Lecture Notes in Mathematics

Edited by A. Dold and B. Eckmann

556

Approximation Theory

Proceedings of an International Colloquium Held at Bonn, Germany, June 8–11, 1976

Edited by R. Schaback and K. Scherer

Springer-Verlag
Berlin · Heidelberg · New York 1976

Editors
Robert Schaback
Lehrstühle für Numerische und
Angewandte Mathematik
Universität Göttingen
Lotzestraße 16–18
3400 Göttingen/BRD

Karl Scherer
Institut für Angewandte Mathematik
Universität Bonn
Wegelerstraße 6
5300 Bonn/BRD

Library of Congress Cataloging in Publication Data
Main entry under title:

Approximation theory.

(Lecture notes in mathematics ; 556)
English or German.
"Vom 8. bis 11. Juni 1976 veranstaltete der Sonderforschungsbereich 72 am Institute für Angewandte Mathematik der Universität Bonn ein internationales Kolloquium über Approximationstheorie."
1. Approximation theory—Congresses. 2. Spline theory—Congresses. 3. Numerical analysis—Congresses. I. Schaback, Robert. II. Scherer, Karl. III. Sonderforschungsbereich Zweiundsiebzig Approximation und Optimierung. IV. Series: Lecture notes in mathematics (Berlin) ; 556.
QA3.L28 no. 556 [QA221] 510'.8s [511'.4] 76-50618

AMS Subject Classifications (1970): 41XX, 42XX, 42A08, 42A24, 42A92, 65DXX, 65NXX, 65N30

ISBN 3-540-08001-5 Springer-Verlag Berlin · Heidelberg · New York
ISBN 0-387-08001-5 Springer-Verlag New York · Heidelberg · Berlin

Printed in Germany
Printing and binding: Beltz Offsetdruck, Hemsbach/Bergstr.

V o r w o r t

Vom 8. bis 11. Juni 1976 veranstaltete der Sonderforschungsbereich 72 am Institut für Angewandte Mathematik der Universität Bonn ein Internationales Kolloquium über Approximationstheorie. Besondere Berücksichtigung fanden die Teilgebiete Spline-Approximation, Konvergenzverhalten und numerische Methoden der Approximation.

Durch die großzügige Unterstützung der Deutschen Forschungsgemeinschaft, der an dieser Stelle nochmals herzlich gedankt sei, konnte neben vielen inländischen auch eine größere Anzahl von Gästen aus dem Ausland eingeladen werden. Die Veranstalter der Tagung möchten sich ferner bei allen Mitgliedern des Sonderforschungsbereichs bedanken, die zu deren organisatorischem Gelingen beigetragen haben, sowie den Tagungsteilnehmern für ihre Vorträge und Beiträge zu den Diskussionen.

Göttingen und Bonn, den 31. 8. 1976

R. Schaback

K. Scherer

Inhaltsverzeichnis Seite

Rationale Approximierbarkeit singulärer Funktionen über $[0,\infty]$.

Hans-Peter Blatt

In letzter Zeit gewinnen rationale Approximationen auf unbeschränkten Intervallen an Interesse. Einmal verwendet man rationale Approximationen von e^{-x} über $[0,\infty]$ zur numerischen Lösung von Wärmeleitungsgleichungen [4], zum anderen treten Approximationsprobleme auf unbeschränkten Intervallen in der Elektrotechnik auf, beispielsweise um Dämpfungsforderungen beim Entwurf elektrischer Schaltungen zu erfüllen. Wir beschäftigen uns hier mit Fragen der Approximationsgüte rationaler Näherungen an Funktionen auf $[0,\infty]$. Wir beweisen einen Weierstraßschen Approximationssatz für Funktionen mit Polen als Singularitäten und Resultate vom Jackson- bzw. Bernstein-Typ.

1. Rationale Approximation auf $[0,\infty]$.

Im Intervall $[0,\infty)$ sind L reelle Punkte $(L \geq 1)$

$$0 \leq x_1 < x_2 < \dots < x_L$$

vorgegeben, denen nichtnegative ganze Zahlen

$$\beta_1 , \beta_2 , \dots , \beta_L$$

zugeordnet sind. Wir setzen

$$\beta := \sum_{i=1}^{L} \beta_i ,$$

$$w(x) := \prod_{i=1}^{L} (x-x_i)^{\beta_i}$$

und betrachten eine Funktion f mit den Eigenschaften:

f ist in jedem Punkt x_i $(1 \leq i \leq L)$ $3\beta_i$-mal differenzierbar, (1.1)

$f = w\cdot\hat{f}$, wobei $\hat{f}$ über $[0,\infty)$ stetig, reellwertig und $\neq 0$ ist, (1.2)

$$\lim_{x\to\infty} \frac{1}{f(x)} = c \geq 0. \tag{1.3}$$

Wir wollen jetzt für $n \geq 3\beta$ die Funktion $\frac{1}{f}$ bezüglich

$$V_n := \{\frac{1}{q} \mid q \in \Pi_n\}$$

approximieren, d.h. wir minimieren die Tschebyscheff-Norm

$$||\frac{1}{f} - \frac{1}{q}|| := \sup_{x\in[0,\infty]} |\frac{1}{f(x)} - \frac{1}{q(x)}|$$

bezüglich $q \in \Pi_n$. Dabei ist Π_n die Menge der Polynome vom Grad $\leq n$.

Wir nennen $\rho_{V_n}(f)$ die Minimalabweichung zu $\frac{1}{f}$ bezüglich V_n.
Dazu bestimmen wir $\tilde{p} \in \Pi_{2\beta-1}$ so, daß

$$\tilde{p}^{(j)}(x_i) = f^{(j)}(x_i) \tag{1.4}$$

für $i = 1,2, \ldots ,L$ und $j = 0,1, \ldots ,2\beta_i - 1$ gilt und definieren

$$\tilde{V}_n := \left\{v = \frac{1}{q} = \frac{1}{\tilde{p}+w^2p} \;\middle|\; \begin{array}{l} p \in \Pi_{n-2\beta},\ q(x) \neq 0 \text{ für} \\ \text{alle } x \in [0,\infty) \text{ mit } x \neq x_i \end{array}\right\}. \tag{1.5}$$

Man beweist ohne Schwierigkeit den

Satz 1:
Für $n \geq 3\beta$ ist $\tilde{V}_n \neq \emptyset$, und die Minimallösung zu $\frac{1}{f}$ bezüglich V_n existiert und liegt in $\tilde{V}_n$, also $\rho_{V_n}(f) = \rho_{\tilde{V}_n}(f)$.

Ähnlich wie bei der rationalen Approximation einer Funktion über $[0,\infty]$, die keine Singularitäten besitzt, läßt sich die Minimallösung durch eine Alternante charakterisieren.

Satz 2: $\tilde{V}_n$ sei $\neq \emptyset$ und $\rho_{V_n}(f) > c$.
Dann ist $v_o = \frac{1}{q_o}$ Minimallösung zu $\frac{1}{f}$ bezüglich V_n.
$\Longleftrightarrow$ (1) Falls $q_o \in \Pi_n - \Pi_{n-1}$, dann existiert in $[0,\infty)$ eine Alternante der Länge $n - 2\beta + 2$ zu $\frac{1}{f} - v_o$.

(2) Falls $q_o \in \Pi_{n-1}$, dann existiert in $[0,\infty)$ eine Alternante der Länge $n - 2\beta + 1$ zu $\frac{1}{f} - v_o$ und der größte Alternantenpunkt ist ein (+) - Punkt von $\frac{1}{f} - v_o$.

Der Beweis ergibt sich durch die gleiche Beweistechnik wie in [2].

Im Fall (2) liegt somit nicht mehr die übliche Anzahl von Alternantenpunkten vor. Dieser mögliche Ausartungsfall wird sich gerade als der Ursprung für viele Schwierigkeiten bei Approximationsgüteaussagen herausstellen. Satz 2 impliziert auch die Eindeutigkeit der Minimallösung.

Wir setzen jetzt $\lim_{x\to\infty} f(x) = \infty$ voraus und betrachten

$$f_r(x) = \begin{cases} f(x) & \text{für } x \le r \\ f(r) & \text{für } x \ge r \end{cases} \tag{1.6}$$

Für $r > x_L$ ergibt sich bei der Approximation von $\frac{1}{f}$ über $[0,\infty]$:

$$|\rho_{V_n}(f) - \rho_{V_n}(f_r)| \le \sup_{x\ge r}\left|\frac{1}{f(x)} - \frac{1}{f_r(x)}\right|$$

Ist weiterhin r so gewählt, daß

$$f(r) = \min_{x\ge r} f(x) , \tag{1.7}$$

so erhält man

$$\rho_{V_n}(f) \le \frac{1}{f(r)} + \rho_{V_n}(f_r) . \tag{1.8}$$

Bezeichnen wir mit

$$\rho_n(f,r) := \inf_{q_n \in \Pi_n} \left|\left| \frac{1}{f} - \frac{1}{q_n} \right|\right|_{[0,r]}$$

die Minimalabweichung von $\frac{1}{f}$ bezüglich V_n über $[0,r]$, so gilt das

Lemma:

(a) Es gibt ein $r^* \ge 0$ und eine Teilfolge $\{n_j\}$, so daß

$$\rho_{V_n}(f) \leq \max\left(\frac{2}{f(r)}, \frac{1}{f(r)} + \rho_n(f,r)\right) \qquad (1.9)$$

für alle $n = n_j$ $(f = 1,2,\ldots)$ und alle $r \geq r^*$ mit der Eigenschaft (1.7).

(b) Existiert eine natürliche Zahl $\tilde{n}$, so daß für $n \geq \tilde{n}$ alle Minimallösungen zu $V_n \setminus V_{n-1}$ gehören, so gilt die obige Aussage für alle $n \geq \tilde{n}$.

(c) Existiert eine Teilfolge $\{n_j\}$, so daß die Minimallösungen zu $\frac{1}{f}$ bezüglich V_{n_j+1} in V_{n_j} liegen, so gilt die Ungleichung (1.9) für $n = n_j (j = 1,2,\ldots)$ and alle $r > x_L$ mit (1.7).

Beweis:

Wir zeigen zunächst (b):

Wegen (1.8) müssen wir nur für $\rho_{V_n}(f_r) > \frac{1}{f(r)}$ zeigen, daß $\rho_{V_n}(f_r) = \rho_n(f,r)$ für $r \geq r^*$ ist. Die Minimallösungen $v_n = \frac{1}{q_n}$ zu $\frac{1}{f}$ bezüglich V_n liegen für $n \geq \tilde{n}$ nicht in V_{n-1}. Also existiert für $\tilde{n}$ eine Alternante der Länge $n - 2\beta + 2$, deren größter Alternantenpunkt ζ ein (-)-Punkt von $\frac{1}{f} - \frac{1}{q_{\tilde{n}}}$ ist. Sei $r^* > \max(\zeta, x_L)$ so gewählt, daß

$$q_{\tilde{n}}'(x) > 0 \quad \text{für} \quad x \geq r^* \text{ ist.}$$

Dann gilt für $r \geq r^*$ mit (1.7) und $n = \tilde{n}$, falls $\rho_{V_n}(f_r) > \frac{1}{f(r)}$ ist:

$$\rho_{V_n}(f_r) = \rho_n(f,r)\,, \qquad (1.10)$$

$$\frac{1}{q_n} \text{ ist Minimallösung zu } \frac{1}{f_r} \text{ bezüglich } V_n \text{ mit } q_n'(x) > 0 \text{ für } x \geq r\,. \qquad (1.11)$$

Wir zeigen (1.10) und (1.11) durch Induktion für alle $n \geq \tilde{n}$ mit $\rho_{V_n}(f_r) > \frac{1}{f(r)}$:

Die Behauptung sei für n richtig und es gelte $\rho_{V_{n+1}}(f_r) > \frac{1}{f(r)}$.
Dann hat die Minimallösung $\frac{1}{q_n}$ zu $\frac{1}{f_r}$ bezüglich V_n eine Alternante

$$0 \leq y_1 < \dots < y_{n-2\beta+2}$$

und der größte Alternantenpunkt $y_{n-2\beta+2}$ ist ein (-)-Punkt der Fehlerfunktion, denn andernfalls wäre $\frac{1}{q_n}$ Minimallösung zu $\frac{1}{f_r}$ bezüglich V_{n+1} und wegen (1.7) und (1.11) auch Minimallösung zu $\frac{1}{f}$ bezüglich V_{n+1}, was unserer Voraussetzung widerspräche.
Nun ist $y_{n-2\beta+2} \leq r$ und mit $q_n = \tilde{p} + w^2 p_n = w(\hat{\tilde{p}} + wp_n)$ gilt für die Minimallösung $\frac{1}{q^*_{n+1}}$ zu $\frac{1}{f_r}$ bezüglich V_{n+1} :

$$\frac{1}{q^*_{n+1}} - \frac{1}{q_n} = \left(\frac{1}{f_r} - \frac{1}{q_n}\right) - \left(\frac{1}{f_r} - \frac{1}{q^*_{n+1}}\right)$$

$$= \frac{p_n - p^*_{n+1}}{(\hat{\tilde{p}}+wp_n)\,(\hat{\tilde{q}}+wp^*_{n+1})} .$$

Da $q^*_{n+1} \notin \Pi_n$, hat $p_n - p^*_{n+1}$ an den Stellen y_i abwechselndes Vorzeichen und somit $n - 2\beta + 1$ Nullstellen ξ_i :

$$0 < \xi_o < \xi_1 < \dots < \xi_{n-2\beta} < r .$$

Wegen $p_n - p^*_{n+1} \in \Pi_{n-2\beta+1}$ folgt damit

$$p^*_{n+1}(x) = p_n(x) + A(x-\xi_o) \dots (x-\xi_{n-2\beta}) \qquad (1.12)$$

mit $A > 0$. Nach Induktionsvoraussetzung ist $q_n'(x) > 0$ für $x \geq n$, also ist auch wegen (1.12) $q^{*\prime}_{n+1}(x) > 0$ für $x \geq r$.
Dann ist aber $\frac{1}{q^*_{n+1}}$ Minimallösung zu $\frac{1}{f}$ bezüglich V_{n+1} über $[0,r]$, $\rho_{V_{n+1}}(f_r) = \rho_{n+1}(f,r)$ und $q^*_{n+1} = q_{n+1}$.
Falls die Voraussetzung von (b) nicht erfüllt ist, so existiert eine Teilfolge $\{n_j\}$ und die Minimallösungen $\frac{1}{q_n}$ zu $\frac{1}{f}$ haben für $n = n_j$ Alternanten der Länge $n - 2\beta + 2$ und der größte Alternanten-

punkt ist ein (+)-Punkt von $\frac{1}{f} - \frac{1}{q_n}$. Ist $r > x_L$ mit (1.7)
und $\rho_{V_n}(f_r) > \frac{1}{f(r)}$, dann liegt dieser größte Alternantenpunkt in $[0,r]$. Also ist $\rho_{V_n}(f_r) = \rho_n(f,r)$ für $n = n_j$ und damit (c) bewiesen.

Die Behauptung (a) ergibt sich aus (b) und (c).

2. Ein Weierstraßscher Approximationssatz.

Mit Hilfe des vorherigen Lemmas können wir für Funktionen des beschriebenen Typs einen Approximationssatz in V_n herleiten.

Satz 3: f sei eine Funktion wie in (1.1) - (1.3) mit c = 0.
Dann ist

$$\lim_{n\to\infty} \rho_{V_n}(f) = 0 .$$

Beweis: Nach Teil (a) des Lemmas gibt es ein $r^* \geq 0$ und eine Teilfolge $\{n_j\}$ mit

$$\rho_{V_n}(f) \leq \max\left(\frac{2}{f(r)}, \frac{1}{f(r)} + \rho_n(f,r)\right) \tag{2.1}$$

für $n = n_j$ und alle $r \geq r^*$ mit der Eigenschaft (1.7).

Wir bestimmen $p \in \Pi_{3\beta-1}$ mit

$$p^{(j)}(x_i) = f^{(j)}(x_i)$$

für $j = 0,1,\ldots,3\beta_i-1$ und $i = 1,2,\ldots,L$. Also gilt

$$f - p = w^3 \cdot F$$

mit einer stetigen Funktion F über $[0,\infty)$. Sei nun p_n für $n \geq 3\beta$ die Minimallösung zu F bezüglich $\Pi_{n-3\beta}$ über $[0,r]$ mit $r \geq r$:

$$||F - p_n||_{[0,r]} = E_{n-3\beta}(r) . \tag{2.2}$$

Wir setzen

$$q_n := p + w^3(p_n + E_{n-3\beta}(r))$$
$$= w(Q_n + w^2\, E_{n-3\beta}(r)) = w\cdot\hat{q}_n$$

mit Polynomen Q_n und $\hat{q}_n$. Da $f(x) = w\cdot\hat{f}(x)$ ist mit $\hat{f}(x) > 0$ für $x \in [0,\infty)$, folgern wir:

$$|\hat{f}(x) - Q_n(x)| = |\frac{1}{w(x)}(f(x) - p(x) - w^3(x)\, p_n(x))|$$
$$= w^2(x)|F(x) - p_n(x)|$$
$$\leq w^2(x)\, E_{n-3\beta}(r)$$

oder

$$\hat{q}_n(x) = |Q_n(x) + w^2(x)\, E_{n-3\beta}(r)|$$
$$\geq \hat{f}(x) + w^2(x)\, E_{n-3\beta}(r) - |\hat{f}(x) - Q_n(x)|$$
$$\geq \hat{f}(x)$$

für alle $0 \leq x \leq r$. Damit erhalten wir für $0 \leq x \leq r$:

$$\left|\frac{1}{f(x)} - \frac{1}{q_n(x)}\right| = \frac{|f(x) - q_n(x)|}{w^2(x)\,\hat{f}(x)\,\hat{q}_n(x)}$$
$$\leq \frac{w(x)|F(x) - p_n(x) - E_{n-3\beta}(r)|}{|\hat{f}(x)|^2}$$
$$\leq K(r)\cdot E_{n-3\beta}(r) \qquad (2.3)$$

mit $K(r) = 2\cdot\left|\left|\frac{w}{\hat{f}^2}\right|\right|_{[0,r]}$.

Sei $\varepsilon > 0$ vorgegeben. Wir bestimmen ein $r \geq r^*$ mit der Eigenschaft (1.7) und $\frac{1}{f(r)} < \varepsilon$, weiterhin $k \in \mathbb{N}$ mit

$$K(r)\, E_{n_k-3\beta}(r) < \varepsilon.$$

Dann folgt aus (2.1) und (2.3):

$$\rho_{V_{n_k}}(f) < 2\varepsilon .$$

3. Differenzierbare Funktionen

Um die Voraussetzungen der folgenden Aussagen nicht zu kompliziert zu gestalten, beschränken wir uns auf $\beta = 0$, die zu approximierende Funktion f hat somit keine Nullstellen in $[0,\infty)$.

Satz 4: f sei k-mal stetig differenzierbar in $[0,\infty)$ und es gebe Konstanten $A > 0$, $\theta > 0$ und $r_o > 0$ so daß $f^{(k)} \in \mathrm{Lip}_{M_r}\,\alpha$ für jedes $r \geq r_o$ im Intervall $[0,r]$ mit

$$M_r \leq A(f(r))^{\theta}\, r^{-k-\alpha}.$$

Dann existiert eine Teilfolge $\{n_j\}$ der natürlichen Zahlen mit

$$\rho_{V_{n_j}}(f) = O(n_j^{-p}) \quad \text{und} \quad p = \frac{k+\alpha}{\theta+1} .$$

Falls $\rho_{V_n}(f) \neq \rho_{V_{n+1}}(f)$ für fast alle n gilt, oder eine Teilfolge $\{n_j\}$ existiert mit $1 \leq n_{j+1}/n_j \leq \kappa$ (mit festem κ) und $\rho_{V_{n_j}}(f) = \rho_{V_{n_j+1}}(f)$, so gilt sogar:

$$\rho_{V_n}(f) = O(n^{-p}) .$$

Beweis: Es sei

$$\rho_{V_n}(f) \leq \max\left\{\frac{2}{f(r)}\, ,\, \frac{1}{f(r)} + \rho_n(f,r)\right\}$$

und r mit der Eigenschaft (1.7). Dann folgt wie im Beweis von Satz 3:

$$\rho_{V_n}(f) \leq \max\left\{\frac{2}{f(r)}\, ,\, \frac{1}{f(r)} + K\cdot E_n(r)\right\} \tag{3.1}$$

mit $K = 2 \cdot \max_{x \geq 0} \frac{1}{f(x)^2}$, $E_n(r) = \inf_{p \in \Pi_n} ||f-p||_{[0,r]}$.

Wir transformieren $0 \leq x \leq r$ durch $\frac{r(1+t)}{2} = x$ auf $-1 \leq t \leq +1$ und setzen

$$g(t) := f(\frac{r(1+t)}{2})$$

Dann ist $g^{(k)}(t) = (\frac{r}{2})^k \cdot f^{(k)}(x)$

und

$$\begin{aligned}|g^{(k)}(t) - g^{(k)}(\tilde{t})| &= (\frac{r}{2})^k \, |f^{(k)}(x) - f^{(k)}(\tilde{x})| \\ &\leq (\frac{r}{2})^k \, A(f(r))^\theta \, r^{-k-\alpha} |x-\tilde{x}|^\alpha \\ &= 2^{-k-\alpha} \cdot A(f(r))^\theta |t-\tilde{t}|^\alpha.\end{aligned}$$

Nach einem Satz von Jackson (Meinardus [5]) folgt dann:

$$\min_{g \in \Pi_n} ||g-p||_{[-1,1]} = E_n(r) \leq B(f(r))^\theta \, n^{-k-\alpha}$$

mit einer Konstanten B. Wir wählen r so, daß (1.7) und

$$f(r)^{\theta+1} = n^{k+\alpha}$$

erfüllt ist und erhalten aus (3.1):

$$\rho_{V_n}(f) \leq C \cdot n^{-\frac{k+\alpha}{\theta+1}} \quad \text{mit einer Konstanten C.}$$

Mit dem Lemma erhalten wir schließlich die asymptotischen Aussagen.

In gewissem Sinn läßt sich Satz 4 umkehren, wir definieren dazu wie im klassischen Fall für ein Intervall $[a,b]$ die Klasse

$$W := \{g \in C[a,b] \mid \omega(\delta) = O(\delta \cdot \log \frac{1}{\delta}) \text{ für } \delta \to 0\}$$

Satz 5: f sei eine reellwertige, stetige Funktion mit endlich vielen Nullstellen in $[0,\infty)$. Ist $\rho_{V_n}(f) = O(n^{-p-\alpha})$ mit $0 < \alpha \leq 1$

und p eine nichtnegative, ganze Zahl, dann existieren in $(0,\infty)$ die Ableitungen von f bis zur p-ten Ordnung und für jedes Intervall $[a,b]$ mit $a > 0$ gilt

$$f^{(p)} \in \mathrm{Lip}\,\alpha \quad \text{für} \quad \alpha > 1\,,$$

$$f^{(p)} \in W \quad \text{für} \quad \alpha = 1\,.$$

Ist weiterhin $p \geq 2k$, dann existieren in $[0,\infty)$ die Ableitungen f bis zur Ordnung k und es gilt im Intervall $[0,r]$

$$||f^{(k)}||_{[0,r]} \leq D||f||^2_{[0,r]} \cdot r^{-k}$$

mit einer Konstanten $D > 0$, die unabhängig von f und r ist.

Beweis: Wir fixieren ein $r > 0$. Es existiert ein $n_o \in N$ und ein Polynom p_n mit

$$||\frac{1}{f} - \frac{1}{p_n}|| \leq A \cdot n^{-p-\alpha}$$

oder

$$|f(x) - p_n(x)| \leq A \cdot n^{-p-\alpha}|f(x)||p_n(x)|$$

für ein geeignetes $A > 0$ und alle $n \geq n_o$.
Außerdem ist für $f(x) \neq 0$:

$$|\frac{1}{p_n(x)}| \geq |\frac{1}{f(x)}| - A\, n^{-p-\alpha}.$$

Mit einem geeigneten $n_1 \geq n_o$ gilt dann für $n \geq n_1$ und $x \in [0,r]$:

$$|p_n(x)| \leq 2\,\|f\|_{[0,r]}$$

Insgesamt erhält man für $n \geq n_1$:

$$||f - p_n||_{[0,r]} \leq 2\cdot A||f||^2_{[0,r]}\, n^{-p-\alpha}.$$

Die Transformation $x = \frac{r(1+t)}{2}$ von $[0,r]$ in $[-1,1]$ liefert nach (3.2):

$$||g - q_n||_{[-1,1]} \leq 2 \cdot A||f||^2_{[0,r]} n^{-p-\alpha}$$

mit $q_n(t) = p_n(\frac{r(1+t)}{2})$.

Wie beim klassischen Beweis (Meinardus [5]) setzt man

$$v_o(t) = q_{n_1+1}(t)$$

$$v_\nu(t) = q_{n_1+2^\nu}(t) - q_{n_1+2^{\nu-1}}(t).$$

Dann gilt

$$q_{n_1+2^m}(t) = \sum_{\nu=0}^{m} v_\nu(t)$$

und der erste Teil der Aussage ergibt sich wie üblich aus der Reihendarstellung

$$g(t) = \sum_{\nu=0}^{\infty} v_\nu(t).$$

Was die zweite Aussage betrifft, so folgt aus der Markoffschen Ungleichung:

$$\left|v_\nu^{(k)}(t)\right| \leq B\cdot 2^{2\nu k}||v_\nu||_{[-1,1]}$$
$$\leq C\cdot 2^{2\nu k-\nu(p+\alpha)}||f||^2_{[0,r]}$$

mit Konstanten B und C. Somit konvergiert die Reihe $\sum_{\nu=0}^{\infty} v_\nu^{(k)}(t)$ gleichmäßig in $[-1,1]$ gegen $g^{(k)}$ und es gilt mit von f und r unabhängigen Konstanten $\tilde{D}$ bzw. D:

$$||g^{(k)}||_{[-1,1]} \leq \tilde{D}||f||^2_{[0,r]}$$

oder

$$||f^{(k)}||_{[0,r]} \leq D||f||^2_{[0,r]} r^{-k}.$$

Beispiele:

(1) $f(x) = \sqrt{x} + 1$:

Ist $\frac{1}{p_n}$ Minimallösung zu $\frac{1}{f}$ bezüglich V_n und $p_n \in V_n \setminus V_{n-1}$ dann muß der größte Alternantenpunkt ein (+)-Punkt sein, da $f^{(n+1)}(x)$ in $(0,\infty)$ einerlei Vorzeichen besitzt. Somit ist p_n für ungerades n aus $V_n \setminus V_{n-1}$ und $\frac{1}{p_n}$ ist für solche n auch Minimallösung bezüglich V_{n+1}. Außerdem gehört f in $[0,r]$ zu $\mathrm{Lip}_1 \frac{1}{2}$ und

$$M_r = 1 \le \frac{f(r)}{\sqrt{r}} .$$

Also gilt

$$\rho_{V_n}(f) = O(n^{-\frac{1}{4}}) \text{ für } n \to \infty.$$

Setzt man jedoch $x = \xi^2$, so ist die ursprüngliche Aufgabe äquivalent zur Approximation von $\frac{1}{x+1}$ in $[0,\infty]$ bezüglich gerader Funktionen aus V_{2n} oder zur Approximation von $\frac{1}{|x|+1}$ bezüglich V_{2n} über $[-\infty,+\infty]$. Überträgt man die obige Theorie auf die Approximation über $[-\infty,\infty]$ (Blatt [2]), so erhält man sogar:

$$\rho_{V_n}(f) = O(n^{-\frac{1}{2}}) \text{ für } n \to \infty .$$

(2) $f(x) = e^x + \log(x+1)$:

Für jedes $k \in \mathbb{N}$ sind die Voraussetzungen von Satz 4 mit $\alpha = 1$ und $\theta = 2$ erfüllt. Es existiert somit zu jedem $p \in \mathbb{N}$ eine Teilfolge $\{n_j\}$ mit

$$\rho_{V_{n_j}}(f) = O(n_j^{-p}).$$

4. Ganze Funktionen

Ausgangspunkt der Untersuchungen über das asymptotische Verhalten der Minimalabweichungen in den letzten Jahren waren Beziehungen zwischen der Holomorphie und der geometrischen Konvergenz der Minimalabweichungen ([4], [6], [7]).
Dazu betrachtet man zu gegebenem $r > 0$ und $s > 1$ die abgeschlos-

sene Ellipse $\mathcal{E}(r,s)$ der komplexen Ebene mit Brennpunkten in 0 und r und der Summe $r\cdot s$ beider Achsen. Ist f eine ganze Funktion, so setzen wir

$$M_f(r,s) := \max_{z\in\mathcal{E}(r,s)} |f(z)| \qquad (4.1)$$

Meinardus, Reddy, Taylor und Varga [7] bewiesen

Satz 6: Sei f(x) eine reelle, stetige Funktion über $[0,\infty)$ mit höchstens endlich vielen Nullstellen in $[0,\infty)$, $\{p_n\}$ eine Folge reeller Polynome mit $p_n \in \Pi_n$ für jedes $n \geq 0$ und q eine reelle Zahl mit

$$\overline{\lim_{n\to\infty}} \left\|\frac{1}{f} - \frac{1}{p_n}\right\|^{\frac{1}{n}} = \frac{1}{q} < 1 .$$

Dann gibt es eine ganze Funktion F(z) mit F(x) = f(x) für alle $x \geq 0$ und F ist von endlicher Ordnung. Außerdem gibt es für jedes $s > 1$ Konstanten $K > \theta$, $\theta > 0$ und $r_o > 0$ mit

$$M_f(r,s) \leq K(\|f\|_{[0,r]})^{\theta} \text{ für alle } r \geq r_o .$$

Mit Hilfe unseres obigen Lemmas können wir eine teilweise Umkehrung dieses Satzes erreichen.

Satz 7: Sei f ein ganze transzendente Funktion mit reellen Taylorkoeffizienten. Gibt es reelle Zahlen $s > 1$, $K > 0$, $\theta > 0$ und $r_o > 0$ mit

$$M_f(r,s) \leq K(f(r))^{\theta} \text{ für alle } r \geq r_o, \qquad (4.2)$$

dann gilt:

$$\underline{\lim_{n\to\infty}} \rho_{V_n}(f)^{\frac{1}{n}} \leq s^{-\frac{1}{1+\theta}} < 1 .$$

Beweis: Wegen (4.2) gilt: $\lim_{x\to\infty} f(x) = \infty$.

Sei $\{n_j\}$ eine Teilfolge und r wie in Teil (a) des Lemmas.

Dann gilt für $n = n_j$:

$$\rho_{V_n}(f) \le \max \{ \frac{2}{f(r)}, \frac{1}{f(r)} + \rho_n(f,r) \},$$

mit den Bezeichnungen wie im Beweis von Satz 3:

$$\rho_{V_n}(f) \le \max \{ \frac{2}{f(r)}, \frac{1}{f(r)} + K(r)\, E_{n-3\beta}(r) \}$$

mit

$$K(r) = 2 \, || \frac{w^3}{f^2} ||_{[0,r]} .$$

Nun ist für $x \ge r_o$:

$$\left| \frac{w^3(x)}{f^2(x)} \right| \le \frac{||w^3||_{[0,x]}}{f^2(x)} \le \tilde{K} \frac{||w^3||_{[0,x]}}{[M_f(x,s)]^{2/\theta}} .$$

Da f eine ganze transzendente Funktion, gilt mit einer geeigneten Konstanten K^*

$$K(r) \le K^*$$

für alle r. Nach einem Satz von Bernstein ([5], S. 112) ist

$$E_{n-3\beta}(r) \le \frac{M_F(r,s)}{(s-1)\, s^{n-3\beta}} .$$

Wegen $F = \frac{f-p}{w^3}$ existiert ein $r_1 > r_o$ mit

$$M_F(r,s) \le 2\, M_f(r,s)$$

für alle $r \ge r_1$. Also gilt für solche r:

$$E_{n-3\beta}(r) \le \frac{2Ks^{3\beta}}{s-1} \frac{(f(r))^\theta}{s^n} .$$

Es existiert wegen $\lim_{x\to\infty} f(x) = \infty$ ein j_o, so daß man für jedes $j \ge j_o$ eine Zahl $r = r(n_j) \ge \max(r_1, r^*)$ bestimmen kann, die (1.7) erfüllt und

$$f(r) = s^{n_j/(1+\theta)} .$$

Dann ist

$$\rho_{V_{n_j}}(f) \le C \cdot s^{-n_j/(1+\theta)}$$

mit einer Konstanten $C > 0$.

Falls f die Voraussetzungen von Satz 6 erfüllt und nur endlich viele negative Taylorkoeffizienten hat, dann folgt durch Verwendung von Teil (b) des Lemmas sogar:

$$\overline{\lim_{n\to\infty}}\ \rho_{V_n}(f)^{\frac{1}{n}} < 1 .$$

Literatur

1. H.-P. Blatt:
Rationale Approximation auf [0,∞], ZAMM 53(1973), T 181-182.

2. H.-P. Blatt:
Rationale Tschebyscheff-Approximation über unbeschränkten Intervallen, Habilitationsschrift Universität Erlangen-Nürnberg, 1974.

3. D. Brink, G. D. Taylor:
Chebyshev Approximation by Reciprocals of Polynomials on [0,∞), J. Approximation Theory 16, 142-149 (1976).

4. W. J. Cody, G. Meinardus, R. S. Varga:
Chebyshev Rational Approximations to e^{-x} in [0,+∞) and Applications to Heat-Conduction Problems,
J. Approximation Theory 2 (1969), 50-65.

5. G. Meinardus:
Approximation of Functions: Theory and Numerical Methods, Springer-Verlag, Berlin, 1967.

6. G. Meinardus, R. S. Varga:
Chebyshev Rational Approximations to Certain Entire Functions in [0,∞),
J. Approximation Theory 3 (1970), 300-309.

7. G. Meinardus, A. R. Reddy, G. D. Taylor, R. S. Varga:
Converse Theorems and Extensions in Chebyshev Rational Approximation to Certain Entire Functions in [0,∞),
Trans. Amer. Math. Soc. 170 (1972), 171-185.

A DEFECT CORRECTION METHOD FOR FUNCTIONAL EQUATIONS

K. Böhmer*)

Summary: We want to solve numerically the functional equation $F(y)=o$. For that purpose we use a discretization method with the property that the global discretization error admits an asymptotic expansion. We combine this with Newton's method and find numerical methods which are related to Pereyra's technique [8]. The first step of these methods have been given for the special case of initial value problems for ordinary differential equations by Zadunaisky [14,15] and Stetter [12].

1. Asymptotic error expansion and Newton's method

In addition to the original problem

(1.1) $\quad F(y)=o, \qquad F:D \subseteq E \to E^o;\ E,E^o$ Banach spaces,

we deal with the discretized problem ($h \in (o,h_o] \subset \mathbb{R}_+$)

(1.2) $\quad \Phi_h(\eta_h)=o, \quad \Phi_h:D_h \subseteq E_h \to E_h{}^o;\ E_h,E_h{}^o$ Banach spaces.

We assume that

(1.3) $\quad$ (1.1) and (1.2) have unique solutions $y \in D$ and $\eta_h \in D_h$.

Further we use the equivalent notations

$$x_1=x_2 + O(h^r),\ r\in \mathbb{R}_+ \ \text{ iff } \ \|x_1-x_2\| = O(h^r).$$

Here $\|\cdot\|$ means any of the norms of $E,E^o,E_h,E_h{}^o$.

(1.1) and (1.2) are correlated by linear bounded discretization operators

---

*) This report was partially supported by a grant of the Volks-Wagen-Foundation.

$$(1.4)\quad \begin{cases} \Delta_h : E \to E_h,\ \Delta_h^{\,o} : E^o \to E_h^{\,o},\ \Delta_h, \Delta_h^{\,o} \text{ linear bounded,} \\ \Phi_h(\Delta_h u) = \Delta_h^{\,o}\{F(u) + \Lambda_h(u)\} \quad \text{for } u \in E. \end{cases}$$

In many cases the "local error mapping" Λ_h admits an asymptotic expansion up to the order ν_q, that is

$$(1.5)\quad \Delta_h^{\,o}\,\Lambda_h(u) = \Delta_h^{\,o}\,\{\sum_{\iota=1}^{q} h^{\nu_\iota} f_\iota(u) + O(h^{\nu_{q+1}})\}, 0<\nu_1<\nu_2<\ldots<\nu_{q+1}$$

$$\text{for } u \in D_q \subseteq D,\ f_\iota : D_q \to E^o, f_\iota \ \text{ independent of } h.$$

If $y \in D_q$ is the solution of (1.1) and (1.2) is consistent to (1.1) of order ν_p then the "local discretization error" λ_h satisfies

$$(1.6)\quad \lambda_h := \Delta_h^{\,o}\Lambda_h(y) = \Phi_h(\Delta_h y) = \Delta_h^{\,o}\{\sum_{\iota=p}^{q} h^{\nu_\iota} f_\iota(y) + O(h^{\nu_{q+1}})\}.$$

Very important for numerical applications is the question,whether (1.6) carries over to the "global discretization error"

$$(1.7)\quad \gamma_h := \eta_h - \Delta_h y,\ \eta_h, y \text{ solutions of (1.1), (1.2).}$$

Gragg [7] had studied this question first for initial value problems of ordinary differential equations. Stetter [10,11] generalized Graggs result to functional equations. In these papers we always have $\nu_\iota = \iota$. Several difficult special cases were treated directly (see for example Benson [1]).We don't discuss the conditions for an asymptotic error expansion, but we assume for the further that

$$(1.8)\quad \begin{cases} \gamma_h = \eta_h - \Delta_h y = \Delta_h\ \{\sum_{\iota=p}^{q} h^{\nu_\iota} g_\iota(y) + O(h^{\nu_{q+1}})\} \text{ where} \\ \mathbb{N}_* := \{\nu_1, \nu_2, \ldots, \nu_q, \nu_{q+1}\}\ ,\ 0 < \nu_1 < \nu_2 < \ldots < \nu_{q+1} \text{ and} \\ \nu_i, \nu_j \in \mathbb{N}_* \Rightarrow \nu_i + \nu_j \in \mathbb{N}_* \text{ or } \nu_i + \nu_j > \nu_{q+1}\ , \\ g_\iota \text{ independent of } h. \end{cases}$$

We will proceed combining this asymptotic expansion with Newton's method (see for example [2]).

Theorem 1: *Let us assume that for* $\rho \in \mathbb{R}_+$ F' *exists and is equicontinuous,* $(F')^{-1}$ *exists and* $\|(F')^{-1}\| \leqq K \in \mathbb{R}_+$ *in* $K_\rho(y) := \{u \in D \mid \|u-y\| < \rho\}$, *y solution of* (1.1), *and let* $L \in (0,1)$ *be given. Then there is a* $\sigma = \sigma(\rho, K, L)$ *such that*

$$(1.9) \qquad y_o,\ c \in K_\sigma(y),\ y_{\nu+1} := y_\nu - F'(c)^{-1}F(y_\nu),\ \nu = 0,1,\ldots$$

implies

$$(1.10) \qquad \begin{cases} \lim\limits_{\nu\to\infty} y_\nu = y \quad \text{and} \\ \|y_{\nu+1}-y\| \leqq L\|y_\nu - y\| \leqq \frac{L}{1-L}\|y_{\nu+1}-y_\nu\| . \end{cases}$$

2. Method of defect-correction

The idea of combining asymptotic expansions of the global discretization error with defect equations to obtain error estimations was suggested in two different versions by Stetter [12] and one version has been treated by Frank and Ueberhuber [4,5] for several concrete examples: Runge-Kutta methods for ordinary differential equations and some boundary value problems. Here we use the second version for general functional equations (1.1).

We solve (1.1) numerically by a discrete problem (1.2) and find $\eta_{h,o} := \eta_h$ as solution of (1.2). Let T_{q_o} be an operator such that

$$(2.1) \qquad T_{q_o} : \begin{cases} E_h \to E \\ \eta_{h,o} \mapsto y_o := T_{q_o}\eta_{h,o}, \quad p \leqq q_o \leqq q \end{cases}$$

and let T_{q_o} essentially reproduce the asymptotic expansions, that is

$$(2.2) \qquad \eta_{h,o} = \Delta_h \{y + \sum_{\iota=p_o}^{q} h^{\nu_\iota} g_\iota(y) + O(h^{\nu_{q+1}})\}$$

implies

$$(2.3)\quad \begin{cases} T_{q_o}\eta_{h_o} = y_o = y + \sum\limits_{\iota=p_o}^{q_o} h^{\nu_\iota} g_\iota(y) + O(h^{\nu^*_{q_o+1}}) \\ \\ \text{and} \quad \nu_{q_o} < \nu^*_{q_o+1} \leqq \nu_{q+1}\ . \end{cases}$$

Usually T_{q_o} will be interpolation operators and, since often $\|\cdot\|_E = \|\cdot\|$ includes derivatives, one has $q_o \leqq q$, instead of $q_o = q$. In Theorem 1 we generally have $c = y_o$. But since we want to discuss the step $\nu = o$ and $\nu > o$ at the same time we assume an expansion for c, differing a little bit of y_o,

$$(2.4)\quad \begin{cases} c = y + \sum\limits_{\iota=r_o}^{r_1} h^{\nu_\iota}\hat{g}_\iota(y) + O(h^{\nu_{r_1+1}}) \\ \nu_{r_o} \leqq \nu_{p_o},\ \nu_{r_1+1} + \nu_{p_o} \geqq \nu^*_{q_o+1},\ \hat{g}_\iota \ \text{independent of } h. \end{cases}$$

Now we have as a consequence of Theorem 1

<u>Theorem 2:</u> *Let* y_o *and* c *be of the form* (2.3) *resp.* (2.4). *Further let* F *be at least* $\kappa := -[-\nu^*_{q_o+1}/\nu_{r_o}]$*- times differentiable,* $(F')^{-1}$ *exist and* $F^{(\kappa)}$ *and* $(F')^{-1}$ *be bounded in* $K_\rho(y)$. *Define* y_1 *by*

$$(2.5)\qquad F'(c)(y_1-y_o) = -F(y_o)\ .$$

Then y_1 *admits an asymptotic expansion*

$$(2.6)\quad \begin{cases} y_1 = y + \sum\limits_{\iota=p_1}^{q_o} h^{\nu_\iota} g_{\iota 1}(y) + O(h^{\nu^*_{q_o+1}}) \\ \textit{and}\ \ \nu_{p_1} = \nu_{p_o} + \nu_{r_o} \ \textit{for}\ \nu_{p_1} < \nu^*_{q_o+1} \end{cases}$$

resp.

$$(2.7)\quad y_1 = y + O(h^{\nu^*_{q_o+1}}) \ \textit{for}\ \nu_{p_1} \geqq \nu^*_{q_o+1}\ .$$

<u>Proof:</u> With (2.3), (1.9), $\bar{\nu} := \nu^*_{q_o+1}$ and $\vartheta, \vartheta_1 \in (0,1)$ we have

$$- F'(c)(y_1-y_o) = F(y_o) = F(y + \sum_{\iota=p_o}^{q_o} h^{\nu_\iota} g_\iota(y) + O(h^{\bar{\nu}}))$$

$$= F(y) + \sum_{j=1}^{\kappa-1} \frac{F^{(j)}(c)+F^{(j)}(y)-F^{(j)}(c)}{j!}\{\sum_{\iota=p_o}^{q_o} h^{\nu_\iota} g_\iota(y)+O(h^{\bar{\nu}})\}^j$$

$$+ F^{(\kappa)}(y + \vartheta \sum_{\iota=p_o}^{q_o} h^{\nu_\iota} g_\iota(y) + O(h^{\bar{\nu}}))\{\sum_{\iota=p_o}^{q_o} h^{\nu_\iota} g_\iota(y) + O(h^{\bar{\nu}})\}^\kappa .$$

Now

$$F^{(j)}(y)-F^{(j)}(c) = - \sum_{\ell=j+1}^{\kappa-1} \frac{F^{(\ell)}(y)}{(\ell-j)!}(c-y)^{\ell-j} - \frac{F^{(\kappa)}(y+\vartheta_1(c-y))}{(\kappa-j)!}(c-y)^{\kappa-j}$$

$$= \sum_{\iota=r_o}^{q_o} h^{\nu_\iota}\tilde{g}_{\iota 1}(y) + O(h^{\bar{\nu}}) - \frac{F^{(\kappa)}(y+\vartheta_1(c-y))}{(\kappa-j)!}(c-y)^{\kappa-j}$$

with suitable $\tilde{g}_{\iota 1}$ independent of h. Since

$$\bar{\nu} = \nu^*_{q_o+1} \leqq \kappa\nu_{p_o} \text{ and } j\nu_{p_o} + (\kappa-j)\nu_{r_o} \geqq \kappa\nu_{r_o} \geqq \bar{\nu}$$

we further have

$$F^{(\kappa)}(y+\vartheta(y_o-y))(y_o-y)^\kappa = O(h^{\bar{\nu}}) \text{ and}$$

$$F^{(\kappa)}(y+\vartheta_1(c-y))(c-y)^{\kappa-j}(y_o-y)^j = O(h^{\bar{\nu}}).$$

So we have altogether with $\nu_{p_1} = \nu_{p_o} + \nu_{r_o} \leqq 2\nu_{p_o}$

$$(2.8) \qquad F(y_o) = F'(c)\left\{\sum_{\iota=p_o}^{p_1-1} h^{\nu_\iota} g_\iota(y) + \sum_{\iota=p_1}^{q_o} h^{\nu_\iota}\hat{g}_{\iota 1}(y) + O(h^{\bar{\nu}})\right\}$$

with $\hat{g}_{\iota 1}$ independent of h. Finally (2.3) implies, with $\hat{g}_{\iota 1}$ independent of h,

$$y_1 = y_o - \sum_{\iota=p_o}^{p_1-1} h^{\nu_\iota} g_\iota(y) - \sum_{\iota=p_1}^{q_o} h^{\nu_\iota}\hat{g}_{\iota 1}(y) + O(h^{\bar{\nu}}) =$$

$$= y + \sum_{\iota=p_1}^{q_o} h^{\nu_\iota} g_{\iota 1}(y) + O(h^{\bar{\nu}}) . \quad \square$$

We use here the notations of Stetter [11]: A discretization method $\mathfrak{M}$ for a given problem $\mathcal{P} = \{E, E^o, F\}$ (see (1.1)) consists of an infinite sequence $\{E_h, E_h^o, \Delta_h, \Delta_h^o, \phi_h\}_{h \in (o, h_o]}$ (see (1.2), (1.4)). Here

$$\text{(2.9)} \qquad \phi_h : \begin{cases} (E \to E^o) \to (E_h \to E_h{}^o) \\ F \mapsto \Phi_h := \phi_h(F). \end{cases}$$

Since usually (2.5) is not solvable exactly, we have to apply this discretization method to our linear problem (2.5). The crucial question is, whether the approximate solution reflects the asymptotic expansion of y_1. To guarantee that we need the properties given in

Definition 3: *In* (1.1) *let* $Fu = F_1u + d$ *with a linear continuous* F_1 *and* d *independent of* u *and let* ϕ_h, *applied to* F_1u+d, *have the following properties*

$$\text{(2.10)} \qquad \begin{cases} \phi_h(F_1u+d) = \phi_h(F_1)\Delta_h u + \Delta_h{}^o d \\ \textit{where} \\ \Phi_h(F_1, \Delta_h u) := \phi_h(F_1)\Delta_h u \ \textit{linear, continuous in } \Delta_h u. \end{cases}$$

If the discretization method $\mathfrak{M}$ *satisfies* (2.10) *we call it a* *linear discretization method* *for the problem* $\mathcal{P} = \{E, E^o, F\}$. *Further we call* $\Phi_1(F_1, \Delta_h u) + \Delta_h{}^o d = o$ *a* *linear discretization* *of* $F_1u + d = o$.

Let

$$\text{(2.11)} \qquad \Phi_h{}^{*}(\Delta_h c)(\eta_{h1} - \eta_{ho}) = -\Delta_h{}^o F(y_o)$$

be a linear discretization of equation (2.5). In many cases $\Phi_h{}^{*}$ is the Frechet derivative of Φ_h from (1.2), but that is not true in every case. We need a second property for discretizations Φ_h (see [9]):

Definition 4: *If there is a* $c \in \mathbb{R}_+$, c *independent of h, such that*

$$\text{(2.12)} \quad \|\eta_1 - \eta_2\| \leqq c\|\Phi_h(\eta_1) - \Phi_h(\eta_2)\| \ \textit{for} \ \eta_1, \eta_2 \in E_h, \ h \in (o, h_o],$$

then Φ_h *is called uniformly stable.*

Generally it is not possible to solve (2.5) exactly. What happens if (2.5) is approximatively solved by a discretization method (see Stetter [11])? The crucial question is whether the approximate solution reflects the asymptotic expansion of y_1. To guarantee that we us

Definition 3: *In* (1.1) *let* $Fu=F_1u+d$ *with* F_1 *continuous,* d *independent of* u *and let the discretized problem to* F_1u+d

(2.9) $\Phi_h(\Delta_h u) = \Phi_h(\Delta_h u, F_1u+d) = \Delta_h^o\{F_1u+d+\Lambda_h(F_1u+d)\}$

have the following properties

$$
(2.10)\quad \begin{cases}
\Lambda_h(F_1u+d) = \Lambda_h F_1 u + \Lambda_h d \text{ that means} \\
\Phi_h(\Delta_h u, F_1u+d) = \Delta_h^o\{F_1u+\Lambda_h F_1 u+d+\Lambda_h d\} \\
\text{and if } d = d_o = \alpha_1 d_1 + \alpha_2 d_2 \text{ then} \\
\Delta_h(d_o) = \Lambda_h(\alpha_1 d_1+\alpha_2 d_2) = \alpha_1\Lambda_h d_1+\alpha_2\Lambda_h d_2+O(h^{\nu^*}) \\
\text{where } \nu^* = \min\{\mu_o,\mu_1,\mu_2\}\ ,\ \mu_j \text{ being the order of} \\
\text{consistency of } \Phi_h(\Delta_h u, F_1u+d_j) = 0 \text{ for } F_1u+d_j=0, j=0,1,2.
\end{cases}
$$

If (2.10) *is satisfied* (2.9) *is called a linear discretization method for the original problem* $F_1u + d = 0$.

We need a second property for the operators Φ_h (see [9]):

Definition 4: *If there is a* $c \in \mathbb{R}_+$, c *independent of* h, *such that*

(2.11) $\|\eta_1-\eta_2\| \leq c\|\Phi_h(\eta_1) - \Phi_h(\eta_2)\|$ *for* $\eta_1,\eta_2 \in E_h$, $h\in (o,h_o]$

then Φ_h *is called uniformly stable.*

Let

(2.12) $\Phi_h^*(\Delta_h c)(\eta_{h1}-\eta_{ho}) = -\Delta_h^o F(y_o)$

be a discrete form of equation (2.5). In many cases one may choose Φ_h^* to be the Frechet derivative of Φ_h from (1.2), but that is not possible in every case.

Theorem 5: *In addition to the conditions of Theorem 2 let* (2.11) *be a uniformly stable linear discretization of* (2.5) *and let the* g_ι *in* (2.2) *be such that*

$$(2.13)\quad \begin{cases} \Phi_h^{*}(\Delta_h c)\varepsilon_\iota = \Delta_h^{o} F'(c) g_\iota(y) \text{ implies} \\ \varepsilon_\iota = \Delta_h\big(g_\iota(y) + \sum_{\ell=r_o}^{s_\iota} h^{\nu_\ell}\hat{k}_{\iota\ell}(y) + O(h^{\nu_{s_\iota+1}})\big) \\ \text{with } \nu_{s_\iota} + \nu_\iota = \nu_{q_o},\ \nu_{s_\iota+1} + \nu_\iota \geq \nu^{*}_{q_o+1} \end{cases}$$

Then η_{h1} *from* (2.12) *satisfies* ($\tilde{g}_{\iota 1}$ *independent of* h)

$$(2.14)\qquad \eta_{h1} = \Delta_h\{y + \sum_{\iota=p_1}^{q_o} h^{\nu_\iota}\tilde{g}_{\iota 1}(y) + O(h^{\nu^{*}_{q_o+1}})\}\ \text{for } \nu_{p_1} < \nu^{*}_{q_o+1}$$

resp.

$$\eta_{h1} = \Delta_h\{y + O(h^{\nu^{*}_{q_o+1}})\}\ \text{for } \nu_{p_1} \geq \nu^{*}_{q_o+1}.$$

Proof: From (2.8) we find

$$-F'(c)(y_1-y_o)=F(y_o)=F'(c)\{\sum_{\iota=p_o}^{p_1-1} h^{\nu_\iota}g_\iota(y)+\sum_{\iota=p_1}^{q_o} h^{\nu_\iota}\hat{g}_{\iota 1}(y)+O(h^{\bar{\nu}})\}$$

and therefore

$$(2.15)\quad \Phi_h^{*}(\Delta_h c)(\eta_{h1}-\eta_{ho})=-\Delta_h^{o}F(y_o)=-\Delta_h^{o}\{F'(c)\ [\sum_{\iota=p_o}^{p_1-1} h^{\nu_\iota}g_\iota(y) + \sum_{\iota=p_1}^{q_o} h^{\nu_{\iota 1}}\hat{g}_{\iota 1}(y)+O(h^{\bar{\nu}})]\}.$$

Since Φ_h^{*} is a linear discretization we have with (2.10)-(2.13), (2.15)

$$\eta_{h1}-\eta_{ho} = -\Delta_h\{\sum_{\iota=p_o}^{p_1-1} h^{\nu_\iota}(g_\iota(y) + \sum_{\ell=r_o}^{s_\iota} h^{\nu_\ell}\hat{k}_{\iota\ell}(y) + O(h^{\nu_{s_\iota+1}})) + O(h^{\nu^{*}_{q_o+1}})\}$$

$$= -\Delta_h\{\sum_{\iota=p_o}^{p_1-1} h^{\nu_\iota}g_\iota(y) + \sum_{\iota=p_1}^{q_o} h^{\nu_\iota}\tilde{k}_\iota(y) + O(h^{\nu^{*}_{q_o+1}})\},$$

and therefore

$$\eta_{h1}=\eta_{ho} - \Delta_h \{\sum_{\iota=p_o}^{p_1-1} h^{\nu_\iota} g_\iota(y) + \sum_{\iota=p_1}^{q_o} h^{\nu_\iota}\tilde{k}_\iota(y) + O(h^{\nu^*_{q_o+1}})\}$$

$$= \Delta_h\{y + \sum_{\iota=p_1}^{q_o} h^{\nu_\iota}\tilde{g}_{\iota 1}(y) + O(h^{\nu^*_{q_o+1}})\}$$

resp. $\eta_{h1}=\Delta_h\{y + O(h^{\nu^*_{q_o+1}})\}$. □

In numerical applications it is often impossible to compute $F(y_o)$ exactly. So it is important to know, how well $F(y_o)$ has to be evaluated.

Theorem 6: *Let the conditions of Theorem 5 be satisfied and*

$$(2.16) \qquad \psi(y_o) = \Delta_h^o\{F(y_o) + \sum_{\iota=p_1}^{q_o} h^{\nu_\iota}\hat{f}_\iota(y) + O(h^{\nu^*_{q_o+1}})\} .$$

Then η^*_{h1} *from*

$$(2.17) \qquad \Phi_h^*(\Delta_h c)(\eta^*_{h1}-\eta_{ho}) = -\psi(y_o)$$

admits an asymptotic expansion, corresponding to (2.14),

$$(2.18) \qquad \eta^*_{h1}=\Delta_h\{y+\sum_{\iota=p_1}^{q_o} h^{\nu_\iota} g^*_{\iota 1}(y)+O(h^{\nu^*_{q_o+1}})\},$$ *for* $\nu_{p_1}<\nu^*_{q_o+1}$ *and* $g^*_{\iota 1}$ *independent of* h

resp.

$$\eta^*_{h1}=\Delta_h\{y+O(h^{\nu^*_{q_o+1}})\} \text{ for } \nu_{p1} \geq \nu^*_{q_o+1} .$$

Proof: In (2.15) we introduce (2.16) and using (2.8), (2.17)

$$\Phi_h^*(\Delta_h c)(\eta^*_{h1}-\eta_{ho})=-\Delta_h^o\{F(y_o)+\sum_{\iota=p_1}^{q_o} h^{\nu_\iota}\hat{f}_\iota(y)+O(h^{\nu^*_{q_o+1}})\}$$

$$= -\Delta_h^o\{F'(c)\sum_{\iota=p_o}^{p_1-1} h^{\nu_\iota} g_\iota(y)+\sum_{\iota=p_1}^{q_o} h^{\nu_\iota}\tilde{f}_\iota(y)+O(h^{\nu^*_{q_o+1}})\}$$

and in the same manner like above we have (2.18). □

So the essential idea of this new approach is the following

1.) Discretization

2.) Computation of the defect (exactly enough and with asymptotic expansion for the error)

3.) Newton-step, go back to 1.)

So, similar to the Fox-Pereyra technique of iterated deferred corrections ([6,8,9]) we can improve our approximations by <u>iterated defect corrections</u>.
Here we have generalized one version of Stetters [12] approach to initial value problems for ordinary differential equations. With definitions 3 and 4 it is possible to generalize the other version, too.

3. Application

We want to apply the preceeding results to a special case: Fredholms integral equation of the second kind

$$(3.1) \qquad y(t) - \int_a^b K(s,t,y(s))ds = 0.$$

We get the discrete problem in approximating the integral by a quadrature formula. To have asymptotic expansions we use the trapecoidal rule and find

$$(3.2) \quad \begin{cases} \eta_h(t) - h\,\{ \sum\limits_{j=0}^{N} \frac{1}{2}(2 - \delta_{j0} - \delta_{jN})K(a+jh,t,\eta_h(a+jt))\} = 0, \\ \text{with } N > 1 \text{ and } h := (b-y)/N. \end{cases}$$

If K is smooth enough all our assumptions in 2 are fulfilled. Especially it is not necessary to evaluate $\int_a^b K(s,t,y_o(s))ds$ exactly, but we can use Richardson extrapolation and it is enough, to have $F(y_\nu)$ to the order $2\nu+2$.

Our method works for linear and nonlinear equations. Here we solve numerically the equations

$$(3.3) \qquad y(t) - \lambda\int_o^1 e^{t-s}y(s)ds - f(t) = o$$

with the exact solutions

(3.4) $$y(t) = f(t) - \frac{\lambda}{\lambda-1} e^{t} \int_{o}^{1} e^{-s} f(s)ds \quad \text{for } \lambda \neq 1.$$

(3.2) reduces to a system of linear equations: With $t_j := \frac{j}{N}$ and $u_j := \eta_h(t_j), j = 0(1)N$, we have

(3.5) $$\begin{cases} \sum_{j=o}^{N} u_j\{-\lambda h e^{(t_i-t_j)}(1 - \frac{\delta_{jo}}{2} - \frac{\delta_{jN}}{2}) + \delta_{ij}\} = f(t_i) \\ \qquad i=o(1)N, \ \delta_{ij} = \begin{cases} o \text{ for } i \neq j \\ 1 \text{ for } i=j. \end{cases} \end{cases}$$

We solve (3.3) resp. (3.5) for

$\lambda := 0.1, 0.5,$

$f(t) = f_i(t), i=1(2)3, f_1(t):=\exp(-10\cdot(x-0.5)^2), f_2(t):=\sin t,$

$f_3(t):=\exp(x^2)$

starting with N = 10 and use for T_{q_o} interpolation operators, namely interpolation by polynomials of degree N resp. splines of degree 9 with incidence vectors $(5,1,1,\ldots,1,5)^T \in \mathbb{R}^{N+1}$. To find approximations for the 4 derivatives in o resp. 1 we take Lagrange-polynomials interpolating in the first resp. last 5 points and use the derivatives of these polynomials as approximations for the derivatives of the $T_{q_o}\eta_h$ (see Swartz-Varga [13]. To check if N is appropriately chosen we compare the defects computed for N and 2 N. For the ν-th step of iteration the quotient should behave approximatively like $2^{2^{(\nu+1)}}$. If that is not the case we go from N to 2 N and start again. If the corrections are smaller than a certain tolerance we stop.

We made some numerical experiments on a UNIVAC 1108 with normal precision and compared the results gained by just one correction. The numbers given in the following table are the maximal relative errors obtained for the corresponding cases.

	λ = 0.1				λ = 0.5			
	polynomial of degree N		spline of order 10 with N knots		polynomial of degree		spline of order 10 with N knots	
	10	20	N=10	N=20	10	20	N=10	N=20
f_1	10^{-4}	x *)	6.10^{-5}	x	$1.5.10^{-4}$	x	$9 \cdot 10^{-5}$	x
f_2	x	+)	x		x		x	
f_3	x		x		x		x	

*) usually exact and approximate solution are the same, in some few cases the last digit differs by 1

+) since the tolerance was reached in the step before, the program was stopped.

It is clear that increasing the degree of the interpolating polynomial does not automatically improve the approximation (for instance f_1 and N=40 give very bad values !).

Acknowledgements: This report was supported by the Volks-Wagen Foundation. I am grateful to Dr. R. Weiss, TU Wien, for interesting discussions about this paper, to Dipl.-Math. P. Kürschner, TU Karlsruhe, for his help in programming the examples and to Mrs. M. Zahn for typing the manuscript.

LITERATUR

[1] Benson, M.: Errors in numerical quadrature for certain singular integrands, and the numerical solution of Abel integral equations, dissertation, University of Wisconsin, Madison 1973.

[2] Böhmer, K.: Über die Mittelwerteigenschaft eines Operators und ihre Anwendung auf die Newton'schen Verfahren, Interner Bericht Nr. 75/4 des Instituts für Praktische Mathematik der Universität Karlsruhe.

[3] Brakhage, H.: Über die numerische Behandlung von Integralgleichungen nach der Quadraturformelmethode, Num. Math. 2, 183-196 (1960).

[4] Frank, R.: The method of iterated defect correction and its application to two-point boundary value problems, Part I, Num. Math. 25, 409-419 (1976).

[5] Frank, R. and Ueberhuber, C.W.: Iterated defect corrections to Runge Kutta methods, Rep. Nr. 14/75, Institut für Numerische Mathematik, Technische Universität Wien.

[6] Fox, L. and E.T. Goodwin: Some new methods for numerical integration of ordinary differential equations, Proc. Comb. Phil. Soc. 45, 373-388 (1949).

[7] Gragg, W.: Repeated extrapolation to the limit in the numerical solution of ordinary differential equations, dissertation UCLA (1963).

[8] Pereyra, V.: On improving an approximate solution of a functional equation by deferred corrections, Num. Math. 8, 376-391 (1966).

[9] Pereyra, V.: Iterated deferred corrections for nonlinear operator equations, Num. Math. 10, 316-323 (1967).

[10] Stetter, H.J.: Asymptotic Expansions for the Error of Discretization Algorithms for Non-linear Functional Equations, Num. Math. 7, 18-31 (1965).

[11] Stetter, H.J.: Analysis of Discretization Methods for Ordinary Differential Equations, Springer-Verlag Berlin, Heidelberg, New York (1973).

[12] Stetter, H.J.: Economical global error estimation, in Stiff Differential Systems, Ed. R.A. Willoughby, New York 1974.

[13] Swartz, B.K. and Varga, R.S.: Error bounds for spline and L-spline interpolation, J. Approx. Theory 6, 6-49 (1972).

[14] Zadunaisky, P.E.: A Method for the Estimation of Errors Propagated in the Numerical Solution of a System of Ordinary Differential Equations, in the Theory of orbits in the solar system and in stellar systems, Proc. of Intern. Astronomical Union, Symp. 25, Thessaloniki 1964, Ed. G. Contopoulos.

[15] Zadunaisky, P.E.: On the Accuracy in the Numerical Computation of Orbits, in Periodic Orbits, Stability and Resonances, 216-227, Ed. G.E.O. Giacaglia, Dordrecht-Holland, 1970.

K. Böhmer
Institut für Praktische Mathematik
der Universität Karlsruhe
Postfach 6380
7500 Karlsruhe

Odd-degree spline interpolation at a biinfinite knot sequence

Carl de Boor*

1. Introduction. Let $\underline{t} := (t_i)_{-\infty}^{+\infty}$ be a biinfinite, strictly increasing sequence, set

$$t_{\pm\infty} := \lim_{i\to\pm\infty} t_i ,$$

let $k = 2r$ be a positive, even integer, and denote by $\$_{k,\underline{t}}$ the collection of spline functions of order k (or, of degree $< k$) with knot sequence $\underline{t}$. Explicitly, $\$_{k,\underline{t}}$ consists of exactly those $k-2$ times continuously differentiable functions on

$$I := (t_{-\infty}, t_{\infty})$$

which, on each interval (t_i, t_{i+1}), coincide with some polynomial of degree $< k$, i.e.,

$$\$_{k,\underline{t}} := \mathbb{P}_{k,\underline{t}} \cap C^{k-2} \quad \text{on} \quad I = (t_{-\infty}, t_{\infty}) .$$

We are particularly interested in *bounded* splines

$$m\$_{k,\underline{t}} := \$_{k,\underline{t}} \cap m(I),$$

i.e., in splines s for which

$$\|s\|_\infty := \sup_{t\in I} |s(t)|$$

is finite. It is obvious that the *restriction map*

$$R_{\underline{t}} : \$_{k,\underline{t}} \to \mathbb{R}^{\mathbb{Z}} : s \mapsto s|_{\underline{t}} := (s(t_i))_{-\infty}^{\infty}$$

carries $m\$_{k,\underline{t}}$ into the space $m(\mathbb{Z})$ of bounded, biinfinite sequences. We are interested in inverting this map, i.e., in interpolation. We consider the

Bounded Interpolation Problem: *To construct, for given* $\alpha \in m(\mathbb{Z})$, *some* $s \in m\$_{k,\underline{t}}$ *for which* $s|_{\underline{t}} = \alpha$.

*Sponsored by the United States Army under Contract DAAG29-75-C-0024

we will say that the B.I.P. is correct (for the given knot sequence $\underline{t}$) if it hasexactly one solution for every $\alpha \in m(\mathbb{Z})$.

We consider under what conditions on $\underline{t}$ the B.I.P. is correct. We also discuss the continuity properties of the map $\alpha \mapsto s_\alpha$ in case the B.I.P. is correct. We establish the following theorem.

Theorem 1. If the global mesh ratio

$$M_{\underline{t}} := \sup_{i,j} \Delta t_i / \Delta t_j$$

is finite, then $I = (-\infty, \infty)$, and $R_{\underline{t}}$ maps $m\$_{k,\underline{t}}$ faithfully onto $m(\mathbb{Z})$, i.e., for every bounded, biinfinite sequence α, there exists one and only one bounded spline $s_\alpha \in \$_{k,\underline{t}}$ for which $s_\alpha(t_i) = \alpha_i$, all i. Moreover,

$$\|s_\alpha\|_\infty \le \text{const } \|\alpha\|_\infty, \quad \text{all } \alpha \in m(\mathbb{Z}) , \tag{1.1}$$

with const depending only on k and $M_{\underline{t}}$.

We note in passing the following immediate corollary.

Corollary. Denote by $\overset{\circ}{C}[a,b]$ the space of continuous (b-a)-periodic functions on $\mathbb{R}$. Given $\underline{\tau} := (\tau_i)_0^n$ with $a = \tau_0 < \ldots < \tau_n = b$, let $\underline{t} = (t_i)_{-\infty}^{\infty}$ be its "(b-a)-periodic extension", i.e.,

$$t_{i+nj} := \tau_i + n(b-a) \text{ for } i=1,\ldots,n \text{ and all } j \in \mathbb{Z}.$$

Denote by $\overset{\circ}{\$}_{k,\underline{\tau}}$ the (b-a)-periodic functions in $\$_{k,\underline{t}}$. Then (as is well known), for every $f \in \overset{\circ}{C}[a,b]$, there exists exactly one $s_f \in \overset{\circ}{\$}_{k,\underline{\tau}}$ which agrees with f at $\tau_0, \tau_1, \ldots, \tau_n$. Further, for some const depending only on the global mesh ratio $M_{\underline{\tau}} = \max_{i,j} \Delta\tau_i/\Delta\tau_j$,

$$\|s_f\|_\infty \le \text{const } \|f\|_\infty, \quad \text{all } f \in \overset{\circ}{C}[a,b].$$

Indeed, if $s_f \in \$_{k,\underline{t}}$ agrees with $f \in \overset{\circ}{C}[a,b]$ at $\underline{t}$, then so does its translate $s_f(\cdot - (b-a))$ which is also in $\$_{k,\underline{t}}$, and therefore must equal s_f, by the uniqueness of the interpolating spline. This shows that s_f is the interpolating spline in $\overset{\circ}{\$}_{k,\underline{\tau}}$ for f, and so $\|s_f\| \le \text{const } \|f\|$ from (1.1).

For the case of uniform $\underline{t}$, $\underline{t} = \mathbb{Z}$ say, the problem of bounded interpolation has been solved some time ago by Ju. Subbotin [17]. In this case, the interpolation conditions $s_\alpha|_{\underline{t}} = \alpha$ establish a one-to-one and continuous correspondence between bounded splines and bounded sequences. Subbotin came upon the interpolating spline as a solution of the extremum problem of finding a function s with $s|_{\underline{t}} = \alpha$ and smallest possible (k-1)st derivative, measured in the supremum norm. Later, I.J. Schoenberg investigated the B.I.P. once more, this time as a special case of cardinal spline interpolation to sequences α which do not grow too fast at infinity [15], [16].

Little is known for more general knot sequences. The simplest case, $k = 2$, of piecewise linear interpolation is, of course, trivial. The next simplest case, $k = 4$, of cubic spline interpolation has been investigated in [6] where the above theorem can be found for this case.

The basic tool of the investigation in [6] is the exponential decay or growth of <u>nullsplines</u>. Nullsplines are therefore the topic of Section 2 of this paper, if only to admit defeat in the attempt to generalize the approach of [6]. We are more successful, in Section 3, in identifying, for each knot sequence $\underline{t}$ and each i, a particular <u>fundamental spline</u> L_i, i.e., a spline with $L_i(t_j) = \delta_{ij}$, which must figure in the solution of the B.I.P., if there is one at all (see Lemmas 1 and 2). The argument is based on an idea of Douglas, Dupont and Wahlbin [12] as used in [7] and further clarified, simplified and extended by S.Demko [10]. It is also shown (in Lemma 3 and its corollary) that the r-th derivative of a nontrivial nullspline must increase exponentially in at least one direction. The exponential decay of the fundamental spline L_i is used in Section 4 to prove Theorem 1. That section also contains a proof of the fact (Theorem 4) that the B.I.P. is solvable in terms of exponentially decaying fundamental splines, if it is correct at all. This fact is closely connected with S.Demko's results [10].

<u>2. Nullsplines and fundamental splines.</u> It is clear that the problem of finding, for an arbitrary given biinfinite sequence α, <u>some</u> spline $s \in \$_{k,\underline{t}}$ for which $s|_{\underline{t}} = \alpha$, always has solutions. In other words, it is clear that $R_{\underline{t}}$ maps $\$_{k,\underline{t}}$ <u>onto</u> $\mathbb{R}^{\mathbb{Z}}$. To see this, start with a polynomial p_0 of order k which satisfies $p_0(t_0) = \alpha_0$, $p_0(t_1) = \alpha_1$, and set $s = p_0$ on $[t_0, t_1]$. Now suppose that we have s already determined on some interval $[t_i, t_j]$ and let p_{j-1} be the polynomial which coincides with s on $[t_{j-1}, t_j]$. Then

$$p_j(t) := p_{j-1}(t) + (\alpha_{j+1} - p_{j-1}(t_{j+1}))\left(\frac{t - t_j}{t_{j+1} - t_j}\right)^{k-1}$$

is the <u>unique</u> polynomial of order k which takes on the value α_{j+1} at t_{j+1} and agrees with p_{j-1} (k-1)-fold at t_j. The definition

$$s = p_j \quad \text{on } [t_j, t_{j+1}]$$

therefore provides an extension of s to $[t_i, t_{j+1}]$, and, in fact, the only one possible. The extension to $[t_{i-1}, t_{j+1}]$ is found analogously. In this way, we find a solution inductively.

The argument shows that we can freely choose the interpolating spline on the interval $[t_0, t_1]$ from the k-2 dimensional linear manifold

$$\{p \in \mathbb{P}_k : p(t_0) = \alpha_0,\ p(t_1) = \alpha_1\}$$

and that, with this choice, the interpolating spline is otherwise uniquely determined. In particular, the set of solutions for $\alpha = 0$, i.e., the kernel or nullspace of the restriction map $R_{\underline{t}}$, is a k-2 dimensional linear space, whose elements we call <u>nullsplines</u>. In other words, nullsplines are splines which vanish at all their knots.

The difficulty with the B.I.P. is therefore not the construction of some interpolating spline. Rather, the problem is interesting because we require an interpolating spline with certain additional characteristics or "side conditions", viz. that it be bounded. Nullsplines can be made to play a major role in the analysis of this problem.

For instance, the question of how many bounded solutions there are is equivalent to the question of how many bounded nullsplines there are. More interestingly, a well known approach to the construction of interpolants consists in trying to solve first the special problem of finding, for each i, a <u>fundamental spline</u>, i.e., a spline $L_i \in \$_{k,\underline{t}}$ for which

$$L_i(t_j) = \delta_{i-j}, \text{ all } j.$$

Such a spline consists (more or less) of two nullsplines joined together smoothly at t_i. Therefore, if one could prove that both nullsplines decay exponentially away from t_i, i.e.,

$$\|L_i\|_{(t_j, t_{j+1})} \leq \text{const}_k \lambda^{|i-j|}, \text{ all } j,$$

at a rate $\lambda \in [0,1)$ which is independent of i, then it would follow that the series

$$s_\alpha := \sum_{i=-\infty}^{\infty} \alpha_i L_i \tag{2.1}$$

converges uniformly on compact subsets of I and gives a solution s_α to the B.I.P.. In fact, s_α then depends continuously on α, i.e.,

$$\|s_\alpha\|_\infty \leq \text{const}_{k,\lambda} \|\alpha\|_\infty, \text{ all } \alpha \in m(\mathbb{Z})$$

for some $\text{const}_{k,\lambda}$ which does not depend on α.

The hope for such exponentially decaying fundamental functions is really not that farfetched. Such functions form the basis for Schoenberg's analysis in the case of equidistant knots, and they occur implicitly already in Subbotin's work. Further, a very nice result of S.Demko [10] to be elaborated upon in the next section (see also C.Chui's talk at this conference) shows that the bounded spline interpolant s_α to bounded data α is necessarily of the form (2.1) with exponentially decaying L_i in case s_α depends continuously on α.

In a rather similar way, nullsplines also occur in the discussion of interpolation error. If f is sufficiently smooth, and s_f is its spline interpolant, i.e., $s_f|_{\underline{t}} = f|_{\underline{t}}$, then one gets, formally at first, that

$$(2.2) \qquad f(t) - s_f(t) = \int_{t_{-\infty}}^{t_{\infty}} K(t,s)\, f^{(k)}(s)\, ds \,.$$

Here, the Peano kernel $K(t,\cdot)$ is a spline function of order k with knots $\underline{t}$ and an additional knot at the point t, and vanishes at all the knots $\underline{t}$. Hence, $K(t,\cdot)$ is again a function put together from two nullsplines. The exponential decay of these two nullsplines away from t is desirable here, since only with such a decay can (2.2) actually be verified for interesting functions f. But, I won't say anything more about this here.

Based on my experience with [6], I had at one time considerable hope that the exponential decay of nullsplines could be proved with the help of the following considerations. A nullspline $s \in \$_{k,\underline{t}}$ is determined on the interval $[t_i, t_{i+1}]$ as soon as one knows the vector

$$\hat{s}_i := (s'(t_i), \ldots, s^{(k-2)}(t_i)/(k-2)!)$$

since one knows that $s(t_i) = s(t_{i+1}) = 0$. One can therefore compute $\hat{s}_{i+1}$ from $\hat{s}_i$ in a linear manner. Specifically,

$$\hat{s}_{i+1} = -A(\Delta t_i)\, \hat{s}_i \,,$$

with A(h) the matrix of the form

$$A(h) := \operatorname{diag}(1, h^{-1}, \ldots, h^{-k+3})\, A\, \operatorname{diag}(1, h, \ldots, h^{k-3})$$

and $A = A(1)$ the matrix

$$A := \left(\binom{k-1}{i} - \binom{j}{i} \right)_{i,j=1}^{k-2} .$$

This means that A(h) has many nice properties. For instance, $A^{-1}(h) = A(-h)$, and A(h) is an oscillation matrix in the sense of Gantmacher and Krein.

In the special cubic case, $k = 4$, A(h) has the simple form

$$A(h) = \begin{pmatrix} 2 & h \\ 3/h & 2 \end{pmatrix}$$

and allows therefore the conclusion that $\hat{s}_i$ grows exponentially either for increasing or else for decreasing index i, at a rate of at least 2. This observation goes back to a paper by Birkhoff and the author [1].

The transformation A(h) has been studied in much detail in the case of equidistant knots in a paper by Schoenberg and the author [8], and also, in more generality, by C. Micchelli [14]. But, such exponential decay or growth for nullsplines on an arbitrary knot sequence has so far not been proved. S. Friedland and C. Micchelli [13] have obtained from such considerations results concerning the maximal allowable <u>local</u> mesh ratio

$$m_{\underline{\underline{t}}} := \sup_{|i-j|=1} \Delta t_i/\Delta t_j .$$

<u>3. Exponential decay of the r-th derivative of fundamental splines and nullsplines of order k = 2r.</u> We base the arguments in this section on the best approximation property of spline interpolation. To recall, the r-th divided difference of a sufficiently differentiable function f at the points $t_i, \ldots, t_{i+r}$ can be represented by

$$[t_i,\ldots,t_{i+r}]f = \int M_i(t)\, f^{(r)}(t)\, dt/r!$$

with $M_i = M_{i,r,\underline{\underline{t}}}$ a B-spline of order r,

$$M_i(t) := r[t_i,\ldots,t_{i+r}](\cdot - t)_+^{r-1},$$

normalized to have unit integral. Further, $\{s^{(r)} : s \in \$_{2r,\underline{\underline{t}}}\} = \$_{r,\underline{\underline{t}}}$ while, by a theorem of Curry and Schoenberg [9],

$$\$_{r,\underline{\underline{t}}} = \left\{ \Sigma_i \beta_i M_i : \beta \in \mathbb{Z}^{\mathbb{R}} \right\} \text{ on } I ,$$

where we take the biinfinite sum <u>pointwise</u>, i.e.,

$$(\Sigma_i \beta_i M_i)(t) := \Sigma_i \beta_i M_i(t), \quad \text{all } t \in \mathbb{R}.$$

This makes good sense since

$M_i(t) \geq 0$ with strict inequality iff $t_i < t < t_{i+r}$.

Lemma 1. Let $\mathcal{L}_i := \{L \in \$_{2r,\underline{t}} : L(t_j) = \delta_{i-j}$, all $j\}$. Then $\mathcal{L}_i$ has exactly one element in common with $\mathbb{L}_2^{(r)}(I)$. We denote this element by

$$L_i$$

and call it the i-th fundamental spline for the knot sequence $\underline{t}$. Further, with the abbreviations

$$\text{(3.1)} \qquad \bar{h} := \sup_j \Delta t_j, \quad \underline{h} := \inf_j \Delta t_j ,$$

we have

$$\text{(3.2)} \qquad \|L_i^{(r)}\|_2 \leq \text{const}_r \bar{h}^{1/2}/\underline{h}^r$$

for some constant const_r depending only on r.

Proof. We first prove that $\mathcal{L}_i$ contains at most one element in $\mathbb{L}_2^{(r)}(I) = \{f \in C^{r-1}(I) : f^{(r-1)}$ abs.cont., $f^{(r)} \in \mathbb{L}_2(I)\}$. Since $\mathcal{L}_i - \mathcal{L}_i = \ker R_{\underline{t}}$, it is sufficient to prove that the only nullspline in $\mathbb{L}_2^{(r)}$ is the trivial nullspline. For this, let $s \in \ker R_{\underline{t}} \cap \mathbb{L}_2^{(r)}(I)$. Then, by the introductory remarks for this section,

$$s^{(r)} = \Sigma_j \beta_j M_j \text{ for some } \beta \in \mathbb{R}^{\mathbb{Z}},\ s^{(r)} \in \mathbb{L}_2, \text{ and } \int M_j s^{(r)} = 0 \text{ for all } j.$$

But, by a theorem in [3], there exists a positive constant D_r which depends only on r so that, for $1 \leq p \leq \infty$, and for all $\gamma \in \mathbb{R}^{\mathbb{Z}}$,

$$\text{(3.3)} \qquad D_r^{-1} \|\gamma\|_p \leq \|\Sigma_j \gamma_j ((t_{j+r}-t_j)/r)^{1-1/p} M_j\|_p \leq \|\gamma\|_p .$$

Here, $\|\gamma\|_p := (\Sigma_j |\gamma|^p)^{1/p}$, while, for f on I, $\|f\|_p := (\int_I |f|^p)^{1/p}$. This shows that the sequence $(\hat{M}_j)$ given by

$$\text{(3.4)} \qquad \hat{M}_j := ((t_{j+r}-t_j)/r)^{1/2} M_j, \text{ all } j,$$

is a Schauder basis for $\$_{r,\underline{t}} \cap \mathbb{L}_2$. Therefore, $\Sigma_j \gamma_j \hat{M}_j$ converges $\mathbb{L}_2$ to the spline function in $\mathbb{L}_2$ it represents. But this means that our particular spline $s^{(r)}$ is in the $\mathbb{L}_2$-span of (M_i), yet orthogonal to every one of the M_i, which means that $s^{(r)}$ vanishes identically. But then, since s vanishes more than r times, s itself must vanish identically.

Next, we prove that $\mathcal{L}_i$ contains <u>at least</u> one element in $\mathbb{L}_2^{(r)}(I)$. For this, we recall from [5] that there exists, for any given $\alpha \in \mathbb{R}^{\mathbb{Z}}$, a function g which is locally in $\mathbb{L}_2^{(r)}$ and satisfies $g|_{\underline{t}} = \alpha$, and whose r-th derivative satisfies

$$(3.5) \qquad \|g^{(r)}\|_2 \leq D_r\Big(\sum_j (t_{j+r}-t_j)([t_j,\ldots,t_{j+r}]\alpha)^2\Big)^{1/2},$$

with D_r the same constant mentioned in (3.3). Here, the number $[t_j,\ldots,t_{j+r}]\alpha$ stands for the rth divided difference at the points t_j, $\ldots$, t_{j+r} of any function f for which $f|_{\underline{t}} = \alpha$. In this way, we obtain for the specific sequence $\alpha = (\delta_{i-j})_{j=-\infty}^{\infty}$ a function $g \in \mathbb{L}_2^{(r)}$ for which

$$g(t_i) = \delta_{i-j}, \text{ all } j,$$

while $\|g^{(r)}\|_2$ is bounded by the right side of (3.5). Note that, for the specific sequence $\alpha = (\delta_{i-j})$, this bound becomes

$$\|g^{(r)}\|_2 \leq D_r\Big(\sum_{j=i-r}^{i} (t_{j+r}-t_j)[1/\prod_{\substack{n=j\\ n\neq i}}^{j+r}(t_i-t_n)]^2\Big)^{1/2}$$

$$\leq \text{const}_r(\bar{h})^{\frac{1}{2}}/\underline{h}^r \ .$$

Now let $\hat{g}$ be any element in $\mathbb{L}_2^{(r)}$ so that $\hat{g}^{(r)}$ is the $\mathbb{L}_2$-approximation to $g^{(r)}$ from $\$_{r,\underline{t}} \cap \mathbb{L}_2$. This makes sense since (3.3) insures that $\$_{r,\underline{t}} \cap \mathbb{L}_2$ is a closed subspace of $\mathbb{L}_2(I)$. Then

$$[t_j,\ldots,t_{j+r}]\hat{g} = \int M_j\hat{g}^{(r)}/r! = \int M_j g^{(r)}/r! = [t_j,\ldots,t_{j+r}]g,$$

all j, while $\|\hat{g}^{(r)}\|_2 \leq \|g^{(r)}\|_2$. But this means that, for an appropriate polynomial p of order r,

$$(\hat{g} + p)(t_j) = g(t_j) = \delta_{i-j}, \text{ all } j,$$

while still $\|(\hat{g} + p)^{(r)}\|_2 \leq \|g^{(r)}\|_2 \leq \text{const}_r\bar{h}^{1/2}/\underline{h}^r$. This shows that $L := \hat{g} + p$ is a function of the desired kind. |||

We continue to use the inequality (3.3) and the abbreviation $\hat{M}_j = ((t_{j+r}-t_j)/r)^{1/2}M_j$, and come now to what I consider to be the main point of this paper.

Lemma 2. If β is the sequence of coefficients for $L_i^{(r)}$ with respect to the basis $(\hat{M}_j)$ for $\$_{r,\underline{t}}$, i.e., if $L_i^{(r)} = \Sigma_j \beta_j \hat{M}_j$, and

$$\beta_j^{(n)} := \begin{cases} 0 \ , & |j-i| < n \\ \beta_j, & |j-i| \geq n \end{cases} \ , \quad n=0,1,2, \ldots,$$

then there exist const_r and $\lambda_r \in [0,1)$ depending only on r so that

$$\|\beta^{(n)}\|_2 \leq \mathrm{const}_r \|\beta\|_2 \lambda_r^n \ , \ n=0,1,2, \ldots . \tag{3.6}$$

The inequalities (3.3) allow us to conclude from Lemma 2 the exponential decay of $L_i^{(r)}$ in the following form.

Corollary. For some const_r, and some $\lambda_r \in [0,1)$ depending only on r, and for all i and n,

$$\|L_i^{(r)}\|_{2,(t_{-\infty},t_{i-n})} + \|L_i^{(r)}\|_{2,(t_{i+n},t_{\infty})} \leq \mathrm{const}_r \|L_i^{(r)}\|_2 \lambda_r^n .$$

Proof of Lemma 2. Let

$$A := (\int \hat{M}_i \hat{M}_j)$$

be the Gram matrix for our appropriately normalized B-spline basis of $\$_{r,\underline{t}}$. A proof of the lemma can be obtained directly from the fact that the elements of the inverse matrix for A decay exponentially away from the diagonal at a rate which can be bounded in terms of r and independently of $\underline{t}$. This is proved in [7] with the aid of a nice inequality due to Douglas, Dupont and Wahlbin [12]. But, between the time I proved Lemma 2 this way and the delivery of this talk, S. Demko wrote a paper [10] in which he demonstrated that such arguments use actually very little specific information about splines. Using the inequality of Douglas, Dupont and Wahlbin, he proved the following nice

Theorem (S. Demko). Let $A := (a_{ij})$ be an invertible band matrix (of finite order). Explicitly, assume that, for some m, $a_{ij} = 0$ whenever $|i-j| > m$, and that, for some positive $\underline{K}$ and $\overline{K}$, and some $p \in [1,\infty]$,

$$\underline{K} \|x\|_p \leq \|Ax\|_p \leq \overline{K} \|x\|_p, \quad \text{all } x.$$

Then the entries of the inverse $A^{-1} =: (b_{ij})$ *satisfy*

$$|b_{ij}| \leq \text{const}\ \lambda^{|i-j|}, \quad \underline{\text{all}}\ i,j,$$

for some const *and some* $\lambda \in [0,1)$ *which depend only on* m, p, $\underline{K}$ *and* $\bar{K}$. *In particular, these constants do not depend on the order of the matrix* A.

The interested reader will have no difficulty in proving this theorem after a study of the following proof of Lemma 2, a proof which makes essential use of Demko's ideas, even though the inequality of Douglas, Dupont and Wahlbin fails to make an explicit appearance. In the bargain, the reader will thereby obtain explicit estimates for const and λ (which Demko did not bother to compute).

We note that the specific matrix $A = (\int \hat{M}_i \hat{M}_j)$ is a band matrix, of band width $m = r-1$ in the sense that $\int \hat{M}_i \hat{M}_j = 0$ for $|i-j| > r-1$. Also, we conclude from (3.3) that the sequence-to-sequence transformation

$$\alpha \longmapsto A\alpha$$

induces a linear map on $\ell_2(\mathbb{Z})$ to $\ell_2(\mathbb{Z})$ which we also call A and which is bounded and boundedly invertible. Specifically, one obtains from (3.3) that

$$\kappa := \|A\|_2 \|A^{-1}\|_2 \leq D_r^2 . \tag{3.7}$$

Here, $\|B\|_2 := \sup \{\|B\alpha\|_2 / \|\alpha\|_2 : \alpha \in \ell_2(\mathbb{Z})\}$, as usual.

We now claim that,

$$\text{for all } n \geq 2r, \quad \|\beta^{(n)}\|_2^2 \leq (\kappa^2/(1+\kappa^2)) \|\beta^{(n-2m)}\|_2^2 \tag{3.8}$$

which, with the $\underline{t}$-independent estimate (3.7) for κ, establishes the lemma (with $\lambda_r \leq (\kappa/(1+\kappa^2)^{1/2})^{1/2m}$).

For the proof of (3.8), we consider without loss of generality only the specific function L_0. We note that

$$\begin{aligned}(A\beta)_i &= \int \hat{M}_i L_0^{(r)} = r!((t_{i+r}-t_i)/r)^{1/2}[t_i,\ldots,t_{i+r}]L_0 \\ &= 0 \ \text{ unless } \ t_i \leq t_0 \leq t_{i+r} .\end{aligned}$$

Therefore,

$$\text{(3.9)} \qquad \operatorname{supp} A\beta \subseteq [-r,0] ,$$

where, for any biinfinite sequence α, we use the abbreviation

$$\operatorname{supp} \alpha := \{i \in \mathbb{Z} : \alpha_i \neq 0\}.$$

We claim that, for $n \geq m$,

$$\text{(3.10)} \qquad \operatorname{supp} A\beta^{(n)} \subseteq (-n-m, n+m) \setminus (-n+m, n-m) .$$

Indeed, $\operatorname{supp} (\beta^{(n)} - \beta) \subseteq (-n,n)$, hence $\operatorname{supp} A(\beta^{(n)} - \beta) \subseteq (-n-m, n+m)$ whbh also contains $\operatorname{supp} A\beta = [-r,0]$, therefore

$$\operatorname{supp} A\beta^{(n)} \subseteq (-n-m, n+m) .$$

On the other hand, $\operatorname{supp} \beta^{(n)} \subseteq \mathbb{Z} \setminus (-n, n)$, therefore also

$$\operatorname{supp} A\beta^{(n)} \subseteq \mathbb{Z} \setminus (-n+m, n-m) .$$

It follows from (3.10) that, for $n \geq 2r$,

$$\text{(3.11)} \qquad \operatorname{supp} A\beta^{(n)} \cap \operatorname{supp} A\beta^{(n-2m)} = \emptyset ,$$

therefore

$$\|A\beta^{(n)}\|_2^2 \leq \|A\beta^{(n)}\|_2^2 + \|A\beta^{(n-2m)}\|_2^2 = \|A(\beta^{(n)} - \beta^{(n-2m)})\|_2^2 .$$

But then

$$\|A^{-1}\|_2^{-1} \|\beta^{(n)}\|_2 \leq \|A\beta^{(n)}\|_2 \leq \|A(\beta^{(n)} - \beta^{(n-2m)})\|_2$$
$$\leq \|A\|_2 \|\beta^{(n)} - \beta^{(n-2m)}\|_2 ,$$

i.e.,

$$\|\beta^{(n)}\|_2^2 \leq \kappa^2 \|\beta^{(n)} - \beta^{(n-2m)}\|_2^2$$
$$= \kappa^2 (\|\beta^{(n-2m)}\|_2^2 - \|\beta^{(n)}\|_2^2)$$

which proves our earlier claim (3.8). |||

It is clear that the argument provides the exponential decay of the form (3.6) and with $\lambda \leq (\kappa/(1+\kappa^2)^{1/2})^{1/2m}$ for any sequence β in $\ell_2(\mathbb{Z})$ for which $A\beta$ has finite support. In particular, one obtains such exponential decay for the sequence $\gamma^{(i)}$ for which $A\gamma^{(i)} = (\delta_{i-j})$, i.e., for the i-th row of the matrix inverse of A. Further, it is clear that (3.11) implies $\|A\beta^{(n)}\|_p^p \leq \|A\beta^{(n)}\|_p^p + \|A\beta^{(n-2m)}\|_p^p$ for any $1 \leq p < \infty$, hence, the argument carries at once from $\ell_2(\mathbb{Z})$ over to any $\ell_p(\mathbb{Z})$ with $1 \leq p < \infty$. Demko obtains such exponential decay also for $p = \infty$ by considering the transposed matrix A^T for which then automatically

$$\|A^T\|_1 \|(A^T)^{-1}\|_1 = \|A\|_\infty \|A^{-1}\|_\infty$$

due to the finite order of the matrix he considers. This switch requires a word or two in the infinite case, as follows. As one easily checks, if a (bi)infinite matrix (a_{ij}) gives rise to a bounded linear map A on ℓ_∞, then its transpose gives a bounded linear map B on ℓ_1, and the adjoint of B is then necessarily A itself. This implies that, if a matrix (a_{ij}) gives rise to a bounded linear map on ℓ_∞ which is boundedly invertible, then its inverse can also be represented by a matrix, viz. the transpose of the matrix which represents the inverse of the linear map on ℓ_1 given by the transpose of (a_{ij}). Of course, exponential decay away from the diagonal is unchanged when going over to the transpose.

These comments establish the following

<u>Theorem 2. Let</u> M <u>be a finite, infinite or biinfinite "interval" in</u> $\mathbb{Z}$, <u>let</u> $1 \leq p \leq \infty$, <u>and let</u> $q := \min\{p, p/(p-1)\}$. <u>Let</u> $(a_{ij})_{i,j\in M}$ <u>be a matrix with band width</u> $m := \sup\{|i-j| : a_{ij} \neq 0\}$, <u>and assume that</u> (a_{ij}) <u>induces a bounded linear map</u> A <u>on</u> $\ell_p(M)$. <u>If</u> A <u>is boundedly invertible, then</u> A^{-1} <u>is also given by a matrix,</u> (b_{ij}) <u>say, and</u>

$$|b_{ij}| \leq \text{const } \lambda^{|i-j|}, \quad \underline{\text{all}}\ i,j,$$

<u>with</u>

$$\lambda := (\kappa/(1+\kappa^q)^{1/q})^{1/2m}, \quad \text{const} \le \|A^{-1}\|_p/\lambda^{2m}, \quad \kappa := \|A\|_p\|A^{-1}\|_p \;.$$

We add one more remark. With the appropriate interpretation of "bandedness", the above argument carries through even for matrices which are not banded in the straightforward sense. As a typical example, consider the Gram matrix for a local support basis of some space of functions of <u>several</u> variables. Then, there is no ordering of that basis for which the corresponding Gram matrix is appropriately banded. But, if we follow the geometry of the underlying problem and think of the Gram "matrix" as acting on functions on some <u>multidimensional</u> index set M having an appropriate metric $|\cdot|$ (instead of on $\mathbb{Z}$), then the statement and the proof of Theorem 2 go through otherwise unchanged. We do not pursue this point here further, but alert the reader to Descloux's fine paper [11] in which such considerations can be uncovered once one knows what to look for.

We finish this section with the observation that the r-th derivative of a nontrivial nullspline must increase exponentially in at least one direction. The argument is rather similar to the proof of Lemma 2. We continue to denote by A the specific matrix $(\int \hat{M}_i \hat{M}_j)$ and recall

(3.7) $$\kappa := \|A\|_2\|A^{-1}\|_2 \le D_r^{\,2} \;.$$

<u>Lemma 3</u>. <u>If</u> $\Sigma_i \beta_i \hat{M}_i$ <u>is the</u> r-<u>th derivative of a nullspline in</u> $\$_{2r,\underline{t}}$ <u>and</u> $i \le j$ <u>are arbitrary indices, then</u>

$$(1+\kappa^2)\sum_{\nu=i}^{j}|\beta_\nu|^2 \le \kappa^2 \sum_{\nu=i-2m}^{j+2m}|\beta_\nu|^2 \;.$$

<u>Proof</u>. Define β', β'' by

$$\beta'_\nu := \begin{cases}\beta_\nu, & i\le\nu\le j\\ 0, & \text{otherwise}\end{cases}, \quad \beta''_\nu := \begin{cases}\beta_\nu, & i-2m\le\nu\le j+2m\\ 0, & \text{otherwise}\end{cases},$$

so that the inequality to be proved reads

(3.12) $$(1+\kappa^2)\|\beta'\|_2^2 \le \kappa^2\,\|\beta''\|_2^2 \;.$$

We have

$$\text{supp } A\beta' \subseteq [i-m, j+m]$$

while

$$\text{supp } (\beta - \beta'') \subseteq \mathbb{Z} \setminus [i-2m, j+2m],$$

therefore, with $A\beta = 0$,

$$\text{supp } A\beta'' = \text{supp } A(\beta - \beta'') \subseteq \mathbb{Z} \setminus [i-m, j+m] \subseteq \mathbb{Z} \setminus \text{supp } A\beta'.$$

Consequently,

$$\|A^{-1}\|_2^{-1} \|\beta'\|_2 \leq \|A\beta'\|_2 \leq \|A(\beta' - \beta'')\|_2 \leq \|A\|_2 \|\beta' - \beta''\|_2 ,$$

or, with $\kappa = \|A\|_2 \|A^{-1}\|_2$,

$$\|\beta'\|_2^2 \leq \kappa^2 \|\beta' - \beta''\|_2^2 = \kappa^2 (\|\beta''\|_2^2 - \|\beta'\|_2^2)$$

which implies (3.12). |||

Corollary. Let $\Sigma_i \beta_i \hat{M}_i$ be the r-th derivative of a nullspline s in $\$_{2r,\underline{t}}$ and set

$$a_j := \sum_{2mj<i\leq 2m(j+1)} |\beta_i|^2, \quad \text{all } j \in \mathbb{Z} ,$$

with m := r-1, as before. Then

(3.13) $$\sum_{i<\nu<j} a_\nu \leq \kappa^2 (a_i + a_j), \text{ for all } i < j.$$

Therefore, for all μ, and either for all $i > \mu$ or for all $i < \mu$,

$$a_i \geq \text{const}_\mu \Lambda^{|i-\mu|}$$

with

$$\text{const}_\mu := \tfrac{1}{2} a_\mu / (\kappa^2 \Lambda)$$

and

$$\Lambda := (1 + \kappa^2)/\kappa^2 > 1 .$$

Proof. Assertion (3.13) follows at once from the lemma. The second assertion of the corollary is less obvious. For its proof, assume without loss that $\mu = 0$. From (3.13),

$$\Lambda^{i-1}a_0 \le \sum_{-i<\nu<i} a_\nu, \; i=1,2,3, \ldots .$$

Therefore,

(3.14) $$\Lambda^{i}c \le \sum_{-i<\nu<i} a_\nu, \; i=1,2,3, \ldots$$

with

$$c := a_0/\Lambda .$$

Let now $\text{const}_0 = \frac{1}{2} c/\kappa^2$, as defined above, and assume that the inequality

$$a_i \ge \text{const}_0 \Lambda^{|i|}$$

is violated for some $i > 0$ while also

(3.15) $$a_{-j} < \text{const}_0 \Lambda^j$$

for some positive j which we assume without loss of generality to be no less than i. Then, we can also assume that j is the smallest index $\ge i$ for which (3.15) holds. We obtain from (3.13) that

(3.16) $$\sum_{-j<\nu<i} a_\nu \le \kappa^2(a_{-j} + a_i) < \kappa^2 \text{const}_0(\Lambda^j + \Lambda^i) = \tfrac{1}{2} c(\Lambda^j + \Lambda^i).$$

On the other hand, by (3.13) and by the choice of j,

$$\begin{aligned}\sum_{-j<\nu<i} a_\nu &= \sum_{-j<\nu\le -i} a_\nu + \sum_{-i<\nu<i} a_\nu \\ &\ge \text{const}_0(\Lambda^j - 1 - (\Lambda^i - 1))/(\Lambda - 1) + c\Lambda^i \\ &= \tfrac{1}{2}c\,(\Lambda^j - \Lambda^i) + c\,\Lambda^i = \tfrac{1}{2}c\,(\Lambda^j + \Lambda^i)\end{aligned}$$

which contradicts (3.16), and so finishes the proof. In the second last equality, we used the fact that $\Lambda - 1 = (\kappa^2+1)/\kappa^2 - 1 = 1/\kappa^2$. |||

<u>Remark.</u> It is easy to see that, in the corollary, $a_{\mu-1} + a_\mu \ne 0$ for any μ in case the nullspline s is not trivial. For if, e.g., $a_{-1} = a_0 = 0$, then $s^{(r)}$ would vanish on $[t_{-2m+1+(r-1)}, t_{2m-(r-1)}] = [t_{2-r}, t_{r-1}]$,

hence s would be a polynomial of degree < r on that interval and vanish 2(r-1) times there, therefore would have to vanish identically there. But then, we would have s = 0 by the considerations in Section 2. We can therefore conclude from the corollary that, for a nontrivial null-spline s,

$$a_i \geq \text{const}\, \Lambda^{|i|}$$

either for all $i > 1$ or else for all $i < -1$, with a_i and Λ as in the corollary and $\text{const} := \frac{1}{2}\max\{a_{-1}, a_0\}/(\kappa\Lambda)^2 > 0$.

The argument for this corollary would have been simpler had I been able to prove that every β with $A\beta = 0$ can be written as a sum $\beta = \beta' + \beta''$ with $\Sigma_{i\geq 0}|\beta_i'|^2 < \infty$ and $\Sigma_{i\leq 0}|\beta_i''|^2 < \infty$, and $A\beta' = A\beta'' = 0$.

A minor variation of the arguments for Lemma 3 and its corollary allow the following conclusion of independent interest in the study of linear difference equations.

Theorem 3. Let $A = (a_{ij})$ be a biinfinite matrix which represents a linear map, also denoted by A, on $\ell_p(\mathbb{Z})$ for some $p \in [1, \infty)$ which is bounded and bounded below, i.e., there exist positive $\underline{K}$ and $\overline{K}$ so that

$$\underline{K}\,\|\alpha\|_p \leq \|A\alpha\|_p \leq \overline{K}\,\|\alpha\|_p \quad \text{for all } \alpha \in \ell_p(\mathbb{Z}).$$

If A is a band matrix, i.e., if

$$m := \sup\{|i-j| : a_{ij} \neq 0\} < \infty,$$

then any nontrivial sequence β for which $A\beta = 0$ must increase exponentially either for increasing or for decreasing i. Explicitly, there exist an index μ and a positive $\text{const}_{\beta,\mu}$ so that, either for all $i > \mu$ or else for all $i < \mu$,

$$\sum_{2mi<j\leq 2m(i+1)} \beta_j^p \geq \text{const}_{\beta,\mu}\, \Lambda^{|i-\mu|}$$

with

$$\Lambda := (1+\kappa^p)/\kappa^p \quad \text{and} \quad \kappa := \underline{K}\overline{K}.$$

Thanks are due to Allan Pinkus for questioning the necessity of an additional assumption in an earlier version of this theorem.

<u>4. Exponential decay of the fundamental spline.</u> Assume that the knot sequence is such that the B.I.P. is correct, i.e., has exactly one solution $s_\alpha \in m\$_{k,\underline{t}}$ for every $\alpha \in m(\mathbb{Z})$. This means that the restriction map $R_{\underline{t}}$, when restricted to $m\$_{k,\underline{t}}$, is one-one, onto, and clearly bounded with respect to the sup-norm. One verifies directly (else see (4.2) below) that $m\$_{k,\underline{t}}$ is a closed subspace of $m(I)$, hence complete. The Open Mapping Theorem therefore provides the conclusion that $R_{\underline{t}}$ is <u>boundedly</u> invertible. This means the existence of some const so that

$$\|s_\alpha\|_\infty \leq \text{const}\ \|\alpha\|_\infty\ , \text{ all } \alpha \in m(\mathbb{Z}). \tag{4.1}$$

Let $N_i = N_{i,k,\underline{t}}$ be the i-th **B-spline of order k for the knot sequence** $\underline{t}$, normalized so that

$$N_i(t) \ := \ ([t_{i+1},\ldots,t_{i+k}] - [t_i,\ldots,t_{i+k-1}])(\cdot - t)_+^{k-1}$$

and so , comparing with the B-splines introduced at the beginning of Section 3,

$$N_{i,k,\underline{t}} \ = \ ((t_{i+k}-t_i)/k)\ M_{i,k,\underline{t}}\ .$$

From (3.3), or already from [2],

$$D_k^{-1}\|\beta\|_\infty \leq \|\Sigma_i \beta_i N_i\|_\infty \leq \|\beta\|_\infty\ , \ \text{ all } \beta \in m(\mathbb{Z}), \tag{4.2}$$

for some positive constant D_k depending only on k and not on $\underline{t}$.

Since $(N_i)_{-\infty}^{\infty}$ is a basis for $\$_{k,\underline{t}}$ (in the sense described in the preceding section), it follows that $s \in \$_{k,\underline{t}}$ satisfies $s|_{\underline{t}} = \alpha$ if and only if its B-spline coefficient sequence β satisfies

$$\Sigma_j N_j(t_i)\beta_j \ = \ \alpha_i,\ \text{ all } i\ , \tag{4.3}$$

while $s \in \$_{k,\underline{t}}$ is bounded if and only if its corresponding B-spline sequence β is bounded, by (4.2). We conclude that the B.I.P. has exactly one solution for every $\alpha \in m(\mathbb{Z})$ iff the matrix

$$A \ := \ (N_j(t_i))$$

maps ℓ_∞ faithfully onto ℓ_∞. We collect these facts in the following

Theorem 4. The bounded interpolation problem is correct if and only if the matrix

$$A = (N_j(t_i))$$

provides a faithful linear map from $\ell_\infty(\mathbb{Z})$ *onto* $\ell_\infty(\mathbb{Z})$. *If one or the other of these conditions holds, then* A, *being trivially bounded, is boundedly invertible. Since* A *is also a band matrix, of band width* $m := k-1$, *it then follows from Theorem 2 that the inverse of* A *is also given by a matrix,* (b_{ij}) *say, and that*

$$|b_{ij}| \leq \text{const } \lambda^{|i-j|}, \quad \underline{\text{all}}\ i,j,$$

with

$$\lambda := (\kappa/(1+\kappa))^{1/2m}, \quad \text{const} \leq \kappa/\lambda^{2m}, \; \kappa := \|A^{-1}\|_\infty$$

since $\|A\|_\infty = 1$. *In particular, for all* i, *the function*

$$L_i := \Sigma_j \, b_{ij} \, N_j$$

is then a fundamental spline which decays exponentially at the rate λ, *and the solution* s_α *of the B.I.P. for given* $\alpha \in m(\mathbb{Z})$ *is given by*

$$s_\alpha = \Sigma_j \alpha_j L_j \, ,$$

a series which converges uniformly on compact subsets of I.

We do not know conditions which are both necessary and sufficient for the correctness of the B.I.P. . Since correctness implies boundedness of the map $\alpha \mapsto s_\alpha$, we obtain from [4; Lemma of Section 2] the necessary condition that the local mesh ratio

$$m_{\underline{\underline{t}}} = \sup_{|i-j|=1} \Delta t_i / \Delta t_j$$

be finite. If the local mesh ratio is indeed finite, then a simple sufficient condition for uniqueness of the interpolating bounded spline is the condition that

$$I = (-\infty, \infty) \, . \tag{4.4}$$

This is connected with the fact that, with $k = 2r$, the r-th derivative of any nontrivial nullspline grows exponentially in at least one direction, as described in Lemma 3 and its corollary. Precisely, we have the following

Lemma 4. If $m_{\underline{t}} := \sup_{|i-j|=1} \Delta t_i/\Delta t_j < \infty$, and there exists a bounded nontrivial nullspline s in $\$_{k,\underline{t}}$, then either $t_{-\infty} > -\infty$ or $t_{\infty} < \infty$.

Proof. Let $s = \Sigma_i \gamma_i N_{i,k}$ be the nontrivial bounded nullspline in $\$_{k,\underline{t}}$. Its j-th derivative is then $s^{(j)} = \Sigma_i \gamma_i^{(j)} N_{i,k-j}$, with

$$\gamma_i^{(j)} := \begin{cases} \gamma_i & , j=0 \\ (k-j)(\gamma_i^{(j-1)} - \gamma_{i-1}^{(j-1)})/(t_{i+k-j}-t_i), & j>0 \end{cases} .$$

This implies the estimate

$$(4.5) \qquad |\gamma_i^{(j)}| \le \frac{k!}{(k-j-1)!} 2^j \max\{|\gamma_{i-j}|,\ldots,|\gamma_i|\}/(t_{i+k-j}-t_i)^j$$

(see, e.g., [4], for similar considerations). Write now the r-th derivative of s in terms of the somewhat differently normalized B-splines $\hat{M}_i := (r/(t_{i+r}-t_i))^{1/2} N_{i,r,\underline{t}}$ introduced in Section 2,

$$s^{(r)} = \Sigma_i \beta_i \hat{M}_i .$$

Then $\beta_i = \gamma_i^{(r)}((t_{i+r}-t_i)/r)^{1/2}$, so that, from (4.5),

$$(4.6) \qquad |\beta_i| \le \text{const}_r \|\gamma\|_\infty/(t_{i+r}-t_i)^{r-1/2} .$$

By the corollary to Lemma 3 (in Section 3), we may assume, without loss of generality, the existence of a positive const so that, with $m = r-1$,

$$\sum_{2mj<i\le 2m(j+1)} |\beta_i|^2 \ge \text{const } \Lambda^j , \quad j=2,3,\ldots$$

where $\Lambda := (1+\kappa^2)/\kappa^2 > 1$ and $\kappa \le D_r^2$, the latter a certain constant independent of $\underline{t}$. In conjunction with (4.6), this implies that

$$\begin{aligned} \text{const } \Lambda^j &\le \text{const}_{r,\gamma} \max\{(t_{i+r}-t_i)^{1-2r} : 2mj<i\le 2m(j+1)\} \\ &\le \text{const}_{r,\gamma} (m_{\underline{t}})^{r^2} \min\{(t_{i+r}-t_i)^{1-2r} : 2mj<i\le 2m(j+1)\}, \end{aligned}$$

where we have used the fact that

$$m_{\underline{\underline{t}}}^{-|i-j|} \le (t_{i+r}-t_i)/(t_{j+r}-t_j) \le m_{\underline{\underline{t}}}^{|i-j|} .$$

It now follows that

$$t_{i+r} - t_i \le \text{const}\, \Lambda^{-i/(2r)}, \quad i=2r, 2r+1, \ldots$$

and therefore

$$t_{\infty} = t_{2r} + \sum_{i=0}^{\infty} (t_{(i+1)r} - t_{ir}) < \infty . \quad |||$$

We note in passing that the argument also establishes uniqueness in case either $t_{-\infty}$ or t_{∞} is finite as long as the local mesh ratio is $< \varrho$ for some ϱ which is greater than 1 and depends on Λ.

We are now ready to prove Theorem 1.

Proof of Theorem 1. **Since** the global mesh ratio $M_{\underline{\underline{t}}} = \sup_{i,j} \Delta t_i/\Delta t_j$ is finite, then, in particular, $I = (-\infty, \infty)$ and Lemma 4 implies that $R_{\underline{\underline{t}}}$ maps $m\$_{k,\underline{\underline{t}}}$ one-one to $m(\mathbb{Z})$.

Next, we prove that, for each i, the fundamental spline function L_i introduced in Lemma 1 decays exponentially away from t_i, i.e., for all j and all $x \in [t_j, t_{j+1}]$,

$$(4.7) \qquad |L_i(x)| \le \text{const}\, \lambda^{|i-j|}$$

for some const depending only on k and $M_{\underline{\underline{t}}}$, and some $\lambda \in [0,1)$ which depends only on k. It suffices to consider $j \ge i$. We have $L_i(t_n) = 0$ for $n \ne i$, therefore

$$\begin{aligned} L_i(x) &= (x-t_{j+1})\ldots(x-t_{j+r})[x, t_{j+1}, \ldots, t_{j+r}]L_i \\ &= \prod_{n=1}^{r}(x-t_{j+n}) \int r[x, t_{j+1}, \ldots, t_{j+r}](\cdot - t)_+^{r-1} L_i^{(r)}(t)dt/r! . \end{aligned}$$

By Hölder's inequality,

$$\int r[x, t_{j+1}, \ldots, t_{j+r}](\cdot - t)_+^{r-1} L_i^{(r)}(t)\, dt \le (r/(t_{j+r}-x))^{1/2} \|L_i^{(r)}\|_{2,[x,t_{j+r}]} ,$$

making use of (3.3), so that, from the corollary to Lemma 2,

$$|L_i(x)| \leq \text{const}_r\,(\bar{h}^r/\underline{h}^{1/2})\,\text{const}_r\,\|L_i^{(r)}\|_2\,\lambda^{j-i}$$

with $\lambda \in [0,1)$ depending only on k, and

$$\bar{h} := \sup_n \Delta t_n, \quad \underline{h} := \inf_n \Delta t_n .$$

But now, from Lemma 1,

$$\|L_i^{(r)}\|_2 \leq \text{const}_r\,\bar{h}^{1/2}/\underline{h}^r ,$$

and (4.7) follows.

The exponential decay of all fundamental splines L_i at a rate which does not depend on i now allows us to construct an interpolant s_α in $\$_{k,\underline{t}}$ for arbitrary $\alpha \in m(\mathbb{Z})$, in the form

$$s_\alpha = \Sigma_i \alpha_i L_i ,$$

which satisfies

$$\|s_\alpha\|_\infty \leq \text{const}\,\|\alpha\|_\infty$$

and therefore is in $m\$_{k,\underline{t}}$. |||

It is clear that the argument for Theorem 1 shows the existence of a number $\varrho > 1$ (which depends on k and on the λ of Lemma 2) so that the conclusions of Theorem 1 hold even if we only know that the <u>local</u> mesh ratio is less than ϱ. A quick analysis of the constants involved shows this provable ϱ to converge to 1 very fast as k increases.

References

1. G. Birkhoff & C. de Boor, Error bounds for spline interpolation, J.Math.Mech. 13 (1964) 827-836

2. C. de Boor, On uniform approximation by splines, J.Approximation Theory 1 (1968) 219-235

3. " , The quasi-interpolant as a tool in elementary polynomial spline theory, in "Approximation Theory", G.G. Lorentz ed., Academic Press, New York, 1973, 269-276

4. " , On bounding spline interpolation, J.Approximation Theory 14 (1975) 191-203

5. " ,How small can one make the derivatives of an interpolating function?, J.Approximation Theory 13 (1975) 105-116

6. " , On cubic spline functions that vanish at all knots, MRC TSR 1424, May 1974; Adv.Math. 20 (1976) 1-17

7. " , A bound on the $\mathbb{L}_\infty$-norm of $\mathbb{L}_2$-approximation by splines in terms of a global mesh ratio, MRC TSR 1597, Aug. 1975; Math. Comp. 30 (1976)

8. " & I.J. Schoenberg, Cardinal interpolation and spline functions VIII. The Budan-Fourier theorem for splines and applications, in "Spline functions, Karlsruhe 1975", Springer Lecture Notes in Mathematics 501, Springer, Berlin, 1976, 1-79

9. H.B. Curry & I.J. Schoenberg, On Pólya frequency functions IV. The fundamental spline functions and their limits, J.d'Analyse Math. 17 (1966) 71-107

10. S. Demko, Inverses of band matrices and local convergence of spline projections, Feb. 1976; to appear in SIAM J.Numer.Anal.

11. J. Descloux, On finite element matrices, SIAM J.Numer.Anal. 9 (1972) 260-265

12. J.Douglas, Jr., T. Dupont & L. Wahlbin, Optimal $\mathbb{L}_\infty$-error estimates for Galerkin approximations to solutions of two point boundary value problems, Math.Comp. 29 (1975) 475-483

13. S. Friedland & C. A. Micchelli, Bounds on the solution of difference equations and spline interpolation at knots, to appear

14. C. A. Micchelli, Oscillation matrices and cardinal spline interpolation, in "Studies in spline functions and approximation theory" by S.Karlin, C. Micchelli, A.Pinkus and I.J.Schoenberg,Academic Press, New York, 1976, 163-201

15. I.J. Schoenberg, Cardinal interpolation and spline functions, J. Approximation Theory 2 (1969) 167-206

16. " , Cardinal interpolation and spline functions II., J.Approximation Theory 6 (1972) 404-420

17. Ju. N. Subbotin, On the relations between finite differences and the corresponding derivatives, Proc.Steklov Inst.Mat. 78 (1965) 24-42; Engl.Transl. by Amer.Mathem.Soc. (1967) 23-42

Mathematics Research Center
610 Walnut Street
Madison, WI. 53706
USA

Zur numerischen Stabilität des Newton-Verfahrens bei der nichtlinearen Tschebyscheff-Approximation

Dietrich Braess

When the nonlinear approximation problem is treated by using Newton's method, at each iteration step the solution of a linear approximation problem is required. If we are concerned with nonlinear Chebyshev approximation, the (linear) auxiliary problem is also non trivial. Thus for generating a more effective algorithm the latter problem is solved only on a finite point set. However, then we must not only choose reference points like in Remez-type algorithmus; the reference set has to be augmented in order to take care of numerical stability.

Wohl jeder Numeriker weiß, daß es im allgemeinen ein Kunstfehler ist, die Eigenwerte einer Matrix über das charakteristische Polynom zu berechnen. Die Aufspaltung des Problems in die zwei Teile: 1. Berechnung des charakteristischen Polynoms, 2. Bestimmung der Nullstellen des Polynoms ist hier unzulässig. - Bei einer üblichen Fassung des Newton-Verfahrens für die nichtlineare Tschebyscheff-Approximation beobachtet man numerische Instabilitäten. Die Analyse zeigt, daß man im Prinzip den gleichen Fehler gemacht hat, nämlich eine unzulässige Entkopplung der verschiedenen Ebenen vorgenommen hat. Diese Erkenntnis zeigt den Weg zu einem stabilen Verfahren. Naturgemäß geht die Stabilisierung zu Lasten der Rechengeschwindigkeit.

Zur Beschreibung des Newtonschen Verfahrens benötigen wir zwei Begriffe der nichtlinearen Approximationstheorie [4].

Definition. Sei E ein normierter Raum und $G\subset E$. Dann heißt $h\in E$ Tangentenvektor bei g_o an G, wenn eine Kurve $\{g_\lambda,\ 0\leqslant\lambda<1\}\subset G$ existiert, so daß

$$\|g_\lambda-g_o-\lambda h\|=o(\lambda) \quad \text{für } \lambda\longrightarrow 0.$$

Die Menge der Tangentenvektoren bilden den Tangentialkegel $C_{g_o}G$.

Ist insbesondere G eine n-dimensionale C^1-Untermannigfaltigkeit von E und $F\colon \mathbb{R}^n\supset D\longrightarrow G$ eine Parametrisierung für $g=F(0)$, dann ist

$$C_g G=\{d_oF\ a,\ a\in\mathbb{R}^n\}.$$

Ferner beschreibt $g_\lambda=F(\lambda a)$ eine zum Tangentenvektor $h=d_oFa$ passende Kurve.

Definition. Sei E normierter Raum und $G\subset E$. Dann heißt $g\in G$ kritischer Punkt zu f in G, wenn 0 eine beste Approximation zu (f-g) in C_gG ist.

Die Bedeutung der kritischen Punkte zeigt das
Lemma (über die 1. Variation). Jede lokal beste Approximation zu f in G ist kritischer Punkt.

Der 3-Zeilen-Beweis ist konstruktiv und auch Grundlage des Newtonschen Verfahrens.

Beweis. Sei g nicht kritisch. Dann existiert ein $h\in C_gG$ mit

$$\alpha:=\|f-g\|-\|f-g-h\|>0. \tag{1}$$

Ein solches h kann man insbesondere bei linearem oder konvexem Tangentialkegel mittels eines linearen bzw. konvexen Approximationsproblems bestimmen. Für jede zugeordnete Kurve gilt dann die Abschätzung:

$$\begin{aligned}\|f-g_\lambda\| &\leqslant \|f-g-\lambda h\|+\|g_\lambda-g-\lambda h\|\\ &\leqslant \|(1-\lambda)(f-g)\|+\lambda\|f-g-h\|+o(\lambda)\\ &\leqslant \|f-g\|-\alpha\lambda-o(\lambda)<\|f-g\|-\frac{1}{2}\alpha\lambda\end{aligned} \tag{2}$$

für hinreichend kleine λ. Also existiert in jeder Umgebung eine bessere Approximation.

Mit dem Beweis ist das allgemeine Schema für die Verfahren vom Newton-Typ vorgezeichnet.

Algorithmus.

0. Wähle einen Startwert g_o. Setze $\nu=0$.
1. Bestimme zu g_ν einen Tangentenvektor $h=h_\nu \in C_g G$ mit

$$\|f-g_\nu-h\| < \|f-g_\nu\|.$$

(Möglichkeiten s.u.)
2. Sei $g_{\nu,\lambda}$ eine zu g_ν, h_ν zugeordnete Bahnkurve. Bestimme näherungsweise einen Minimalwert der eindimensional Funktion

$$\lambda \longrightarrow \|f-g_{\nu,\lambda}\|,$$

etwa nach der Halbierungsmethode. Die Lösung sei $g_{\nu+1}$. Fahre bei 1. fort.

Es gibt eine Reihe von Algorithmen, die sich in dieses Schema einordnen. Wir betrachten zunächst zwei Spezialfälle und setzen der Einfachheit halber voraus, daß die Tangentialkegel linear sind.

1. Newton-Verfahren im engeren Sinne. Es wird mit jedem Schritt das lineare Approximationsproblem

$$\|f-g_\nu-h\| = \min_{h \in C_g G}! \tag{3}$$

vollständig gelöst. Die Lösung ergibt das im Schritt 1 des Algorithmus genannte h_ν.

Die Bezeichnung "Newton-Verfahren" rührt von der Tatsache her, daß man im Fall eines linearen Ausgangsproblems bereits mit einem einzigen Schritt die Lösung erhält. In der Literatur wird es auch als Osborn-Watson-Verfahren [10] bezeichnet. Quadratische Konvergenz bei guter Ausgangsnormierung wurde in [7] unter relativ schwachen Voraussetzungen gezeigt (vergl. auch [1]).

Das Verfahren hat jedoch bei der nichtlinearen Tschebyscheff-Approximation den Nachteil, daß das Hilfsproblem wiederum mit einem Iterationsverfahren gelöst werden muß. Ein vergleichsweise großer Rechenaufwand zeigt sich, selbst wenn das lineare Problem nicht über einem Intervall, sondern über einer diskreten Punktmenge gelöst wird [9], vergl. auch [2].

2. Newton-Verfahren mit Referenzen. (Es wird vorausgesetzt, daß der Tangentialkegel einen n-dimensionalen Haarschen Unterraum von C(I) bildet, wobei I ein Intervall ist)

Es wird das linearisierte Approximationsproblem jedesmal nur über n+1 Punkten $t_i = t_i^{(\nu)}$; $i=0,1,\ldots,n$ gelöst:

$$\max_{0\leq i\leq n} |(f-g_\nu-h)(t_i)| = \min_{h\in C_g G}! \tag{4}$$

Dabei werden die Referenz-Punkte $t_0, t_1, \ldots t_n$ bei jedem Schritt in Abhängigkeit von der Fehlerkurve $\varepsilon = \varepsilon_\nu := f-g_\nu$ gewählt. Man zerlege das Intervall I so in Teilintervalle, $I_1, I_2 \ldots$ (Vorzeichenintervalle), daß ε in jedem dieser Intervalle nicht das Vorzeichen wechselt, im benachbarten jedoch das entgegengesetzte Vorzeichen annimmt.

Fall 1. Es gibt höchstens n+2 Vorzeichenintervalle und in jedem nimmt $\varepsilon(t)$ sein absolutes Maximum genau einmal an. Die Menge der Extrema liefert nach eventueller Ergänzung weiterer Punkte die Referenz.

Fall 2. Es gibt mehr als n+1 Vorzeichenintervalle, aber eine Auswahl von n+1 der Intervalle verhält sich wie in Fall 1 angegeben. In den anderen Intervallen sei $|\varepsilon(t)| < \|\varepsilon\|$. Die Extrema der ausgewählten n+1 Intervalle bilden die Referenz.

Fall 3. Alle sonstigen Fälle. Es wird gewissermaßen als Notschritt ein passender Tangentialvektor wie beim Newton-Verfahren im engen Sinne ausgeführt.

In den Fällen 1 und 2 braucht man insbesondere nur ein lineares Gleichungssystem zu lösen. Das Verfahren ist also weniger aufwendig.

Bei dem 2. Verfahren wird also für jeden Newton-Schritt gewissermaßen nur ein Schritt des Remez-Algorithmus ausgeführt. Wie im Spezialfall der rationalen Approximation in [11] und der Exponentialapproximation in [3] läßt sich allgemein zeigen, daß die Iterationsfolgen bei beiden Varianten gegen einen kritischen Punkt konvergieren. Der entscheidende Punkt des Konvergenzbeweises ist das folgende. Wenn h eine Lösung von (3) ist, dann gilt wegen der Wahl der Referenzpunkte für hinreichend kleines λ

$$\|f-g_\nu-\lambda h\| < \|f-g_\nu\|. \tag{5}$$

Für die Herleitung von (2) kommt man mit der im Vergleich zu (1) schwächeren Abschätzung (5) aus.

Die Praxis zeigt jedoch ein Bild, das nicht im Einklang mit dem theoretischen Konvergenzbeweis ist. In einem großen Prozentsatz hat man eine extrem langsame Konvergenz, so daß man vom praktischen Standpunkt nicht mehr von Konvergenz sprechen kann. Die Schwierigkeiten scheinen ähnlich gelagert zu sein, wie sie Krabs von dem in [8] geschilderten Verfahren berichtete. Dort wird ein passender Tangentenvektor h_ν aus Überlegungen gewonnen, die sich an das Kolmogorov-Kriterium anlehnen.

Die in der Praxis auftretenden Komplikationen zeigen, daß das Verfahren numerisch instabil ist. Woran liegt diese Instabilität?

Im Grunde genommen führt man einen zweistufigen Prozeß nach folgendem Schema durch:

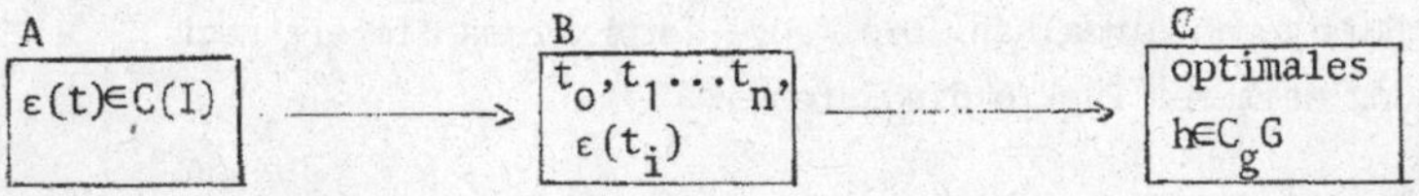

Beim Newton-Verfahren im engeren Sinne gewinnt man aus der Fehlerfunktion, also den Daten im Kasten A direkt einen Tangentialvektor. Die zugeordnete Abbildung von A nach C ist im allgemeinen recht stabil. Beim 2. Verfahren reduziert man die Information A in einem Zwischenschritt nach B so, daß die Zuordnung B ––> C numerisch leicht berechenbar ist. Aber letztere ist wesentlich empfindlicher gegenüber geringfügigen Änderungen der Daten.

Der Ausweg ist jetzt vorgezeichnet. Man muß nach B mehr Informationen übertragen und ein etwas aufwendigeres Verfahren für den Schritt von B nach C in Kauf nehmen. Ein Gewinn gegenüber der ersten Version bleibt immer noch.

<u>3. Stabilisiertes Newton-Verfahren.</u> Man fixiert vor dem Start der Iteration eine Menge von wenigstens (n+1) Punkten; die man äquidistant oder besser an den Intervallenden etwas dichter wählt. Dabei genügen n+1 Punkte bei gutartigen Problemen, während bei weniger gutartigen 2(n+1) Punkte empfohlen werden. - Bei jedem Iterationsschritt bildet man T_ν als Vereinigung der fixierten Punktmenge und der Referenzpunkte, die jeweils wie bei der 2. Variante bestimmt werden. Dann löst man das lineare Programm

$$\max_{t\in T_\nu} |(f-g_\nu-h)(t)| = \min_{h\in C_{g_\nu} G}! \qquad (6)$$

Es ist also ein diskretes Tschebyscheff-Approximationsproblem zu lösen. Dies kann z.B. mit dem Stiefelschen Austauschverfahren ähnlich wie beim Gaußschen Eliminationsverfahren erfolgen.

Mit der Stabilisierung hat der Autor tatsächlich praktische Fortschritte erzielt. Ohne diese ließen sich bei der Exponentialapproximation nur Probleme mit 2 nichtlinearen Parametern sicher lösen. Die Stabilisierung ermöglichte die Behandlung mit 5 (und teilweise mit 6) nichtlinearen Parametern. Der Vollständigkeit halber sei bemerkt, daß der Ansatz ebensoviele lineare Parameter enthielt und daß diese Information ebenfalls verwertet wurde [5].

An dieser Stelle erscheint eine Abgrenzung gegenüber dem nichtlinearen Remez-Algorithmus [1,9] angebracht. Dort wird zunächst eine Referenz ausgewählt und dann zur Lösung eines nichtlinearen Gleichungssystems häufig das Newton-Verfahren herangezogen. Das Newtonsche Verfahren liegt hier gegenüber der Auswahl der Referenz in der untergeordneten Verfahrensebene. Die Situation ist also anders als in dem zentralen Verfahren dieser Arbeit. Wir benötigen im Gegensatz zum Remez-Algorithmus keine à priori Aussage über die Längen von Alternanten.

Im Mittelpunkt dieser Arbeit standen die Diskussionen zweier Ebenen des Newton-Verfahrens. Abschließend sei bemerkt, daß man in diesem Sinne bei der Exponentialapproximation 5 Verfahrensebenen unterscheiden kann. Sie sind schematisch in der folgenden Tabelle dargestellt. Zwecks Vereinfachung besteht die Tendenz, die einzelnen Ebenen getrennt zu betrachten. Weil dies nur bedingt möglich ist, bereitet das Problem auch solche Schwierigkeiten.

Historisch gesehen, entfiel zunächst die oberste Ebene aufgrund der Theorie von Rice. Alle auftretenden Schwierigkeiten schob man auf numerische Instabilitäten in den unteren Ebenen. Später konzentrierten sich die Bemühungen naturgemäß auf die oberste Ebene. Die Probleme dieser Ebene sind seit einigen Wochen grundsätzlich gelöst [6]. Insbesondere scheint es so,

	Verfahrensebene	Reduktion	typische Fragestellung
1	Iterationen mit verschiedenen Startwerten	global ⟶ lokal	Wie wählt man Startwerte, damit die Folgen das globale Minimum erreichen?
2	Newton-Schritt	nichtlinear ⟶ linear	Wie wird garantiert, daß das Abstiegsverfahren einen kritischen Punkt erreicht?
3	Lineare (bzw. konvexe) Tschebyscheff-Approximation	kontinuierliches Problem ⟶ diskretes Problem (lineares Programm oder lineare Gleichungen)	Wie vollständig muß man das lineare T-Problem lösen ?
4	Lineares Gleichungssystem	Gleichungssystem ⟶ elementare Matrizenmanipulation	Wie wählt man die Basis des Tangentialraums, damit die Matrizen gut konditioniert sind?
5	Arithmetrische Operationen	⟶|	Ist doppelte Genauigkeit erforderlich?

Tab. Numerische Behandlung der Exponentialapproximation.

als ob man sich bei der numerischen Behandlung "fast immer" auf Tangentialräume beschränken und die Auswertung von Tangentialkegeln umgangen werden kann. (Dies bedeutet leider nicht, daß der Numeriker die Charakterisierung mittels Tangentialkegel nicht zu kennen braucht.) Das Interesse kann sich jetzt wieder verstärkt auf die anderen Ebenen konzentrieren.

Literatur

1. R.B.Barrar and H.L.Loeb, On the Remez Algorithm for nonlinear families. Numer. Math. 15, 382-391 (1970)

2. A.Baumann, Die Konstruktion bester Tschebyscheff-Approximationen mit Splines bei freien Knoten. Schriftenreihe des Rechenzentrums der Universität Münster Nr. 11 (1974)

3. D.Braess, Die Konstruktion der Tschebyscheff-Approximierenden bei der Anpassung mit Exponentialsummen. J.Approximation Theory 3, 261-273 (1970)

4. -, Kritische Punkte bei der nichtlinearen Tschebyscheff Approximation. Math. Z. 132, 327-341 (1973)

5. -, Eine Möglichkeit zur Konvergenzbeschleunigung bei Iterationsverfahren für bestimmte nichtlineare Probleme. Numer. Math. 14, 468-475 (1970)

6. -, Chebyshev approximation by γ-polynomials III. (in preparation)

7. L.Cromme, Eine Klasse von Verfahren zur Ermittlung bester nichtlinearer Tschebyscheff-Approximationen. Numer. Math. 25, 447-459 (1976)

8. W.Krabs, Ein Verfahren zur Lösung gewisser nichtlinearer diskreter Approximationsprobleme. ZAMM 50, 359-368 (1970)

9. C.M.Lee and F.D.K.Roberts, A comparison of algorithms for rational ℓ_∞ - approximation. Math. Comp. 27, 111-121 (1973)

10. M.R.Osborne and G.A.Watson, An algorithm for minimax approximation in the nonlinear case. Comp.J.12, 63-68 (1969)

11. H.Werner, Die konstruktive Ermittlung der Tschebyscheff-Approximierenden im Bercich der rationalen Funktionen. Arch. Rational Mech. Anal. 11, 368-384 (1962)

Prof. Dr. Dietrich Braess
Ruhr-Universität Bochum
Institut für Mathematik

Universitätsstr. 15o

4630 Bochum

ZUR STETIGEN ABHÄNGIGKEIT DER MENGE DER MINIMALPUNKTE BEI GEWISSEN MINIMIERUNGSAUFGABEN

von

Bruno Brosowski

1. EINLEITUNG

Zahlreiche Aufgaben der mathematischen Optimierung und aus der Theorie der besten Approximation lassen sich als Minimaxaufgaben formulieren. In der vorliegenden Arbeit benutzen wir diesen Ansatz, um die stetige Abhängigkeit der Minimalpunkte von einem Parameter bei Optimierungs- und Approximationsaufgaben zu untersuchen. Wir verwenden diesen Ansatz in der folgenden Form:

Es seien X ein metrischer Raum, Y ein normierter Vektorraum, T ein kompakter Hausdorff-Raum und

$$\Phi : X \times Y \times T \to \mathbb{R} \cup \{\infty\}$$

eine Abbildung. Mit diesen Größen definieren wir die Funktion

$$\varphi : X \times Y \to \mathbb{R} \cup \{\infty\}$$

durch

$$\varphi(x,y) := \sup_{t \in T} \Phi(x,y,t) .$$

Jedem Element x aus X ordnen wir die Menge

$$P_\Phi(x) := \{v_o \in Y \mid \varphi(x,v_o) = \inf_{v \in Y} \varphi(x,v)\}$$

zu. Auf diese Weise erhalten wir eine mengenwertige Abbildung, die den metrischen Raum X in die Potenzmenge $POT(Y)$ von Y abbildet. Im folgenden geben wir Bedingungen für die Ober- und Unterhalbstetigkeit von P_Φ an. Diese Stetigkeitsbegirffe werden in der folgenden Definition gegeben.

DEFINITION 1. Es seien X,Y topologische Räume. Eine Abbildung

$$P : X \to POT(Y)$$

heißt *oberhalbstetig* bzw. *unterhalbstetig in* x_o *aus* X, wenn es zu jeder offenen Menge W aus Y mit $P(x_o) \subset W$ bzw. $P(x_o) \cap W \neq \emptyset$ eine Umgebung U von x_o gibt, so daß $P(x) \subset W$ bzw. $P(x) \cap W \neq \emptyset$ für alle x aus U gilt.

Wir erläutern diesen Ansatz an einigen Beispielen.

BEISPIEL 2. Es seien X ein normierter Vektorraum, $Y:=X$ und T die Einheitskugel im stetigen Dual von X versehen mit der $\sigma(X^*,X)$-Topologie. Ferner sei V eine nichtleere Teilmenge von X. Setzt man

$$\Phi(x,y,x^*) := \begin{cases} x^*(x-y) & \text{falls } y \in V \\ \infty & \text{sonst} \end{cases},$$

so erfaßt man durch diesen Ansatz die beste Approximation eines Elementes x aus X durch Elemente aus V. Man nennt in diesem Fall P_Φ die *metrische Projektion auf* V und bezeichnet sie mit P_V.

BEISPIEL 3. Es seien Y ein normierter Vektorraum, T die Einheitskugel im stetigen Dual von Y versehen mit der $\sigma(Y^*,Y)$-Topologie und V eine nichtleere Teilmenge von Y. Ferner seien y_o aus Y und v_o eine beste Approximation an y_o bezüglich V. Schließlich sei noch

$$X := \{v_o + \lambda(y_o + v_o) \in Y \mid \lambda \geq 0\}.$$

Die Abbildung Φ sei definiert durch

$$\Phi(x,y,x^*) := \begin{cases} x^*(x-y) & \text{falls } y \in V \\ \infty & \text{sonst} \end{cases}.$$

Durch diesen Ansatz erfaßt man die sogenannte *radiale Stetigkeit* der metrischen Projektion, die von BROSOWSKI und DEUTSCH [1972, 1974 a,b] untersucht wurde.

BEISPIEL 4. Es seien X ein normierter Vektorraum, $Y:=X, p: X \to \mathbb{R}$ ein stetiges sublineares Funktional und T das Subdifferential von p in Θ. Ferner sei V eine nichtleere Teilmenge von Y. Setzt man

$$\Phi(x,y,x^*) := \begin{cases} x^*(x-y) & \text{falls } y \text{ aus } V \\ & , \\ \infty & \text{sonst} \end{cases}$$

so erfaßt man durch diesen Ansatz die Minimierung eines sublinearen Funktionals auf einer Teilmenge V von Y.

BEISPIEL 5. Gegeben seien ein metrischer Raum X, ein normierter Vektorraum Y, ein kompakter Hausdorff-Raum T_1 sowie Abbildungen $p:Y \to \mathbb{R}$, $q:Y \times T_1 \to \mathbb{R}$ und $e:X \times T_1 \to \mathbb{R}$. Nun betrachten wir die Aufgabe: Minimiere $p:Y \to \mathbb{R}$ unter den Nebenbedingungen

$$(*) \qquad q(y,t) \leqq e(x,t) \qquad \text{für jedes } t \in T_1.$$

Zur Zurückführung dieser parametrischen Minimierungsaufgabe auf eine ebensolche Minimaxaufgabe setzen wir

$$d_p(x) := \inf_{y \in Y} p(y),$$

wobei das Infimum unter den Nebenbedingungen (*) zu bilden ist. Ferner sei $T := T_1 \cup \{\omega\}$ mit $\omega \notin T_1$ und T versehen mit der Topologie erzeugt durch die offenen Mengen aus T_1 und $\{\omega\}$ offen. Nun definieren wir die Abbildung

$$\Phi(x,y,t) := \begin{cases} q(y,t) - e(x,t) & \text{für } t \in T_1 \\ & . \\ p(y) - d_p(x) & \text{für } t = \omega \end{cases}$$

Dann sind die Minimalpunkte der ursprünglichen Aufgabe genau die Minimalpunkte der neuen Minimaxaufgabe. Auf diese Weise haben wir eine Optimierungsaufgabe mit einem Parameter im Restriktionsvektor zurückgeführt auf eine parametrische Minimaxaufgabe. Ein ähnlicher Ansatz führt zum Ziel, wenn ein Parameter in der Zielfunktion oder in einer der Funktionen q vorliegt.

Für das weitere werden wir stets annehmen, daß die Abbildung $\Phi: X \times Y \times T \to \mathbb{R} \cup \{\infty\}$ bei festem (x,y) bezüglich der Variablen t stetig ist. Ferner setzen wir voraus, daß $P_\Phi(x)$ für alle x aus X nicht leer ist.

2. BEDINGUNGEN FÜR DIE OBERHALBSTETIGKEIT VON P_Φ BEI LINEARER PARAMETERABHÄNGIGKEIT.

In diesem Abschnitt betrachten wir die folgende speziellere Minimaxaufgabe: Es sei Y ein normierter Vektorraum, V eine nichtleere Teilmenge von Y, T eine $\sigma(Y^*,Y)$-kompakte Teilmenge von Y^* und y_o ein Element aus Y. Mit diesen Größen definieren wir die Abbildung $\Psi: Y \times T \to \mathbb{R} \cup \{\infty\}$ gemäß

$$\Psi(y,x^*) := \begin{cases} x^*(y_o - y) & \text{falls } y \in V \\ \infty & \text{sonst} \end{cases}$$

Es sei v_o ein Minimalpunkt des Funktionals

$$\psi(y) := \max_{x^* \in T} \Psi(y,x^*).$$

Ferner seien X ein Kegel eines normierten Vektorraumes Z und $\eta: Z \to Y$ eine lineare Abbildung mit

$$IM(\eta|X) \supset \{\lambda(y_o - v_o) \in Y \mid \lambda \geq 0\}.$$

Nun betrachten wir das durch die Abbildung

$$\Phi(x,y,x^*) := \Psi(y,x^*) + x^* \circ \eta(x)$$

definierte parametrische Minimaxproblem.
Ist V ein linearer Teilraum von Y, so gilt der

<u>Satz 6.</u> *Ist P_Φ an der Stelle Θ aus X oberhalbstetig und ist $P_\Phi(\Theta)$ sternförmig bzgl. v_o, beschränkt und abgeschlossen, so ist $P_\Phi(\Theta)$ kompakt.*

<u>BEWEIS.</u> Falls $P_\Phi(\Theta)$ nicht kompakt ist, so gibt es eine Folge (v_n) aus $P_\Phi(\Theta)$ ohne Häufungspunkt in $P_\Phi(\Theta)$, wobei o.B.d.A. $v_n \neq v_o$ angenommen werden kann. Mit Hilfe der Folge (v_n) definieren wir die reellen Zahlen

$$\beta_n := \sup\{\beta \in \mathbb{R} \mid v_o + \beta(v_n - v_o) \in P_\Phi(\Theta)\},$$

für die wegen der Beschränktheit von $P_\Phi(\Theta)$ die Beziehung $1 \leq \beta_n < \infty$ gilt. Außerdem gilt

$$v_o + \frac{n+1}{n}\beta_n(v_n-v_o) \notin P_\phi(\Theta).$$

Wegen

$$IM(\eta|X) \supset \{\lambda(y_o-v_o) \in Y \,|\, \lambda \geqq 0\}$$

gibt es ein x_1 aus X mit $\eta(x_1)=y_o-v_o$. Wir behaupten

$$v_o + \frac{n+1}{n}\beta_n(v_n-v_o) \in P_\phi(\frac{1}{n}\bar{x}).$$

Es gilt nämlich mit geeigneten $x_1^*, x_2^* \in T$ und beliebigem $v \in V$

$$\varphi\left(\frac{1}{n}x_1, v_o + \frac{n+1}{n}\beta_n(v_n-v_o)\right)$$

$$= x_1^*\left(y_o-\left(v_o + \frac{n+1}{n}\beta_n(v_n-v_o)\right)\right) + x_1^*\left(\frac{1}{n}(y_o-v_o)\right)$$

$$= \frac{n+1}{n}x_1^*(y_o-(v_o+\beta_n(v_n-v_o)))$$

$$\leqq \frac{n+1}{n}\varphi(\Theta, y_o-(v_o+\beta_n(v_n-v_o)))$$

$$\leqq \frac{n+1}{n}\varphi\left(\Theta, y_o-\left(\left(1-\frac{n}{n+1}\,v_o\right)+\frac{n}{n+1}v\right)\right)$$

$$= \frac{n+1}{n}x_2^*\left(y_o-\left(\left(1-\frac{n}{n+1}v_o\right)+\frac{n}{n+1}v\right)\right)$$

$$= x_2^*(y_o-v)+x_2^*\left(\frac{1}{n}(y_o-v_o)\right)$$

$$= x_2^*(y_o-v)+x_2^*\circ\eta\left(\frac{1}{n}\,x1\right)$$

$$\leqq \varphi\left(\frac{1}{n}\,x_1, v\right).$$

Wegen $\frac{n+1}{n}\beta_n \geqq 1$ hat mit (v_n) auch $\left\{v_o+\frac{n+1}{n}\beta_n(v_n-v_o)\right\}$ keinen Häufungspunnkt. Daher ist die Menge

$$\Gamma := \left\{v_o + \frac{n+1}{n}\beta_n(v_n-v_o)\right\}$$

abgeschlossen und somit $U:=Y\setminus\Gamma$ offen. Wegen $U \supset P_\phi(\Theta)$, P_ϕ oberhalbstetig in Θ und $\frac{1}{n}x_1 \to \Theta$ für $n \to \infty$ existiert ein n_o aus $\mathbb{N}$, so daß für alle $n \geqq n_o$ die Inklusion

$$U \supset P_\phi(\frac{1}{n}\,x_1)$$

gilt. Dies ist jedoch wegen

$$v_o + \frac{n+1}{n}\beta_n(v_n - v_o) \in P_\Phi(\frac{1}{n} x_1)$$

nicht möglich. Folglich ist $P_\Phi(\Theta)$ kompakt. ▨

Aus Satz 6 und Beispiel 2 ergibt sich das folgende Ergebnis von SINGER [1972]:

Zusatz 7. *Es sei V ein proximinaler linearer Teilraum eines normierten Vektorraumes Y mit oberhalbstetiger metrischer Projektion P_V. Dann ist für jedes Element y aus Y die Menge $P_V(y)$ kompakt.*

Von BROSOWSKI und DEUTSCH [1974 b] wurde das Singersche Ergebnis verallgemeinert: Ist V eine proximinale Teilmenge aus Y mit $P_V(y)$ konvex für alle $y \in Y$ und P_V *ORU*-stetig, so ist $P_V(y)$ kompakt für alle $y \in Y$.

Dieses Ergebnis läßt sich jedoch nicht aus Satz 6 herleiten.

3. BEDINGUNGEN FÜR DIE UNTERHALBSTETIGKEIT VON P_Φ.

Zur Herleitung von Bedingungen für die Unterhalbstetigkeit von P_Φ ist der von REINDLMEIER [1975] eingeführte Begriff der *P*-Abbildung wichtig.

DEFINITION 8. Es seien X ein topologischer Raum und Y ein topologischer Vektorraum über $\mathbb{R}$. Eine Abbildung

$$\Psi : X \times Y \to \mathbb{R} \cup \{\infty\}$$

heißt *P-Abbildung* an der Stelle x_o aus X, wenn für alle y_o aus Y und alle w aus $Y \setminus \{\Theta\}$ gilt:

Ist $\Psi(x_o, y_o + w) \leq \Psi(x_o, y_o)$, so gibt es Umgebungen $U(x_o), U(y_o)$ und ein $\varepsilon > 0$, so daß für alle x aus $U(x_o)$ und alle y aus $U(y_o)$ die Beziehung

$$\Psi(x, y + \varepsilon w) \leq \Psi(x, y)$$

gilt.

Zur Formulierung des Satzes benötigen wir noch die

DEFINITION 9. Es seien X,Y topologische Räume. Eine mengenwertige Abbildung

$$\psi: X \to POT(Y)$$

heißt *graphenkompakt* in x_o aus X, wenn es für jede Folge (x_n) aus X mit $x_n \to x_o$ und jede Folge (y_n) aus Y mit $y_n \in \psi(x_n)$ eine konvergente Teilfolge (y_{n_k}) gibt mit

$$\lim_{k\to\infty} y_{n_k} = y_o \quad \text{aus } \psi(x_o).$$

Der folgende Satz ist eine geringfügige Verallgemeinerung eines Ergebnisses von REINDLMEIER [1975], der diesen Satz unter der Voraussetzung "konvex" statt "quasikonvex" bewies.

Satz 10. Gegeben sei eine Abbildung

$$\Phi: X \times Y \times T \to \mathbb{R} \cup \{\infty\}$$

mit den Voraussetzungen der Einleitung.
Es seien $\varphi: X \times Y \to \mathbb{R} \cup \{\infty\}$ in x_o graphenkompakt, $P_\Phi: X \to POT(Y)$ eine P-Abbildung in x_o und φ bezüglich der zweiten Variablen quasikonvex.
Dann ist P_Φ an der Stelle x_o unterhalbstetig.

BEWEIS. Ist P_Φ nicht unterhalbstetig in x_o, so gibt es eine Folge (x_n) aus X mit $x_n \to x$ und ein Element y_1 aus $P_\Phi(x_o)$ und eine Umgebung $U(y_1)$ mit

$$P_\Phi(x_n) \cap U(y_1) = \emptyset$$

für jedes n aus $\mathbb{N}$. Nun wähle man aus jedem $P_\Phi(x_n)$ ein Element v_n. Wegen der Graphenkompaktheit von P_Φ in x_o gibt es eine Teilfolge von (v_n), die gegen ein Element y_2 aus $P_\Phi(x_o)$ konvergiert. Es gilt $y_1 \neq y_2$. Zur Abkürzung setzen wir

$$[y_1,y_2] := \{\lambda y_1 + (1-\lambda) y_2 \in Y \mid \lambda \in [0,1]\},$$

$$\Gamma := \{v \in V \mid \text{Für jede Umgebung } U(v) \text{ gilt } U(v) \cap P_\Phi(x_n) \neq \emptyset \text{ für unendlich viele } n\}$$

und

$$M := [y_1,y_2] \cap \Gamma.$$

Es gilt $y_1 \notin M$ und $y_2 \in M \subset P_\Phi(x_o)$. Folglich ist M nicht leer. Da M abgeschlossen ist, gibt es ein maximales t aus $[0,1]$ mit

$$y_o := y_2 + t(y_1 - y_2) \qquad \text{aus } M.$$

Wegen $y_1 \notin M$ ist $t<1$ und daher

$$w := (1-t)(y_1 - y_2) \neq \Theta.$$

Somit ist $y_o + w = y_1$ aus $P_\Phi(x_o)$. Also gilt

$$\varphi(x_o, y_o + w) = \varphi(x_o, y_o).$$

Da φ nach Voraussetzung eine P-Abbildung ist, gibt es Umgebungen $U(x_o)$, $U(y_o)$ und eine reelle Zahl $\varepsilon > 0$, so daß für alle $x \in U(x_o)$ und für alle $y \in U(y_o)$ die Beziehung

$$\varphi(x, y + \varepsilon w) \leqq \varphi(x, y)$$

gilt. Wegen y_o aus M gibt es eine Teilfolge (x_{n_i}) und Elemente u_i aus $P_\Phi(x_{n_i})$ mit $u_i \to y_o$ für $i \to \infty$. Folglich existiert ein i_o derart, daß für alle $i \geqq i_o$ die Beziehungen

$$x_{n_i} \text{ aus } U(x_o) \quad \text{und} \quad u_i \text{ aus } U(y_o)$$

gelten. Also gilt

$$\varphi(x_{n_i}, u_i + \varepsilon w) \leqq \varphi(x_{n_i}, u_i).$$

Wegen u_i aus $P_\Phi(x_{n_i})$ ist $\varphi(x_{n_i}, u_i)$ minimal und es gilt daher

$$\varphi(x_{n_i}, u_i + \varepsilon w) = \varphi(x_{n_i}, u_i)$$

oder

$$u_i + \varepsilon w \text{ aus } P(x_{n_i}).$$

Ferner gilt $u_i + \varepsilon w \to y_o + \varepsilon w$ für $i \to \infty$. Daher folgt für jede Umgebung $U(y_o + \varepsilon w)$ die Beziehung

$$U(y_o + \varepsilon w) \cap P_\Phi(x_{n_i}) \neq \emptyset$$

für unendlich viele i. Also gilt

$$y_o + \varepsilon w \in \Gamma.$$

Die Quasikonvexität von $\varphi(x,v)$ bezüglich v impliziert, daß aus der Ungleichung

$$\varphi(x,y+\varepsilon w) \leqq \varphi(x,y)$$

die Ungleichung

$$\varphi(x,y+\delta w) \leqq \varphi(x,y)$$

für alle δ mit $0 < \delta \leqq \varepsilon$ folgt. Daher ist $\varepsilon>0$ so wählbar, daß gilt

$$y_o + \varepsilon w \in [y_1,y_2].$$

Insgesamt haben wir

$$y_o + \varepsilon w \in [y_1,y_2] \cap \Gamma = M,$$

also

$$\begin{aligned} y_o + \varepsilon w &= y_2+t(y_1-y_2) + \varepsilon(1-t)(y_1-y_2) \\ &= y_2+[t+\varepsilon(1-t)](y_1-y_2) \end{aligned}$$

aus M. Dies ist aber nicht möglich, da t maximal gewählt war. Folglich ist P_Φ unterhalbstetig im Punkte x_o. ▨

4. LITERATUR

BROSOWSKI, B.; F.DEUTSCH

1972 Some new continuity concepts for metric projections. Bull.Am.Math.Soc. 78, 974-978.

1974 a On some geometrical properties of suns. J.Approx.Theory 10, 245-267.

1974 b Radial Continuity of Set-valued metric projections. J.Approx.Theory 11, 236-253.

REINDLMEIER, J.

1975 Zur stetigen Parameterabhängigkeit der Menge der Minimalpunkte bei Minmax-Aufgaben. Dissertation. Universität Göttingen.

SINGER, I.

1972 On set-valued metric projections.
in "Linear Operators and Approximation", ISNM Vol.20, 217-233.

ANSCHRIFT:

Prof.Dr.Bruno Brosowski
Fachbereich Mathematik
Johann Wolfgang Goethe-Universität
Robert Mayer Str. 6-10
D-6000 Frankfurt

PIECEWISE POLYNOMIAL APPROXIMATION, EMBEDDING THEOREM AND RATIONAL APPROXIMATION

Ju. A. Brudnyi

1°. The aim of this lecture is to expose some results of the author in piecewise polynomial and rational approximation of functions of several variables. In proofs we use facts of the local approximatien theory, that is why it will be useful to begin with a brief survey of this theory.

2°. Let $M(S)$ be a quasi-Banach space* of functions measurable on a bounded set $S \subset R^n$ with a monotone permutation-invariant norm, e.g., L_p - space, $0 < p \leq \infty$, Orlicz space, Lorentz space L_{pq}, $0 < p, q \leq \infty$. Let π be a finite cube's packing, i.e., a finite set of non-overlaping cubes in R^n with centers in S . Finally denote the space of piecewise polynowial functions by $\mathcal{P}_k(\pi)$, namely a function f defined on S belongs to $\mathcal{P}_k(\pi)$ iff the trace $f|_{Q \cap S}$ coinsides on $Q \cap S$ with polynomial of degree** $\leq k-1$ for every cube $Q \in \pi$.

Definition of the local approximation.

$$E_k(f;\pi) := \inf\{\|f-p\|_{M(S)} \,;\, p \in \mathcal{P}_k(\pi)\} \quad (1)$$

In the case $\pi = \{Q\}$ we define (1) by $E_k(f;Q)$. The important aim of the theory is to ascertain relations between the behavior of (1) (as the function of π) and properties of f .

To formulate the first result us remind that k - modulus of continuity of the function $f \in M(S)$ is

$$\omega_k(f;t) := \sup\{\|\Delta_h^k f\|_{M(S_h)} \,;\, h \in R^n,\ |h| \leq \tau\}$$

* i.e. the triangle inequlity is fulfilled with a constant $C \geq 1$ in the right part. The space is Banach if $C = 1$.

** a polynomial of negative degree is considered to be zero.

Here S_h is the domain of the function $x \mapsto \Delta_h^k(f;x)$ and $\Delta_h^k := \Delta_h(\Delta_h^{k-1})$ with $\Delta_h(f;x) = f(x+h) - f(x)$.

Theorem 1. If S is a region with the Lipschitz boundary, $M(S)$ is Banach space with Fatou property, then for $f \in M(S)$

$$E_k(f;\pi) \leq \gamma\omega_k(f;|\pi|), \ \forall \pi. \qquad (2)$$

Here $|\pi| := \sup\{\operatorname{diam} Q; Q \in \pi\}$ and $\gamma = \gamma(k,S)$.

Inversely, there exists the packing π_t consisting of equal cubes of the diameter t such that

$$\omega_k(f;t) \leq \gamma(k)E_k(f;\pi_t), \ \forall t > 0. \qquad (3)$$

Remarks. 1) Inequality (3) is true in more general situation. 2) The condition on the boundary S is exact.

Corollary 1.

$$\omega_k(f;t) \approx \sup_{\pi} E_k(f;\pi), \ \forall t > 0. \qquad (4)$$

Here $\sup$ is taken with respect to π consisting of equal cubes of the diameter t.

Definition of approximation k -modulus of continuity.

$$\tilde{\omega}_k(f;t) := \sup_{\pi} E_k(f;\pi),$$

where $\sup$ is as in (4).

Thus $\tilde{\omega}_k(f;t) \approx \omega_k(f;t)$ in the situation of theorem 1; in general the inequality $\omega_k(f;t) \lesssim \tilde{\omega}_k(f;t)$ is only true; see (3).

The corollary of theorem 1 in the case $M(S) = L_\infty(0,1)$ and $\pi = \{[0,1]\}$ is Whitney's theorem [32]; special cases of theorem 1 are proved in [3], [4], [7], p.79, the general case is contained in [10].

3°. Now consider the relations between differential properties of f at a point and the behavior of it local approximation in a neighbourhood of this point.

Denote by $\| f; \tilde{S} \|^*$, $\tilde{S} \subset S$, the number $\| f \chi \|_{M(S)} / \| \chi \|_{M(S)}$, where χ is the characteristic function of $\tilde{S}$. The denominator of this fraction is equal (by definition) to $\varphi_M(\mathrm{mes}\, \tilde{S})$, where φ_M is the fundamental function of M. Let

$$E_k^*(f; Q) = E_k(f; Q) / \varphi_M(\mathrm{mes}\, Q).$$

Following in the essential Calderon and Zygmund [17], say that f from $M(S)$ belongs to $T_x^\lambda M(S)$, $x \in S$, $-\infty < \lambda < \infty$, if there exists (Taylor) polynomial P of degree $< \lambda$ such that

$$\| f - P; Q \cap S \|^* = O\{(\mathrm{mes}\, Q)^{\lambda/n}\} \tag{5}$$

for every cube Q with the center x.

The definition of the space $t_x^\lambda M(S)$ can be obtained if we suppose that o is in the right part of (5) and the degree of P is less than or equal to λ.

Remark. Calderon-Zygmund spaces $T_p^\lambda(x)$ and $t_p^\lambda(x)$ are realized in case $M = L_p$, $1 \le p \le \infty$, and $S = R^n$.

Taylor polynomial is not unique in general, but uniqueness is in the case:

$$\inf_Q \frac{\mathrm{mes}(Q \cap S)}{\mathrm{mes}\, Q} > 0 \tag{6}$$

where Q is a cube with the center x and of volume ≤ 1.

Theorem 2. Suppose that $\lambda \le k$, $\lambda \notin \{0, 1, \dots, k-1\}$ and (6) is fulfilled. Then

$$f \in T_x^\lambda M(S) \iff \overline{\lim_{Q \to x}} \frac{E_k^*(f; Q)}{(\mathrm{mes}\, Q)^{\lambda/n}} < +\infty \tag{7}$$

If moreover $\lambda \neq \kappa$, then the equality of the limit in (7) to zero is equivalent to the fact that f belongs to $t_x^\lambda M(S)$.

Theorem 3. Suppose that $M = L_p$, $0 < p \leq \infty$, and $\tilde{S}$ is a measurable subset of S; then the limit (7) with $\lambda = \kappa$ is finite for almost every $x \in \tilde{S}$ iff $f \in t_x^\kappa L_p(S)$ for almost every $x \in \tilde{S}$.

Corollary 2. If $f \in T_p^\kappa(x)$ for $x \in \tilde{S}$, then $f \in t_p^\kappa(x)$ for almost every $x \in \tilde{S}$.

Theorem 2 strengthens the corresponding theorem from [3], theorem 3 is presented in [9]. Corollary 2 in case $1 \leq p \leq \infty$ is the classical result of Rademacher-Stepanoff-Calderon-Zygmund.

4°. Let us consider the question of existence of derivatives of f in subsets of S. Denote $E_\kappa^*(f;Q)$ by $E_r^*(x)$ in the case when Q has the center x and the volume r^n and suppose $0 < \lambda \leq \kappa$ and $\lambda \notin \{1, \ldots, \kappa-1\}$.

Theorem 4. If (6) takes place iniformly with respect to $x \in \tilde{S} \subset S$ and

$$\sup_{x \in \tilde{S}} E_r^*(x) = O(r^\lambda) \tag{8}$$

then the trace $f|_{\tilde{S}}$ is extended up to the function from $Lip(\lambda; R^n)$.

If moreover $\lambda \neq \kappa$ and the right side of (8) is $o(r^\lambda)$ then the trace $f|_{\tilde{S}}$ is extended to the function from $lip(\lambda; R^n)$.

Suppose now that (6) is fulfilled iniformly with respect to $x \in \tilde{S}$ and instead of (8) the following condition takes place:

For certain integer m, $0 \leq m \leq \kappa-1$, and certain quasi-Banach space $N(\tilde{S})$

$$h(r) = \int_0^r \frac{\|E_r^*\|_{N(\tilde{S})}}{r^{m+1}} dr < +\infty \tag{9}$$

Under these assumptions is true the following

<u>Theorem 5</u>. The trace $f|_{\tilde{S}} = \tilde{f}$ has in $N(\tilde{S})$ strong derivatives* of order $\le m$ and

$$\omega_{k-m}(\mathcal{D}^m \tilde{f}; \tau)_{N(\tilde{S})} \le \gamma h(5\tau),$$

where γ does not depend on f and τ.

<u>Remark</u>. It is also possible to evaluate ω_p, $0 < p < k-m$, for this it is necessary to add $\gamma \tau^p \{ \int_\tau^d \frac{h(s)}{s^{p+1}} ds + \| f \|_{N(S)} \}$ to the right part, where $d = diam\, S$ and γ does not depend on f and τ.

Theorem 4 is contained in [5], theorem 5 for typical special cases is in [3], [7], pp.85-87.

5°. Another important goal of the local approximation theory is the study of approximative spaces of the local type (LA -spaces). Examples of such spaces are BMO space of Iohn-Nirenberg, $\mathcal{L}$ -spaces, studied in works of Campanato, Stampacchia, Spanne, V.P. Il'yin and others, and also spaces of functions of the bounded variation (in different sences). The general approach to the theory of LA -spaces on the base of the local approximation theory has developed in [7].

Let Ω be a measurable space of which points are packings π (not necessary all). Denote by X a quasi-Banach space of measurable functions on Ω with the monotone norm.

<u>Definition of LA -space.</u>

$$\mathcal{L}^k_M(X; S) := \{ f \in M(S);\ \| E_k(f; *) \|_X < +\infty \}$$

Indicate only one general assertion about these spaces.

<u>Teorem 6.</u> Suppose the measure on Ω is σ -finite and

---

* A strong derivative $\mathcal{D}_i f$ is defined as a limit in $N(\tilde{S})$ of the corresponding difference ratio ; $\mathcal{D}^\alpha = \mathcal{D}_1^{\alpha_1} \cdots \mathcal{D}_n^{\alpha_n}$

the sets* $\{\pi \in \Omega ; \pi < \pi_0\}$ and $\{\pi \in \Omega ; \pi > \pi_0\}$ are measurable in Ω. Finally, suppose the spaces X_i, M_i are Banach and more-over M_i is separable and has the Fatou property, $i = 0, 1$. Then

$$[\mathcal{Z}_0, \mathcal{Z}_1]_\theta \subset \mathcal{Z}^\kappa_M (X; S).$$

Here $\mathcal{Z}_i = \mathcal{Z}^\kappa_{M_i}(X_i; S)$, $i = 0, 1$, $M = [M_0, M_1]_\theta$, $X = X_0^{1-\theta} X_1^\theta$.

Remarks. a) For the real method of interpolation we have, without the conjectures of theorem 6 but with $M_0 = M_1 = M$

$$(\mathcal{Z}_0, \mathcal{Z}_1)_{\theta q} \subset \mathcal{Z}^\kappa_M (X; S)$$

where $X = (X_0, X_1)_{\theta q/p}$ and $p = p_M \le 1$ ($p_M = 1$ if M Banach space).

b) Theorem 1 is a generalisation of theorem 1 in [7], p. 126; however in this theorem the condition $B_\theta = [B_0, B_1]_\theta$ is omitted and the proof of this theorem is not correct.

6°. Some families of LA-spaces play the essential role for applications. So $\tilde{\Lambda}^\kappa_M(X; S)$ spaces play the first fiddle in studying imbedding theorems and $V^{\kappa\lambda}_{pq}(S)$ spaces - in studying differential properties of functions of several variables. Consider the family $\tilde{\Lambda}^\kappa_M$ at first. Suppose X is a quasi-Banach space of measurable functions defined on R_+ (with the measure $x^{-1} dx$) and the monotone quasinorm.

Definition of Lipschitz space

$$\tilde{\Lambda}^\kappa_M (X; S) := \{ f \in M(S) ; \| \tilde{\omega}_\kappa (f; *) \|_X < +\infty \}.$$

---

* $\pi < \pi_0$ iff each cube of π is contained in a cube of π_0.

To formulate the main result define the operator Γ transforming functions of a single variable into functions of two variables according to formula

$$\Gamma(f;t,s) := \int_t^{2s^n} \frac{f(\sqrt[n]{u})}{\varphi_M(u)} \frac{du}{u}, \quad 0 < t \le s^n \le \operatorname{mes} S;$$

Γ is equal to zero for other values of variables $(t,s) \in R_+^2$

Let further $N \circ X$ be the quasi-Banach space on R_+^2 defined by the quasinorm

$$\|f\|_{N \circ X} := \| \, \|f(*,t)\|_N \|_X .$$

Theorem 7. If (6) holds uniformly on S * and $\Gamma : X \to N \circ Y$ then it is the continious imbedding

$$\tilde{\Lambda}_M(X;S) \subset \tilde{\Lambda}_N(Y;S)$$

Remark. The right part coincides with $N(S)$ when $Y = L_\infty$. Note some important corollaries

A. Consider the family of quasi-Banach spaces $\Phi_{q\theta}$, $0 < \theta, q \le \infty$

$$\|f\|_{\Phi_{\theta q}} := \left\{ \int_0^\infty \left| \frac{f(u)}{u^q} \right|^\theta \frac{du}{u} \right\}^{1/\theta}$$

If S is a region with the Lipschitz boundary and M is a Banach space with Fatou property then

$$W_M^k(S) \subset \tilde{\Lambda}_M^k(\Phi_{k\infty}; S) \quad ** \tag{10}$$

With these assumptions for S and M

$$B_M^{\lambda\theta}(S) = \tilde{\Lambda}_M^k(\Phi_{\lambda\theta}; S) \tag{11}$$

---

* Such sets will be named regular farther.

** It is equality for reflexive M .

There are Sobolev and Nikolskiy-Besov spaces formed on the base $M(S)$ in the left part of (10), (11).

<u>Remark</u>. The spaces are denoted by W_p^k and $B_p^{\lambda\theta}$ (and also $B_{p\theta}^{\lambda}$) in the case $M = L_p$.

The relation (11) gives the base for the definition

$$\mathcal{B}_p^{\lambda\theta}(S) := \tilde{\Lambda}_{L_p}^{k}(\Phi_{\lambda\theta}; S), \; 0 < \lambda < k. \qquad (12)$$

Owing to (11) it is a correct widening of the scale of spaces for the set S different from the region, and for p or θ is less than unity.

We get as a corollary of the theorems 5 and 7.

<u>Theorem 8</u>. If S is a regular set then

a) k can be replaced by an arbitrary $s > \lambda$ in (12);

b) function of the space (12) has strong derivatives of order $m < \lambda$ and m -th derivatives belong to $\mathcal{B}_p^{\lambda-m,\theta}(S)$;

c) if $1 \leq p \leq \infty$ and λ is non integer then

$$\mathcal{B}_p^{\lambda}(S) = B_p^{\lambda}(R^n)|_S \; ;$$

d) if $\frac{\lambda-\mu}{n} = \frac{1}{p} - \frac{1}{q}$, $p < q \leq \infty$, then

$$\mathcal{B}_p^{\lambda\theta}(S) \subset \mathcal{B}_q^{\mu\theta}(S);$$

e) if $\frac{\lambda}{n} = \frac{1}{p} - \frac{1}{q}$, $p < q \leq \infty$, then*

$$\mathcal{B}_p^{\lambda\theta}(S) \subset L_{q\theta}(S)$$

The right part can be replaced by BMO space in the case $\theta = \infty$, $q = \infty$ and the space $C(\bar{S})$ in the case $\theta = 1$, $q = \infty$.

---

* The quasinorm of $L_{q\theta}(S)$ is defined in the case $q = \infty$ by $\left\{ \int_0^{a/2} \left| \frac{f^*(u)}{\ln a/u} \right|^{\theta} \frac{du}{u} \right\}^{1/\theta}$, $a = 2 \operatorname{mes} S$.

f) if $\frac{\lambda}{n} = \frac{1}{p} - \frac{1}{\theta} + \frac{1}{q}$, $\theta < q < \infty$, then the operator $(f,g) \mapsto f \cdot g$ maps continuously $\mathcal{B}_p^{\lambda\theta}(S) \times L_{q\infty}(S)$ into $L_\theta(S)$.

We obtain already known and some new properties of $B_p^{\lambda\theta}$ spaces (e.g. the assertion f) and the assertions e), $q = \infty$). Note, the imbedd ing e) contains also the following inequality

$$\omega_K(f;\tau)_{L_{q\theta}(S)} \leq \gamma \left\{ \int_0^\tau \left| \frac{\omega_K(f;u)_{L_p(S)}}{u^\lambda} \right|^\theta \frac{du}{u} \right\}^{1/\theta} ;$$

here S is the region with the Lipschitz boundary and λ, p, q as in e). The inequality was obtained by P.L.Ul'janov in the case $S = [0,1]$, $\theta = q$ in other way [31].

In the case of Sobolev spaces I restrict myself by the following result. Let

$$\psi_a(\tau) = \int_\tau^a \frac{u^{K/n - 1}}{\varphi_M(u)} du$$

and $\psi_a(\tau) = 0$ if $\tau > a$. Further M , N are Banach spaces, M has Fatou property and S is region with the Lipschits boundary.

Theorem 9. If $\psi_a \in N(0,1)$ then

$$W_M^K(S) \subset N(S)$$

If $\psi_a \to 0$ in N , then this imbedding is compact.

Theorem 7 and its corollaries are presented in [8] , [11] , [12] .

B. Point out some applications of theorem 8 to stating the structural theorems for $\mathcal{L}$ -spaces; the spaces were studied in works of Campanato, Meyers, Stampacchia, Peetre, Spanne, V.P.Il'in etc.

Definition of $\mathcal{L}_p^{\lambda\theta}(S)$ space.

$$\mathcal{L}_p^{\lambda\theta}(S) := \{ f \in L(S); \left(\int_S |E_r^*(x)|^p dx \right)^{1/p} \in \Phi_{\lambda\theta} \} \qquad (13)$$

Here $E_n^*(x)$ equals, as in 4°, $E_k(f;Q)_{L(S)}/mes\,Q$, where cube Q has the center x and the volume r^n.

Theorem 10. If S is a regular set, $0<\lambda<k$ and $1\le p\le\infty$ then

$$\mathcal{L}_p^{\lambda\theta}(S)=\mathcal{B}_p^{\lambda\theta}(S)$$

Corollary 3. If λ is non integer then*

$$\mathcal{L}_p^{\lambda p}(S)=B_p^{\lambda}(R^n)|_S$$

The space (13) is isomorphic to BMO space in the case $\lambda=0$, $\theta=p=\infty$ and it is isomorphic to Morrey space in the case $\lambda<0$, $p=\theta=\infty$. So the following imbedding is a generalisation of Morrey's criterion of compactness

$$W_1^k\mathcal{L}_p^{\lambda\theta}(S)\subset B_p^{k+\lambda,\theta}(S),\ k+\lambda>0;$$

here S is a region with Lipschitz boundary.

The typical particular case of theorem 10 was presented in [7] , p.88. Corollary 3 is obtained from teorem 4 in the case $p=\infty$ and from theorem 5 and the extension theorem of [22] in the case $1\le p<\infty$.

7°. Finally consider just one more family of LA-spaces. Define a space of functions of a cube Var_p by finiteness of p-variation, i.e. $\mathcal{I}\in Var_p$ iff

$$Var_p\,\mathcal{I}:=\sup_{\pi}\left\{\sum_{Q\in\pi}|\mathcal{I}(Q)|^p\right\}^{1/p}<+\infty$$

Definition of $V_{pq}^{k\lambda}(S)$ space.

---

* It is $Lip(\lambda;R^n)|_S$ in the right part in the case $p=\infty$

$$V_{pq}^{\kappa\lambda}(S) := \{ f \in L_q(S) ;\ \mathcal{I}_f \in Var_p \}. \qquad (14)$$

Here $\mathcal{I}_f(Q) = (mes Q)^{-\lambda/n} E_\kappa(f;Q)_{L_q(S)}$ and $0 < p, q \le \infty$, $-\infty < \lambda < \infty$. The closure of $C^\infty(R^n)$ in the space (14) is denoted by $AC_{pq}^{\kappa\lambda}(S)$.

The most important delineation of functions of the space (14) is its "smoothness" that will be defined below. Functions of negative smoothness have no interesting properties; the following theorems give the description of differential properties of functions of positive smoothness.

Let Λ^σ and M^σ, $0 < \sigma \le n$, be the Hausdorff (outer) σ-measure and the Hausdorff σ-capasity respectively. Thus

$$\Lambda^\sigma(S) := \lim_{\varepsilon \to 0} \Lambda^\sigma_{(\varepsilon)}(S) ;\ M^\sigma(S) := \Lambda^\sigma_{(\infty)}(S),$$

where

$$\Lambda^\sigma_{(\varepsilon)}(S) := \inf \{ \sum (mes Q_i)^{\sigma/n} ;\ S \subset \bigcup Q_i ,\ mes Q_i \le \varepsilon^n \}.$$

Denote furthermore by $T_x^{\lambda-0} M(S)$ the space defined as $T_x^\lambda M(S)$, see (5), but with adding $\ln(\frac{1}{mes Q})$ to the right part. Define the space $Lip(\lambda - 0; R^n)$ similarly, adding the logarithmic multiplier to the right part.

<u>Theorem 11</u>. Suppose $f \in V_{pq}^{\kappa\lambda}(S)$, the set S is regular and the smoothness of the function with respect to the measure Λ^σ, i.e. the number $\mu = \lambda + \frac{\sigma}{p} - \frac{n}{q}$, is greater than zero. Then there are the following assertions

1) if $\sigma = n$ and μ is an integer $\ge \kappa$, then $f \in t_x^\kappa L_q(S)$, a.e. Moreover for every $\varepsilon > 0$ there exists such a set S_ε of

the Lebegue measure $<\varepsilon$ that the trace of f on it's complement extends on R^n as a function of $C^k(R^n)$.

2) If $0<\sigma\leq n$ and μ is non integral, then $f\in T_x^\mu L_q(S)$ for $x\in S\setminus S_0$ where $\Lambda^\sigma(S_0)=0$. Moreover for every $\varepsilon>0$ there exists such a set S_ε of the Hausdorff σ - capacity $<\varepsilon$ that the trace of the function f on its complement extends on R^n as a function of $Lip(\mu;R^n)$.

3) If $0<\sigma\leq n$ and μ is integral $<k$, then there are the assertions of 2) with replacing T^μ by $T^{\mu-0}$ and $Lip\,\mu$ by $Lip(\mu-0)$.

Theorem 12. With the conjectures and the notations above let moreover $f\in AC_{pq}^{k\lambda}(S)$ and μ is not integral, $0<\mu<k$. Then $f\in t_x^\mu L_q(S)$ for $x\in S\setminus S_0$ where $\Lambda^\sigma(S_0)=0$ and for every $\varepsilon>0$ there exists such a set S_ε of the Hausdorff σ - capacity $<\varepsilon$ that the trace of f on its complement extends on R^n as a function of $lip(\mu;R^n)$.

The following lemma is used in the proofs of theorem 12.

Lemma 1. If $\mathfrak{I}\in Var_p$ then $\overline{\lim\limits_{Q\to x}}\, |\mathfrak{I}(Q)|/(mes\,Q)^{\sigma/np}<+\infty$ for $x\in S\setminus S_0$ where $\Lambda^\sigma(S_0)=0$.

Using the lemma with $\mathfrak{I}=\mathfrak{I}_f$, see (14), and then successively theorems 2,3,4 we obtain all the assertions of the theorem 12. We use the following assertion instead of lemma 1 in order to prove theorem 12 in the case $\sigma=n$.

Lemma 2. Let $\mathfrak{I}\in Var_p$ and the function $j(x)=\overline{\lim\limits_{Q\to x}}\frac{|\mathfrak{I}(Q)|}{(mes\,Q)^{1/p}}$ measurable, then

$$\|j\|_{L_p(R^n)}\leq Var_p\,\mathfrak{I}.$$

If a function $f_\varepsilon\in C^\infty(R^n)$ is in ε -neighbourhood of f in $V_{pq}^{k\lambda}$, then let $j_\varepsilon(x)=\overline{\lim\limits_{Q\to x}}\frac{E_k(f-f_\varepsilon;Q)_{L_q}}{(mes\,Q)^\gamma}$, $\gamma=\frac{\lambda}{n}+\frac{1}{p}$.

By lemma 2 we obtain $\|j_\varepsilon\|_{L_p} < \varepsilon$, but as $f_\varepsilon \in C^\infty$ and $\mu < S$ we have $\overline{\lim_{Q\to x}} \frac{E_\kappa(f;Q)_{L_q}}{(mes\,Q)^\gamma} \le j_\varepsilon(x)$.

Thus, $\lim_{Q\to x} \frac{E_\kappa(f;Q)_{L_q}}{(mes\,Q)^\gamma} = 0$ a.e. as ε is arbitrary. From Egorov's theorem this convergence can be made uniform in the complement of a set of Lebesgue measure $< \varepsilon$. From here and from theorems 2,3,4 we obtain the assertion of theorem 12 for the case $\sigma = n$.

In the case $0 < \sigma < n$ we use a simpler result .

Lemma 3. If $\mathcal{J} \in Var_p$ and $|\mathcal{J}|^p$ is absolutely continuous, then $\lim_{Q\to x} \frac{|\mathcal{J}(Q)|}{(mes\,Q)^{\sigma/np}} = 0$ for $x \in S \setminus S_0$ with $\Lambda^\sigma(S_0) = 0$. Furthermore, this convergence can be made uniform in the complement of an open set S_ε with $M_\sigma(S_\varepsilon) < \varepsilon$.

The proof that follows is obvious. The proofs of all the lemmas use the standard technique based upon Besicovitch's and Vitali's covering theorems.

The assertions of theorems 11,12 in a less complete form are presented in [7] , p. 107 and in [9] .

Theorems 11, 12 have many corollaries. Point out some of them.

A. $V^{1,0}_{p,\varphi}(0,1)$ space coinsides obviously with V_p space of functions of bounded p -variation. Thus the assertion 1) of theorem 11 in the case $p = 1$ leads to classical Lebesgue theorem of differtiation of functions of bounded variatian. Theorem 11 however gives some new results even in this case.

Further $V^{1,0}_{1,q}([0,1]^n)$, $q = \frac{n}{n-1}$, coinsides with BV - space of functions of bounded variation in Tonneli sense. Differential properties of functions of this space were studied by Tonneli, Stepanov, Deny-Lions, Calderon-Zygmund, Vol'pert ets (see [19]).

Hovever in this case theorem 11 gives new facts too.

B. If $0<\mu<\kappa$ and S is a regular set then

$$\mathcal{B}_p^{\mu}(S)\subset AC_{pq}^{\kappa\lambda}(S),$$

where $\lambda>0$, $0<p\leq q\leq\infty$, $\mu=\lambda+\frac{n}{p}-\frac{n}{q}$ (see [7], p. 117). Thus we obtain the differential properties of functions of $\mathcal{B}_p^{\mu}$ (and B_p^{μ}, in particular); all these properties were not known before.

C. If $L_p^{\mu}(S)$ is Bessel potentials space (when μ is integer we replace it by W_p^{μ}), then

$$L_p^{\mu}(S)\subset AC_{pq}^{\kappa\lambda}(S),$$

where $\lambda>0$, $1\leq p\leq q\leq\infty$, $\mu=\lambda+\frac{n}{p}-\frac{n}{q}$, $0<\mu<\kappa$ (see [7], p. 112 for μ integer; the general case obtain by using theorem 6). Differential properties of functions of L_p^{μ} were studied by many authors; the basic results for μ integer have been received by Calderon and Zygmund [17], the most complete results have been established recently by Bagby and Ziemer, see [1] where preceding works are presented also. The second assertion of theorem 12 for this case is nevertheless new; other assertions of theorems 11, 12 give new proofs contained in [1] results.

8°. Let now $S=Q_0=[0,1]^n$, $M=L_p$, $1\leq p\leq\infty$, with C instead of L_∞ and everywhere below $\mathfrak{T}$ denotes partition of Q_0 into cubes.

Denote by π_N the partition of Q_0 into equal cubes of volume N^{-n} and consider the question of the properties of the functions for which

$$E_\kappa(f;\pi_N)_{L_p}\leq\varphi(N^{-1}),\ \forall N.$$

Theorem 13. For some constant $\gamma = \gamma(k)$

$$E_k(f;\pi_N)_{L_p} \le \gamma\,\omega_k(f;N^{-1})_{L_p}$$

In the case $n = 1$ and $k \geq 2$ the estimate is right for approximation by spline-functions*.

The first statement is directly following from theorem 1 (cf. also [7] , p. 79). For obtaining the second we can use

Lemma 4. Let $\tilde{\pi}$ denote the partition of $[a,b]$ into $k-1$ equal intervals, q is a polynomial of degree $\le k-1$ and $f \in \mathcal{P}_k(\tilde{\pi}) \cap C^{k-2}$ such a spline function that $f^{(s)}(a) = 0$, $f^{(s)}(b) = q^{(s)}(b)$, $s = 0,1,\dots,k-2$. Then

$$\|f\|_{L_p(a,b)} \le \gamma(k)\,\|q\|_{L_p(a,b)}$$

It is enough to consider the case $a = 0$, $b = 1$. Further, for the finding f we can write the linear system of equations which has the only solutions and therefore

$$\max_{0 \le x \le 1} |f| \le \gamma(k) \sum_{s=0}^{k-2} |q^{(s)}(1)|$$

Making use of Markov inequality we obtain

$$\max_{0 \le x \le 1} |f| \le \tilde{\gamma}(k) \max_{0 \le x \le 1} |q| \;;$$

hence and from relation $\|q\|_{L_p(0,1)} \approx \max_{0 \le x \le 1} |f|$ lemma follows.

Let now Q_i be $[i/N, \, (i+1)/N]$, $\tilde{Q}_i = Q_i \cup Q_{i+1}$ and q_i is a polynomial of degree $\le k-1$ deviating least from f in $L_p(\tilde{Q}_i)$. We construct the spline ,

---

* i.e. the functions from $\mathcal{P}_k(\pi_N) \cap C^{k-2}$.

$\tilde{N} = (k-1)(N-2)+2$, on setting it equal to $q_0(x)$ for $x \in Q_0$ and to $q_{N-2}(x)$ for $x \in Q_{N-1}$ and define it in Q_i , $1 \le i \le N-2$ like in lemma 4 but with another boundary conditions:

$$s^{(j)}(i/N) = q_{i-1}^{(j)}(i/N),\ s^{(j)}(i+1/N) = q_i^{(j)}(i+1/N),\ j = 0, 1, \dots, k-2.$$

Then

$$\|f - S\|_{L_p(0,1)} \le \Big\{\sum_{i=0}^{N-2} E_k(f;\tilde{Q}_i)_{L_p}^p\Big\}^{1/p} + \Big\{\sum_{i=1}^{N-2} \|S - q_{i-1}\|_{L_p(Q_i)}^p\Big\}^{1/p}$$

From lemma 4, the second sum is majorized by the first. Hence at last

$$\|f - S\|_{L_p(0,1)} \le \tilde{\gamma}(k)\Big\{\sum_{i=0}^{N-2} E_k(f;\tilde{Q}_i)_{L_p}^p\Big\}^{1/p} =$$

$$= \tilde{\gamma}(k)\Big\{\sum_{i=0,1} E_k(f;\pi^i)_{L_p}^p\Big\}^{1/p},$$

where $\pi^0 = \{\tilde{Q}_i\ ;\ i = 0, 2, 4, \dots\}$, $\pi^1 = \{\tilde{Q}_i\ ;\ i = 1, 3, 5, \dots\}$

Every from two items in the right side is majorized by $\omega_k(f; N^{-1})_{L_p}^p$ by theorem 1 so the theorem proves. Converse theorem is established for the present in the case $p = \infty$; for $p < \infty$ the given results are not complete.

<u>Theorem 14</u>. Let, for $f \in C(Q)$,

$$E_k(f;\pi_N)_C \le \varphi(N^{-1}),\ \forall N, \tag{15}$$

where f is monotone and satisfies Δ_2 -condition. Then

$$\omega_k(f;t)_C \le \gamma(k)\varphi(t),\ \forall t > 0$$

<u>Proof</u>. In consequence of the second statement of theorem 1, we have

$$\omega_k(f;t)_C \le \gamma(k)\sup_Q E_k(f;Q)_C$$

where $Q \subset Q_0$ and its volume $\leq t^n$; and so it is sufficient to estimate the number $E_\kappa(f;Q)_c$ in the case $mes\, Q \leq t^n$.

Let $N = N(t)$ satisfy the condition $(N+1)^{-1} \leq t < N$ and Q^a , $1 \leq a \leq 2^n$, are the cubes from π_N with a common vertex so that $Q \subset \bigcup_a Q^a$. We shall consider that cubes with next numbers have a common $(n-1)$ -dimensional face and set $U_s = \bigcup_{a \leq s} Q^a$

Lemma 5. For some constant $\gamma = \gamma(\kappa)$

$$E_\kappa(f;U_s)_c \leq \gamma\{E_\kappa(f;U_{s-1})_c + E_\kappa(f;Q_s)_c\},$$

where Q_s is a cube from a certain partition $\pi_{\tilde{N}}$ with $\tilde{N} \leq 2N$.

Let the projection of common face of cubes Q^{s-1} and Q^s on to the perpendicular to this face coordinate axis Γ be a number ν/N; we set $\tilde{N} = N + [\frac{N}{2\nu}]$ and take as Q_s such a cube from $\pi_{\tilde{N}}$ wich is contained in $Q^{s-1} \cup Q^s$ and is projected on to segment $[\nu/\tilde{N}, \nu+1/\tilde{N}]$ in Γ . It is easy to show that

$$\min\left\{\frac{mes(Q^{s-1} \cap Q^s)}{mes\, Q^{s-1}}, \frac{mes(Q^s \cap Q_s)}{mes\, Q_s}\right\} \geq \gamma,$$

where γ depends only on dimension. Therefore the lemma follows from theorem 2 in [4].

Consecutively applying the lemma for $s = 2, 3, \ldots, 2^n$ we obtain together with (15)

$$E_\kappa(f;Q)_c \leq E_\kappa(f;U_{2^n})_c \leq \gamma\varphi(N^{-1});$$

that completes the proof.

Corollary 4. The condition

$$E_\kappa(f;\pi_N)_c = O(N^{-\lambda}),\ 0 < \lambda \leq \kappa,$$

is necessary and sufficient for belonging f to $Lip(\lambda;Q_0)$ in the case λ is noninteger or $\lambda = k$. In the case λ is integer this condition is equivalent belonging f to $C^{\lambda-1}(Q_0)$ and satisfying of Zygmund condition for derivatives of higher order of f.

Corollary 5. If $E_k(f;\pi_N) = o(N^{-k})$, $N \to \infty$, then f is a polynomial of degree $\leq k-1$.

Remarks. a) In the case L_p, $p < \infty$, the arguments of theorem 14 give only estimates $\omega_k(f;t)_{L_p} \leq \varphi(t)|\ln t|^{1/p}$ and very likely the logarithmis multiplier is unnecessary.
b) Making use the results from the works of Gajer [21] and Butler-Richards [16], we can nevertheless get results similar corollaries 4, 5 in the case

It should be noted that theorems 13, 14 and corollaries 4, 5 in the case of $C[0,1]$ space had been obtained yet in 1961 in the article [14]; particular cases of these results were obtained later in [26], [27], [29], [30], [25].

The second statement of theorem 13 uses the construction which the author applied in the proof of theorem 7 in [6]. A more sofistical method of spline-approximation was given indepedently in [20].

We note yet another method of piecewise polynomial approximation, namely, let

$$P_N^k(f;L_p) := \sup_{\pi} E_k(f;\pi)_{L_p}, \tag{16}$$

where sup is taken for these partition π which contains of cubes with volumes less or equal than N^{-n}. Then from theorem 1 follows immediately the relation $P_N^k(f;L_p) \asymp \omega_k(f;N^{-1})_{L_p}$ so in this case analogue of the corollaries 4, 5 are faithful.

The more careful investigation shows that we could use in the converse statement

$$\tilde{P}_N^K(f;L_p) := \max\{E_K(f;\pi_N)_{L_p};\ E_K(f;\pi_{N+\frac{1}{2}})_{L_p}\},$$

where partition $\pi_{N+\frac{1}{2}}$ contains of the two kinds of cubes: those adjoined to the bound Q_0 have the volume $(2N)^{-n}$ and the nest have volume N^{-n}. The approximation (16) in the case $K=2$ and $C[0,1]$ space have been studied in the work of Nitsche [26]; all the results are the consequences of theorems from [14]. The general case have been considered in [6]; spline approximation in the simil~~l~~ar situation have been independently investigated by Scherer [28].

9°. We consider another way of piecewise polynomial approximation of function from $L_p(Q_0)$; namely, we set

$$S_N^K(f;L_p) := \inf\{E_K(f;\pi_N)_{L_p};\ \operatorname{card}\pi \le N\}$$

The complete description of class

$$\{f\in L_p(0,1);\ S_N^K(f;L_p) = O(\varphi(N^{-1}))\}$$

is given in [10], [13]. Here we consider the functions of several variables.

We assume of $f \in L_p(0,1)$ the following:

a) for certain $\lambda > 0$

$$\Omega(t) = t^\lambda \int_0^t \frac{\omega_K(f;s)_{L_p}}{s^{\lambda+1}}\,ds < +\infty;$$

b) number q, $p < q \le \infty$, satisfying the inequaty

$$\frac{\lambda}{n} > \frac{1}{p} - \frac{1}{q}$$

Then it is true

Theorem 15. For every natural N

$$\delta_N^{k+1}(f;L_q) \le \gamma\,\Omega(N^{-1/n}),$$

where γ does not depend on f and N.

In order to formulate lemma 1 we define "homogeneous" Besov space $\mathring{B}_p^\lambda$ by the finitness of seminorm

$$|f|_{\mathring{B}_p^\lambda} := \left\{\int_0^1 \left|\frac{\omega_k(f;s)_{L_p}}{s^\lambda}\right|^p \frac{ds}{s}\right\}^{1/p}.$$

Lemma 6 (Birman-Solomiak [2]). If $f \in \mathring{B}_p^\lambda$, λ is noninteger and N is natural, then piesewise polynomial function $g_N(f) \in \mathcal{P}_k(\pi_N)$ exists such as that

$$\|f - g_N(f)\|_{L_q} \le \gamma N^{-\lambda/n} |f|_{\mathring{B}_p^\lambda}.$$

Here π_N is a diadic partition* of Q_0 into N cubes and γ does not depend on f and N.

To formulate the second lemma we define the space $H_p^k(\omega)$ by the finitness of seminorm

$$|f|_{H_p^k(\omega)} := \sup_{\tau>0} \frac{\omega_k(f;\tau)_{L_p}}{\omega(\tau)}.$$

Here ω is monotone function satisfying the Δ_2-condition. Further, define the apace $\mathcal{K}$ by seminorm

$$|f|_{\mathcal{K}} := \sup_{t>0} \frac{t^\lambda K(t;f;\mathring{B}_p^\lambda,\mathring{B}_p^\mu)}{\psi(t)}$$

there is K - functional Peetre in the right part, $0<\lambda<\mu<k$, and

---

* i.e., π_N is constructing in a such a way. First we devide Q_0 into 2^n equal cubes; then some of them are once more divided into 2^n equal cubes etc.

$$\psi(t) = t^{\lambda}\int_0^t \frac{\omega(s)}{s^{\lambda+1}}\,ds + t^{\mu}\int_t^1 \frac{\omega(s)}{s^{\mu+1}}\,ds.$$

Lemma 7. There is the continious embedding

$$H_p^k(\omega) \subset \mathcal{K}$$

For the proof one uses the usual considerations of theory of interpolation spaces (see,f.e., [15]).

Let now be f_0 , f_1 such as that $f = f_0 + f_1$ and

$$K(t; f; \mathring{B}_p^{\lambda}, \mathring{B}_p^{\mu}) \approx |f_0|_{\mathring{B}_p^{\lambda}} + t\,|f_1|_{\mathring{B}_p^{\mu}}$$

for $t = N^{-\frac{\mu-\lambda}{n}}$. We set $\omega(t) = \omega_{k+1}(f;t)_{L_p}$ and choose μ out of the condition $k < \mu < k+1$. Then $f \in H_p^{k+1}(\omega)$ and by lemma 6

$$S_{2N}^{k+1}(f;L_q) \le \|f_0 - g_N(f_0)\|_{L_q} + \|f_1 - g_N(f_1)\|_{L_q} \le$$

$$\le \gamma N^{-\lambda/n}\{|f_0|_{\mathring{B}_p^{\lambda}} + N^{-\frac{\mu-\lambda}{n}}|f_1|_{\mathring{B}_p^{\mu}}\}.$$

Since the way we choose f_0 , f_1 and according to lemma 7 the right side is less than or equal $\gamma\psi(N^{-1/n})$. But

$$\psi(t) \le 2\Omega(t) + 2\int_t^1 \frac{\omega_k(f;s)_{L_p}}{s^{\mu+1}}\,ds$$

As $s^{-k}\omega_k(f;s)_{L_p}$ is nearly decreasing and $\mu > k$ the second item is majorezed by first. So,

$$S_{2N}^{k+1}(f;L_q) \le \gamma\,\Omega(N^{-1/n})$$

and it establishes the theorem.

We state now the converse theorem.

Theorem 16. If $f \in L_p(Q_0)$ and for some $\lambda > 0$ and $q = \min\{(\lambda + \frac{1}{p})^{-1}, 1\}$ we have

$$\sum_{N=1}^{\infty} N^{-1} | N^{\lambda/n} S_N^{\kappa}(f; L_p)|^{q} < +\infty$$

then $f \in V_{\bar{q}p}^{\kappa \bar{\lambda}}$ with $\bar{\lambda} = -(n-1)\lambda$, $\bar{q} = (\lambda + \frac{1}{p})^{-1}$. Therefore all statements of theorem 11 are true for the functions with $\mu = -(n-1)\lambda + \frac{\sigma}{\bar{q}} - \frac{n}{p}$ and such a σ that $\mu > 0$ *.

For the proof we use the following

Lemma 8. If $f = f_2 + f_1$, $f_i \in \mathcal{P}_{\kappa}(\Pi_i)$, $i = 1, 2$, and

$$\mathrm{card}(\Pi_1 \cup \Pi_2) \le N$$

then

$$\| f \|_{V_{\bar{q}p}^{\kappa \bar{\lambda}}} \le \gamma N^{\lambda/n} \| f \|_{L_p}$$

It is sufficient to estimate

$$K(\Pi) = \sum_{Q \in \Pi} \left| \frac{E_{\kappa}(f; Q)_{L_p}}{(\mathrm{mes}\, Q)^{\bar{\lambda}}} \right|^{\bar{q}}$$

Cubes, containing in $Q_1 \cap Q_2$ for some $Q_i \in \Pi_i$, are the zero contibution in the right part. Remain cubes we divide into two sets: the first contains of the cubes having a vertex of, at least, one cubes from Π_i, $i = 1, 2$. There are no more than $2^n N$ such cubes. For the rest of them the following estimate is true:

$$\sum (\mathrm{mes}\, Q)^{\frac{n-1}{n}} \le \gamma(n) N^{1/n}$$

By these estimates and Gölder inequality we evaluate $K(\Pi)$.

Selecting, now, $g_j \in \mathcal{P}_{\kappa}(\Pi_j)$, $\mathrm{card}\, \Pi_j \le 2^j$, so as that

$$S_{2^j}^{\kappa}(f; L_p) \asymp \| f - g_j \|_{L_p}$$

and estimating by lemma 8 every item of the sum $g_0 + \sum_{j=0}^{\infty} (g_{j+1} - g_j) = f$

---

* in particular, we have $\mu = \lambda$ for $\sigma = n$.

we get the desired result.

Remark. From the theorem we can get a statement about class of saturation for this approximation. The more exact assertion is known only in a particular situation ([6] , corollary 3); this assertion is based on inequality

$$\operatorname*{Var}_{[0,1]} f \leq \underline{\lim}\, N S_N'(f;C)$$

which is true for $f \in C(0,1)$.

The approximation of this paragraph was first studied by Kahane [23] in the case $K = 1$ and $C[0,1]$ space and Birman-Solomiak [2] in general. Theorem 15 strengthens their result.

10°. Consider shortly the case of rational approximation. Denote by $R_N(f;L_p)$ distance in $L_p(Q_0)$ from f to set of rational functions of degree $\leq N$ depending rationally no more than $2N$ parameters. Making use of certain considerations from the proof of theorem 15 we can establish following

Theorem 17. In the assumptions and the notations of theorem 15

$$R_N(f;L_q) \leq \gamma\, \Omega(N^{-1/n})\beta_N$$

where γ does not depend on f and N and β_N increases to infinity with speed no more than iteration logarithm any order.

References

1. Th.Bagby and W.P.Ziemer. Pointwise differentiability and absolute continuity. Trans.Amer.Math.Soc. 191 (1974), 129-148.

2. M.S.Birman and M.Z.Solomjak. Piecewise-polynomial approximations of functions of classes W_p^α. Mat.Sbornik 73, No. 3 (1967), 331-355 = Math.USSR Sb.2 (1967), 295-317.

3. Ju.A.Brudnyi. On local best approximations. Dokl.Akad.Nauk SSSR, 161, No.4 (1965), 746-749 (Russian).

4. Ju.A.Brudnyi. A multidimensional analog of a theorem of Whitney. Mat.Sb. 82 (124) (1970), 175-191 = Math.USSR sb. 11(1970), 157-170.

5. Ju.A.Brudnyi. An extension theorem. Func.Anal. and Appl., 4, No. 3 (1970), 97-98 (Russian).

6. Ju.A.Brudnyi. Piecewise polynomial approximation and local approximation. Dokl.Akad.Nauk SSSR 201 (1971), 16-18 = Soviet Math.Dokl. 12 (1971), 1591-1594.

7. Ju.A.Brudnyi. Spaces defined by means of local approximation. Trudy Moskov.Mat.Obsc. 24 (1971), 69-132 = Trans.Moskow Math. Soc. 24 (1971), 74-139.

8. Ju.A.Brudnyi. On permutation of smooth function. Uspehi Mat.Nauk 27, No. 2 (1972), 165-166 (Russian).

9. Ju.A.Brudnyi. Local approximation and differential properties of functions of several variables. Uspehi Mat.Nauk 29, No. 4 (1974), 163-164,(Russian).

10. Ju.A.Brudnyi. Spline approximation and functions of bounded variation. Dokl.Akad.Nauk SSSR, 215, No. 3 (1974), 511-513 = Soviet Math.Dokl. 15, No. 2 (1974), 518-521.

11. Ju.A.Brudnyi. On scale of $\mathcal{L}_{pq}^\lambda$ spaces and exact embedding

theorems. Proc. Conference of Embedding Theorems (Alma-Ata, 1973), Alma-Ata, 1975 (Russian).

12. Ju.A.Brudnyi. On extension theorem for some family of functional spaces. Zap.naucn.Semin. LOMI., 56 (1976), 170-173 (Russian).

13. Ju.A.Brudnyi. Some nonlinear methods of best approximation. Proc. Intern.Conf.Appr.Theory (Kalouga, 1975). Moskow, 1977 (in preparation).

14. Ju.A.Brudnyi and I.E.Copengauz. Approximation by piecewise polynomial functions. Uzv.Akad.Nauk SSSR; Ser.Mat. 27 (1963), 723-746 (Russian).

15. P.L.Butzer and H.Berens. Semi-Groups of Operators and Approximation. Sringer-Verlag, Berlin, 1967.

16. G.J. Butler and F.B.Richrds. On L_p saturation theorem for spline. Can.J.Math., 24, No. 5 (1972), 957-966.

17. A.P.Calderon and A.Zygmund. Local properties of solutions of elliptic partial differential equations. Studia Math., 20 (1961), 171-225.

18. S.Campanato. Proprieta di una famiglia di spazi funzionali. Ann. Scuola Norm. Sup.Pisa, 18 (1964), 137-160.

19. H.Federer. Geometric Measure Theory. Sringer-Verlag. New York, 1969.

20. G.Freud and V.A.Popov. Some questions of approximation by spline-functions and polynomials. Studia Sci.Math.Hung. 5(1970), 161-171 (Russian).

21. D.Gajer. Saturation bei Spline-Approximation und Quadratur. Numer. Math. 16 (1970), 129-140.

22. A.Jonsson and H.Wallin. A Whithey extension theorem in L_p and Besov space. Dep.Math.Univ.Umea (Publ), 1975, No. 5, 60 pp.

23. J.-P.Kahane. Theoria constructiva de functiones. Cursos y Semin.Math.Univ.Buenos-Aires, No. 5, 1961.

24. C.B.Morrey. Functions of several variables and absolute continuity. Duke Math.J., 6 (1940), 187-215.

25. D.Newman.The Zygmund condition for polygonal approximation. Proc.Amer.Math.Soc., 45, No. 2 (1974), 303-305.

26. I.Nitsche. Sätze vou Jackson-Bernstein-Tyr für die approximationen mit splines-funktionen. Math.Z., 109, No. 2 (1969), (97-106).

27. F.B.Richards. On the saturation class for spline functions. Proc.Amer.Math.Soc., 33, No. 2 (1972), 471-475.

28. K.Scherer. Über die beste approximation von L_p funktionen durch splines. Proc.Intern.Conf. (Varna). Sofia, (1972), 277-286.

29. O.Shisha. Characterization of functions having Zygmund's property. I.Appr.Theory, 9, No. 9 (1973), 395-397.

30. O.Shisha. Characterization of smoothness properties of functions by means of their degree of approximation by splines. I.Appr.Theory, 12, No. 4 (1974), 365-371.

31. P.L.Ulianov. Imbedding theorems and correlations between best approximations for different metrics. Mat.Sb.,81, No.1 (1971), 104-131 (Russian).

32. H.Whitney. On functions with bounded n -th differences. J.Math.Pures and Appl., 9, No. 36 (1957), 67-95.

UN ALGORITHME GENERAL POUR L'APPROXIMATION AU SENS DE TCHEBYCHEFF DE FONCTIONS BORNEES SUR UN ENSEMBLE QUELCONQUE.

C. CARASSO
Université de Saint-Etienne

P.J. LAURENT
Université de Grenoble

Abstract

An algorithm is given for the calculation of a best Chebyshev approximation of a bounded function defined on an arbitrary set by a linear combination of functions of the same type (not necessarily independant). The formulation of the problem includes in fact the problem of linear approximation in an arbitrary normed linear space. The convergence holds without Haar condition.

Résumé

On propose un algorithme pour calculer une meilleure approximation au sens de Tchebycheff d'une fonction bornée sur un ensemble quelconque par une combinaison linéaire de fonctions du même type (non nécessairement indépendantes). La formulation du problème contient en fait le problème de l'approximation linéaire dans un espace normé arbitraire. La convergence est assurée sans hypothèse de Haar.

Introduction

En 1967, une généralisation de l'algorithme de Rémès a été proposée [12] pour la construction du meilleur approximant d'un élément dans un sous-espace vectoriel V d'un espace normé quelconque. Cet algorithme nécessitait la connaissance d'une base de V et surtout sa définition aussi bien que sa convergence étaient fondées sur une hypothèse de type Haar. Ces conditions sont très fortes. La seconde en particulier n'est jamais vérifiée dans le cas de l'approximation au sens de Tchebycheff de fonctions de plusieurs variables.

Dans [2] une première tentative a été faite pour remplacer la condition de Haar par une hypothèse beaucoup plus faible sur le déroulement effectif de l'algorithme (hypothèse d'itérativité). Enfin, dans le cadre plus général de l'optimisation avec contraintes, un nouvel algorithme a été proposé dans [2], [7] et [8] pour éviter l'hypothèse d'itérativité. Nous décrivons ici d'un point de vue pratique ce dernier algorithme dans le cas particulier d'un problème d'approximation. Pour la convergence nous renvoyons à [6].

1. Enoncé du problème d'approximation

On désigne par E l'espace Euclidien de dimension n et on note $< x,x' >$ le produit scalaire ordinaire de x et x' dans E.

Soit T un ensemble quelconque, on note $\mathcal{B}(T)$ l'espace des fonctions réelles bornées définies sur T. Etant données n+1 fonctions $c,b_1,\ldots,b_n$ appartenant à $\mathcal{B}(T)$ (non nécessairement indépendantes) on cherche à approcher c uniformément sur T par une fonction de la forme $\sum_{i=1}^{n} x_i\, b_i$, où $x \in E$ vérifie éventuellement des relations linéaires. Définissons pour cela la variété affine :

$$W = \{x \in E \mid < x,\beta(t) > = \gamma(t) \ ,\ t \in S_o\}$$

où les $\beta(t)$, $t \in S_o$ sont des éléments linéairement indépendants de E ; et les $\gamma(t)$, $t \in S_o$ des nombres réels. On note V le sous-espace de dimension k_o engendré par $v(t)$, $t \in S_o$. La variété W est parallèle au sous-espace vectoriel $V^{\perp}$ orthogonal à V et dont la dimension est égale à $n-k_o$.

Si l'on pose :

$$f(x) = \operatorname*{Sup}_{t\in T} \left| \sum_{i=1}^{n} x_i\, b_i(t) - c(t) \right| ,$$

le problème consiste à minimiser $f(x)$ pour $x \in W$.

Posons :

$$(P) \qquad \alpha = \operatorname*{Min}_{x\in W} f(x)$$

On sait qu'il existe des solutions, c'est-à-dire des éléments $\bar{x} \in W$ vérifiant $\alpha = f(\bar{x})$.

On appellera <u>solution à ε près</u> (pour $\varepsilon > 0$) tout élément $\tilde{x} \in W$ vérifiant $f(\tilde{x}) \leq \alpha + \varepsilon$.

Pour $\varepsilon > 0$ donné, l'algorithme que nous allons décrire fournira en un nombre fini d'itérations une solution à ε près. Si η est un nombre positif tel que $\eta < \frac{\varepsilon}{2^{n_o}}$, la mise en oeuvre de cet algorithme demandera seulement que pour tout $\hat{\varepsilon}$ vérifiant $\eta \leq \hat{\varepsilon} \leq \varepsilon$ et tout $x \in W$ on soit capable de déterminer $\hat{t} \in T$ vérifiant :

$$\left| \sum_{i=1}^{n} x_i\, b_i(\hat{t}) - c(\hat{t}) \right| \geq f(x) - \hat{\varepsilon} \ .$$

Remarque

La formulation du problème précédent englobe en fait le problème de l'approximation dans une variété affine de dimension finie d'un espace vectoriel normé quelconque.

Soit en effet Y un espace vectoriel normé réel dont la norme est notée $\|y\|$ pour $y \in Y$. Etant donnés n+1 éléments $y, y_1, \ldots, y_n$ de Y (non nécessairement indépendants) on cherche à approcher y par un élément de la forme $\sum_{i=1}^{n} x_i\, y_i$, avec $x \in W$ (variété affine de E comme plus haut). On est donc amené à minimiser :

$$f(x) = \|\sum_{i=1}^{n} x_i\, y_i - y\| \qquad \text{pour} \quad x \in W .$$

Notons Y' le dual topologique de Y et (y,y') la valeur en y d'une fonctionnelle linéaire continue $y' \in Y'$. Soit $T \subset Y'$ un sous-ensemble tel que :

$$\|y\| = \operatorname*{Sup}_{y' \in T} |(y,y')| \qquad \text{pour tout} \quad y \in Y .$$

On pourra prendre, par exemple, pour ensemble T la boule unité S' de Y' ou encore l'ensemble $\mathcal{E}(S')$ des points extrémaux de S'.

On a alors :

$$f(x) = \operatorname*{Sup}_{y' \in T} \left| \sum_{i=1}^{n} x_i\, (y_i,y') - (y,y') \right| .$$

On est donc ramené à la formulation précédente en posant $b_i(y') = (y_i,y')$ et $c(y') = (y,y')$ pour $y' \in T$.

2. Support minimal

Soit $\mathcal{V}$ un sous-espace vectoriel de dimension d de E engendré par une famille finie d'éléments v(t), $t \in D$ (non nécessairement linéairement indépendants). Pour la commodité, on supposera que D et T sont disjoints, ce qui ne restreint pas la généralité.

Dans la suite, on notera b, l'application de T dans E définie par :

$$b(t) = [b_1(t), \ldots, b_n(t)] \quad .$$

2.1. Support relatif à un sous-espace

Un sous-ensemble fini non vide S de T sera dit **support relativement à** $\mathcal{V}$ s'il existe des coefficients $\lambda(t)$, $t \in S$, non tous nuls, tels que $\sum_{t\in S} \lambda(t)b(t) \in \mathcal{V}$.

Il résulte directement de la définition que tout sous-ensemble de T comportant un nombre de points supérieur ou égal à n-d+1 est automatiquement un support.

2.2. Support minimal relatif à un sous-espace

Un support S relativement à $\mathcal{V}$ sera dit minimal s'il n'existe pas de support relativement à $\mathcal{V}$ qui soit strictement contenu dans S.

On montre facilement qu'un sous-ensemble $S = \{t_1,\ldots,t_{k+1}\}$ comportant k+1 points de T ($k \geq 0$) est un support minimal relativement à $\mathcal{V}$ si et seulement s'il existe des coefficients $\lambda(t)$, $t \in S$, tous non nuls tels que $\sum_{t\in S} \lambda(t)b(t) \in \mathcal{V}$ et le sous-espace engendré par $v(t)$, $t \in D$ et $b(t)$, $t \in S$ est de dimension d+k.

Il résulte directement de la définition qu'un support minimal comporte au plus n-d+1 points et que tout support contient un support minimal.

2.3. Coefficients associés à un support minimal

Si S est un support minimal, alors il existe des coefficients $\lambda_S(t)$, $t \in D \cup S$ tels que :

$$\sum_{t\in D} \lambda_S(t)v(t) + \sum_{t\in S} \lambda_S(t)b(t) = 0 \quad ,$$

$$\sum_{t\in S} |\lambda_S(t)| = 1 \quad .$$

Les coefficients $\lambda_S(t)$, $t \in S$ sont uniques au signe près (on peut multiplier tous les $\lambda_S(t)$ par -1) et sont tous non nuls.

2.4. Approximation relative à un support minimal

Soit $S = \{t_1,\ldots,t_{k+1}\} \subset T$ un support minimal relativement à $\mathcal{V}$. Considérons le problème de l'approximation sur S de l'élément c par un élément de la forme $\sum_{i=1}^{n} x_i b_i$ avec $x \in \mathcal{W}$, où $\mathcal{W}$ est une variété affine parallèle à $\mathcal{V}^\perp$, l'orthogonal de $\mathcal{V}$ dans E, définie par :

$$\mathcal{W} = \{x \in E \mid \langle x, v(t) \rangle = e(t),\ t \in D\}$$

où les $e(t)$, $t \in D$ sont des nombres réels.

On définit la fonctionnelle d'écart associée à S par :

$$f_S(x) = \max_{t \in S} \left| \sum_{i=1}^{n} x_i \, b_i(t) - c(t) \right|$$

et on pose :

$$\alpha_S = \min_{x \in \mathcal{W}} f_S(x) \quad .$$

On notera $\mathcal{W}_S$ l'ensemble des solutions, i.e. des éléments $\bar{x} \in \mathcal{W}$ tels que $\alpha_S = f_S(\bar{x})$.

Désignons par $\lambda_S(t)$, $t \in D \cup S$, une famille de coefficients associés à S comme en 2.3.. On a alors le résultat suivant :

Théorème

Si l'on pose

$$z_S = - \sum_{t \in S} \lambda_S(t) c(t) - \sum_{t \in D} \lambda_S(t) e(t) ,$$

alors on a :

$\alpha_S = |z_S|$

et

$\mathcal{W}_S = \{x \in \mathcal{W} \mid \langle x, b(t) \rangle - c(t) = \text{signe}\,(\lambda_S(t)) z_S,\ t \in S\}$

qui est une variété affine parallèle au sous-espace $\mathcal{V}_S^{\perp}$ *, où* $\mathcal{V}_S$ *est le sous-espace de dimension d+k engendré par* $v(t)$*,* $t \in D$ *et* $b(t)$*,* $t \in S$ *.*

Démonstration

Considérons les restrictions $\tilde{b}$ et $\tilde{c}$ des fonctions b et c au sous-ensemble S. L'ensemble S étant fini, on peut appliquer le théorème classique de caractérisation (3.3.7. p. 91 de [15]) : un élément $\bar{x}$ est solution si et seulement s'il existe h+1 points ($h \leq n-d$) $s_1, \ldots, s_{h+1}$ de S , des coefficients $\rho_1, \ldots, \rho_{h+1}$ positifs ($\sum_{i=1}^{h+1} \rho_i = 1$) et des entiers $\varepsilon_i = \pm 1$, $i=1,\ldots,h+1$, tels que :

a) $\varepsilon_j (\langle \bar{x}, b(s_j) \rangle - c(s_j)) = f_S(\bar{x}) \quad , \quad j=1,\ldots,h+1 \quad ,$

b) $\sum_{j=1}^{h+1} \rho_j \, \varepsilon_j \, b(s_j) \in \mathcal{V}$.

La condition b) exprime que l'ensemble $\{s_1, \ldots, s_{h+1}\}$ est un support relativement à $\mathcal{V}$. Comme S est minimal on a h = k et $\{s_1, \ldots, s_{k+1}\} = S$.

Les coefficients $\lambda_S(t)$, $(t \in S)$ sont uniques au signe près ; on a donc :

$$\lambda_S(s_j) = \varepsilon\, \rho_j\, \varepsilon_j \quad , \quad j=1,\dots,k+1 \quad \text{avec} \quad \varepsilon = \pm 1 \ ,$$

et, en introduisant des coefficients $\lambda_S(t)$, $t \in D$, b) peut s'écrire :

$$(*) \qquad \sum_{t\in S} \lambda_S(t) b(t) + \sum_{t\in D} \lambda_S(t) v(t) = 0 \ .$$

L'élément $\bar{x}$ étant solution, on a $f_S(\bar{x}) = \alpha_S$. Comme $\sum_{j=1}^{h+1} \rho_j = 1$ on a :

$$\alpha_S = f_S(\bar{x}) = \sum_{j=1}^{h+1} \rho_j\, \varepsilon_j (< \bar{x}, b(s_j) > - c(s_j))$$

$$= \varepsilon < \bar{x}, \sum_{t\in S} \lambda_S(t) b(t) > - \varepsilon \sum_{t\in S} \lambda_S(t) c(t)$$

et en utilisant $(*)$ et le fait que $\bar{x} \in \mathcal{W}$:

$$\alpha_S = - \varepsilon \left(\sum_{t\in D} \lambda_S(t) < \bar{x}, v(t) > + \sum_{t\in S} \lambda_S(t) c(t) \right)$$

$$= - \varepsilon \left(\sum_{t\in S} \lambda_S(t) c(t) + \sum_{t\in D} \lambda_S(t) e(t) \right) = \varepsilon \ z_S$$

d'où $\alpha_S = |z_S|$.

En remarquant que $\varepsilon_j = \varepsilon$ signe $\lambda_S(s_j)$, la condition a) devient :

$< \bar{x}, b(t) > - c(t) = \text{signe}\,(\lambda_S(t)) z_S$, pour $t \in S$.

Q.E.D.

L'algorithme que nous allons décrire va nous fournir, pour un nombre $\varepsilon > 0$, donné, une séquence finie $S^1, S^2, \dots, S^\mu$ de supports minimaux relatifs à V et une séquence associée $x^1, \dots, x^\mu$ d'éléments de W tels que :

$$\alpha^\nu = - \sum_{t\in S^\nu} \lambda_{S^\nu}(t) c(t) - \sum_{t\in S_o} \lambda_{S^\nu}(t) \gamma(t)$$

forme une séquence non décroissante avec $f(x^\mu) - \alpha^\mu \leq \varepsilon$, ce qui entrainera simultanément $\alpha - \alpha^\mu \leq \varepsilon$ et $f(x^\mu) - \alpha \leq \varepsilon$, donc en particulier que x^μ est une solution à ε près du problème.

3. Chaîne de supports minimaux

Soit S_1 un support minimal relatif à V. Appelons $f_1 = f_{S_1}$ la fonctionnelle d'écart qui lui est associée et considérons la minimisation de f_1 sur W. Notons α_1 le montant du minimum, W_1 l'ensemble des solutions et V_1 le sous-espace tel que W_1

soit parallèle à $V_1^{\perp}$.
Si S_1 comporte k_1+1 éléments, le sous-espace V_1 est de dimension k_o+k_1.

On refait la même construction mais relativement à V_1. Si S_2 désigne un support minimal de V_1 comportant k_2+1 éléments, on forme la fonctionnelle d'écart $f_2 = f_{S_2}$ et on note α_2 le montant de son minimum sur W_1, W_2 l'ensemble des solutions et V_2 le sous-espace vectoriel de dimension $k_o + k_1 + k_2$ tel que $V_2^{\perp}$ soit parallèle à W_2. On continue ainsi de proche en proche cette construction.

3.1. Chaîne de supports minimaux

Définition

On appelle chaîne de supports minimaux (en abrégé chaîne) une séquence finie de supports minimaux $\mathcal{C} = \{S_1,\ldots,S_m\}$ obtenue comme ci-dessus pour laquelle $V_m = E$.

En résumé, si l'on pose $V_o = V$, la séquence $\{S_1,\ldots,S_m\}$ est une chaîne si l'on a :

S_i support minimal de V_{i-1} ,

V_i sous-espace vectoriel engendré par $\beta(t)$, $t \in S_o$ et $b(t)$, $t \in \bigcup_{j=1}^{i} S_j$

$i=1,\ldots,m.$

$V_m = E$.

3.2. Solution associée à une chaîne

A une chaîne $\mathcal{C} = \{S_1,\ldots,S_m\}$ on peut associer :

. la séquence des fonctionnelles d'écart $\{f_1,\ldots,f_m\}$,

. la séquence des ensembles de solutions successifs $\{W_1,\ldots,W_m\}$ parallèles à l'orthogonal des sous-espaces vectoriels correspondants $\{V_1,\ldots,V_m\}$, avec $V_m = E$,

. la séquence des montants des minima de f_i sur W_{i-1} , $\{\alpha_1,\ldots,\alpha_m\}$.

Comme $V_m = E$, la variété affine W_m est réduite à un point $x = x_{\mathcal{C}}$ qui sera appelé solution associée à la chaîne $\mathcal{C}$.

Le calcul de $x_{\mathcal{C}}$ et des montants $\alpha_1,\ldots,\alpha_m$ se fait en résolvant un système linéaire de n+m équation à n+m inconnues. A chaque support minimal S_i de la chaîne $\mathcal{C}$, on peut associer comme on l'a vu en 2.3. (où $\mathcal{V}$ est l'espace engendré par $\beta(t)$, $t \in S_o$ et $b(t)$, $t \in S_j$, $j=1,\ldots,i-1$) une famille de coefficients $\lambda_{S_i}(t)$, $t \in S_j$, $j=0,\ldots,i$ (uniques au signe près pour $t \in S_i$) telle que :

$$\sum_{t\in S_o} \lambda_{S_i}(t)\beta(t) + \sum_{j=1}^{i} \sum_{t\in S_j} \lambda_{S_i}(t)b(t) = 0 ,$$

$$\sum_{t\in S_i} |\lambda_{S_i}(t)| = 1 .$$

D'après le théorème 2.4., la solution $x = x_{\mathscr{C}}$ vérifie les n+m équations :

$$(1)\quad \begin{cases} < x, \beta(t) > \; = \gamma(t) \quad , \quad t \in S_o & (k_o \text{ équations}) \\ < x, b(t) > \; - \text{ signe } (\lambda_{S_i}(t)) z_i = c(t) \quad , \quad t \in S_i \; , & (k_i+1 \text{ équations}) \\ \qquad\qquad\qquad\qquad i=1,\ldots,m \; ; & \end{cases}$$

dont les inconnues sont $x_1,\ldots,x_n,z_1,\ldots,z_m$.

On pose alors $\alpha_i = |z_i|$, i=1,...,m .

On note $A_{\mathscr{C}}$ la matrice à n+m lignes et n+m colonnes associée au système linéaire précédent et $c_{\mathscr{C}}$ le vecteur colonne du deuxième membre.

Donnons la structure de ce système dans le cas particulier où n = 5, $k_o = 1$, $k_1 = 2$, $k_2 = k_3 = 1$ en notant $S_i = \{t_{i,1},\ldots,t_{i,k_i+1}\}$ et

$\varepsilon_{i,j} = \text{signe } \lambda_{S_i}(t_{i,j})$ $(i=1,\ldots,3 \; ; \; j=1,\ldots,k_{i+1})$.

Le système $(A_{\mathscr{C}}, c_{\mathscr{C}})$ s'écrit :

$\beta(1)$					x_1		$\gamma(1)$
$b(t_{11})$	$-\varepsilon_{11}$				x_2		$c(t_{11})$
$b(t_{12})$	$-\varepsilon_{12}$				x_3		$c(t_{12})$
$b(t_{13})$	$-\varepsilon_{13}$				x_4		$c(t_{13})$
$b(t_{21})$		$-\varepsilon_{21}$		×	x_5	=	$c(t_{21})$
$b(t_{22})$		$-\varepsilon_{22}$			z_1		$c(t_{22})$
$b(t_{31})$			$-\varepsilon_{31}$		z_2		$c(t_{31})$
$b(t_{32})$			$-\varepsilon_{32}$		z_3		$c(t_{32})$

3.3. Chaîne régulière

Définition

On dira qu'une chaîne $\mathscr{C} = \{S_1,S_2,\ldots,S_m\}$ est <u>régulière</u> si tous les supports minimaux S_i qui la composent sont constitués d'au moins deux éléments, c'est-à-dire si $k_i \geq 1$, i=1,...,m.

Si la chaîne est régulière, V_i a donc une dimension strictement supérieure à celle de V_{i-1}. Ainsi la longueur m d'une chaîne régulière (c'est-à-dire le nombre m de supports minimaux qui la composent) est inférieur ou égale à $n_o = n - k_o$.

Etant donnée une chaîne quelconque $\mathcal{C}$, si l'on supprime tous les supports minimaux réduits à un point, on obtient une nouvelle chaîne $\mathcal{C}'$ régulière. Cette opération ne change pas la solution x associée à la chaîne et les montants α_i qui n'ont pas été supprimés. Elle peut modifier les coefficients $\lambda_{S_i}(t)$ pour tous les supports S_i de la chaîne qui ont un indice i supérieur à un support éliminé. Par contre, S_i étant un support minimal, les coefficients $\lambda_{S_i}(t)$ pour $t \in S_i$ restent valables.

4. Théorème d'échange généralisé

L'algorithme sera basé sur le théorème suivant qui généralise et complète le théorème classique d'échange de Stiefel ([21], [22], [15] p. 117).

4.1. Théorème

Soit $\mathcal{V}$ un sous-espace quelconque de E engendré par les vecteurs v(t), $t \in D$.

Si S_1 est un support minimal relativement à $\mathcal{V}$ de coefficients associés $\lambda_{S_1}(t)$, $t \in S_1 \cup D$ et si S_2 est un support minimal relativement à l'espace $\mathcal{V}_1$ engendré par b(t), $t \in S_1$ et v(t), $t \in D$ de coefficients associés $\lambda_{S_2}(t)$, $(t \in D \cup S_1 \cup S_2)$ alors la bipartition de S_1 en $C_1 = \{t \in S_1 \mid \lambda_{S_2}(t)/\lambda_{S_1}(t) = r\}$ et $B_1 = S_1 \setminus C_1$

(où $r = \min\limits_{t \in S_1} \dfrac{\lambda_{S_2}(t)}{\lambda_{S_1}(t)}$) est telle que :

$\tilde{S}_1 = B_1 \cup S_2$ est un support minimal relativement à $\mathcal{V}$

et $\tilde{S}_2 = C_1$ est un support minimal relativement à l'espace $\mathcal{V}_1$ engendré par b(t), $t \in S_1$ et v(t), $t \in D$.

De nouveaux coefficients $\lambda_{\tilde{S}_1}(t)$, $t \in D \cup \tilde{S}_1$ et $\lambda_{\tilde{S}_2}(t)$, $t \in D \cup \tilde{S}_1 \cup \tilde{S}_2$ peuvent être obtenus au moyen des formules :

$$\lambda_{\tilde{S}_1}(t) = \begin{cases} \frac{1}{m}\lambda_{S_2}(t) & ,\text{si } t \in S_2 \\ \frac{1}{m}(\lambda_{S_2}(t) - r\,\lambda_{S_1}(t)) & ,\text{si } t \in B_1 \cup D \end{cases}$$

$$\lambda_{\tilde{S}_2}(t) = \begin{cases} \frac{1}{m}\lambda_{S_1}(t) & \text{si} \quad t \in S_1 \cup D \\ 0 & \text{si} \quad t \in S_2 \end{cases}$$

avec $m = \sum_{t\in S_2} |\lambda_{S_2}(t)| + \sum_{t\in B_1} |\lambda_{S_2}(t) - r\,\lambda_{S_1}(t)|$ et $p = \sum_{t\in C_1} |\lambda_{S_1}(t)|$.

Démonstration

Puisque S_1 et S_2 sont des supports minimaux relativement à $\mathcal{V}$ et $\mathcal{V}_1$, respectivement, on a :

(1) $$\sum_{t\in S_1} \lambda_{S_1}(t)b(t) + \sum_{t\in D} \lambda_{S_1}(t)v(t) = 0 \quad \text{et} \quad \sum_{t\in S_1} |\lambda_{S_1}(t)| = 1$$

et

(2) $$\sum_{t\in S_2} \lambda_{S_2}(t)b(t) + \sum_{t\in S_1} \lambda_{S_2}(t)b(t) + \sum_{t\in D} \lambda_{S_2}(t)v(t) = 0 \quad \text{et} \quad \sum_{t\in S_2} |\lambda_{S_2}(t)| = 1.$$

Si on multiplie la première équation par $-r$ et si on l'ajoute à la seconde on obtient :

(3) $$\sum_{t\in S_2} \lambda_{S_2}(t)b(t) + \sum_{t\in B_1} (\lambda_{S_2}(t) - r\lambda_{S_1}(t))b(t) + \sum_{t\in D} (\lambda_{S_2}(t) - r\lambda_{S_1}(t))v(t) = 0,$$

ce qui montre que $\tilde{S}_1 = B_1 \cup S_2$ est un support de $\mathcal{V}$. Ce support est minimal : s'il n'en était pas ainsi on pourrait trouver $\tilde{t} \in B_1$ et des coefficients $\lambda'(t)$ tels que :

$$\sum_{t\in S_2} \lambda_{S_2}(t)b(t) + \sum_{t\in B_1\setminus\{\tilde{t}\}} \lambda'(t)b(t) + \sum_{t\in D} \lambda'(t)v(t) = 0$$

et en retranchant cette équation à (3) on verrait que S_1 n'est plus minimal relativement à $\mathcal{V}$.

En divisant par m, on obtient les coefficients $\lambda_{\tilde{S}_1}(t)$ associés à $\tilde{S}_1$ et D.

La formule (1) peut aussi s'écrire :

$$\sum_{t\in C_1} \lambda_{S_1}(t)b(t) + \sum_{t\in B_1} \lambda_{S_1}(t)b(t) + \sum_{t\in D} \lambda_{S_1}(t)v(t) = 0$$

ce qui montre, les coefficients $\lambda_{S_1}(t)$, $t \in C_1$ étant uniques au signe près, que $C_1 = \tilde{S}_2$ est un support minimal de $\mathcal{V}_1$, les coefficients $\lambda_{\tilde{S}_2}(t)$ étant obtenus en divisant (1) par p.

Q.E.D.

Remarque

Supposons que $\mathcal{V}$ soit de dimension d et que S_1 comporte exactement n-d+1 éléments; alors $\mathcal{V}_1 = E$ et ainsi tout ensemble $S_2 = \{\hat{t}\}$ réduit à un seul élément $\hat{t} \in T$ est évidemment un support minimal de $\mathcal{V}_1$. Le théorème d'échange nous indique la partie C_1 de S_1 que l'on peut échanger avec $\hat{t}$ de sorte que :

$$\tilde{S}_1 = (S_1 \setminus C_1) \cup \{\hat{t}\}$$

soit encore un support minimal de $\mathcal{V}$ et C_1 un support minimal de l'espace engendré par $b(t)$, $t \in \tilde{S}_1$ et une famille génératrice de $\mathcal{V}$.

Lorsque C_1 se compose d'un seul élément t_o on a alors simplement échangé $\hat{t}$ et t_o de sorte que l'ensemble $\tilde{S}_1 = (S_1 \setminus \{t_o\}) \cup \{\hat{t}\}$ forme un support minimal de $\mathcal{V}$.

4.2. Opération d'échange sur une chaîne

Définition

Etant donnée une chaîne $\mathcal{C} = \{S_1, \ldots, S_m\}$ on dira que l'on "échange" S_{j-1} et S_j si l'on remplace ces deux supports par $\tilde{S}_{j-1}$ et $\tilde{S}_j$ selon le théorème d'échange précédent de façon à obtenir une nouvelle chaîne.

En reprenant les notations du paragraphe 3.1., S_{j-1} est un support minimal de $\mathcal{V} = V_{j-2}$ et S_j est un support minimal de $\mathcal{V}_1$, espace vectoriel engendré par la famille génératrice de V_{i-2} et $b(t)$, $t \in S_{j-1}$. Le théorème d'échange nous permet de construire une bipartition de S_{j-1} en B_{j-1} et $C_{j-1} \neq \emptyset$ telle que, si l'on pose $\tilde{S}_{j-1} = B_{j-1} \cup S_j$ et $\tilde{S}_j = C_{j-1}$ alors :

$$\tilde{\mathcal{C}} = \{S_1, S_2, \ldots, S_{j-2}, \tilde{S}_{j-1}, \tilde{S}'_j, S_{j+1}, \ldots, S_m\}$$

constitue à nouveau une chaîne.

On remarque que, même si $\mathcal{C}$ est une chaîne régulière, il n'en est pas forcément de même pour $\tilde{\mathcal{C}}$ car le support $\tilde{S}_j$ peut être réduit à un point.

Si $\lambda_{S_i}(t)$, $(t \in \bigcup_{j=0}^{i} S_j)$, $i=1,\ldots,m$, désigne une famille de coefficients associée à $\mathcal{C}$, on a :

$$\sum_{j=1}^{i} \sum_{t \in S_j} \lambda_{S_i}(t) b(t) + \sum_{t \in S_o} \lambda_{S_i}(t) \beta(t) = 0 \quad \text{et} \quad \sum_{t \in S_i} |\lambda_{S_i}(t)| = 1 \ , \ i=1,\ldots,m.$$

L'échange entre S_{j-1} et S_j ne modifie donc pas les coefficients $\lambda_{S_i}(t)$ pour $i=1,\ldots,j-2,j+1,\ldots,m$. Les nouveaux coefficients $\lambda_{\tilde{S}_{j-1}}(t)$ et $\lambda_{\tilde{S}_j}(t)$ sont calculés suivant le théorème d'échange (les coefficients $\lambda_{S_i}(t)$ pour $i=j+1,\ldots,m$ doivent simplement être réordonnés pour $t \in \tilde{S}_j \cup \tilde{S}_{j-1}$).

5. Algorithme

L'algorithme va consister en la construction d'une suite de chaînes régulières :

$$\mathscr{C} = \{S_1^\nu, S_2^\nu, \ldots, S_{m^\nu}^\nu\}$$

telle que la suite des montants associés :

$$\{\alpha_1^\nu, \ldots, \alpha_{m^\nu}^\nu\}$$

soit lexicographiquement strictement croissante, c'est-à-dire que pour tout ν , il existe un entier ℓ^ν $(1 \le \ell^\nu \le m^\nu)$ tel que :

$$\alpha_i^{\nu+1} = \alpha_i^\nu \quad , \quad i=1,\ldots,\ell^\nu-1 \; ,$$

$$\alpha_{\ell^\nu}^{\nu+1} > \alpha_{\ell^\nu}^\nu$$

Etant donné un nombre positif ε_1, arbitrairement petit, après un nombre fini μ d'itérations on obtiendra une chaîne $\mathscr{C}^\mu$ et une solution $x^\mu \in W$ telle que l'on ait :

$$f(x^\mu) - \alpha_1^\mu \le \varepsilon_1$$

ce qui entraine :

$$\alpha - \alpha_1^\mu \le \varepsilon_1 \quad \text{et} \quad f(x^\mu) - \alpha \le \varepsilon_1 \quad ,$$

donc que x^μ est une solution à ε_1 près du problème.

Si ε_1 est la précision à atteindre, on se donne des nombres positifs $\varepsilon_2, \ldots, \varepsilon_{n-k_o+1}$ tels que :

$$\varepsilon_{i+1} < \varepsilon_i/2 \quad , \quad i=1,\ldots,n-k_o \; .$$

5.1. Description de l'algorithme

On posera dans toute la suite $S_o^\nu = S_o$.

Supposons qu'à l'itération ν on ait une chaîne régulière $\mathscr{C}^\nu = \{S_1^\nu, \ldots, S_{m^\nu}^\nu\}$, des coefficients $\lambda_{S_i^\nu}(t)$, $(t \in \bigcup_{j=0}^{i} S_j^\nu)$, $(i=1,\ldots,m^\nu)$ associés, la solution x^ν ainsi que les montants $\alpha_1^\nu, \ldots, \alpha_{m^\nu}^\nu$ qui lui correspondent. On note $A^\nu = A_{\mathscr{C}^\nu}$ la matrice de dimension $n+m^\nu$ associée à la chaîne et $c^\nu = c_{\mathscr{C}^\nu}$ le vecteur colonne du deuxième membre, (cf. paragraphe 3.2.).

Déterminons un élément $t^\nu \in T$ tel que :

$$f(x^\nu) - |< x^\nu, b(t^\nu) > - c(t^\nu)| \le \varepsilon_{m^\nu+1}$$

et posons :

$$S^\nu_{m^\nu+1} = \{t^\nu\}$$

$$z^\nu_{m^\nu+1} = < x^\nu, b(t^\nu) > - c(t^\nu) \quad ; \quad \alpha^\nu_{m^\nu+1} = |z^\nu_{m^\nu+1}|$$

$$\lambda_{S^\nu_{m^\nu+1}}(t^\nu) = 1 .$$

Les coefficients $\lambda_{S^\nu_{m^\nu+1}}(t)$, pour $t \in \bigcup_{j=0}^{m^\nu} S_j$, sont calculés en résolvant le système linéaire :

$${}^tA^\nu \lambda = u \quad ,$$

où u est le vecteur colonne dont les n premiers éléments sont formées des éléments du vecteur $-b(t^\nu)$, les autres étant nuls.

(On remarque que $S^\nu_{m^\nu+1}$ peut être considéré comme un support minimal de $V_{m^\nu} = E$).

Soit :

$$J^\nu = \{j \in \{1,\ldots,m^\nu+1\} \mid \alpha^\nu_{m^\nu+1} + \varepsilon_{m^\nu+1} \le \alpha^\nu_j + \varepsilon_j\}$$

On voit que l'ensemble J^ν contient au moins l'indice $m^\nu + 1$. Notons :

$$j^\nu = \min (j \mid j \in j^\nu) \quad .$$

On distinguera trois cas suivant la valeur de j^ν :

1er cas : $j^\nu = 1$.

On a alors $f(x^\nu) \le \alpha^\nu_{m^\nu+1} + \varepsilon_{m^\nu+1} \le \alpha^\nu_1 + \varepsilon_1$, donc $f(x^\nu) - \alpha^\nu_1 \le \varepsilon_1$, ce qui signifie que x^ν est solution à ε_1 près de (P) ; on arrête donc le calcul.

2ème cas : $2 \le j^\nu \le m^\nu$.

On échange alors $S^\nu_{j^\nu-1}$ et $S^\nu_{j^\nu}$ au sens qui a été précisé au paragraphe 4.2., c'est-à-dire qu'on les remplace par $\tilde{S}^\nu_{j^\nu-1}$ et $\tilde{S}^\nu_{j^\nu}$ et qu'on remplace les coefficients

$\lambda_{S^{\nu}_{j^{\nu}-1}}(t)$ et $\lambda_{S^{\nu}_{j^{\nu}}}(t)$ par $\lambda_{\tilde{S}^{\nu}_{j^{\nu}-1}}(t)$ et $\lambda_{\tilde{S}^{\nu}_{j^{\nu}}}(t)$ calculés selon le théorème d'échange.

(Il faut également "réordonner" les coefficients $\lambda_{S^{\nu}_{j}}(t)$ pour $j > j^{\nu}$ et $t \in \tilde{S}^{\nu}_{j^{\nu}-1} \cup \tilde{S}^{\nu}_{j^{\nu}}$).

On obtient ainsi une nouvelle chaîne $\tilde{\mathscr{C}}^{\nu}$:

$$\tilde{\mathscr{C}}^{\nu} = \{S^{\nu}_{1},\dots,S^{\nu}_{j^{\nu}-2},\tilde{S}^{\nu}_{j^{\nu}-1},\tilde{S}^{\nu}_{j^{\nu}},S^{\nu}_{j^{\nu}+1},\dots,S^{\nu}_{m^{\nu}}\}$$

(le support $S^{\nu}_{m^{\nu}+1} = \{t^{\nu}\}$ n'est pas introduit dans $\tilde{\mathscr{C}}^{\nu}$).

On remarque que $\tilde{S}^{\nu}_{j^{\nu}-1}$ ne peut être réduit à un point, car il contient $S^{\nu}_{j^{\nu}}$ et la chaîne $\mathscr{C}^{\nu}$ a été supposée régulière. Par contre $\tilde{S}^{\nu}_{j^{\nu}}$ peut être réduit à un point ; si c'est le cas on le supprime ; on aboutit ainsi à une chaîne $\mathscr{C}^{\nu+1}$ qui est régulière et dont le nombre de niveaux est égal soit à m^{ν} soit à $m^{\nu} - 1$. La suppression d'un support $\tilde{S}^{\nu}_{j^{\nu}}$ réduit à un point nécessite de modifier les coefficients

$\lambda_{S^{\nu}_{j}}(t)$, $(t \in \bigcup_{i=0}^{j^{\nu}-1} S_{i})$ pour $j > j^{\nu}$.

La modification se fait de la façon suivante: Posons $\tilde{S}^{\nu}_{j^{\nu}} = \{\hat{t}\}$ et $\lambda_{\tilde{S}^{\nu}_{j^{\nu}}} = \hat{\varepsilon}$ (où $\hat{\varepsilon}$ est égal à +1 ou -1 par construction). On a :

$$(1) \qquad \sum_{t \in S_{o}} \lambda_{\tilde{S}^{\nu}_{j^{\nu}}}(t)\beta(t) + \sum_{i=1}^{j^{\nu}-2} \sum_{t \in S^{\nu}_{i}} \lambda_{\tilde{S}^{\nu}_{j^{\nu}}}(t)b(t) + \sum_{t \in \tilde{S}^{\nu}_{j^{\nu}-1}} \lambda_{\tilde{S}^{\nu}_{j^{\nu}-1}}(t)b(t) + \hat{\varepsilon}\, b(t) = 0$$

$$(2) \qquad \sum_{t \in S_{o}} \lambda_{S^{\nu}_{j}}(t)\beta(t) + \sum_{\substack{i=0 \\ i \neq j^{\nu}-1 \\ i \neq j^{\nu}}}^{j} \sum_{t \in S^{\nu}_{i}} \lambda_{S^{\nu}_{j}}(t)b(t) + \sum_{t \in \tilde{S}^{\nu}_{j^{\nu}-1}} \lambda_{S^{\nu}_{j}}(t)b(t) + \lambda_{S^{\nu}_{j}}(\hat{t})b(\hat{t}) = 0 \qquad j=j^{\nu}+1,\dots,m .$$

On tire alors $b(\hat{t})$ de l'équation (1) pour le porter dans (2). On obtient ainsi de nouveaux coefficients :

$$\lambda'_{S^{\nu}_{j}}(t) = \lambda_{S^{\nu}_{j}}(t) - \hat{\varepsilon}\lambda_{\tilde{S}^{\nu}_{j^{\nu}}}(t) \text{ , pour } t \in \bigcup_{i=0}^{j^{\nu}-2} S^{\nu}_{i} \cup \tilde{S}^{\nu}_{j^{\nu}-1} \text{ et } j=j^{\nu}+1,\dots,m^{\nu} .$$

On décale les supports S^{ν}_{j} , pour $j=j^{\nu}+1,\dots,m^{\nu}$.

$$S^{\nu}_{j} = S^{\nu}_{j+1} \quad , \quad j=j^{\nu},\dots,m^{\nu}-1$$

et les coefficients :

$$\lambda_{S_j^\nu}(t) = \lambda'_{S_{j+1}^\nu}(t)\ ,\ \text{pour } t \in \bigcup_{i=0}^{j^\nu-2} S_i^\nu \cup \tilde{S}_{j^\nu-1}^\nu \quad \text{et } j=j^\nu,\dots,m^\nu-1\ .$$

<u>$3^{\text{ème}}$ *cas*</u> : $j^\nu = m^\nu + 1$.

Partant de la chaîne :

$$\{S_1^\nu,S_2^\nu,\dots,S_{m^\nu}^\nu,\{t^\nu\}\}$$

on détermine le plus petit indice i^ν ($1 \leq i^\nu \leq m^\nu + 1$) tel que :

$$\{S_1^\nu,S_2^\nu,\dots,S_{i^\nu-1}^\nu,\{t^\nu\},S_{i^\nu}^\nu,\dots,S_{m^\nu}^\nu\}$$

soit encore une chaîne. Si l'on pose :

$$I^\nu = \{i \in \{1,\dots,m^\nu+1\} \mid b(t^\nu) \in V_{i-1}\}$$

on a : $i^\nu = \min\,(i \mid i \in I^\nu)$.

Si l'on a $i^\nu = 1$, cela signifie que $\{t^\nu\}$ est un support de V et que le montant correspondant :

$$\alpha_{m^\nu+1}^\nu = |\langle x^\nu,b(t^\nu)\rangle - c(t^\nu)|$$

vérifie :

$$f(x^\nu) - \alpha_{m^\nu+1}^\nu \leq \varepsilon_{m^\nu+1} \leq \varepsilon_1 \ ,$$

donc que x^ν est une solution à ε_1 près de (P). On arrête donc le calcul.

Si l'on a $2 \leq i^\nu \leq m^\nu + 1$, on échange $S_{i^\nu-1}^\nu$ et $\{t^\nu\}$, ce qui donne $\tilde{S}_{i^\nu-1}^\nu$ et $\tilde{S}_{i^\nu}^\nu$. Le support minimal $\tilde{S}_{i^\nu-1}^\nu$ ne peut être réduit à un point car ce point serait t^ν et on a $b(t^\nu) \notin V_{i^\nu-2}$. Si $\tilde{S}_{i^\nu}^\nu$ est réduit à un point , on le supprime et on modifie les coefficients $\lambda_{S_j^\nu}(t)$ pour $j > i^\nu$ comme dans le deuxième cas.

<u>*Dans les trois cas*</u>, on aboutit donc (sauf si le calcul s'arrête, la précision étant atteinte) à une nouvelle chaîne régulière $\mathcal{C}^{\nu+1}$. On calcule alors la nouvelle solution $x^{\nu+1}$ et les montants $\alpha_1^{\nu+1},\dots,\alpha_{m^{\nu+1}}^{\nu+1}$ qui lui sont associés.

<u>*Remarques*</u>

1) Les opérations effectuées dans le troisième cas peuvent être remplacées par une succession d'opérations d'échange (ce sera le cas dans l'organigramme ci-dessous) :

Organigramme

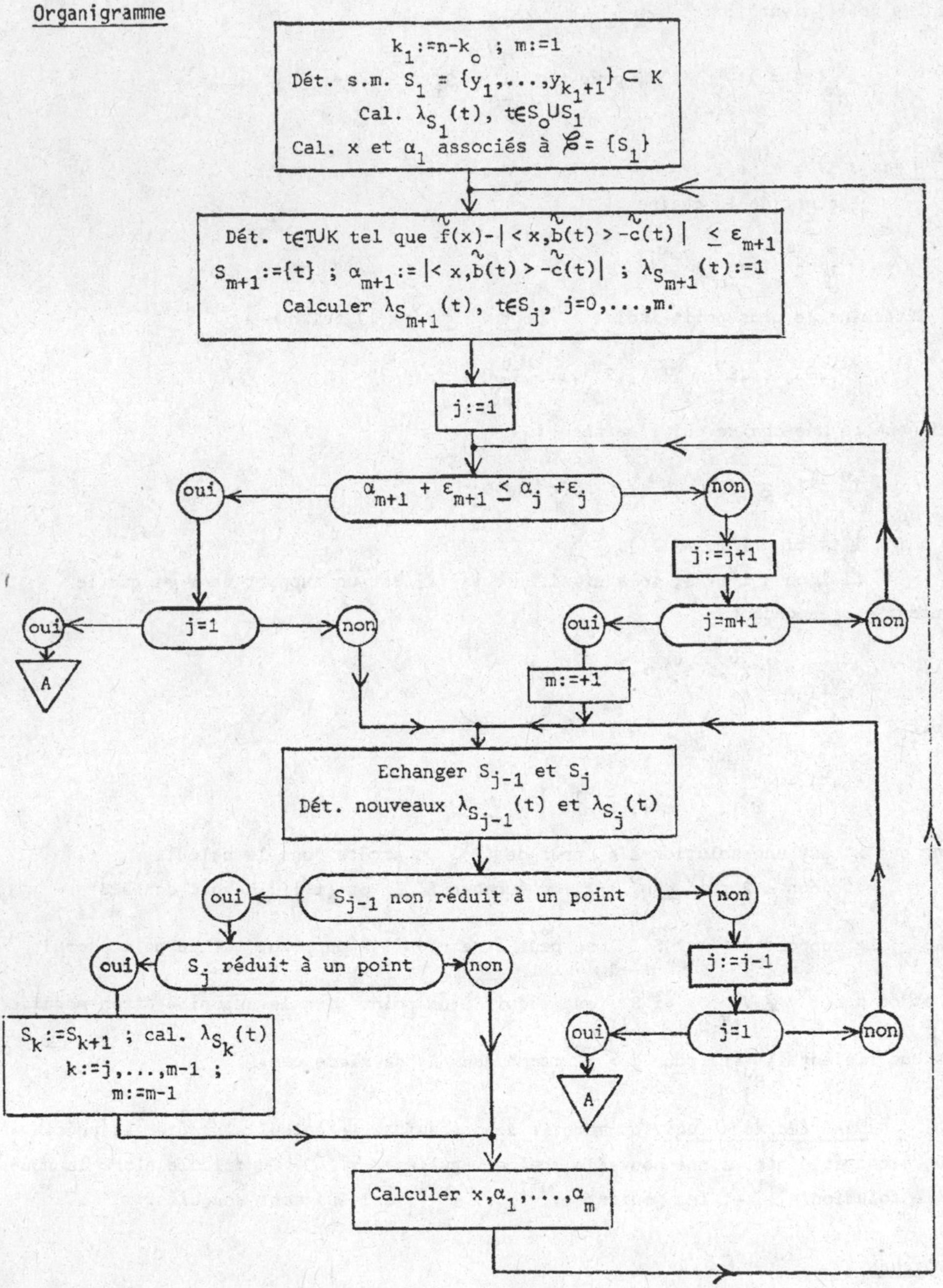

Partant de la chaîne :

$$\{S_1^\nu, S_2^\nu, \dots, S_{m^\nu}^\nu, \{t^\nu\}\}$$

on échange $S_{m^\nu}^\nu$ et $\{t^\nu\}$, ce qui donne $\tilde{S}_{m^\nu}^\nu$ et $\tilde{S}_{m^\nu+1}^\nu$. Si $\tilde{S}_{m^\nu}^\nu$ est réduit à un point, cela signifie que $\tilde{S}_{m^\nu}^\nu = \{t^\nu\}$ et que $\tilde{S}_{m^\nu+1}^\nu = S_{m^\nu}^\nu$, c'est-à-dire qu'on a la chaîne :

$$\{S_1^\nu, S_2^\nu, \dots, S_{m^\nu-1}^\nu, \{t^\nu\}, S_{m^\nu}^\nu\}.$$

On échange alors $S_{m^\nu-1}^\nu$ et $\{t^\nu\}$ et ainsi de suite jusqu'à ce qu'on obtienne :

soit $\{\{t^\nu\}, S_1^\nu, \dots, S_{m^\nu}^\nu\}$ et on arrête le calcul

soit $\{S_1^\nu, S_2^\nu, \dots, \tilde{S}_{i^\nu-1}^\nu, \tilde{S}_{i^\nu}^\nu, S_{i^\nu+1}^\nu, \dots, S_{m^\nu}^\nu\}$

où $S_{i^\nu-1}^\nu$ contient t^ν et n'est pas réduit à ce point. Si $\tilde{S}_{i^\nu}^\nu$ est réduit à un point, on le supprime comme plus haut.

2) La démonstration de la convergence de l'algorithme précédent est faite dans un cadre plus général dans [7].

5.2. Initialisation de l'algorithme

La détermination d'une chaîne initiale régulière $\mathcal{C}^\circ = \{S_1^\circ, \dots, S_m^\circ\}$ peut être difficile, voire impossible. Ainsi, si la dimension de l'espace vectoriel engendré par les fonctions $b_1, \dots, b_n$ est strictement inférieur à n et si $V = E$ ($k_o = 0$) on ne peut trouver une chaîne régulière.

Afin de permettre une initialisation aisée de l'algorithme on modifie la fonction f à minimiser de façon à ce que toute solution du nouveau problème soit encore solution du problème initial et que l'on puisse déterminer une chaîne initiale $\mathcal{C}^\circ = \{S_1^\circ\}$ avec S_1° comportant exactement $n-k_o+1$ éléments. On suppose que α est positif. Soit r un scalaire positif tel qu'il existe une solution $\bar{x}$ vérifiant $\|\bar{x}\| \leq r$ et soit $\eta > 0$ tel que $\eta \leq f(\bar{x}) = \alpha$. On considère la fonction :

$$\tilde{f}(x) = \text{Max } [f(x) ; \frac{\eta}{r} \|x\|]$$

où $\|x\|$ désigne la norme euclidienne de $x \in E$.
La fonction $x \to \frac{\eta}{r} \|x\|$ peut aussi s'écrire :

$$\frac{\eta}{r} \|x\| = \sup_{y \in K} < x, y >$$

où K est l'ensemble de E défini par :

$$K = \{y \in E \mid \|y\| \leq \eta r^{-1}\}$$

La fonction $\tilde{f}$ peut alors s'écrire :

$$\tilde{f}(x) = \underset{t \in T \cup K}{\text{Max}} |< x, \tilde{b}(t) > - \tilde{c}(t)|$$

$\tilde{b}$ et $\tilde{c}$ désignant les prolongements de b et de c à K définis par

$$\tilde{b}(t) = \begin{cases} b(t) & \text{si } t \in T \\ t & \text{si } t \in K \end{cases} \quad \text{et} \quad \tilde{c}(t) = \begin{cases} c(t) & \text{si } t \in T \\ 0 & \text{si } t \in K \end{cases}$$

On considère alors le problème $(\tilde{P})$ de la minimisation de $\tilde{f}(x)$, pour $x \in W$. Par le choix de r et η, on a :

$$\underset{x \in W}{\text{Inf}}\, f(x) = \underset{x \in W}{\text{Inf}}\, \tilde{f}(x)$$

et l'ensemble des solutions de $(\tilde{P})$ est exactement égal à l'ensemble (non vide) des solutions du problème initial dont la norme est inférieure ou égale à $\frac{\alpha}{\eta} r$.

L'initialisation du problème $(\tilde{P})$ est aisée ; il suffit de construire k_1+1 éléments t_i de K avec $k_1 = n-k_o$ tels que $S_1^o = \{t_1,\ldots,t_{k_1+1}\}$ soit un support minimal de V. La chaîne initiale est alors $\mathcal{C}^o = \{S_1^o\}$. Cette technique d'initialisation permet, notamment, de résoudre le cas où les fonctions b_i, $(i=1,\ldots,n)$ sont linéairement dépendantes. Dans ce cas le support minimal S_1^μ de V tel que x^μ soit solution à ε près du problème contient toujours au moins un élément de K.

6. Exemple

Nous prendrons un exemple volontairement simple afin de voir comment se font les échanges successifs.

On considère le problème d'approximation avec n=2, $k_o=0$, $T = [0,4]$ défini par:

$$c(t) = \begin{cases} 0 & \text{si } t < 1 \\ t-1 & \text{si } t \geq 1 \end{cases} \qquad b_1(t) = \begin{cases} \sin(2\pi t) & \text{si } t < 1 \\ 0 & \text{si } t \geq 1 \end{cases}$$

$$b_2(t) = \begin{cases} \cos(2\pi t) & \text{si } t < 1 \\ 1 & \text{si } 1 \leq t < 2 \\ 2 - \frac{t}{2} & \text{si } 2 \leq t < 3 \\ 5 - \frac{3t}{2} & \text{si } 3 \leq t \end{cases}$$

Pour initialiser l'algorithme on prendra $\eta = 10^{-2}$, $r = 10$.
On choisit pour former le support initial :

$$y_1 = (10^{-3},0) \quad y_2=(-7\times 10^{-4},7\times 10^{-4}) \quad \text{et} \quad y_3 = (0,-10^{-3}).$$

La suite des chaînes construites et des systèmes linéaires de matrices $(A_{\mathcal{C}},c_{\mathcal{C}})$ s'écrit :

étape 1 : $m^1 = 1$

$$\begin{bmatrix} 10^{-3} & 0 & -1 \\ -7\times 10^{-4} & 7\times 10^{-4} & -1 \\ 0 & -10^{-3} & -1 \end{bmatrix} \times \begin{bmatrix} x_1 \\ x_2 \\ z_1 \end{bmatrix} = \begin{bmatrix} 0 \\ 0 \\ 0 \end{bmatrix}$$

Le point de $T \cup K$ sélectionné est $\hat{t} = 4 \in T$ et on a $b(\hat{t}) = (0,-1)$, $c(\hat{t}) = 3$.
On échange $b(\hat{t})$ avec (y_1,y_2).

étape 2 : $m^2 = 2$

$$\left[\begin{array}{cc|c|c} 0 & -10^{-3} & -1 & 0 \\ 0 & -1 & -1 & 0 \\ \hline 10^{-3} & 0 & 0 & -1 \\ -7\times 10^{-4} & 7\times 10^{-4} & 0 & -1 \end{array}\right] \times \begin{bmatrix} x_1 \\ x_2 \\ z_1 \\ z_2 \end{bmatrix} = \begin{bmatrix} 0 \\ 3 \\ 0 \\ 0 \end{bmatrix}$$

$\hat{t} = 2$; $b(\hat{t}) = (0,1)$; $c(\hat{t}) = 1$.
On échange $b(\hat{t})$ et S_2.

étape 3 : $m^3 = 3$

$$\left[\begin{array}{cc|c|c|c} 0 & -10^{-3} & -1 & 0 & 0 \\ 0 & -1 & -1 & 0 & 0 \\ \hline 0 & 1 & 0 & -1 & 0 \\ \hline 10^{-3} & 0 & 0 & 0 & -1 \\ -7\times 10^{-4} & 7\times 10^{-4} & 0 & 0 & -1 \end{array}\right] \times \begin{bmatrix} x_1 \\ x_2 \\ z_1 \\ z_2 \\ z_3 \end{bmatrix} = \begin{bmatrix} 0 \\ 3 \\ 1 \\ 0 \\ 0 \end{bmatrix}$$

On se trouve dans la situation indiquée dans la remarque du paragraphe 5.1.
On échange $\hat{t}$ et S_1.

étape 4 : $m^4 = 3$

$$\left[\begin{array}{cc|c|c|c} 0 & 1 & -1 & 0 & 0 \\ 0 & -1 & -1 & 0 & 0 \\ 0 & -10^{-3} & 0 & -1 & 0 \\ \hline 10^{-3} & 0 & 0 & 0 & -1 \\ \hline -7\times10^{-4} & 7\times10^{-4} & 0 & 0 & -1 \end{array}\right] \times \begin{bmatrix} x_1 \\ x_2 \\ z_1 \\ z_2 \\ z_3 \end{bmatrix} = \begin{bmatrix} 1 \\ 3 \\ 0 \\ 0 \\ 0 \end{bmatrix}$$

On élimine S_2 qui ne contient qu'un élément.

étape 5 : $m^5 = 2$

$$\left[\begin{array}{cc|c|c} 0 & 1 & -1 & 0 \\ 0 & -1 & -1 & 0 \\ \hline 10^{-3} & 0 & 0 & -1 \\ -7\times10^{-4} & 7\times10^{-4} & 0 & -1 \end{array}\right] \times \begin{bmatrix} x_1 \\ x_2 \\ z_1 \\ z_2 \end{bmatrix} = \begin{bmatrix} 1 \\ 3 \\ 0 \\ 0 \end{bmatrix}$$

$\hat{t} = 3$; $b(\hat{t}) = (0,0.5)$; $c(\hat{t}) = 2$.

On échange $b(\hat{t})$ et S_2 .

étape 6 : $m^6 = 3$

$$\left[\begin{array}{cc|c|c|c} 0 & 1 & -1 & 0 & 0 \\ 0 & -1 & -1 & 0 & 0 \\ \hline 0 & 0.5 & 0 & -1 & 0 \\ \hline 10^{-3} & 0 & 0 & 0 & -1 \\ -7\times10^{-4} & 7\times10^{-4} & 0 & 0 & -1 \end{array}\right] \times \begin{bmatrix} x_1 \\ x_2 \\ z_1 \\ z_2 \\ z_3 \end{bmatrix} = \begin{bmatrix} 1 \\ 3 \\ 2 \\ 0 \\ 0 \end{bmatrix}$$

On échange $S_2 = \{\hat{t}\}$ et S_1 ; $\tilde{S}_2$ est alors réduit à un élément et on le supprime (étape 7). On a alors :

étape 8 :

$$\left[\begin{array}{cc|c|c} 0 & 0.5 & -1 & 0 \\ 0 & -1 & -1 & 0 \\ \hline 10^{-3} & 0 & 0 & -1 \\ -7\times10^{-4} & 7\times10^{-4} & 0 & -1 \end{array}\right] \times \begin{bmatrix} x_1 \\ x_2 \\ z_1 \\ z_2 \end{bmatrix} = \begin{bmatrix} 2 \\ 3 \\ 0 \\ 0 \end{bmatrix}$$

Ce dernier système nous donne une solution du problème d'approximation :

$x_1 = -0.275$ $x_2 = -0.667$

L'erreur est $\alpha_1 = |z_1| = 2.333$.

Remarques

1) L'exemple précédent ne vérifie pas la condition de Haar et ne pourrait pas être résolu par l'algorithme de Rémès classique. Il n'est même pas "itératif" au sens défini dans [3].

2) Si on remplace $T = [0,4]$ par $T = [1,4]$ les fonctions b_1 et b_2 deviennent linéairement dépendantes (car $b_1 \equiv 0$). L'algorithme se déroule cependant de la même façon et donne une solution $x_1 = -0.275$ et $x_2 = -0.667$ identique.

Références

[1] Brosowski, B. : Über Tschebyscheffsche Approximationen mit linearen Nebenbedingungen. Math. Zeitschr., 88, (1965), 105-128.

[2] Carasso, C. : L'algorithme d'échange en optimisation convexe. Thèse Grenoble, (1973).

[3] Carasso, C. : Etude de l'algorithme de Rémès en l'absence de conditions de Haar. Num. Math., 20, (1972), 165-178.

[4] Carasso, C. : Densité des hypothèses assurant la convergence de l'algorithme de Rémès, R.A.I.R.O., R3, (1972), 69-84.

[5] Carasso, C. : Un algorithme de minimisation de fonctions convexes avec ou sans contraintes : "l'algorithme d'échange". Prépublication N° 3, Math., Univ. de Saint-Etienne, 7th IFIP Conference on Optimization Techniques, Springer-Verlag (1975).

[6] Carasso, C. et Laurent, P.J.: Un algorithme pour la minimisation d'une fonctionnelle convexe sur une variété affine. Séminaire d'Analyse Numérique, Grenoble, (18 octobre 1973).

[7] Carasso, C. et Laurent, P.J. : Un algorithme de minimisation en chaîne en optimisation convexe. Séminaire d'Analyse Numérique, N° 242, (29 janvier 1976) à paraître.

[8] Carasso, C. et Laurent, P.J. : An algorithm of successive minimization in convex programming. IX. Intern. Symp. on Mathematical Programming, Budapest, 23-27 août 1976.

[9] Cheney, E.W., and Goldstein, A.A. : Newton's method for convex programming and Tchebyscheff approximation. Num. Math. 1, (1959), 253-268.

[10] Descloux, J. : Dégénérescences dans les approximations de Tschebycheff linéaires et discrètes. Num. Math. 3, (1961), 180-187.

[11] Goldstein, A.A. : Constructive real analysis. Harper's series in modern mathematics, Harper and Row, (1967).

[12] Laurent, P.J. : Approximation uniforme de fonctions continues sur un compact avec contraintes de type inégalité. Rev. Franç. d'Inf. et de Rech. Opér. 5, (1967), 81-95.

[13] Laurent, P.J. : Théorèmes de caractérisation d'une meilleure approximation dans un espace normé et généralisation de l'algorithme de Rémès. Num. Math. 10, (1967), 190-208.

[14] Laurent, P.J. : Charakterizierung und Gewinnung einer besten Approximation in einer konvexen Teilmenge eines normierten Raumes. ISNM 12, (1968), 91-102, Birkhäuser Verlag.

[15] Laurent, P.J. : Approximation et Optimisation, Hermann, Paris, (1972).

[16] Laurent, P.J. : Exchange algorithm in convex analysis. Conf. on Approximation Theory, the Univ. of Texas, Austin, (1973). Acad. Press.

[17] Laurent, P.J. and Pham Dinh Tuan :Global approximation of a compact set by elements of a convex set in a normed space. Num. Math. 15, (1970), 137-150.

[18] Rémès, E. : Sur le calcul effectif des polynomes d'approximation de Tchebycheff. C.R.A.S., Paris, 199, (1934), 337-340.

[19] Rémès, E. : General computational methods for Chebyshev approximation. Problems with real parameters entering linearly. Izdat. Akad. Nauk Ukrainsk. SSR, Kiev (1957), Atomic Energy Commission Translations 4491.

[20] Rockafellar, R.T. : Convex analysis, Princ. Univ. Press (1970).

[21] Stiefel, E.L. : Über diskrete und lineare Tschebyscheff-Approximationen. Num. Math. 1, (1959), 1-28.

[22] Stiefel, E.L. : Numerical Methods of Tschebycheff Approximation. In "On Numerical Approximation", R. Langer Ed., Univ. of Wisconsin,(1959),217-232.

[23] Stiefel, E.L. : Note on Jordan Elimination, Linear Programming and Tchebycheff Approximation. Num. Math. 2, (1960), 1-17.

[24] Töpfer, H.J. : Über die Tschebyscheffsche Approximationsaufgabe bei nicht erfüllter Haarscher Bedingung. Berichte des Hahn-Meitner-Instituts Berlin, HMI-B40(1965).

[25] Töpfer, H.J. : Tschebyscheff-Approximation bei nicht erfüllter Haarscher Bedingung. Z.A.M.M., 45, (1965), T81-T82.

[26] Töpfer, H.J. : Tschebyscheff-Approximation und Austauschverfahren bei nicht erfüllter Haarscher Bedingung. Tagung, Oberwolfach (1965). ISNM7, Birkhäuser Verlag (1967), 71-89.

Remerciements

Les auteurs remercient vivement J. AZEMA, de l'Université de Saint-Etienne, pour la mise en oeuvre et la programmation de l'algorithme.

ON THE RANGE OF CERTAIN LOCALLY DETERMINED SPLINE PROJECTIONS

C. K. Chui, P. W. Smith and J. D. Ward

1. Introduction.

Let $H^{\ell,2}$ be the linear space of functions which are ℓ-fold integrals of $L_2(R)$ functions. Many authors, including Schoenberg [11], Jerome-Schumaker [9], Golomb [8], Smith [12, 13], and de Boor [4], have treated problems centered about the question: What can one say about the solution $s \in H^{\ell,2}$ to the problem

$$\inf\{\int_R (D^\ell f)^2 : f \in H^{\ell,2},\ f(t_i) = Y_i,\ -\infty < i < \infty\},$$

concerning existence, characterization, continuity with respect to the data, etc.? The solution s is called an $H^{\ell,2}$-spline (interpolating the data) and is known to be a piecewise polynomial of order 2ℓ (degree $2\ell - 1$). In this paper we derive sufficient conditions to characterize when an $H^{\ell,2}$-spline is in $L_2(R)$, thus continuing the study initiated in [13].

In the course of deriving these results, we develop certain properties of banded matrices and their inverses in section 2. These results are then coupled with certain results on locally determined spline projections in section 3 to obtain necessary and/or sufficient conditions on the data for spline projection operators to produce an $L_p(R)$ spline. Also in this section an interesting variant of Helly's Theorem is stated and proved. Section 4 contains examples related to spline interpolation including the treatment of $H^{\ell,2}$-splines.

2. Preliminaries.

In this section, we will derive some preliminary results concerning banded matrices and their inverses. It turns out that if a banded matrix A and its inverse A^{-1} exist as operators on a Banach space X, then it very often happens that both A and A^{-1} exist on certain "related" Banach spaces. The results we obtain here will be applied to determine which data sequences can be interpolated by splines of a certain order in $L_p(R)$.

The following proposition is a modification of a recent result of Demko [7]. Since Demko's proof essentially applies here, we just state the result without proof.

PROPOSITION 2.1. *Let* $A = (a_{ij})$ *be a bi-infinite matrix such that*

(a) *there exists a nonnegative integer* m *so that* $a_{ij} = 0$ *for all* i *and* j *with* $|i - j| > m$, *and*

(b) *considered as an operator on some* ℓ_q, $1 \le q \le \infty$, A *satisfies* $\|A\|_q \le 1$ *and* $\|A^{-1}\|_q \le 1/\mu$ *for some* $\mu > 0$. *Set* $A^{-1} = (\alpha_{ij})$. *Then there exists a positive constant and an* r, $0 \le r \le 1$, *depending only on* μ *and* m, *such that* $|\alpha_{ij}| \le Kr^{|i-j|}$ *for all* i, j.

In fact, it can be shown (cf. de Boor [6]) that

$$r \le (K_q/(1+K_q^q)^{1/q})^{1/2m}$$

where $K_q = \|A\|_q \|A^{-1}\|_q$.

Let S be the collection of all doubly infinte sequences which tend to zero at $\pm\infty$ faster than the reciprocal of any polynomial. We may, and will, define a countable family $P = \{P_n\}$, $n = 1,2,\ldots$, of seminorms on S by

$$P_n(y) = \sup_i \{(1+i^2)^n |y_i|\}.$$

Then P defines a locally convex topology on S, and in fact,

S becomes a Fréchet space (see [10], p. 46). By a standard argument, it is clear that the dual space S' of S can be identified with the space of all sequences $\{b_i\}$, $-\infty < i < \infty$, which have at most power growth; that is, $|b_i| \leq |p(i)|$ for some polynomial p. We have the following result.

PROPOSITION 2.2. Let A satisfy hypotheses (a) and (b) in Proposition 2.1. Then A is a continuous, one-one and onto operator from S' to S' with the weak star topology.

Proof. Let A^T be the formal adjoint of A; that is $A^T=(\beta_{ij})$ with $\beta_{ij} = a_{ji}$ for all i,j. The proof will be accomplished by showing that A^T is a continuous bijective map from S to S. In turn, it suffices to show that A^T and $(A^{-1})^T$ are both continuous from S into itself. Now since A satisfies (a) and (b) in Proposition 2.1., both A^T and $(A^{-1})^T$ are continuous and well-defined on some ℓ_p space. Hence, as linear (but not necessarily bounded) operators on S, A^T and $(A^{-1})^T$ exist as mappings from S to ℓ_p. Thus, Proposition 2.2. will be established if we can prove:

PROPOSITION 2.3. Let A satisfy hypotheses (a) and (b) in Proposition 2.1. and let $B = (A^{-1})^T$. Then both A^T and B are continuous operators from S into itself.

Proof. We only give a proof for the operator B since the proof for A^T is clear. Take $y = (y_i) \in S$. Then for each $k \geq 0$, $P_{k+1}(y) < \infty$. Fix a positive integer k. By Proposition 2.1., we have

$$(1 + j^2)^k (By)_j = (1 + j^2)^k \sum_i c_{ji}\, r^{|i-j|} y_i$$

where $|c_{ij}| \leq K$ for all i, j. Therefore

$$|(1 + j^2)^k (By)_j| \leq K \sum_i r^{|i-j|} y_i (1 + j^2)^k$$

$$\leq KP_{k+1}(y) \sum_i r^{|i-j|}(1+i^2)^{-k}(1+j^2)^k(1+i^2)^{-1}$$

$$\equiv KP_{k+1}(y)\{\sum_{|i|<|j|} + \sum_{|i|\geq|j|}\}.$$

Here and throughout we have omitted the summands for simplicity. For $|i| < |j|$, $0 < |j| - |i| \leq |i - j|$, so that $r^{|i-j|} \leq r^{|j|-|i|}$, and this implies

$$\sum_{|i|<|j|} \leq \sum_{|i|<|j|} r^{|j|}\{r^{-|i|}(1+i^2)^{-k}\}(1+j^2)^k(1+i^2)^{-1}.$$

To estimate $r^{-|i|}(1+i^2)^{-k}$, let us consider the function

$$f(x) = r^{-x}(1+x^2)^{-k}, \; x > 0.$$

Since $f'/f = \log(1/r) - 2kx/(1+x^2)$, we have $f'(x) > 0$ for all large x, and this gives

$$r^{-|i|}(1+i^2)^{-k} \leq r^{-|j|}(1+j^2)^{-k}$$

for all large $|i|$ with $|i| < |j|$. Pick $|j|$ large enough. Then

$$\sum_{|i|<|j|} \leq C_1 \sum_{|i|<|j|} (1+i^2)^{-1} \leq 2C_1 \sum_{i=0}^{\infty} \frac{1}{1+i^2} \equiv C_2$$

for some $C_1 \geq 1$ <u>independent</u> of j.

On the other hand, for $|i| \geq |j|$, we have $0 \leq |i| - |j| \leq |i - j|$, so that $r^{|i-j|} \leq r^{|i-j|}$, and this gives

$$\sum_{|i|\geq|j|} \leq \sum_{|i|\geq|j|} [r^{-|j|}(1+j^2)^k][r^{|i|}(1+i^2)^{-k}](1+i^2)^{-1}$$

$$\leq \sum_{|i|\geq|j|} (1+i^2)^{-1} < 2\sum_{i=0}^{\infty} \frac{1}{1+i^2} = C_3 .$$

Here, we have noted that $\{r^{|i|}(1+i^2)^{-k}\}$, $i = 1,2,\ldots$, is a monotone decreasing sequence.

Hence, $|(1+j^2)^k(By)_j| \leq K(C_2 + C_3)P_{k+1}(y) \equiv C_4 P_{k+1}(y)$

for all j, so that we have $P_k(By) \leq C_4 P_{k+1}(y)$. This completes our proof of the proposition.

PROPOSITION 2.4. *Let* A *satisfy hypotheses* (a) *and* (b) *in Proposition* 2.1. *Then* A *is a continuous, one-one operator from* ℓ_p *onto* ℓ_p *for any* p, $1 \leq p \leq \infty$.

Proof. By hypothesis (b), A is a continuous, one-one and onto operator on some ℓ_q space. Thus, as linear operators, A and A^{-1} are defined on ℓ_p for $1 \leq p \leq q$. Using an argument similar to that in the proof of Proposition 2.2., it is readily seen that A and A^{-1} are continuous on ℓ_p into ℓ_p for $1 \leq p \leq q$, and hence the required result follows for $1 \leq p \leq q$. Now by appealing to adjoints, we see that A^T and $(A^{-1})^T$ satisfy the conclusions of our proposition for $1 \leq p \leq \infty$. Appealing to adjoints once again completes the proof.

3. Local spline projections.

Let $k \geq 2$ be a positive integer and $\underline{\underline{t}} = \{t_i\}$, $-\infty < i < \infty$, be a bi-infinite knot sequence satisfying $\ldots \leq t_i \leq t_{i+1} \leq \ldots$, $t_{i+k-1} > t_i$ for all i, and $\lim_{i\to\infty} t_i = -\lim_{i\to\infty} t_{-i} = \infty$. We will denote by $S^k_{\underline{\underline{t}}}$ the space of all splines of order k with the knot sequence $\underline{\underline{t}}$ (see [1]). The Banach space of all real-valued bounded continuous functions on the real line R with the supremum norm $\|\cdot\|_\infty$ will be denoted by $BC(R)$, and we will let $BS^k_{\underline{\underline{t}}}$ be the subspace that consists of all bounded spline functions in $S^k_{\underline{\underline{t}}}$. That is, $BS^k_{\underline{\underline{t}}} = S^k_{\underline{\underline{t}}} \cap BC(R)$.

We also consider a sequence $\{\phi_i\}$, $-\infty < i < \infty$, of continuous linear functionals on $BC(R)$ that satisfies the following two conditions:

(a) $\|\phi_i\| = 1$ and $\text{supp}\,(\phi_i) \subset [t_i, t_{i+k}]$ for all i. Here, by $\text{supp}\,(\phi)$, $\phi \in [BC(R)]^*$, we mean the complement of the largest open set G so that for every $f \in [BC(R)]$ with compact support in G, we have $\phi(f) = 0$.

(b) There is a positive number Γ such that

$$(3.1)\quad \Gamma^{-1} \sum_{i=-\infty}^{\infty} |a_i| \leq \| \sum_{i=-\infty}^{\infty} a_i \phi_i \|_\infty \leq \sum_{i=-\infty}^{\infty} |a_i|$$

for all sequences $\{a_i\}$ in ℓ_1.

Let P be a linear map from $BC(R)$ into $BS_{\underline{\underline{t}}}^k$ defined by $\phi_i(Pf) = \phi_i(f)$, $f \in BC(R)$, for all i. We will say that P is the (unique) linear projector determined by $\{\phi_i\}$ onto $BS_{\underline{\underline{t}}}^k$ if P is bounded. By using the spanning property of the normalized B-splines (see [2]), we may, and will, realize P as the inverse of the banded matrix A defined by

$$(3.2)\quad A = (a_{ij}),\ a_{ij} = \phi_j(N_{i,k})$$

as long as A maps ℓ_∞ onto ℓ_∞. To this end we have the following theorem which implies that A is onto.

THEOREM 3.1. <u>Let</u> $\{\phi_i\}$ <u>be a bi-infinite sequence of continuous linear functionals on</u> $BC(R)$ <u>that satisfies conditions</u> (a) <u>and</u> (b) <u>and determines a linear projector</u> P <u>from</u> $BC(R)$ <u>onto</u> $BS_{\underline{\underline{t}}}^k$. <u>Then for every data sequence</u> $\{\beta_i\}$ <u>in</u> ℓ_∞, <u>there is a spline</u> $s \in BS_{\underline{\underline{t}}}^k$ <u>such that</u> $\phi_i(s) = \beta_i$ <u>for all</u> $i = 0, \pm 1, \ldots$.

<u>Proof</u>. Without loss of generality, we may assume $\|\{\beta_i\}\|_\infty = 1$. By Helly's Theorem (cf. [10], p. 111) which is valid because of (3.1), there exists an $x_n \in BC(R)$ with $\|x_n\| \leq \Gamma + \varepsilon$ where $\varepsilon > 0$, so that $\phi_i(x_n) = \beta_i$ for $|i| \leq n$. Let

$$s_n = P(x_n) = \sum_{i=-\infty}^{\infty} \gamma_{i,n} N_{i,k} .$$

Then $\|s_n\| \leq \|P\|(\Gamma + \varepsilon)$ and $|\gamma_{i,n}| \leq M$ for some M and all i, n. Choose subsequences $\gamma_{i,n_\ell} \to \gamma_i^*$ where $|\gamma_i^*| \leq M$ for all i. Now set $s = \sum \gamma_i^* N_{i,k}$. It is clear that $s \in BS_{\underline{\underline{t}}}^k$ and $\phi_i(s) = \beta_i$ for all i. This completes the proof of the theorem.

REMARKS. (1) We have just shown that A maps ℓ_∞ onto ℓ_∞. Since A is known to be injective and bounded, we conclude that A^{-1} exists as a bounded mapping on ℓ_∞. Thus, by the results of the previous section, if $\{\beta_i\}$, $-\infty < i < \infty$, is of power growth (i.e. in S'), then there is a unique $s = \sum \alpha_i N_{i,k}$ in $S_{\underline{\underline{t}}}^k$ satisfying $\phi_i(s) = \beta_i$ for all i and $\{\alpha_i\} \in S'$. We will say more about this later.

(2) Since A satisfies the conditions in Proposition 2.1, there is a number $r \in [0,1)$, and we will denote it by $r(A)$, which bounds the rate of decay of the elements in A^{-1}.

(3) Recalling $A = (a_{ij})$ as in (3.2), we will write $A^{-1} = (\alpha_{ij})$; and in view of Proposition 2.1, we will also write $\alpha_{i,j} = c_{i,j} r^{|i-j|}$ where $r = r(A)$ and $|c_{i,j}| \leq K$ for all i and j.

LEMMA 3.1. _Let_ $\phi_i \in [BC(R)]^*$, $-\infty < i < \infty$, _satisfy the hypotheses in Theorem_ 3.1, _and let_ $r(A) = r \in [0,1)$. _Suppose that the knot sequence_ $\underline{\underline{t}} = \{t_i\}$ _satisfies_

$$(3.3) \qquad \sum_{j=-\infty}^{\infty} r^{|i-j|} \left[\frac{t_{i+k}-t_i}{t_{j+k}-t_j}\right]^{q/p} \leq D < \infty$$

for all i, $-\infty < i < \infty$, _where_ $1/p + 1/q = 1$, $1 \leq p \leq \infty$. _Then for any_ $f \in BC(R)$, _the data sequence_ $\{x_i\}$, $x_i = \phi_i(f)$, _projects onto a spline in_ $L_p(R)$ _if_ $\{x_i(t_{i+k}-t_i)^{1/p}\} \in \ell_p$.

Proof. It has been shown by de Boor [2] that there exist absolute constants C_1 and C_2 such that

$$(3.4) \qquad C_1 \|\{y_i(t_{i+k}-t_i)^{1/p}\}\|_{\ell_p} \leq \left\|\sum_{i=-\infty}^{\infty} y_i N_{i,k}\right\|_p \leq C_2 \|\{y_i(t_{i+k}-t_i)^{1/p}\}\|_{\ell_p} ,$$

$1 \leq p \leq \infty$. Thus, it is sufficient to prove that

$$(3.5) \qquad \{(\sum_{j=-\infty}^{\infty} \alpha_{ij} x_j)(t_{i+k}-t_i)^{1/p}\} \in \ell_p .$$

We will only verify the case where $1 < p < \infty$. The cases $p = 1, \infty$ can be proved similarly. Assuming hypothesis (3.3) on the spacing of knots, we now give the following estimate to arrive at (3.5).

$$\sum_{i=-\infty}^{\infty} \left| \sum_{j=-\infty}^{\infty} \alpha_{ij} x_j \right|^p (t_{i+k} - t_i)$$

$$= \sum_{i=-\infty}^{\infty} (t_{i+k} - t_i) \left| \sum_{j=-\infty}^{\infty} c_{ij} r^{|i-j|/p} x_j (t_{j+k} - t_j)^{1/p} r^{|i-j|/q} (t_{j+k} - t_j)^{-1/p} \right|^p$$

$$\leq K^p \sum_{i=-\infty}^{\infty} (t_{i+k} - t_i) \{ \sum_{j=-\infty}^{\infty} r^{|i-j|} |x_j|^p (t_{j+k} - t_j) \} \{ \sum_{j=-\infty}^{\infty} r^{|i-j|} (t_{j+k} - t_j)^{-q/p} \}^{p/q}$$

$$\leq K^p D^{p/q} \sum_{i=-\infty}^{\infty} \sum_{j=-\infty}^{\infty} r^{|i-j|} |x_j|^p (t_{j+k} - t_j)$$

$$= K^p D^{p/q} \sum_{j=-\infty}^{\infty} |x_j|^p (t_{j+k} - t_j) \{ \sum_{i=-\infty}^{\infty} r^{|i-j|} \}$$

$$= K^p D^{p/q} (1 + \frac{2r}{1-r}) \| \{ x_j (t_{j+k} - t_j)^{1/p} \} \|_{\ell_p}^p .$$

In fact, we have proved that

$$\| \{ (\sum_{j=-\infty}^{\infty} \alpha_{ij} x_j) (t_{i+k} - t_i)^{1/p} \} \|_{\ell_p}$$
$$\leq K D^{1/q} (1 + \frac{2r}{1-r})^{1/p} \| \{ x_j (t_{j+k} - t_j)^{1/p} \} \|_{\ell_p} .$$

A partial converse to Lemma 3.1 also holds as in the following

LEMMA 3.2. Let $\phi_i \in [BC(R)]^*$, $-\infty < i < \infty$, satisfy the hypotheses in Theorem 3.1 and let the knot sequence $\underline{t} = \{t_i\}$ satisfy

$$(3.6) \qquad \sum_{j=i-k+1}^{i+k-1} \frac{t_{j+k} - t_j}{t_{i+k} - t_i} \leq E < \infty$$

for all i, $-\infty < i < \infty$. *Suppose* $1 \le p \le \infty$ *and* $s \in S^k_{\underline{t}} \cap L_p(R)$. *Then* $\{\phi_i(s)(t_{i+k}-t_i)^{1/p}\} \in \ell_p$.

We remark that (3.3) trivially implies (3.6) but the converse does not necessarily hold.

Proof of Lemma 3.2. Since $s = \sum y_i N_{i,k} \in L_p(R)$, we have $\{y_i(t_{i+k}-t_i)^{1/p}\} \in \ell_p$ from (3.4). Now

$$\phi_j(s) = \sum_{i=-\infty}^{\infty} y_i \phi_j(N_{i,k}) = \sum_{i=-\infty}^{\infty} y_i a_{ij}$$

where $|a_{ij}| \le 1$ for all i, j and $a_{ij} = 0$ if $|i-j| \ge k$. Again, we only give the proof for $p < \infty$. In this case,

$$\sum_{j=-\infty}^{\infty} |\phi_j(s)|^p (t_{j+k}-t_j) = \sum_{j=-\infty}^{\infty} \Big|\sum_{i=-\infty}^{\infty} y_i a_{ij}\Big|^p (t_{j+k}-t_j)$$

$$\le \sum_{j=-\infty}^{\infty} \{\sum_{|i-j|<k} |y_i|\}^p (t_{j+k}-t_j)$$

$$\le \sum_{j=-\infty}^{\infty} (2k-1)^{p/q} \sum_{|i-j|<k} |y_i|^p (t_{j+k}-t_j)$$

$$= (2k-1)^{p/q} \sum_{i=-\infty}^{\infty} |y_i|^p (t_{i+k}-t_i) \sum_{j=i-k+1}^{i+k-1} \frac{t_{j+k}-t_j}{t_{i+k}-t_i}$$

$$\le (2k-1)^{p/q} E \|\{y_i(t_{i+k}-t_i)^{1/p}\}\|^p_{\ell_p} < \infty.$$

The case $p = \infty$ can be treated similarly.

By applying the above two lemmas and the estimates in the proofs, we have the following result.

THEOREM 3.2. *Let* $\phi_i \in [BC(R)]^*$, $-\infty < i < \infty$, *satisfy the hypotheses in Theorem* 3.1. *and* $\underline{t}$ *be a knot sequence satisfying* (3.3). *Then for any* $f \in BC(R)$ *and* $1 \le p \le \infty$, *we have* $P(f) \in L_p(R)$ *if and only if* $\{\phi_i(f)(t_{i+k}-t_i)^{1/p}\} \in \ell_p$. *In fact, there exist positive absolute constants* C' *and* C''

such that

$$C'\|\{\phi_i(f)(t_{i+k}-t_i)^{1/p}\}\|_{\ell_p} \le \|P(f)\|_p \le C''\|\{\phi_i(f)(t_{i+k}-t_i)^{1/p}\}\|_{\ell_p} .$$

4. Applications to spline interpolation.

In this section we consider spline interpolation on the real line. That is, the linear functionals $\phi_i \in [BC(R)]^*$ which determine the projector P as in Section 3 will be point evaluation functionals. The notation used in Section 3 will be retained. A main result in this section is Theorem 4.2 which gives sufficient conditions on the knot and data sequences for an $H^{\ell,2}$-spline interpolant to be in $L_2(R)$. (Here, and throughout, we use the standard notation for $H^{\ell,2}$ and $H^{\ell,2}$-splines; see, for example, [4,12].) We introduce the following notation to describe the spacing for the knot sequence $\underline{t} = \{t_i\}$:

$$\delta_i^m = t_{i+m} - t_i ,$$

(4.1) $$\Delta_i^m = \max\, [\delta_i^m/\delta_{i+1}^m , \delta_{i+1}^m/\delta_i^m] , \text{ and}$$

$$\Pi_i^m = \max\, [\Delta_{i-1}^m , \Delta_i^m] .$$

Let us first record the following result.

PROPOSITION 4.1. *Let* $\underline{t}$ *be a knot sequence satisfying*

(4.2) $$\limsup_{i\to\pm\infty} \Pi_i^k \le \rho$$

and ϕ_i, $-\infty < i < \infty$, *be continuous linear functionals on* $BC(R)$ *that satisfy the hypotheses in Theorem* 3.1. *Let* $1 \le p \le \infty$, $1/p + 1/q = 1$ *and matrix* A *be given as in* (3.2) *such that* $r(A) = r$ *satisfies* $\rho^{q/p} < 1/r$. *Then a function* $f \in BC(R)$ *projects onto a spline in* $L_p(R)$ *if and only if the sequence* $\{\phi_i(f)(t_{i+k}-t_i)^{1/p}\}$ *is in* ℓ_p.

Proof. Since (4.2) implies (3.6), one direction follows from Lemma 3.2. The other direction can also be obtained by showing that the mesh ratio implies (3.5).

Sharper results can be obtained if the ϕ_i's are point evaluation functionals. In this case, conditions (a) and (b) in Section 3 are automatically satisfied (with $\Gamma = 1$) by the ϕ_i's. The only other condition we need is the boundedness of P. To our knowledge the sharpest theorem we can use concerning cubic spline interpolation at the knots is given by de Boor [3]. If we set $k = 4$, $\phi_i(f) = f(t_i)$ and observe that the condition $\limsup_{i \to \pm\infty} \Pi_i^1 < (3 + \sqrt{5})/2$ implies that the sequence $\{\phi_i\}$ uniquely determines the linear projector P as defined in Section 3, then by applying Proposition 4.1, we have the following

THEOREM 4.1. *Suppose that the knot sequence* $\underline{\underline{t}} = \{t_i\}$ *satisfies*

$$\limsup_{i \to \pm\infty} \Pi_i^1 < (3 + \sqrt{5})/2 \quad \text{and} \quad \limsup_{i \to \pm\infty} \Pi_i^4 = 1.$$

Then a function $f \in BC(R)$ *projects onto a spline in* $S_{\underline{\underline{t}}}^4 \cap L_p(R)$, $1 \le p \le \infty$, *if and only if* $\{f(t_i)(t_{i+4} - t_i)^{1/p}\} \in \ell_p$.

We now consider the problem of $H^{\ell,2}$-spline interpolants. Let $\underline{\underline{t}} = \{t_i\}, \ldots < t_i < t_{i+1} < \ldots$, be a knot sequence and $g \in H^{\ell,2}$. Then there is a unique spline $s \in H^{\ell,2}$ satisfying $s(t_i) = g(t_i)$ for all i and

$$(4.3) \quad \|D^\ell s\|_2 = \inf \{\|D^\ell f\|_2 : f(t_i) = g(t_i), -\infty < i < \infty\}.$$

It is well-known (cf. [8,9,11]) that s is a $C^{2\ell-2}$ piecewise polynomial of order $k \equiv 2\ell$ and $s \in S_{\underline{\underline{t}}}^k$. We wish to establish conditions which will guarantee that $s \in L_2(R)$. It was shown by the second author [13] that if the mesh is quasi-uniform (i.e. $\max \delta_i^1 / \min \delta_i^1$ is finite) then $s \in L_2(R)$ if and only if $\{g(t_i)\} \in \ell_2$. We can now prove a much more general result as in the following

THEOREM 4.2. Let $\underline{\underline{t}} = \{t_i\}$ be a knot sequence satisfying $\delta_i^1 \geq h > 0$ for all i and $\{\phi_i\} \subset [BC(R)]^*$ be defined by $\phi_i(f) = f(t_i)$ such that $\{\phi_i\}$ uniquely determines the linear (bounded) projector P as described in Section 3. Let $\underline{\underline{t}}$ satisfy (3.3), $g \in H^{\ell,2}$ and $k = 2\ell$. Then g has an $H^{\ell,2}$-spline interpolant in $L_2(R)$ if and only if $\{g(t_i)(t_{i+k}-t_i)^{1/2}\} \in \ell_2$.

Proof. Suppose that g has an $H^{\ell,2}$-spline interpolant in $L_2(R)$. Then $\{g(t_i)\}$ must be bounded (see [12,13]) and, hence, we can assume $g \in BC(R)$. Theorem 3.2 then yields the result that $\{g(t_i)(t_{i+k}-t_i)^{1/2}\} \in \ell_2$.

Conversely, suppose that $\{g(t_i)(t_{i+k}-t_i)^{1/2}\} \in \ell_2$. Since $\delta_i^1 \geq h > 0$ for all i, it is easy to see that g is in $BC(R)$. Then by Theorem 3.2 and (3.4), the spline interpolant $P(g) \equiv s$ is in $L_2(R)$ and has the form $s = \sum_{i=-\infty}^{\infty} A_i N_{i,k}$ with $\{A_i(t_{i+k}-t_i)^{1/2}\} \in \ell_2$. From [1], we see that

$$s^{(\ell)} = K \sum_{i=-\infty}^{\infty} A_i^{(\ell)} N_{i,\ell}$$

where $K = (k-1)(k-2)\ldots\ell$ and

$$A_i^{(j)} = \begin{cases} A_i & \text{if } j = 0 \\ (A_i^{(j-1)} - A_{i-1}^{(j-1)})/(t_{i+k-j}-t_i) & \text{if } j \geq 1. \end{cases}$$

Hence, to show that s is the $H^{\ell,2}$-spline interpolant to g, we have to verify that $s^{(\ell)}$ is in $L_2(R)$. This is equivalent to showing that $\{A_i^{(\ell)}(t_{i+\ell}-t_i)^{1/2}\}$ is in ℓ_2. It can be done as in the following estimates:

$$|A_i^{(\ell)}|(t_{i+\ell}-t_i)^{1/2} = \left|A_i^{(\ell-1)} - A_{i-1}^{(\ell-1)}\right|/(t_{i+\ell}-t_i)^{1/2}$$

$$\leq (\ell h)^{-1/2}\left|A_i^{(\ell-1)} - A_{i-1}^{(\ell-1)}\right| \leq \cdots$$

$$\leq c_o h^{-\ell+1/2} \sum_{j=0}^{\ell} \binom{\ell}{j} |A_{i-j}|$$

$$= c_o h^{-\ell} \sum_{j=0}^{\ell} \binom{\ell}{j} |A_{i-j}| |h|^{1/2}$$

$$\leq c_o' h^{-\ell} \sum_{j=0}^{\ell} \binom{\ell}{j} |A_{i-j}| (t_{i-j+k} - t_{i-j})^{1/2}$$

where the constants c_o and c_o' depend only on ℓ. Hence, since $\{A_i(t_{i+k}-t_i)^{1/2}\} \in \ell_2$, we also have $\{A_i^{(\ell)}(t_{i+\ell}-t_i)^{1/2}\} \in \ell_2$ and this completes the proof of the theorem.

If we combine Theorem 4.1 with Theorem 4.2 and a result of de Boor [4] we obtain

THEOREM 4.3. _Let_ $\underline{t} = \{t_i\}$ _satisfy the hypotheses in Theorems 4.1 and 4.2. Let_ $\underline{\gamma} = \{\gamma_i\}$ _be a sequence of real numbers. Then there exists an_ $H^{2,2}$_-spline interpolant in_ $L_2(R)$ _of the data_ (t_i, γ_i) _if and only if_

(4.3) $\{\gamma_i(t_{i+4}-t_i)^{1/2}\} \in \ell_2$

and

(4.4) $\{([t_i, t_{i+1}, t_{i+2}]\underline{\gamma})(t_{i+2}-t_i)^{1/2}\} \in \ell_2$.

Proof. Condition (4.4) is a necessary and sufficient condition for the existence of an $H^{2,2}$-spline interpolant to the data, see [4], and condition (4.3) is necessary and sufficient for the existence of a spline interpolating the data $\{(t_i, \gamma_i)\}$ to be in $L_2(R)$ (given the condition in knots). This completes the proof.

In order to make further progress in this direction it is clear that a more thorough study of spline projections on the real line must be made.

References.

1. C. de Boor, On calculating with B-splines, J. Approximation Theory 6 (1972), 50-62.

2. C. de Boor, The quasi-interpolant as a tool in elementary polynomial spline theory, in Approximation Theory, G. G. Lorentz, ed., Academic Press, N.Y., 1973, 269-276.

3. C. de Boor, On cubic spline functions which vanish at all knots. To appear.

4. C. de Boor, How small can one make the derivatives of an interpolating function? J. Approximation Theory 13 (1975), 105-116.

5. C. de Boor, On bounding spline interpolation, J. Approximation Theory 14 (1975), 191-203.

6. C. de Boor, Odd-degree spline interpolation at a bifinite knot sequence. This volume.

7. S. Demko, Inverses of band matrices and local convergence of spline projections. To appear.

8. M. Golomb, $H^{m,p}$-extensions of $H^{m,p}$-splines, J. Approximation Theory 5 (1972), 238-275.

9. J. W. Jerome and L. L. Schumaker, Characterization of functions with higher order derivatives in L^p, Trans. Amer. Math. Soc. 143 (1969), 363-371.

10. R. Larsen, Functional Analysis an Introduction, Marcel Dekker, Inc., New York, 1973.

11. I. J. Schoenberg, Spline interpolation and higher derivatives, Proc. Nat. Acad. Sci. U.S.A. 51 (1964), 24-28.

12. P. W. Smith, Characterization of the function class $W^{m,p}$, in Approximation Theory, G. G. Lorentz, ed., Academic Press, N.Y., 1973, 485-490.

13. P. W. Smith, $W^{r,p}(R)$-splines, J. Approximation Theory 10 (1974), 337-357.

Charles K. Chui, Philip W. Smith and Joseph D. Ward
Department of Mathematics
Texas A&M University
College Station, Texas 77843, USA

Work supported by the U. S. Army Research Office under grant numbers DAHC04-75-G-0816 and DAAG29-75-G-0186

EINIGE ANWENDUNGEN DER NICHTLINEAREN APPROXIMATIONS-THEORIE AUF RANDWERTAUFGABEN

LOTHAR COLLATZ, HAMBURG

Inhalt: Bei Anwendung der Approximationstheorie auf Naturwissenschaften, Technik, Wirtschaftswissenschaften u.a. treten häufig Probleme auf, die aus dem Rahmen der normalerweise betrachteten Approximationsaufgaben herausfallen. In diesem Übersichtsvortrag wird gezeigt, daß man bei gewissen Klassen von Singularitäten, wie sie bei Randwertaufgaben partieller Differentialgleichungen auftreten, in einfach gelagerten Fällen approximationstheoretische Methoden mit Erfolg anwenden und in günstigen Fällen sogar brauchbare exakte Einschließungen für die Lösungen erhalten kann.

I. In den Anwendungen auftretende Problemklassen

Bei Anwendung der Approximationstheorie auf Analysis und außermathematische Wissenschaften tritt häufig eine oder mehrere der im folgenden genannten Erschwerungen gegenüber den gewöhnlich behandelten Approximationsaufgaben auf:

1) Man hat kompliziertere Funktionenklassen, insbesondere mit mehreren unabhängigen Veränderlichen, zur Approximation zu verwenden als die häufig betrachteten einfachen Funktionen wie Polynome, rationale Funktionen, Exponentialfunktionen, trigonometrische Funktionen und dergleichen.

2) Neben die gewöhnliche Approximation treten kompliziertere Arten von Approximation, Simultanapproximation, Combiapproximation, Feldapproximation u.a. (Morsund [68], Bredendiek [69], [76])

3) Die Parameter treten verkettet auf ("Verkettete Approximation", Hoffmann [69] [76], Collatz-Krabs [73] S. 195, Collatz [75])

4) Es tritt Approximation durch Funktionentypen auf. Man hat in der Klasse der Approximationsfunktionen nicht nur endlich viele Parameter, sondern Funktionen zu bestimmen.

5) Es treten lokale Minima auf. Man beobachtet dies bei Verzweigungsaufgaben und allgemein bei nichtlinearen Randwertaufgaben, z.B. auch bei expandierenden Operatoren.

6) Das Problem weist Singularitäten auf, z.B. geometrischer Art bei dem vorliegenden Bereich oder analytischer Art bei den auftretenden Koeffizienten.

II. Zwei Formen des Einschließungssatzes

Bei nichtlinearer Approximation leistet häufig der folgende Einschließungssatz gute Dienste:

Es sei B ein Gebiet des n-dimensionalen Raumes R^n der Vektoren $x=\{x_1, ,,, x_n\}$ und C(B) der Banachraum der auf B stetigen Funktionen g(x) bei der üblichen Maximumnorm

$$(2.1) \quad \|g\| = \sup_{x\varepsilon B} |g(x)|$$

Ferner sei W=w(x,a) eine Teilmenge von C(B), die von einem Parametervektor $a=\{a_1, \;, \;, a_p\}$ abhängt. Sei die Funktion $f\varepsilon C(B)$ fest gewählt, dann heißt

$$(2.2) \quad \rho = \inf_{w\varepsilon W} \|W-f\|$$

der Minimalabstand von f bezüglich W, und ein Element $w\varepsilon W$ mit $\rho = \|w-f\|$ heißt Minimallösung. Dann gelten folgende zwei Formen des Einschließungssatzes (Meinardus [67], Collatz-Krabs [73], Collatz [69]):

1) Es gebe in B zwei Punktmengen M_1 und M_2 der Art, daß für kein Paar $(\hat{w}, \tilde{w})\varepsilon W$ die Differenz $\hat{w}-\tilde{w}$ auf M_1 positiv und auf M_2 negativ ausfällt. Es sei g eine feste Funktion aus W. Wenn dann der Fehler $\varepsilon=g-f$ auf M_1 positiv und auf M_2 negativ ist, so gilt für die Minimalabweichung ϱ bei der Tschebyschew-Approximation der Einschließungssatz

$$(2.3) \quad \mu_1 \le \varrho \le \mu_2; \text{ mit } \mu_1 = \inf_{x\varepsilon H}|\varepsilon|, \; \mu_2 = \sup_{x\varepsilon B}|\varepsilon| .$$

Dabei ist $H=M_1 \cup M_2$. Im Falle $\mu_1 = \mu_2$ ist g eine Minimallösung (Kolmogorow-Kriterium).

Eine Punktmenge H, die die genannten Voraussetzungen erfüllt, wird H-Menge genannt.

2) Bei der zweiten Form des Einschließungssatzes haben f,g,ε dieselbe Bedeutung wie bei der ersten Fassung, aber die Voraussetzungen für die H-Menge werden durch die folgenden ersetzt:

Es sei $\varepsilon=g-f\neq 0$ auf H

und es gebe kein $w\varepsilon W$ mit $\varepsilon(g-w)>0$ auf H. Dann gilt wörtlich wieder die Einschließungsaussage (2.3).

Bei p Parametern wird in vielen Fällen die H-Menge (p+1)Punkte enthalten. Bei linearer Approximation hat W die Form

$$(2.4) \quad w = \sum_{\nu=1}^{p} a_\nu w_\nu(x_1, \ldots x_n), \text{ mit fest gegebenen Funktionen}$$

w_ν, und auch die Differenz $\hat{w}-\tilde{w}$ hat wieder dieselbe Form. Zum Nachprüfen der Voraussetzungen hat man ein System von p+1 linearen Ungleichungen für p Parameter. Dafür stehen fertige Algorithmen zur Verfügung. Will man aber die erste Form des Einschließungssatzes auf eine nichtlineare Approximation mit den Elementen

(2.5) $\hat{w}=w(x_1,\ldots x_n,\hat{a}_\nu)$, $\tilde{w}=w(x_1,\ldots,x_n,\tilde{a}_\nu)$ anwenden, so erhält man nur p+1 Ungleichungen für 2p Parameter $\hat{a}_\nu$, $\tilde{a}_\nu$, aus denen man normalerweise keine Schlüsse ziehen kann. Bei der zweiten Form des Einschließungssatzes, jedoch erhält man wieder p+1 Ungleichungen für nur p Parameter a_ν.

Dieser Vorteil der zweiten Form des Einschließungssatzes werde an folgendem Beispiel erläutert.

Beispiel: Es liege für eine Funktion y(x) die nichtlineare Randwertaufgabe vor

(2.6) $Ty=y''+\frac{1+x^2}{4}\,y^2=0,\ y(\pm 1)=0$

Nun wird für y ein Polynom z als Näherung angesetzt:

(2.7) $y(x)\approx z(x)=1+\sum_{j=1}^{p} a_j(1-x^{2j})$;

dann wird

Abb.1a

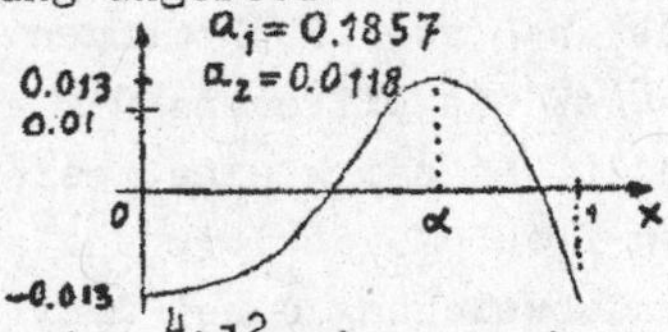

(2.8) $Tz=-2a_1-12a_2x^2+\frac{1+x^2}{4}\,[1+a_1(1-x^2)+a_2(1-x^4)]^2=w(x,a_1,a_2)$,

Es soll $Tz\approx 0$ sein. Das heißt, es ist die Funktion f=0 durch $w(x,a_1,a_2)$ anzunähern. In (2.8) treten sowohl a_1 als auch a_2 an verschiedenen Stellen auf. Es liegt verkettete Approximation vor. Diese Tschebyscheffsche Approximation hat zwei Lösungen. Abb.1a,b zeigt den Fehlerverlauf für diese beiden Lösungen. In beiden Fällen hat man eine Alternante von 3 Punkten. Für die Lösung mit dem kleineren Beträgen der Parameter erhält man die Ungleichungen:

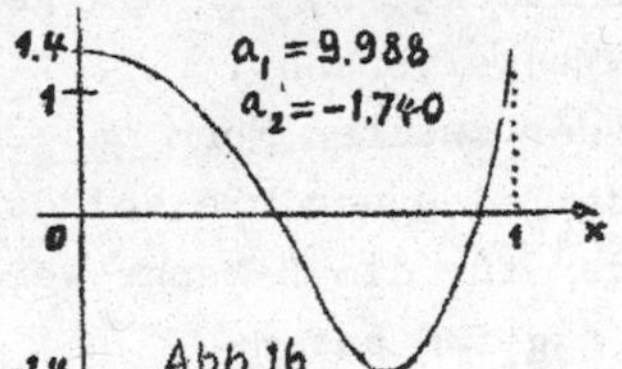

Abb.1b

$$(2.9)\begin{cases} x=0:\ \varphi_1=2a_1-\frac{1}{4}(1+a_1+a_2)^2<0.013\\ \alpha^2=x^2=\beta:\ \varphi_2=-2a_1-12\beta a_2+\frac{1}{4}(1+\beta)\,[1-a_1(1-\beta)+a_2(1-\beta^2]^2<0.013\\ x=1:\ \varphi_3=2a_1+12a_2-\frac{1}{2}<0.013\end{cases}$$

In Abb. 2 sind in einer a_1,a_2-Ebene die Bereiche $\varphi_j > 0$ angedeutet, in welchen die Ungleichungen erfüllt sind. Alle drei Ungleichungen zugleich sind nur in dem Punkte P erfüllt, welche der Lösung entspricht, Diese gehört zum globalen Minimum. Man hat also im Rahmen der Rechengenauigkeit die bestmögliche Approximation an f innerhalb der Klasse $w(x,a_1,a_2)$ erreicht.

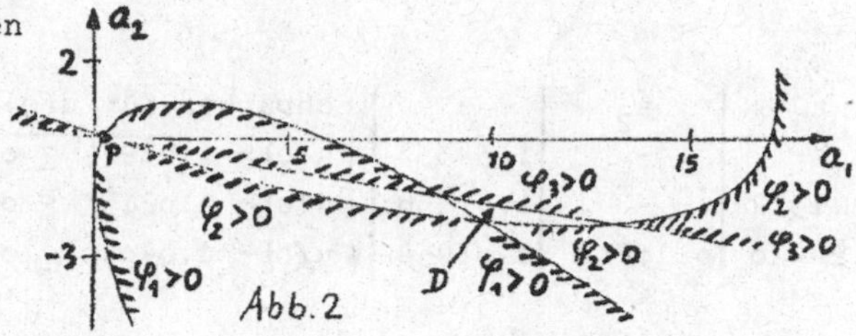

Abb.2

Die Lösung mit den größeren Beträgen der Parameter liegt in Abb.2 in dem kleinen dreieckförmigen Bereich D und entspricht einem lokalen Minimum.

Frau Susanne Böttger und Herrn Rolf Wildhack danke ich für Ausführung numerischer Rechnungen am Computer.

III. <u>Approximation durch Funktionentypen</u>

Bei Hammerstein'schen Integralgleichungen ersetzt Sprekels [75] den Kern K(t,s) durch einen entarteten Kern der Form p(t)q(s). Hierbei variieren t von s in einem Bereich B des n-dimensionalen Punktraumes R^n. Man bildet den Quotienten Q und eine Konstante C nach

$$(3.1)\quad Q(t,s) = \frac{K(t,s)}{p(t)q(s)},\quad C = \frac{\sup_B Q(t,x)}{\inf_B Q(t,x)} \geq 1$$

Dabei sind die Funktionen p(t) und q(s) so zu wählen, daß C möglichst klein ausfällt. Über Approximationsaufgaben von diesem Typ ist noch relativ wenig bekannt. Trotzdem kann man mit der genannten Approximation unter gewissen Voraussetzungen zu punktweisen exakten Schranken für die Lösungsfunktion kommen. Ich danke Herrn Dr. Sprekels für das folgende Beispiel:

$$(3.2)\quad u(t)=Tu(t)= \int_0^1 K(t,s)[u(s)]^3 ds$$

mit $K(t,s)=1+e^{-(1+t)(1+s)}$

Macht man für p(t) und q(s) den Ansatz

$$(3.3)\quad p(t)=q(t)=1+ \sum_{\nu=1}^{N} a_\nu t^\nu$$

so kann man die positive Lösung u(t) von (3.2) in Schranken schließen. Durch Erhöhung des Grades des Polynomes p(t) kann man die Genauigkeit steigern, wie es die folgende Tabelle zeigt.

N	a_1	a_2	c	Schranken für u(o)
o	-	-	1.3433	$\lvert u(o) - 1.0897\rvert \leq 0.4532$
1	-o.1372	-	1.07356	$\lvert u(o) - 1.0215\rvert \leq 0.1084$
2	-o.2442o	o.1o7o2	1.03954	$\lvert u(o) - 1.04010\rvert \leq 0.06043$

IV. Weitere Anwendungen

1) Freie Randwertaufgabe

Als sehr einfaches eindimensionales Modell einer freien Randwertaufgabe sei genannt

(4.1) $Ty=1-[(1+e^x)y']'=0,\ y(o)=1,\ y(a)=y'(a)=0$

Dabei ist a nicht bekannt, wohl aber sei eine eventuell sehr große Schranke A bekannt mit a<A.

Dieser Aufgabe entspricht das Variationsproblem

(4.2) $\Phi(\phi)=\int_o^A[(1+e^x)\phi'^2+2\phi]dx = \text{Min},$

wobei $\phi(x)$ den Bedingungen

(4.3) $\phi(o)=1,\ \phi\epsilon C^1[O,A],\ \phi(x)\geq o$ in $[O,A]$

zu genügen hat. Der einfachste Ansatz

(4.4) $\phi(x)\approx w_1(x)=(1-a_1x)^2$, a>o für $a_1x\leq 1$, $w_1=0$ für $a_1x>1$,

führt bereits zu einer verketteten Approximation. Die Forderung

(4.5) $\Phi(a_1)=8a_1{}^4[e^{1/a_1}-1]-8a_1^3-4a_1^2+\frac{4}{3}a_1+\frac{2}{3a_1} \overset{!}{=} \text{Min}$

führt zu Min Φ=3.18. Der etwas genauere Ansatz

(4.6) $\phi(x)\approx w_2(x)=(1-a_1x-a_2x^2)^2$

führt auf

(4.7) a_1=o.63o, a_2=-o.o94, w'(o)=-1,26, Min Φ =3.14

Die Behandlung komplizierterer zweidimensionaler freier Randwertaufgaben mit Approximationsmethoden soll demnächst veröffentlicht werden.

2) Unsachgemäße Aufgaben

Aus dem weiten Gebiete der inkorrekt gestellten Aufgaben sei nur ein Beispiel genannt. Für die Temperaturverteilung u(x,t) in einem Stabe sei die Wärmeleitungsgleichung

(4,8) $Lu = \frac{\partial u}{\partial t} - \frac{\partial^2 u}{\partial x^2} = 0$ in B $\{0<x<\infty\ ,\ -1<t<0\}$,

die Verteilung von u zur Zeit t=o und die Verteilung am Endpunkt x=o, (4.9) $u(x,0) = f(x) = \frac{1}{1+x}$ für $0 \le x < \infty$

(4.10) $u(0,t) = g(t) = 1$ für $-1 \le t \le 0$

gegeben und man fragt nach der Temperaturverteilung zu der früheren Zeit t=-1. Der Näherungsansatz für u

(4.11) $u \approx w = c\,\mathrm{Re}(e^{k^2t+kx})$ mit $k = a+ ib$, (a,b,c reell)

oder

(4.12) $w = c\, e^{ax+(a^2-b^2)t} \cos(bx+2abt)$

führt auf eine stark verkettete Approximation mit den Werten

a = - 0.37985

b = 0.24672

c = 0.95581

3) Singularitäten

Bei manchen Typen von Singularitäten, insbesondere bei Unstetigkeiten der Randwerte, kann man durch Abziehen der Singularitäten zu Aufgaben gelangen, bei denen die Approximationstheorie sehr gute Resultate liefert. Es sei z.B., nach der stationären Temperaturverteilung u(x,y) in einer homogenen Platte mit der Grundfläche eines Rechteckes $B(|x|<2, |y|<1)$ gefragt, wobei die gegebenen Randwerte unstetig sind:

(4.13) $\Delta u = \frac{\partial^2 u}{\partial x^2} + \frac{\partial^2 u}{\partial y^2} = 0$ in B

(4.14) $u(\pm 2,y) = 1$ für $|y| < 1$, $u(x,\pm 1) = 0$ für $|x| < 2$,

Führt man den Singularitäten entsprechend die Winkel φ_j (j=1,2,3,4) in den vier Ecken, Abb. 3, ein, so kann man u annähern durch w:

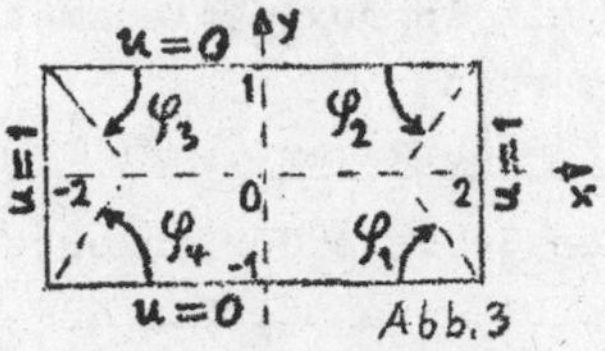

Abb.3

$$(4.15)\quad u \approx w = \frac{2}{\pi}\sum_{j=1}^{4}\varphi_j + \sum_{k=1}^{p} a_k\, w_k(x,y)$$

Dabei sind die w_k Potentialfunktionen, $\Delta w_k = 0$ z.B. $w_1 = 1$, $w_2 = x^2 - y^2$; bei Approximation im Tschebyscheffschen Sinne wurden für die Lösung u die exakten Schranken gewonnen

p=	$\lvert w-u \rvert \leq$
2	0.1141
3	0.0020
4	0.00063
5	0.000114

Es sind auch viele andere Fälle von Singularitäten numerisch mit Approximationsmethoden behandelt worden (z.B. Collatz, Günther, Sprekels [76]) und Approximationsverfahren mit anderen Methoden z.B. der Methode der Finiten Elemente kombiniert worden, z. B. Werner [75], Wetterling-Kothmann [76], Whiteman [73], Strang-Fix [73]u.a.

Literatur

Bredendiek, E. [69] Simultan Approximation, Arch Rat. Mech.Anal., 33(1969), 307-330.

Bredendiek, E. und L. Collatz [76] Simultan Approximation bei Randwertaufgaben, Intern. Ser.Num.Math. Bd.30, 1976, 147-174

Collatz, L. [69] Nichtlineare Approximation bei Randwertaufgaben, V.IKM (Internationaler Kongreß für Anwendungen der Mathematik) Weimar, herausgegeben von H. Matzke, 1969, S. 169-182.

Collatz, L. [75] Bemerkungen zur verketteten Approximation, Intern.Ser.Num.Math.26, 41-45(1975).

Collatz, L , H. Günther, J. Sprekels [76] Vergleiche zwischen Diskretisierungsverfahren und parametrischen Methoden an einfachen Testbeispielen. Z.angew.Math.Mech. 56(1976)1-11.

Collatz, L. und W. Krabs [73] Approximationstheorie, Teubner, Stuttgart, 1973, 208 S.

Hoffmann, K.H. [69] Zur Theorie der nichtlinearen Tschebyscheff Approximation mit Nebenbedingungen, Numer.Math. 14, 24-41(1969).

Hoffmann, K.H. [76] Über verkettete Approximation, erscheint in Intern.Ser.Numer.Math., 1976.

Meinardus, G. [67] Approximation of Functions, Theory and Numerical Methods, Springer Verlag, 1967, 198 S.

Sprekels, J. [75] Existenz und 1. Einschließung bei superlinearen Randwertaufgaben, erscheint demnächst.

Strang, G. und G. Fix [73] An analysis of the Finite Element Method, Prentice Hall 1973, 3o6 S.

Werner, B. [75] Monotonie und Finite Elemente bei elliptischen Differentialgleichungen, Intern.Ser.Numer.Math.27, 3o9-329(1975).

Wetterling, W. und Kothmann, P. [75] Randmaximumsätze bei Gebietszerlegungen, Intern.Ser.Numer.Math. 28, 153-157 (1975).

Whiteman, J.R., [73] The mathematics of Finite Elements and applications, Acad. Press 1973, 52o S.

Lothar Collatz
Institut für Angewandte Mathematik
der Universität Hamburg

ZUR TSCHEBYSCHEFF-APPROXIMATION BEI UNGLEICHUNGSNEBENBEDINGUNGEN IM FUNKTIONENRAUM

Ludwig Cromme

Many practical problems, in particular in ordinary and partial differential equations, lead to approximation problems with pointwise inequality constraints on the approximating functions. The concepts of "extended strong uniqueness" and "solvability condition" play an important role in describing the convergence of the associated numerical procedures. In this paper, we develop some methods for checking these conditions and study some cases where they are satisfied. It turns out that in a certain sense the requirements are minimal in that reasonable convergence properties cannot be expected without them.

Viele praktische Probleme, insbesondere auch bei gewöhnlichen und partiellen Differentialgleichungen, führen auf Approximationsprobleme mit Ungleichungsnebenbedingungen im Funktionenraum. Bei der Beschreibung des Konvergenzverhaltens von hierfür entwickelten numerischen Verfahren, spielen die Begriffe "erweiterte starke Eindeutigkeit" und "Auflösbarkeitsbedingung" eine wichtige Rolle. In der vorliegenden Arbeit soll genauer untersucht werden, wann diese Eigenschaften vorliegen und wie sie nachgeprüft werden können. Dabei ergibt sich, daß diese Forderungen in gewissem Sinne Minimalforderungen sind, ohne die man kein iteratives Verfahren mit vernünftigen Konvergenzeigenschaften erwarten kann.

1. EINLEITUNG

Einseitige Tschebyscheff-Approximation, Approximation mit positiven oder negativen Funktionen zur Gewinnung von Fehlerabschätzungen bei gewöhnlichen oder partiellen Differentialgleichungen und Einschränkungen bei der Approximation von Startwerten für iterative Prozesse sind nur ein paar Stichworte, die den weiten Einsatzbereich der T-Approximation mit Nebenbedingungen durch Schrankenfunktionen deutlich machen. In CROMME [3] wurden für derartige Probleme numerische Methoden entwickelt, bei deren Konvergenzanalyse eine erweiterte starke Eindeutigkeits-Forderung und eine Auflösbarkeitsbedingung eine wichtige Rolle spielten. Da nicht ohne weiteres ersichtlich ist, wann diese Eigenschaften vorliegen, und was sie bedeuten, sollen diese Begriffe in der vorliegenden Arbeit näher untersucht werden. Teilweise kann das Vorliegen dieser Eigenschaften unabhängig von der zu approximierenden Funktion für ganze

Approximationsfamilien nachgewiesen werden. Insgesamt gestatten diese Untersuchungen eine bessere Übersicht über die Anwendbarkeit der in [3] hergeleiteten Konvergenzsätze. Schließlich ergibt sich, daß die erweiterte starke Eindeutigkeit und die Auflösbarkeitsbedingung in gewissem Sinne Minimalforderungen sind, ohne die ein iteratives Verfahren nicht vernünftig arbeiten kann.

Q sei ein kompakter metrischer Raum und C(Q) der Banachraum der reellwertigen stetigen Funktion auf Q. Für einen Banach-Raum B über $\mathbb{R}$ seien durch

$$A \subseteq B \qquad \text{und} \qquad F : A \to C(Q)$$

eine Parametermenge und eine zugehörige Parametrisierung gegeben. Mit festem $u, v, f \in C(Q)$ lautet das Approximationsproblem:

$$(1.1) \quad \begin{cases} \text{Gesucht ist ein } \hat{a} \in A \cap F^{-1}(R) \text{ mit} \\ \|f - F(\hat{a})\|_\infty = \inf\limits_{a \in A \cap F^{-1}(R)} \|f - F(a)\|_\infty, \text{ wo } R \text{ durch} \\ \\ R := \{r \in C(Q) \mid u(x) \le r(x) \le v(x) \quad \forall x \in Q\} \\ \text{gegeben ist.} \end{cases}$$

Die T-Norm sei wie üblich definiert durch

$$\|g\|_\infty := \sup_{x \in Q} |g(x)| \qquad \forall g \in C(Q).$$

2. STARKE EINDEUTIGKEIT

Die (klassische) starke Eindeutigkeit liefert für nicht optimale Approximationen eine Abschätzung des Approximationsfehlers nach unten und besagt anschaulich, daß die Vergrößerung des Approximationsfehlers durch Wegrücken von der besten Approximation mindestens so groß ist wie ein festes Vielfaches des Abstandes von der besten Approximation.

Definition $\hat{w} \in W \subseteq C(Q)$ heißt stark eindeutig beste Approximation von $f \in C(Q)$, wenn es eine Konstante $K > 0$ gibt mit

$$(2.1) \qquad \|f - w\|_\infty \ge \|f - \hat{w}\|_\infty + K\,\|w - \hat{w}\|_\infty \qquad \forall w \in W.$$

Wir sprechen von lokaler starker Eindeutigkeit, wenn (2.1) für alle $w \in W \cap U$ gültig ist, wo U eine geeignete Umgebung von $\hat{w}$ ist.

Lokale starke Eindeutigkeit ist entscheidend für numerisches Wohlverhalten bei Verfahren zur nichtlinearen Tschebyscheff-Approximation, da sie sicherstellt, daß Lösungen von linearisierten Approximationsproblemen hinreichend gute Näherungen für die Lösung des nichtlinearen Problems sind (siehe CROMME [2]). So überrascht es nicht, daß auch bei Verfahren zur Behandlung von Problem (1.1) entsprechende Eigenschaften gefordert werden müssen. Allerdings genügt es hier nicht mehr, nur die Veränderung des Approximationsfehlers bei Störungen der besten Approximation zu betrachten, vielmehr kann es jetzt durchaus Störungen geben, die den Approximationsfehler noch verkleinern, jedoch auf Kosten einer Verletzung der Nebenbedingungen durch Schrankenfunktionen. Ein für Problem (1.1) angemessener starker Eindeutigkeitsbegriff muß also nicht nur die Veränderung des Approximationsfehlers bei Störungen der besten Approximation, sondern auch eine mögliche Verletzung der Nebenbedingungen mit berücksichtigen. Diese Überlegungen führen auf den folgenden erweiterten starken Eindeutigkeitsbegriff, dem wir noch eine Hilfsdefinition voranstellen.

<u>Definition</u> Für $w \in C(Q)$ werde die <u>Verletzung der Nebenbedingungen</u> durch Schrankenfunktionen gemessen durch

$$(2.2) \quad \mathrm{Verl}(w) := \mathrm{Max}\left(\sup_{\substack{x \in Q \\ w(x) < u(x)}} (u(x) - w(x)),\ \sup_{\substack{x \in Q \\ w(x) > v(x)}} (w(x) - v(x)) \right)$$

Die Suprema in (2.1) seien definitionsgemäß gleich Null gesetzt, wenn die Mengen, über die sich die Supremumsbildung erstreckt, leer sind.

<u>Definition</u> $\hat{w} \in W \subseteq C(Q)$ mit $\mathrm{Verl}(\hat{w}) = o$ heißt <u>stark eindeutig beste Approximation von f bei der Approximation mit Ungleichungsnebenbedingungen u und v</u> (kurz: st.e.b.A.Neb.), wenn es ein $K > o$ gibt mit

$$(2.3) \quad \mathrm{Max}\,(\mathrm{Verl}(w),\ \|f-w\|_\infty - \|f-\hat{w}\|_\infty) \geq K \cdot \|w-\hat{w}\|_\infty \quad \forall w \in W$$

Für den zugehörigen lokalen Begriff (l.st.e.b.A.Neb.) beschränkt man sich darauf, (2.3) für alle $w \in W \cap U$ für eine geeignete Umgebung U von $\hat{w}$ zu fordern. Wenn keine Mißverständnisse zu befürchten sind, sprechen wir auch kurz von "erweiterter starker Eindeutigkeit".

Im folgenden soll gezeigt werden, daß bei der Approximation in Tangentialkegeln die erweiterte lokale starke Eindeutigkeit bereits die entsprechende globale Eigenschaft nach sich zieht; außerdem werden wir nachweisen, daß erweiterte starke Eindeutigkeit bei nichtlinearen Parametrisierungen in Tangentialkegeln nachgeprüft werden kann, was den Nachweis solcher

Eigenschaften in der Praxis erleichtert. Die erzielten Ergebnisse stehen teilweise in Analogie zu entsprechenden Ergebnissen bei der klassischen starken Eindeutigkeit und beinhalten teilweise auch Erweiterungen; siehe BRAESS [1] und WULBERT [6].

Lemma 1 $W \subseteq C(Q)$ sei ein Kegel mit der Spitze $\hat{w}$. Es ist $\hat{w} \in W$ (global) stark eindeutig beste Approximation von f bei der Approximation mit Nebenbedingungen u und v genau dann, wenn die erweiterte starke Eindeutigkeitsforderung lokal erfüllt ist.

Beweis Offenbar ist aus der lokalen erweiterten starken Eindeutigkeit das Vorliegen der entsprechenden globalen Eigenschaften zu folgern. Seien also $\varepsilon, K > o$ geeignete Zahlen, sodaß für beliebiges $w \in W$ mit $\|w-\hat{w}\|_\infty \leq \varepsilon$ gilt:

$$\text{(2.4)} \qquad \operatorname{Max}\left(\operatorname{Verl}(w),\ \|f-w\|_\infty - \|f-\hat{w}\|_\infty\right) \geq K\ \|w-\hat{w}\|_\infty \ .$$

Behauptung 1 $\forall w \in W: \ \|w-\hat{w}\|_\infty = \varepsilon \ \ \exists x \in Q:$

$$\text{(2.5)} \qquad w(x) \geq \kappa + \hat{w}(x) \quad \text{und} \quad \begin{cases} (f-\hat{w})(x) = \|f-\hat{w}\|_\infty \ \text{oder} \\ \hat{w}(x) = v(x) \end{cases}$$

oder

$$\text{(2.6)} \qquad w(x) \leq -\kappa + \hat{w}(x) \quad \text{und} \quad \begin{cases} (f-\hat{w})(x) = -\|f-\hat{w}\|_\infty \ \text{oder} \\ \hat{w}(x) = u(x) \end{cases}$$

Dabei sei $\kappa := K \cdot \varepsilon$ gesetzt.

Beweis 1 Sei also $w \in W$ und $\|w-\hat{w}\|_\infty = \varepsilon$. Für die Folge $(w_\nu \mid \nu \in \mathbb{N})$

$$w_\nu := \hat{w} + \frac{1}{\nu}\,(w-\hat{w})$$

folgt aus (2.4) die Ungleichung

$$\text{(2.7)} \qquad \operatorname{Max}\left(\operatorname{Verl}(w_\nu),\ \|f-w_\nu\|_\infty - \|f-\hat{w}\|_\infty\right) \geq K\ \|w_\nu-\hat{w}\|_\infty = K \cdot \varepsilon \cdot \frac{1}{\nu}$$

Durch Fallunterscheidung wird daraus Behauptung 1 hergeleitet. Sei zum Beispiel für unendlich viele $\nu \in \mathbb{N}$

$$\operatorname{Verl}(w_\nu) \geq K \cdot \varepsilon \cdot \frac{1}{\nu}$$

und es werde etwa die obere Schrankenfunktion v verletzt. Dann gibt es also eine Folge

$(x_{\nu_i} \mid i \in \mathbb{N})$ mit

$$w_{\nu_i}(x_{\nu_i}) - v(x_{\nu_i}) \geq K \cdot \varepsilon \cdot \frac{1}{\nu_i}$$

Dann gilt

$$\hat{w}(x_{\nu_i}) + \frac{1}{\nu_i}(w-\hat{w})(x_{\nu_i}) - v(x_{\nu_i}) \geq K \cdot \varepsilon \cdot \frac{1}{\nu_i}$$

$$(w-\hat{w})(x_{\nu_i}) \geq K \cdot \varepsilon + \nu_i (v(x_{\nu_i}) - \hat{w}(x_{\nu_i})) \geq K \cdot \varepsilon$$

und für einen beliebigen Häufungspunkt x von $(x_{\nu_i} \mid i \in \mathbb{N})$ - wegen der Kompaktheit von Q gibt es mindestens einen Häufungspunkt - folgt aus der Stetigkeit der betrachteten Funktionen

$$(w-\hat{w})(x) \geq K \cdot \varepsilon$$

Durch entsprechende Behandlung der anderen Fälle folgt Behauptung 1. Abschließend braucht aus Behauptung 1 also nur noch die globale erweiterte starke Eindeutigkeit gefolgert zu werden, die jedoch jetzt offen auf der Hand liegt:
Sei $w \in W$; Behauptung 1 läßt sich auf

$$\tilde{w} := \hat{w} + \frac{\varepsilon}{\|w-\hat{w}\|_\infty}(w-\hat{w})$$

anwenden und es folgt

$$\mathrm{Max}\,(\mathrm{Verl}(w),\ \|f-w\|_\infty - \|f-\hat{w}\|_\infty) \geq K \cdot \frac{\|w-\hat{w}\|_\infty}{\varepsilon} = K \cdot \|w-\hat{w}\|_\infty .$$

<u>Bemerkung</u> Der Beweis von Lemma 1 zeigt, daß bei der globalen erweiterten starken Eindeutigkeit auf dem Kegel die gleiche Konstante K genommen werden kann wie bei der lokalen Version.

Wir wenden uns jetzt der Frage zu, inwieweit sich erweiterte starke Eindeutigkeit bei nichtlinearen Parametrisierungen durch die Betrachtung von Tangentialkegeln nachweisen läßt, was im allgemeinen eine Vereinfachung bedeutet. Wir beschränken uns nicht auf Tangential<u>räume</u>, sondern legen Tangentialkegel zugrunde, weil diese bei einigen Approximationsklassen, z.B. Exponentialsummen, in natürlicher Weise auftreten. Sei also $K \subseteq B$ ein nicht notwendig konvexer Kegel mit Spitze O; für hinreichend kleines $\varepsilon > 0$ gelte

$$\{\hat{a}+k \mid k \in K,\ \|k\| \leq \varepsilon\} = \{a \in A \mid \|a-\hat{a}\| \leq \varepsilon\}.$$

Schließlich sei

$$F_{\hat{a}}^{K} : K \to C(Q)$$

eine Abbildung mit

$$F_{\hat{a}}^{K}(\lambda\cdot k) = \lambda\cdot F_{\hat{a}}^{K}(k) \quad \text{für } k\in K,\ \lambda\geq o\ ,$$

d.h., $F_{\hat{a}}^{K}$ sei positiv homogen und es gelte

$$(2.8) \qquad \|F(\hat{a}+\Delta a)-F(\hat{a})-F_{\hat{a}}^{K}(\Delta a)\|_{\infty} = o(\Delta a)\ ,\ a+\Delta a\in A.$$

Wir nennen $F_{\hat{a}}^{K}$ regulär, wenn es Konstante $\alpha,\beta>o$ gibt mit

$$(2.9) \qquad \alpha\,\|k\| \leq \|F_{\hat{a}}^{K}(k)\|_{\infty} \leq \beta\,\|k\| \qquad \forall k\in K.$$

<u>Lemma 2</u> Mit den soeben angegebenen Bezeichnungen gilt bei regulärem $F_{\hat{a}}^{K}$:

Für eine genügend kleine Umgebung U von $\hat{a}$ ist $F(\hat{a})$ stark eindeutig beste Approximation von f in F(U) bei Schrankenfunktionen u und v genau dann, wenn o im erweiterten Sinne lokal (und damit nach Lemma 1 auch global) stark eindeutig beste Approximation von $f-F(\hat{a})$ in $F_{\hat{a}}^{K}(K)$ ist.

<u>Beweis</u> Wegen der Regularitätsvoraussetzung (2.9) und (2.8) folgt zunächst die Existenz positiver Zahlen α' und β' mit

$$(2.10) \qquad \alpha'\,\|k\| \leq \|F(\hat{a}+k)-F(\hat{a})\|_{\infty} \leq \beta'\,\|k\| \qquad \forall k\in K\cap V,$$

wo V eine geeignete Nullumgebung ist.

Sei nun o stark eindeutig beste Approximation von $f-F(\hat{a})$ im erweiterten Sinne bezüglich $F_{\hat{a}}^{K}(k)$ mit einer Konstanten K'. Es folgt für geeignetes $\delta>o$

$$\begin{aligned}
& \operatorname{Max}\,(\operatorname{Verl}(F(a)),\ \|f-F(a)\|_{\infty}-\|f-F(\hat{a})\|_{\infty})\\
\geq\ & \operatorname{Max}\,(\operatorname{Verl}(F(\hat{a})+F_{\hat{a}}^{K}(a-\hat{a})),\ \|f-F(\hat{a})-F_{\hat{a}}^{K}(a-\hat{a})\|_{\infty}-\|f-F(\hat{a})\|_{\infty})\\
& -\ \|F(a)-F(\hat{a})-F_{\hat{a}}^{K}(a-\hat{a})\|_{\infty}\\
\geq\ & K'\,\|F_{\hat{a}}^{K}(a-\hat{a})\|_{\infty}-\|F(a)-F(\hat{a})-F_{\hat{a}}^{K}(a-\hat{a})\|_{\infty}\\
\geq\ & K'\,\|F(a)-F(\hat{a})\|_{\infty}-(K'+1)\,\|F(a)-F(\hat{a})-F_{\hat{a}}^{K}(a-\hat{a})\|_{\infty}\\
\geq\ & (K'-\delta\frac{K'+1}{\alpha'})\,\|F(a)-F(\hat{a})\|_{\infty}
\end{aligned}$$

Für a aus einer hinreichend kleinen Umgebung U von $\hat{a}$ folgt daraus die erweiterte starke Eindeutigkeit von $F(\hat{a})$ bei der Approximation in F(U). Durch eine analoge Abschätzung schließt man von der erweiterten starken

Eindeutigkeit bezüglich F(U) auf die lokale und nach Lemma 1 auf die globale erweiterte starke Eindeutigkeit der o im Tangentialkegel $F'_{\hat{a}}(K)$.

Bemerkung Durch Lemma 2 wird der Fall Fréchet-differenzierbarer Parametrisierungen natürlich mit erfaßt.

Die Frage, ob sich erweiterte starke Eindeutigkeit a priori, d.h., bevor man das noch im Verlauf eines Iterationsverfahrens zu berechnende $\hat{a}$ kennt, nachweisen läßt, kann in vielen Fällen positiv beantwortet werden. Dies gilt insbesondere bei Haarsch eingebetteten Tangentialmannigfaltigkeiten, wo sich der Alternationszahl-Begriff aus SCHABACK [4] anwenden läßt. Das liegt daran, daß sich beste Approximationen in Haarschen Fällen bei Nebenbedingungen durch Schrankenfunktionen oft völlig analog zum Fall ohne Nebenbedingungen durch verallgemeinerte Alternanten charakterisieren lassen. So ergibt sich als unmittelbare Folgerung aus TAYLOR [5], SCHABACK [4] und Lemma 2 das

Korollar 3 F sei in $\hat{a}\in\mathring{A}$ Fréchet-differenzierbar, $B=\mathbb{R}^n$, $F'_{\hat{a}}$ sei regulär (2.9) und Haarsch. Schließlich gelte für alle $x\in Q\subseteq I\subseteq\mathbb{R}$

$$u(x)\lneqq v(x) \qquad u(x)\leq f(x)\leq v(x) \quad .$$

Ist $F(\hat{a})$ lokal beste Approximation bei der Approximation mit Nebenbedingungen durch Schrankenfunktionen, dann ist $F(\hat{a})$ bereits im erweiterten Sinne stark eindeutig beste Approximation in F(U) für eine geeignete Umgebung U von $\hat{a}$.

3. DIE AUFLÖSBARKEITSBEDINGUNG

Eine Eigenschaft, die wir in [3] Auflösbarkeitsbedingung genannt haben, spielt bei der Konvergenzanalyse von Verfahren der nichtlinearen T-Approximation ebenfalls eine wichtige Rolle. Von einem Verfahren, das in der Praxis vernünftig arbeiten soll, wird man verlangen, daß es stabil ist gegen kleine Störungen der Eingangsdaten, die aufgrund der beschränkten Darstellbarkeit von reellen Zahlen auf digitalen Rechenanlagen unvermeidlich sind. Insbesondere sollten kleine Störungen der Nebenbedingungen nur geringfügige Änderungen der besten Approximation nach sich ziehen. In diesem Zusammenhang bietet sich die folgende Definition an, die sich von der entsprechenden Definition in [2] nur geringfügig unterscheidet.

Definition $\hat{w}\in W\subseteq C(Q)$ erfüllt die lokale Auflösbarkeitsbedingung, wenn es positive reelle Zahlen ε und K gibt mit

$$\forall u', v' \in C(Q) : \|u-u'\|_\infty \leq \delta < \varepsilon \text{ und } \|v-v'\|_\infty \leq \delta < \varepsilon$$

(3.1)

$$\exists w \in W : \|w-\hat{w}\|_\infty \leq K \cdot \delta \text{ und } u'(x) \leq w(x) \leq v'(x) \quad \forall x \in Q$$

Den Zusammenhang mit der in [3] gegebenen Definition stellt Lemma 4 her.

Lemma 4 F sei in $\hat{a} \in \mathring{A}$ Fréchet-differenzierbar und $F'_{\hat{a}}$ regulär (siehe 2.9). Dann gilt die lokale Auflösbarkeitsbedingung für $F(\hat{a}) \in F(U)$ für eine passende Umgebung U von $\hat{a}$ genau dann, wenn sie für $F(\hat{a})$ in $F(\hat{a}) + F'_{\hat{a}}(B)$ gilt.

Das Lemma zeigt, daß die hier gegebene Definition der Auflösbarkeitsbedingung etwas allgemeiner ist als die in [3] angegebene, jedoch in dem dort betrachteten Zusammenhang dazu äquivalent ist. Der Beweis von Lemma 4 läßt sich mit der gleichen Technik erbringen, die beim Beweis von Lemma 2 benutzt wurde unter Ausnutzung der Tangentialeigenschaften von $F'_{\hat{a}}(B)$ und soll deshalb hier nicht ausgeführt werden.

Zum besseren Verständnis der Auflösbarkeitsbedingung wird im folgenden der Haarsche Fall näher untersucht. In Anlehnung an TAYLOR [5] definieren wir:

$$x \text{ heißt } \underline{\text{positiver Extremalpunkt}} \iff \begin{cases} f(x)-F(a)(x) = \|f-F(a)\|_\infty & \text{oder} \\ F(a)(x) = v(x) \end{cases}$$

$$x \text{ heißt } \underline{\text{negativer Extremalpunkt}} \iff \begin{cases} f(x)-F(a)(x) = -\|f-F(a)\|_\infty & \text{oder} \\ F(a)(x) = u(x) \end{cases}$$

x heißt positiver Nebenextremalpunkt, wenn $F(a)(x)=v(x)$, und negativer Nebenextremalpunkt, wenn $F(a)(x)=u(x)$ gilt.

Q sei im folgenden kompakte Teilmenge eines reellen abgeschlossenen Intervalls $I = [a,b]$. $x_1<x_2<\ldots<x_{m+1}$ heißt verallgemeinerte Alternante der Länge m, wenn die x_i abwechselnd positive und negative Extremalpunkte sind (oder umgekehrt); schließlich sprechen wir von einer Nebenalternante der Länge m, wenn die x_i abwechselnd positive und negative Nebenextremalpunkte sind (bzw. umgekehrt). Dann gilt folgende Charakterisierung der Auflösbarkeitsbedingung.

Lemma 5 F sei in $\hat{a} \in \mathring{A}$ Fréchet-differenzierbar, $B = \mathbb{R}^n$, $F'_{\hat{a}}(\mathbb{R}^n)$ Haarsch von der Dimension n. Es gelte $u(x)<v(x)$ $\forall x \in Q$. $F(\hat{a})$ erfüllt die

lokale Auflösbarkeitsbedingung bezüglich F(U) für eine geeignete Umgebung U von $\hat{a}$ genau dann, wenn es keine Nebenalternante gibt.

Beweis Wegen Lemma 4 können wir uns auf die Betrachtung des Tangentialraumes beschränken.

„$\Longrightarrow$" Die Auflösbarkeitsbedingung sei erfüllt. Wir nehmen an, daß es eine Nebenalternante $x_1<\ldots<x_{n+1}$ gibt und führen diese Annahme zum Widerspruch.

Sei dazu $\varepsilon_0>0$ genügend klein, ε beliebig, $0<\varepsilon<\varepsilon_0$. Wir definieren

$$\left.\begin{aligned} u'(x) &:= u(x) + \varepsilon \\ v'(x) &:= v(x) - \varepsilon \end{aligned}\right\} \forall x \in Q$$

Wegen der Auflösbarkeitsbedingung und aus Stetigkeits- und Kompaktheitsgründen existiert nun eine beste Approximation

$$F(\hat{a}) + \sum_{j=1}^{n} z_j \frac{\partial F}{\partial a_j}\Big|\hat{a}$$

von f bei den Schrankenfunktionen u' und v' auf $F(\hat{a})+F'_{\hat{a}}(\mathbb{R}^n)$. Wegen der modifizierten Schrankenfunktionen u' und v' gilt für passendes Vorzeichen σ

$$\text{Signum}\left[\sum_{j=1}^{n} z_j \frac{\partial F}{\partial a_j}\Big|\hat{a}\ (x_i)\right] = \sigma(-1)^i \quad i=1,\ldots n+1,$$

was $z_j=0$, $j=1,\ldots n$, zur Folge hat, da $F'_{\hat{a}}(\mathbb{R}^n)$ Haarsch ist.

Das führt jedoch auf einen Widerspruch, da zum Beispiel

$$F(\hat{a})(x_1) = v(x_1)>v(x_1)-\varepsilon = v'(x_1)$$

oder

$$F(\hat{a})(x_1) = u(x_1)<u(x_1)+\varepsilon = u'(x_1)$$

gilt. Damit ist die eine Richtung von Lemma 5 bewiesen.

„$\Longleftarrow$" Für ein geeignetes $m\leq n$ gibt es Punkte

$$z_0 := a<z_1<\ldots<z_{m-1}<z_m := b \quad \text{mit}$$

a) z_i ist kein Nebenextremalpunkt $(i=1,\ldots m-1)$

b) $[z_i,z_{i+1}]\cap Q$ enthält nur Nebenextremalpunkte gleicher Signatur, und zwar mindestens einen $(i=0,\ldots m-1)$

c) $[z_i,z_{i+1}]\cap Q$ und $[z_{i+1},z_{i+2}]\cap Q$ enthalten keine Nebenextremalpunkte gleicher Signatur $(i=0,\ldots m-2)$.

Durch Interpolation der O in den Punkten z_i $(i=1,\ldots m-1)$ und durch geeignete Festlegung des Vorzeichens auf $[z_i, z_{i+1}] \cap Q$ (i=0,...m-1) erhält man offenbar die Möglichkeit, die aus kleinen Änderungen der Schrankenfunktionen u und v resultierenden Verletzungen der Nebenbedingungen durch eine proportional dazu gewählte Modifikation der Parameter zu kompensieren; damit ist Lemma 5 bewiesen.

Im Kontext von Lemma 5 ist also die Auflösbarkeitsbedingung eine sehr milde Forderung, die sicherstellt, daß nicht beliebig kleine Störungen der Schrankenfunktionen bereits zur Unlösbarkeit des gestörten Approximationsproblems führen; ohne diese Bedingung wird man kein stabil arbeitendes iteratives Verfahren erwarten. Diese Interpretation läßt sich auf den allgemeinen Fall übertragen, daß Q wieder ein kompakter metrischer Raum und B ein Banachraum über $\mathbb{R}$ ist.

LITERATUR

1. Braess, D.: Kritische Punkte bei der nichtlinearen Tschebyscheff-Approximation. Math. Z. 132, 327-341 (1973)
2. Cromme, L.: Eine Klasse von Verfahren zur Ermittlung bester nichtlinearer Tschebyscheff-Approximationen. Numer. Math. 25, 447-459 (1976)
3. Cromme, L.: Numerische Methoden zur Behandlung einiger Problemklassen der nichtlinearen Tschebyscheff-Approximation mit Nebenbedingungen. Numer. Math., eingereicht.
4. Schaback, R.: On Alternation Numbers in Nonlinear Chebyshev Approximation. J. Approximation Th., erscheint demnächst.
5. Taylor, G. D.: On Approximation by Polynomials Having Restricted Ranges. J. SIAM Numer. Anal. 5, 258-268 (1968)
6. Wulbert, D.: Uniqueness and Differential Characterization of Approximations from Manifolds of Functions. Amer. J. Math. 93, 350-366 (1971)

Dr. L. Cromme

Sonderforschungsbereich 72 und
Institut für Angewandte Mathematik
der Universität Bonn

Wegelerstraße 6

D - 5300 BONN

Schnelle Konvergenz: Charakterisierung der besten Approximation und Entropie

W. Dahmen - E. Görlich

The results announced in this talk will be published under the title:

The characterization problem for best approximation with exponential error orders and evaluation of entropy. Math. Nachrichten (to appear)
W. Dahmen - E. Görlich

CONVERGENCE OF ABSTRACT SPLINES

F.J. DELVOS / W. SCHÄFER / W. SCHEMPP

0. Introduction

In this paper convergence properties of abstract splines in the sense of Atteia [1], Laurent [13], and Sard [17] [18] are discussed. The first result in this direction has been published by Joly [12]. In a recent paper [9] a similar result has been proved within the framework of the theory of spline systems. This approach makes it possible to treat also convergence of splines with transfinite observation (cf. Delvos [6]).
Based on an extension of Sard's method (cf. Delvos-Schempp [9]) we will give a more detailed analysis of the convergence of splines. In particular spline interpolation is characterized as a (transfinite) Ritz approximation (see also Hulme [11], Ciarlet-Varga [4] for concrete results of this type). Thus, the ideas of Nitsche [15] [16] are applicable. The theory of intermediate spaces is used to derive estimates for the error in best approximation of linear functionals. Finally, the Mangeron operator (cf. Birkhoff-Gordon [2]) is considered.

1. On Joly's convergence theorem

Let X be a Hilbert space carrying a graph scalar product

$$(x,y)_1 = (Ux,Uy) + (F_o x,F_o y) ,$$

where $$U : X \to Y \quad \text{and} \quad F_o : X \to Z_o$$

denote continuous linear mappings from X into arbitrary Hilbert spaces Y and Z_o.
Consider now a sequence of (continuous linear) observations (cf. Sard [17])

$$F_n : X \to Z_n \quad (n \in \mathbb{N})$$

such that $$\text{Ker } F_{n+1} \subset \text{Ker } F_n \quad (n \in \mathbb{Z}^+). \qquad (1.1)$$

For simplicity $\{Z_n\}_{n\geq 1}$ is a sequence of Hilbert spaces.

It is well known (cf. Sard [18], Delvos-Schempp [9]) that there exists a unique spline interpolant x_n of $x \in X$ with respect to F_n and U:

$$F_n x_n = F_n x \quad , \quad ||Ux_n|| = \min_{F_n z = F_n x} ||Uz|| \; .$$

The following convergence theorem can be proved (cf. Delvos-Schempp [9]).

THEOREM 1.1

Suppose that $\bigcap_{n=1}^{\infty} \text{Ker } F_n = \{0\}$. Then the relation

$$\lim_{n\to\infty} ||x - x_n||_1 = 0 \tag{1.2}$$

is true for every $x \in X$.

Proof: Let P_n be the orthogonal projector of X which is defined by the equality

$$\text{Ker } P_n = \text{Ker } F_n \; . \tag{1.3}$$

Then

$$x_n = P_n x \; . \tag{1.4}$$

LEMMA 1.1

The set $S = \{P_n x : x \in X, n \in \mathbb{N}\}$ is dense in X.

Proof: Note that (1.1) implies that S is a subspace of X. Suppose that $z \in X$ satisfies

$$(z, P_n x)_1 = 0 \quad (x \in X; n \in \mathbb{N}) \; .$$

Then

$$\begin{aligned} z \in & \bigcap_{n\geq 1} \text{Ker } P_n \\ = & \bigcap_{n\geq 1} \text{Ker } F_n \\ = & \{0\} \; , \end{aligned}$$

i.e. $z = 0$. This proves Lemma 1.1.

Let I denote the identity mapping of X.

LEMMA 1.2

For every $z \in S$ the relation

$$\lim_{n\to\infty} P_n z = Iz = z$$

is true.

Proof: There is a natural number m such that $z = P_m z$. Then

$$P_n z = z \quad (n \geq m)$$

which proves Lemma 1.2.__

Taking into account Lemma 1.1, Lemma 1.2 and $||P_n|| \leq 1$, an application of the Banach-Steinhaus theorem (cf. Laurent [13]) yields the relation (1.2).__

REMARK 1.1

Joly [12] has proved (1.2) for the special observations

$$F_n = \bigtimes_{i=1}^{n+k} L_i \quad (L_i \in X^*,\ i \in \mathbb{N})$$

(dim Ker U = k < ∞).

2. Spline interpolation and Ritz approximation

Under suitable conditions a more detailed analysis of the convergence of $\{x_n\}_{n\geq 1}$ can be given.

Let X be a continuously imbedded, dense linear subspace of Y

$$X \hookrightarrow Y \tag{2.1}$$

such that

$$\overline{\text{Ker } F_o} = Y \tag{2.2}$$

(see Fisher-Jerome [10] for similar conditions). Then the following theorem has been proved (cf. Delvos-Schempp [8]).

THEOREM 2.1

The operator U_o *defined by*

$$U_o x = Ux \quad , \quad x \in \text{Dom } U_o = \text{Ker } F_o$$

is closed in Y, *and*

$$A = U_o^* U_o$$

is the unique positive operator in Y *such that*

$$\text{Dom } A \subset \text{Ker } F_o \ ,$$

$$(Ux, Uy) = (x, Ay) \quad (x \in \text{Ker } F_o,\ y \in \text{Dom } A) \ . \tag{2.3}$$

See also Fisher-Jerome [10]. Because of (1.3) the spline interpolant x_n is obtained by an orthogonal projection of x onto the subspace Ker $F_n^{\perp}$. Since Ker F_o is the energy space of A the <u>spline interpolant</u> x_n of $x \in$ Ker F_o is also the <u>Ritz approximation</u> of x in the subspace H_n = Ker $F_n^{\perp} \cap$ Ker F_o. Therefore we consider the number

$$C_n = \sup \{ ||x|| : F_n x = 0, \ ||x||_1 = 1 \} \qquad (2.4)$$

as the <u>natural error norm of spline interpolation</u>, since

$$||x - x_n|| \le C_n \ ||Ux|| \le C_n \ ||x||_1 \qquad (2.5)$$

(see Nitsche [15]).

It is the main purpose of this section to prove, that under suitable assumptions the relation

$$\lim_{n\to\infty} C_n = 0 \qquad (2.6)$$

is true.

Before doing this we will derive some properties of C_n.

<u>LEMMA 2.1</u>

<u>Let</u> A^{-1} <u>be compact. Then for every</u> $n \in \mathbb{N}$ <u>there is an element</u> $z_n \in X$ <u>such that</u>

$$F_n z_n = 0 \quad , \quad ||z_n||_1 = 1 \quad , \quad ||z_n|| = C_n \quad . \qquad (2.7)$$

<u>Proof</u>: Note, that the compactness of A^{-1} implies the weak (sequential) continuity of the functional

$$x \to ||x - P_n x||$$

on X. By the Alaoglu-Bourbaki theorem the unit ball $\{x \in X : ||x||_1 \le 1\}$ is weak compact. Thus Lemma 2.1 follows from the generalized Weierstraß theorem (cf. Ljusternik-Sobolew [14], p. 140).__

<u>LEMMA 2.2</u>

<u>If</u> $u_n = A^{-1} z_n$, <u>then</u>

$$C_n^2 = ||u_n - P_n u_n||_1 \le ||u_n||_1 \ . \qquad (2.8)$$

<u>Proof</u>: We consider the continuous linear functional on X defined by

$$F(x) = (x - P_n x, z_n) \ .$$

Because of (2.5) we have

$$|F(x)| \leq C_n \; ||x - P_n x|| \leq C_n^2 \; ||x||_1 ,$$

i.e. $||F|| \leq C_n^2$.

Putting $x = z_n$ we obtain $F(z_n) = C_n^2$. Since $||z_n||_1 = 1$, we obtain

$$||F|| = C_n^2 .$$

On the other, we have

$$\begin{aligned} F(x) &= (x-P_n x, z_n) \\ &= (x-P_n x, u_n)_1 \\ &= (x, u_n - P_n u_n)_1 , \end{aligned}$$

which implies

$$||u_n - P_n u_n||_1 = ||F|| = C_n^2 \quad —$$

<u>THEOREM 2.2</u>

<u>Suppose that</u> A^{-1} <u>is compact</u>. <u>If</u> $\bigcap_{n=1}^{\infty} \text{Ker } F_n = \{0\}$ <u>then</u>

$$\lim_{n\to\infty} C_n = 0 .$$

<u>Proof</u>: Since $C_n^2 \leq ||u_n||_1$, it suffices to show that

$$\lim_{n\to\infty} ||u_n||_1 = 0 . \qquad (2.9)$$

It follows from Lemma 2.1 and (1.1) that $P_k z_n = 0 \; (n \geq k)$.
This implies

$$\lim_{n\to\infty} (P_k x, z_n)_1 = \lim_{n\to\infty} (x, P_k z_n) = 0 \quad (x \in X;\; k \in \mathbb{N}) .$$

Since $||z_n||_1 = 1$ $(n \in \mathbb{N})$ and $\{P_k x : x \in X,\; k \in \mathbb{N}\}$ is dense in X (Lemma 1.1), an application of the Banach-Steinhaus theorem shows that

$$\lim_{n\to\infty} (x, z_n)_1 = 0 \quad (x \in X) . \qquad (2.10)$$

Moreover, the assumption $X \hookrightarrow Y$ implies

$$\lim_{n\to\infty} (y, z_n) = 0 \quad (y \in Y) \qquad (2.11)$$

(see Smirnow [20], p. 270). Since A^{-1} is compact we obtain from (2.11)

$$\lim_{n\to\infty} ||u_n|| = 0 . \qquad (2,12)$$

Now, a standard fact concerning weak convergence in Hilbert spaces (cf. Smirnow [20], p. 357) yields

$$\lim_{n\to\infty} ||u_n||_1^2 = \lim_{n\to\infty} (u_n, u_n)_1$$

$$= \lim_{n\to\infty} (u_n, z_n)$$

$$= 0 ,$$

whence Theorem 2.2 is proved.__

3. The degree of optimal approximation

Let $L \in X$ be given. Then

$$L_n = LP_n$$

is the optimal approximation of L with respect to the observation F_n and the coobservation U (cf. Sard [17], Delvos-Posdorf [7]). In this section we will estimate the error norm

$$||L - L_n||$$

in terms of C_n. Put $l = L^*(1)$, $l_n = L_n^*(1)$. Then we have

$$l_n = P_n l \quad , \quad ||L - L_n|| = ||l - l_n||_1$$

(see also de Boor-Lynch [3]).
Therefore, the error norm depends on the error in spline interpolation. We will derive an estimate for

$$||l - l_n||_1$$

which depends on the smoothness of $l = L^*(1)$.
For $\alpha \geq 0$ and $x \in \mathrm{Dom}\,(A^{\alpha/2})$ we introduce the α-norm

$$||x||_\alpha = ||A^{\alpha/2}x|| \ .$$

Then the following theorem has been proved by Schäfer-Schempp [19].

THEOREM 3.1

If $x \in \mathrm{Dom}\,(A^{(1+\beta)/2})$ $(0 \leq \beta \leq 1)$, *then*

$$||x - x_n||_\alpha \leq C_n^{1+\beta-\alpha}\, ||x||_{1+\beta} \quad (0 \leq \alpha \leq 1) \ . \qquad (3.1)$$

REMARK 3.1

For $\alpha = o$, $\beta = o$ we obtain the estimate (2.5)

$$||x - x_n|| \leq C_n \; ||x||_1 \; . \qquad (3.2)$$

The case $\alpha = o$, $\beta = 1$

$$||x - x_n|| \; \leq \; C_n^2 \; ||x||_2 \qquad (3.3)$$

has been proved by Nitsche [16].

Suppose now that A^{-1} is compact. Let $\{\lambda_n\}_{n\geq 1}$ be the increasing sequence of eigenvalues of A. Moreover, $\{e_n\}_{n\geq 1}$ is supposed to be the sequence of corresponding eigenvectors of A:

$$Ae_n = \lambda_n e_n \qquad (n \in \mathbb{N}) \qquad (3.4)$$

THEOREM 3.2

Suppose that $L \in X^*$ satisfies

$$\sum_{m=1}^{\infty} \lambda_m^{\beta-1} \; |L(e_m)|^2 < \infty \qquad (\beta \in [0,1]) \qquad (3.5)$$

Then the relation

$$||L - L_n|| = \mathcal{O}(C_n^{\beta}) \qquad (3.6)$$

holds.

Proof: Put $l = (L - L_o)^*(1)$, $l_n = P_n l$ $(n \geq 1)$. It follows from (1.1) and (1.3) that

$$||L - L_n|| = ||l - l_n||_1 \qquad (n \geq 1) \; . \qquad (3.7)$$

LEMMA 3.1

The condition (3.5) implies

$$l = (L - L_o)^*(1) \in \text{Dom}\,(A^{(1+\beta)/2}) \qquad (3.8)$$

Proof: For every $n \in \mathbb{N}$ we have

$$(l,e_n) = \lambda_n^{-1} \; (l,e_n)_1 = \lambda_n^{-1} \; \overline{L(e_n)} \; .$$

This implies

$$||A^{(1+\beta)/2}l||^2 = \sum_{n=1}^{\infty} \lambda_n^{\beta-1} \ |L(e_n)|^2 < \infty$$

whence (3.8) is proved.__

Taking into account (3.8) and (3.1) we can conclude

$$\begin{aligned} & ||l - l_n||_1^2 \\ &= (l, l - l_n)_1 \\ &= (A^{1/2}l, A^{1/2}(l - l_n)) \\ &= (A^{(1+\beta)/2}l, A^{(1-\beta)/2}(l - l_n)) \\ &\le |||l|||_{1+\beta} \ ||l - l_n||_{1-\beta} \\ &\le C_n^{2\beta} \ |||l|||_{1+\beta}^2 \end{aligned}$$

whence

$$||L - L_n|| \le C_n^{\beta} \ |||l|||_{1+\beta} \ . \tag{3.9}$$

Thus, Theorem 3.2 is proved.__

REMARK 3.2

Note that β doesn't depend on F_n.

As an application we will prove

THEOREM 3.3

For any $y \in Y$ *let* $L_y(x) = (x,y)$ $(x \in X)$. *Then*

$$||L_y - L_y P_n|| \le C_n \ ||y|| \ . \tag{3.10}$$

Moreover, the relation

$$C_n = \max \{||L_y - L_y P_n|| : ||y|| = 1\} \tag{3.11}$$

is true.

Proof: Note that $l_y = (L_y - L_y P_o)^*(1) = A^{-1}y$ implies (3.10) (see (3.9)).

Put $y_n = C_n^{-1} \ z_n$ (see the proof of Lemma 2.2). Then

$$(L_{y_n} - L_{y_n} P_n)(z_n) = ||z_n|| = C_n \ ,$$

whence (3.11) follows.__

4. Application to Mangeron's equation

It is well known that (transfinite) blending interpolation is closely related to Mangeron's equation (cf. Birkhoff-Gordon [2]).

For any $f \in W_2^1(J) \otimes W_2^1(J)$ $(J = [0,1])$ the blended linear interpolant

$$\begin{aligned} f_o(x,y) &= (1-x)f(0,y) + xf(1,y) \\ &+ (1-y)f(x,0) + yf(x,1) \\ &- (1-x)(1-y)f(0,0) - (1-x)yf(0,1) \\ &-x(1-y)f(1,0) - xyf(1,1) \end{aligned}$$

is the unique function in the set of functions $g \in W_2^1(J) \otimes W_2^1(J)$ satisfying

$$g(x,0) = f(x,0) \quad , \quad g(x,1) = f(x,1)$$

$$g(0,y) = f(0,y) \quad , \quad g(1,y) = f(1,y) \qquad (x,y \in J) \; ,$$

which minimizes

$$\int_0^1\int_0^1 |D_1 D_2 \, g(x,y)|^2 \, dxdy$$

(cf. Delvos [5]).

For sufficiently smooth functions $f \in C^{2,2}(R)$ $(R = J \times J)$ f_o is the unique solution of the Dirichlet problem for Mangeron's equation

$$D_1^2 D_2^2 \, f_o = 0 \; ,$$

$$f_o|_{\partial R} = f|_{\partial R} \; .$$

The preceeding theory is applicable by putting

$$U = D_1 D_2 \quad , \quad F_o = |_{\partial R} \; .$$

The operator A is just Mangeron's operator

$$A = D_1^2 D_2^2 \; ,$$

$$\text{Dom } A = \{f \in W_2^2(J) \otimes W_2^2(J) \; : \; f|_{\partial R} = 0\} \; .$$

The eigenvalues of $D_1^2 D_2^2$ are

$$\lambda_{ik} = \pi^4 \cdot i^2 \cdot k^2 \qquad (i,k \in \mathbb{N}) ,$$

and the corresponding eigenfunctions are

$$e_{ik}(x,y) = 2 \sin(\pi ix) \sin(\pi ky) \qquad (i,k \in \mathbb{N}) .$$

Let us define F_n by

$$F_n(f) = ((f,e_{ik})_{i\cdot k\leq n}, f|_{\partial R}) \qquad (n \in \mathbb{N}) .$$

It is well known that

$$c_n = 1/[\pi^2(n+1)] .$$

The estimate (2.5) reads as follows:

THEOREM 4.1

Suppose that $f \in W_2^1(J) \otimes W_2^1(J)$ *satisfies* $f|_{\partial R} = 0$. *Then*

$$||f - \sum_{i\cdot k\leq n} (f,e_{ik})\, e_{ik}||_2 \leq 1/[\pi^2(n+1)]\ ||D_1 D_2 f||_2 .$$

Finally, we present an illustration of the use and the limits of Theorem 3.2.

For any $(s,t) \in \mathring{R}$ the Dirac measure

$$L(f) = f(s,t)$$

is a continuous linear functional on $W_2^1(J) \otimes W_2^1(J)$. It is easily seen that for $\beta > 1/2$ the estimate

$$\sum_{i,k=1}^{\infty} (\pi^2 \cdot i^2 \cdot k^2)^{\beta-1}\ 4 \sin^2(\pi is) \sin^2(\pi kt) < \infty$$

is valid. Putting

$$\beta = 1/2 + \varepsilon \qquad (0 < \varepsilon < 1/2)$$

we obtain from Theorem 3.2

THEOREM 4.2

Let $f \in W_2^1(J) \otimes W_2^1(J)$ satisfy $f|_{\partial R} = 0$. For every $\varepsilon \in (0,1/2)$ there is a positive constant $M(\varepsilon)$ such that

$$|f(s,t) - \sum_{i \cdot k \leq n} (f,e_{ik})\, e_{ik}(s,t)| \leq M(\varepsilon)/[(n+1)^{1/2-\varepsilon}]\; ||D_1 D_2 f||_2 \qquad (n \geq 1).$$

REMARK 4.1

It doesn't follow from the theory that $\sup_{\varepsilon > 0} M(\varepsilon) < \infty$.

References

1. ATTEIA, M.: Généralisationde la définition et des propriétés des "spline-fonctions". C.R. Acad. Sci. Paris 260, 3550 - 3553 (1965).
2. BIRKHOFF, G., GORDON, W.: The draftman's and related equations. J. Approximation Theory 1, 199 - 208 (1968).
3. de BOOR, C.,LYNCH, R.E.: On splines and their minimum properties. J. Math. Mech. 15, 953 - 969 (1966).
4. CIARLET, P.G., VARGA, R.S.: Discrete Variational Green's Function. II. One dimensional problem. Num. Math. 16, 115 - 128 (1970).
5. DELVOS, F.J.: On surface interpolation. J. Approximation Theory 15, 209 - 213 (1975).
6. DELVOS, F.J.: Über die Konstruktion von Spline Systemen. Dissertation. 65 pp. Ruhr-Universität-Bochum 1972.
7. DELVOS, F.J., POSDORF, H.: On optimal tensor product approximation. J. Approximation Theory (to appear).
8. DELVOS, F.J., SCHEMPP, W.: An extension of Sard's method. In "Spline functions, Karlsruhe 1975" (eds.: K. Böhmer, G. Meinardus, W. Schempp), Lecture Notes in Mathematics 501, 80 - 91 (1976).
9. DELVOS, F.J., SCHEMPP, W.: Sard's method and the theory of spline systems. J. Approximation Theory 14, 230 - 243 (1975).

10. FISHER, S.D., JEROME, J.W.: Minimum Norm Extremals in Function Spaces. With Applications to Classical and Modern Analysis. VIII. Lecture Notes in Mathematics 479, 209 pp. (1975).

11. HULME, B.L.: Interpolation by Ritz approximation. J. Math. Mech. 18, 337 - 342 (1968).

12. JOLY, J.L.: Théorémes de convergence pour les fonctions-spline générales d'interpolation et d'ajustement. C.R. Acad. Sci. Paris 264, 126 - 128 (1967).

13. LAURENT, P.J.: "Approximation et optimisation". Hermann. Paris 1972.

14. LJUSTERNIK, L.A., SOBOLEW, W.I.: "Elemente der Funktionalanalysis". Akademie Verlag. Berlin 1965.

15. NITSCHE, J.: Ein Kriterium für die Quasi-Optimalität des Ritzschen Verfahrens. Num. Math. 11, 346 - 348 (1968).

16. NITSCHE, J.: Verfahren von Ritz und Spline-Interpolation bei Sturm-Liouville-Randwertproblemen. Num. Math. 13, 260 - 265 (1969).

17. SARD, A.: Approximation based on Nonscalar Observations. J. Approximation Theory 8, 315 - 334 (1973).

18. SARD, A.: Optimal approximation. J. Functional Analysis 1, 222 - 244 (1967); 2, 368 - 369 (1968).

19. SCHÄFER, W., SCHEMPP, W.: Splineapproximation in intermediären Räumen. In "Spline functions, Karlsruhe 1975" (eds.: K. Böhmer, G. Meinardus, W. Schempp). Lecture Notes in Mathematics 501, 226 - 246 (1976).

20. SMIRNOW, W.I.: Lehrgang der höheren Mathematik. Band V. VEB Deutscher Verlag der Wissenschaften. Berlin 1967.

Dr. F.J. Delvos
Dr. W. Schäfer
Prof. Dr. W. Schempp

Gesamthochschule Siegen
Lehrstuhl für Mathematik I

Hölderlinstr. 3
D-59 Siegen 21

A CONSTRUCTIVE THEORY FOR APPROXIMATION BY SPLINES WITH AN ARBITRARY SEQUENCE OF KNOT SETS

by

R. DEVORE [1]) and K. SCHERER

Introduction. We are interested in the global constructive theory for splines. By global we mean that we consider the global approximation of functions by splines and wish to compare this to the global smoothness of the function as described by properties of the modulus of continuity. Splines are more suited for local approximation and when we study global results, we are forced to accept essentially the worst of the local. Thus, for example, global direct theorems (i.e. estimates of the degree of spline approximation) are given in terms of the maximum distance between adjacent knots. In this paper, we will develope an inverse theory (i.e. a measure of the global smoothness of the function in terms of its degree of approximation) in terms of the minimum distance between adjacent knots. Our main contribution is that we give inverse theorems for completely arbitrary sequences of knot sets whereas prior results require some additional assumptions, usually some sort of mixing condition on the knot sequence.

Let $\Delta: 0 = x_o < x_1 < \dots < x_m = 1$ be a set of knots. If $r > 0$ and $-1 \le \rho \le r-2$, we let $\mathcal{S}_{r,\rho}(\Delta)$ denote the space of splines of order r (degree r-1) with knots Δ and smoothness ρ. That is, $S \in \mathcal{S}_{r,\rho}(\Delta)$ if and only if $S \in C^\rho[0,1]$, and on $(x_{\nu-1}, x_\nu)$ S is a polynomial of degree $\le r-1$, for $\nu = 1,2,\dots,m$. When $\rho = -1$, no continuity is assumed but we do make the convention that S is continuous from the right at each knot and from the left at 0.

The degree of approximation by splines from $\mathcal{S}_{r,\rho}(\Delta)$ can be measured in terms of the upper mesh length $\overline{\Delta} = \max_{1 \le i \le m} |x_{i-1} - x_i|$. Namely, see e.g. [4] if $f \in L_p[0,1]$, then there is a spline $S \in \mathcal{S}_{r,\rho}(\Delta)$ such that

$$(1.1) \qquad \|f - S\|_p \le C\omega_r(f,\overline{\Delta})_p$$

---

1) This author was supported in part by the Alexander von Humboldt Stiftung and the National Science Foundation in Grant GP 19620.

with $\omega_r(f,\cdot)_p$ the r-th order modulus of smoothness as measured in L_p and C depending only on r. If $(\Delta_n)_1^\infty$ is a sequence of knot sets then (1.1) gives that for each $f \in L_p[0,1]$ there are splines $S_n \in \mathcal{S}_{r,\rho}(\Delta_n)$ so that

$$\|f - S_n\|_p \le C\omega_r(f,\overline{\Delta}_n)_p , \quad n = 1,2,\ldots . \tag{1.2}$$

Our interest in this paper is the inverse problem to (1.2). In Section 3, we give a general inverse theorem for approximation by splines from $\mathcal{S}_{r,\rho}(\Delta_n)$. The inverse results are given in terms of the lower mesh lengths $\underline{\Delta}_n = \min_{1\le i\le m_n} |x_i^{(n)} - x_{i-1}^{(n)}|$ with the $x_i^{(n)}$, the knots of Δ_n. The reason for the dependence of inverse theorems on the lower mesh lengths is dicussed somewhat in Section 4.

Our inverse theorem requires no added assumptions on the knot sequence in contrast to most of the known inverse theorems, e.g. [7], [9], [3], [5], [6] where some sort of mixing condition is assumed. Thus, we are dealing in essence with the worst situation. This turns out to be the case of nested knots. Some partial results for nested sequences of knots were given in [6]. The situation for nested knots (and therefore general knot sets as well) is complicated by the fact that there is no saturation phenomenon (see [6,10]).

The inverse theorems for general knot sequences depend on ρ and p as well as r. This is in contrast to the mixed knot sequences where the inverse theorems depend only on r. Let us mention only one interesting case. If the sequence of knot sets has bounded mesh ratios, i.e. $\underline{\Delta}_n^{-1}\,\overline{\Delta}_n \le M$, $n = 1,2,\ldots$ then we show that for $0 < \alpha < \rho+1+1/p$, there are splines $S_n \in \mathcal{S}_{r,\rho}(\Delta_n)$ with $\|f - S_n\|_p = O(\overline{\Delta}_n^\alpha)$ if and only if $\omega_r(f,t) = O(t^\alpha)$. When $\alpha = \rho+1+1/p$, one obtains only that if $\|f - S_n\|_p = O(\underline{\Delta}_n^\alpha)$ then $\omega_r(f,t)_p = O(t^\alpha\,|\log t|)$. This last result cannot be improved as we show with examples in Section 4. The limitation on α given by $\rho+1+1/p$ comes from the inherent smoothness that a spline $S \in \mathcal{S}_{r,\rho}(\Delta_n)$ has in $L_p[0,1]$.

2. <u>Smoothness of splines.</u> The proof of our inverse theorem rests

on estimating the smoothness of the approximating splines S_n. This will give in turn an estimate for the smoothness of f. The smoothness of a spline is controlled by its jumps and jumps in its derivatives. If $S \in \mathcal{S}_{r,\rho}(\Delta)$ then S has continuous derivatives up to order ρ on $[0,1]$. Of course, S has derivatives of all orders on each interval (x_{i-1},x_i), $i=1,2,\dots,m$. For a knot x_i, we define

$$(2.1) \qquad [S^{(\nu)}]_i = S^{(\nu)}(x_i+) - S^{(\nu)}(x_i-).$$

Thus $[S^{(\nu)}]_i$ is the jump in $S^{(\nu)}$ at x_i. It becomes convenient in certain sums to let i range from 0 to m, so we define $[S^{(\nu)}]_i = 0$ when i=0 or m.

L_p estimates for the smoothness of $S^{(\nu)}$ are given in terms of the seminorms

$$(2.2) \qquad J_p(S^{(\nu)}) = \left(\sum_{i=1}^{m-1} |[S^{(\nu)}]_i|^p\right)^{1/p} \qquad 1 \le p < \infty$$

$$J_\infty(S^{(\nu)}) = \max_{1\le i<m} |[S^{(\nu)}]_i|, \quad p = \infty.$$

Namely, we have the following lemma which estimates the smoothness of a spline in terms of its jumps.

Lemma 1. <u>If</u> S <u>is in</u> $\mathcal{S}_{r,\rho}(\Delta)$ and $1 \le p \le \infty$ then

$$(2.3) \qquad \omega_r(S,h)_p \le C(1+h/\underline{\Delta})^{1/p'} h^{1/p} \sum_{\nu=\rho+1}^{r-1} h^\nu J_p(S)$$

<u>with</u> C <u>a constant depending only on</u> r <u>and with</u> p' <u>the conjugate index to</u> p, $1/p'+1/p = 1$.

<u>Proof</u>. We will establish this inequality by showing that for any $2 \le k \le r$ and $\rho < \nu < r$, we have for $h > 0$,

$$(2.4) \qquad \omega_k(S^{(\nu)},h)_p \le C\,\{(1+h/\underline{\Delta})^{1/p'} h^{1/p} J_p(S^{(\nu)}) + h\omega_{k-1}(S^{(\nu+1)},h)_p\}$$

If we apply (2.4) iteratively starting with $\nu=0$ and $k=r$ then we arrive at (2.3).

Now, to show the inequality (2.4), we write $S^{(\nu)} = T_1 + T_2$ with T_1 the spline in $\mathcal{S}_{1,-1}(\Delta)$ which has the same jump at each x_i as $S^{(\nu)}$, i.e. $[T_1]_i = [S^{(\nu)}]_i$, $i=1,2,\ldots,m-1$. Then T_2 is absolutely continuous and $T_2'(x) = S^{(\nu+1)}(x)$, $x \notin \Delta$. Thus,

$$\text{(2.5)} \qquad \begin{aligned} \omega_k(S^{(\nu)},h)_p &\le \omega_k(T_1,h)_p + \omega_k(T_2,h)_p \\ &\le C\,\{\omega_1(T_1,h)_p + h\omega_{k-1}(S^{(\nu+1)},h)_p\} \end{aligned}$$

with C a constant depending only on r. Here, we have used well known properties of moduli of smoothness.
We finish the proof for $1 \le p < \infty$. The case $p = \infty$ is handled similarily.

In order to estimate $\omega_1(T_1,h)_p$, we let $d\mu_p$ denote the purely atomic measure with mass $|[S^{(\nu)}]_i|^p$ at each point x_i. Then, from Holder's inequality we find

$$\begin{aligned} |T_1(x+h)-T_1(x)|^p &\le \Big(\sum_{x\le x_i\le x+h} |[S^{(\nu)}]_i|\Big)^p \\ &\le (1+h/\underline{\Delta})^{p/p'} \sum_{x\le x_i\le x+h} |[S^{(\nu)}]_i|^p \\ &\le (1+h/\underline{\Delta})^{p/p'} \int_0^h d\mu_p(x+t) \end{aligned}$$

where we used the fact that on $[x,x+h]$ there are at most $(1+h/\underline{\Delta})$ knots. Integrating this last inequality and interchanging integrals, we find

$$\begin{aligned} \|T_1(x+h)-T_1(x)\|_{p,[0,1-h]} &\le (1+h/\underline{\Delta})^{1/p'} \Big(\int_0^h \Big[\int_0 |d\mu_p(x)|\Big]\, dt\Big)^{1/p} \\ &\le (1+h/\underline{\Delta})^{1/p'}\, h^{1/p}\, J_p(S^{(\nu)}) \,. \end{aligned}$$

Thus,

$$\omega_1(T_1,h)_p \le (1+h/\underline{\Delta})^{1/p'}\, h^{1/p}\, J_p(S^{(\nu)}) \,.$$

When this is placed in (2.5), we obtain the desired estimate (2.4) and the lemma is proved.

3. Inverse theorems. Let $(\Delta_n)_1^\infty$ be an arbitrary sequence of knot sets with $\lim \underline{\Delta}_n = 0$. By reindexing this sequence if necessary, we can assume that $\underline{\Delta}_{n+1} \le \underline{\Delta}_n$, $n=1,2,\ldots$. We also define $\Delta_o = \{0,1\}$. As we have remarked earlier, the key to proving inverse theorems is to estimate the jumps in the approximating splines and their derivatives. We consider first the case of smooth splines, i.e. $\mathcal{S}_{r,r-2}(\Delta_n)$. This is the simplest case for estimating the jumps and the arguments are the most transparent here.

Lemma 2. Let $1 \le p \le \infty$ and $r > 0$. If there are splines $S_n \in \mathcal{S}_{r,r-2}(\Delta_n)$ with

(3.1) $\qquad \|f-S_n\|_p \le \varepsilon_n, \quad n=1,2,\ldots \quad ,$

then for each $n > 0$,

(3.2) $\qquad J_p(S_n^{(r-1)}) \le C \sum_{k=1}^{n} \underline{\Delta}_k^{-\theta}(\varepsilon_k + \varepsilon_{k-1})$

with C depending only on r, $\theta = r-1+1/p$ and $\varepsilon_o = \|f\|_p$

Proof. Let $1 \le k \le n$. We will show that

(3.3) $\qquad J_p(S_k^{(r-1)}) \le C \underline{\Delta}_k^{-\theta}(\varepsilon_k + \varepsilon_{k-1}) + J_p(S_{k-1}^{(r-1)})$

where for $k = 0$, $S_o = 0$. Then, (3.2) follows from repeated application of (3.3) and the fact that $J_p(S_o^{(r-1)}) = 0$.

Now to prove (3.3), suppose that $x_i^{(k)}$ is a knot from Δ_k, $1 \le i < m_k$. Determine j so that $x_j^{(k-1)} \le x_i^{(k)} < x_{j+1}^{(k-1)}$. When $|x_j^{(k)} - x_j^{(k-1)}| \ge \frac{1}{4}\underline{\Delta}_k$ we let $a_i = x_i^{(k)}$ otherwise we let $a_i = x_j^{(k-1)}$. In any case, a_i is the right end point of an open interval $I_{i,\ell}$ with $|I_{i,\ell}| \ge \frac{1}{4}\underline{\Delta}_k$ and the spline $T_k = S_k - S_{k-1}$ has no knots on $I_{i,\ell}$. Similarily, if $|x_i^{(k)} - x_j^{(k-1)}| \ge \frac{1}{4}\underline{\Delta}_k$, we let $b_i = x_i^{(k)}$, otherwise we let $b_i = x_{j+1}^{(k-1)}$. Then b_i is the left end point of an interval $I_{i,\varkappa}$ of length $\ge \frac{1}{4}\underline{\Delta}_k$ on which T_k has no knots. In our selection of a_i and b_i, either both of these are $x_i^{(k)}$ or one of these is $x_i^{(k)}$ and the other is a knot of S_{k-1}, namely $x_j^{(k-1)}$ or $x_{j+1}^{(k-1)}$. In the latter case, we let $x_{j_i}^{(k-1)}$

denote this knot.

To estimate the jump in $S_k^{(r-1)}$ at $x_i^{(k)}$, we write simply

$$(3.4)\qquad [S_k^{(r-1)}]_i = T_k^{(r-1)}(x_i^{(k)}+) - T_k^{(r-1)}(x_i^{(k)}-) + S_{k-1}^{(r-1)}(x_i^{(k)}+) - S_{k-1}^{(r-1)}(x_i^{(k)}-)$$

$$= T_k^{(r-1)}(b_i+) - T_k^{(r-1)}(a_i-) + [S_{k-1}^{(r-1)}]_{j_i} .$$

Here, we used the fact that $S_{k-1}^{(r-1)}$ and $T_k^{(r-1)}$ are constant on (a_i,b_i). Note that for a given value of i there may be no j_i and then it is understood that the jump in $S_{k-1}^{(r-1)}$ at $x_{j_i}^{(k-1)}$ does not appear. If such a situation would always appear, this would be a property of the knot sets which is essentially equivalent to a mixing condition, cf. the remarks of Section 1.

We complete the proof for the case $1 \le p < \infty$. The case $p = \infty$ is similar and somewhat simpler. Since $T_k^{(r-1)}$ is constant on $I_{i,\ell}$ and $|I_{i,\ell}| \ge \frac{1}{4}\underline{\Delta}_k$, we have

$$(3.5)\qquad |T_k^{(r-1)}(a_i-)|^p = |I_{i,\ell}|^{-1} \int_{I_{i,\ell}} |T_k^{(r-1)}(x)|^p\,dx$$

$$\le 4\,\underline{\Delta}_k^{-1} \int_{I_{i,\ell}} |T_k^{(r-1)}(x)|^p\,dx$$

$$\le C\,\underline{\Delta}_k^{(1-r)p-1} \int_{I_{i,\ell}} |T_k(x)|^p\,dx$$

with C depending only on r. The last inequality is a Markov type inequality for polynomials, see [11,p.236] . The same estimate holds for $T_k^{(r-1)}(b_i+)$ except that now the integral is taken over $I_{i,r}$. Using these estimates back in (3.4) gives

$$J_p(S_k^{(r-1)}) \le \left(\sum_{i=1}^{m-1} |T_k^{(r-1)}(a_i-)|^p\right)^{1/p} + \left(\sum_{i=1}^{m-1} |T_k^{(r-1)}(b_i+)|^p\right)^{1/p} + J_p(S_{k-1}^{(r-1)}$$

$$\leq C\, \underline{\Delta}_k^{-\theta} \|T_k\|_p + J_p(S_{k-1}^{(r-1)})$$

where we used the fact that the intervals $I_{i,\ell}$ and $I_{i,\kappa}$ are disjoint. Our desired inequality (3.3) now follows from the fact that

$$\|T_k\|_p \leq \|f-S_k\|_p + \|f-S_{k-1}\|_p \leq \varepsilon_k + \varepsilon_{k-1} .$$

This completes the proof of the lemma.

We can give a general version of Lemma 2 for arbitrary smoothness ρ. However, the details are more complicated.

Lemma 3. Let $1 \leq p \leq \infty$, $r > 0$ and $-1 \leq \rho \leq r-2$. If there are splines $S_n \in \mathcal{S}_{r,\rho}(\Delta_n)$ with

$$\|f-S_n\|_p \leq \varepsilon_n, \quad n=1,2,\ldots ,$$

then for each $n > 0$ with $\Delta_n \leq \frac{1}{4}$, we have

$$(3.6) \qquad \sum_{\nu=\rho+1}^{r-1} \underline{\Delta}_n^{\nu+1/p} J_p(S_n^{(\nu)}) \leq C\, \Delta_n^{\theta} \sum_{k=1}^{n} \underline{\Delta}_k^{-\theta} (\varepsilon_k + \varepsilon_{k-1})$$

where $\theta = \rho+1+1/p$, $\varepsilon_o = \|f\|$ and C depends only on r.

Proof. Case I. We assume additionally that $\underline{\Delta}_{k+1} \leq 1/2\, \underline{\Delta}_k$, for $1 \leq k < n$. Let $S_o = 0$. We will show that for each $1 \leq k \leq n$, we have

$$(3.7) \qquad \sum_{\nu=1}^{r-1-\rho} \underline{\Delta}_k^{\nu-1} J_p(S_k^{(\rho+\nu)}) \leq C\, \underline{\Delta}_k^{-\theta} (\varepsilon_k + \varepsilon_{k-1})$$

$$+ A_k \sum_{\nu=1}^{r-1-\rho} \underline{\Delta}_{k-1}^{\nu-1} J_p(S_{k-1}^{(\rho+\nu)})$$

where $A_k = 1$ for $2 \leq k \leq n$. Then, (3.6) follows from successive application of (3.7) and the fact that $J_p(S_o^{(\mu)}) = 0$, for all μ.

To prove (3.7), we argue in a similar manner to the proof of (3.3) in Lemma 2. For a knot $x_i^{(k)} \in \Delta_k$, we define a_i, b_i, $I_{i,\ell}$, $I_{i,\kappa}$ and j_i exactly as in the proof of Lemma 2. For each $0 \leq i \leq m_k$ we have looking at the jumps of T_k passing from a_i to b_i

$$(3.8)\quad [S_k^{(\rho+\nu)}]_i = [T_k^{(\rho+\nu)}(b_i+) - T_k^{(\rho+\nu)}(b_i-)] +$$

$$+ [T_k^{(\rho+\nu)}(a_i+) - T_k^{(\rho+\nu)}(a_i-)] + [S_{k-1}^{(\rho+\nu)}]_{j_i}$$

$$= T_k^{(\rho+\nu)}(b_i+) - T_k^{(\rho+\nu)}(a_i-) + [S_{k-1}^{(\rho+\nu)}]_{j_i}$$

$$- (T_k^{(\rho+\nu)}(b_i-) - T_k^{(\rho+\nu)}(a_i+))$$

Here the term $T_k^{(\rho+\nu)}(b_i-)-T_k^{(\rho+\nu)}(a_i+)$ will be 0 if $a_i = b_i$. In the other case, for $1 \le \nu \le r-2-\rho$, the last term in brackets is new compared to our proof of (3.3) and we now work to estimate it. The other terms will be handled similar to that in (3.3).

If Q is any polynomial of degree $\le r$ and I is some interval then $\|Q\|_\infty(I) \le C|I|^{-1/p}\|Q\|_p(I)$ with C a constant depending only on r. The notation indicates that the norms are taken over the interval I. Since the function T_k has no knots on the interval $I_{i,\ell}$, we have that for any $\nu \ge 1$

$$(3.9)\quad \underline{\Delta}_k^{\nu-1}\ |T_k^{(\rho+\nu)}(a_i-)| \le C\ \underline{\Delta}_k^{-\rho-1}\ \|T_k\|_\infty(I_{i,\ell}) \le C\ \underline{\Delta}_k^{-\theta}\|T_k\|_p(I_{i,\ell})$$

where we have also used Markov's inequality for polynomials. Similarily, we have

$$(3.10)\quad \underline{\Delta}_k^{\nu-1}\ |T_k^{(\rho+\nu)}(b_i+)| \le \underline{\Delta}_k^{-\theta}\ \|T_k\|_p(I_{i,r}).$$

Since $a_i \ne b_i$ one of the a_i or b_i is $x_i^{(k)}$ and the other is the knot $x_{j_i}^{(k-1)}$. Let us suppose that $b_i = x_i^{(k)}$. The other case gives the same estimate. We write $T_k^{(\rho+\nu)}$ in its Taylor expansion at a_i+ to find

$$(3.11)\quad |T_k^{(\rho+\nu)}(b_i-)-T_k^{(\rho+\nu)}(a_i+)| \le \sum_{\mu=1}^{s_\nu} \frac{|T_k^{(\rho+\nu+\mu)}(a_i+)|}{\mu!}(b_i-a_i)^\mu$$

$$\le \sum_{\mu=1}^{s_\nu} \frac{|T_k^{(\rho+\nu+\mu)}(a_i+)|}{\mu!}(\tfrac{1}{2}\ \underline{\Delta}_k)^\mu$$

with $s_\nu = r-1-\rho-\nu$. Here we used the fact that $|b_i - a_i| \le 1/2\, \underline{\Delta}_k$. From (3.9) we see that

$$(3.12)\qquad \underline{\Delta}_k^{\nu+\mu-1}\, |T_k^{(\rho+\nu+\mu)}(a_i+)| \le \underline{\Delta}_k^{\nu+\mu-1}\,(|T_k^{(\rho+\nu+\mu)}(a_i-)| + |[S_{k-1}^{(\rho+\nu+\mu)}]_{j_i}|)$$

$$\le C\, \underline{\Delta}_k^{-\theta} \|T_k\|_p(I_{i,\ell}) + \underline{\Delta}_k^{\nu+\mu-1}\, |[S_{k-1}^{(\rho+\nu+\mu)}]_{j_i}|\,.$$

Now let us return to (3.8). If we apply an ℓ_p norm to all sequences as a function of i and multiply by $\underline{\Delta}_k^{\nu-1}$, we find

$$(3.13)\qquad \underline{\Delta}_k^{\nu-1}\, J_p(S_k^{(\rho+\nu)}) \le C\, \underline{\Delta}_k^{-\theta}\, \|T_k\|_p + \underline{\Delta}_k^{\nu-1}\, J_p(S_{k-1}^{(\rho+\nu)})$$

$$+\, C\, \underline{\Delta}_k^{-\theta}\, \|T_k\|_p \sum_{\mu=1}^{s_\nu} \frac{(1/2)^\mu}{\mu!} + \underline{\Delta}_k^{\nu-1} \sum_{\mu=1}^{s_\nu} \frac{(\underline{\Delta}_k/2)^\mu}{\mu!}\, J_p(S_{k-1}^{(\rho+\nu+\mu)})$$

where $T_k^{(\rho+\nu)}(b_i+)$ was estimated by (3.10); $T_k^{(\rho+\nu)}(a_i-)$ was estimated by (3.9); $S_{k-1}^{(\rho+\nu)}$ gives $J_p(S_{k-1}^{(\rho+\nu)})$; and $(T_k^{(\rho+\nu)}(b_i-) - T_k^{(\rho+\nu)}(a_i+))$ was estimated by using (3.11) and (3.12). Summing (3.13) over ν we have

$$\sum_{\nu=1}^{r-1-\rho} \underline{\Delta}_k^{\nu-1}\, J_p(S_k^{(\rho+\nu)}) \le C\, \underline{\Delta}_k^{-\theta} \|T_k\|_p + \sum_{\nu=1}^{r-1-\rho} \underline{\Delta}_k^{\nu-1}\, J_p(S_{k-1}^{(\rho+\nu)})$$

$$+ \sum_1^\infty \frac{(1/2)^\mu}{\mu!} \sum_{\nu=2}^{r-1-\rho} \underline{\Delta}_k^{\nu-1}\, J_p(S_{k-1}^{(\rho+\nu)}).$$

The desired result (3.7) for $k \ge 2$ follows from the facts that $\|T_k\|_p \le \varepsilon_k + \varepsilon_{k-1}$;

$$1 + \sum_1^\infty \frac{(1/2)^\mu}{\mu!} \le e^{1/2} \le 2$$

and

$$\underline{\Delta}_k^{\nu-1} \le (1/2\underline{\Delta}_{k-1})^{\nu-1} \le 1/2\, \underline{\Delta}_{k-1}^{\nu-1}, \quad \nu > 1.$$

This last inequality is where we used the additional assumption of Case I.

Case II. In the general case, we choose for given n a subsequence $(\Delta_{n_k})_0^\lambda$ which satisfies Case I and some additional condition. We start with $n_\lambda = n$. Then

i) if $\Delta_{n_\lambda - 1} \geq 2\,\Delta_{n_\lambda}$ we choose $n_{\lambda-1} = n_\lambda - 1$,
or if this is not the case,

ii) let j be the largest index $\leq n_\lambda$ with $\Delta_{n_\lambda - j} < 2\,\Delta_{n_\lambda}$ and choose $n_{\lambda-1} = n_\lambda - j - 1$.

If we have $j = n_\lambda$ in ii) we terminate the sequence by setting $0 = n_{\lambda-1} = n_0$. Otherwise, and if the resulting $n_{\lambda-1}$ is positive, we replace n_λ by $n_{\lambda-1}$, select $n_{\lambda-2}$ by the above procedure and continue in this way. This selection guaratees not only that $(\Delta_{n_k})_0^\lambda$ satisfies the condition $\Delta_{n_k} \leq (1/2)\,\Delta_{n_{k-1}}$ but also in addition $\Delta_{n_k} \geq (1/2)\,\Delta_{n_{k-1}+1}$.

From Case I we have

$$(3.14)\qquad \sum_{\nu=\rho+1}^{r-1} \Delta_n^{\nu+1/p}\, J_p(S_n^{(\nu)}) \;\leq\; C\,\Delta_n^\theta \sum_{k=1}^{\lambda} \Delta_{n_k}^{-\theta}(\varepsilon_{n_k} + \varepsilon_{n_{k-1}}).$$

Certainly, we have

$$(3.15)\qquad \sum_{k=1}^{\lambda} \Delta_{n_k}^{-\theta}\,\varepsilon_{n_k} \;\leq\; \sum_{k=1}^{n} \Delta_k^{-\theta}\,\varepsilon_k .$$

On the other hand, our selection guarantees that $\Delta_{n_k}^{-\theta} \leq 2^\theta \Delta_{n_{k-1}+1}^{-\theta}$, so that

$$\Delta_{n_k}^{-\theta}\,\varepsilon_{n_{k-1}} \;\leq\; 2^\theta\,\Delta_{n_{k-1}+1}^{-\theta}\,\varepsilon_{n_{k-1}} .$$

Summing, we find

$$(3.16)\qquad \sum_{k=1}^{\lambda} \Delta_{n_k}^{-\theta}\,\varepsilon_{n_{k-1}} \;\leq\; 2^\theta \sum_{k=1}^{n} \Delta_k^{-\theta}(\varepsilon_k + \varepsilon_{k-1}) .$$

The estimates (3.15) and (3.16) when used in (3.14) provide the desired estimate (3.6) and the lemma is proved.

It is now a simple matter to combine Lemmas 1 and 3 to obtain the following general inverse theorem.

Theorem 1. <u>Let</u> $(\Delta_n)_0^\infty$ <u>be a sequence of knot sets with</u> $\underline{\Delta}_{n+1} \le \underline{\Delta}_n$, $n = 0,1,\ldots$ <u>and</u> $\Delta_0 = \{0,1\}$. <u>Suppose that</u> $r > 0$, $-1 \le \rho \le r-2$ <u>and</u> $1 \le p \le \infty$. <u>If for each</u> $n \ge 0$ <u>there is an</u> $S_n \in \mathcal{S}_{r,\rho}(\Delta_n)$ <u>such that</u> $\|f-S_n\|_p \le \varepsilon_n$, <u>then for each</u> $n \ge 0$ <u>with</u> $\underline{\Delta}_n \le 1/4$, <u>we have</u>

$$(3.17)\qquad \omega_r(f,\underline{\Delta}_n)_p \le C\, \underline{\Delta}_n^{\theta} \sum_{k=1}^{n} \underline{\Delta}_k^{-\theta} (\varepsilon_k + \varepsilon_{k-1})$$

<u>where</u> $\theta = \rho+1+1/p$ <u>and</u> C <u>is a constant depending only on</u> r.

<u>Proof</u>. The estimates (2.3) and (3.6) show that

$$\omega_r(S_n,\underline{\Delta}_n)_p \le C\, \underline{\Delta}_n^{\theta} \sum_{k=1}^{n} \underline{\Delta}_k^{-\theta} (\varepsilon_k + \varepsilon_{k-1}).$$

Since $\|f-S_n\|_p \le \varepsilon_n$, we also have $\omega_r(f-S_n,\underline{\Delta}_n)_p \le 2^r \varepsilon_n$. These two inequalities and the fact that $\omega_r(f,\underline{\Delta}_n)_p \le \omega_r(f-S_n,\underline{\Delta}_n)_p + \omega_r(S_n,\underline{\Delta}_n)_p$ combine to give (3.17).

Because of the monotonicity of ω_r, the estimate in (3.17) gives an estimate of $\omega_r(f,t)_p$ for any value of t. In the case that the sequence (Δ_n) is too sparse we loose something. By too sparse we mean if $\underline{\Delta}_{n+1}^{-1} \underline{\Delta}_n$ is not uniformly bounded.

An important special case in Theorem 1 is when the sequence (Δ_n) also satisfies the bounded ratio condition, that is

$$(3.18)\qquad \overline{\Delta}_n \underline{\Delta}_n^{-1} \le C, \quad n = 0,1,\ldots \quad \text{with } C \text{ a constant.}$$

In this case, for example, we can characterize the Lip (α,r,p) = $= \{f: \omega_r(f,t)_p = O(t^\alpha),\ t > 0\}$ in terms of the degree of approximation

$$E_{n,\rho}(f) = \inf_{S \in \mathcal{S}_{r,\rho}(\Delta_n)} \|f-S\|_p .$$

Of course we will still need the condition

$$(3.19)\qquad C\,\underline{\Delta}_n \le \underline{\Delta}_{n+1} \le \underline{\Delta}_n \quad \text{with } C > 0 \text{ a constant,}$$

which prevents the sequence (Δ_n) from being too sparse.

Corollary 1. Let (Δ_n) be a sequence of knot sets with $\lim \underline{\Delta}_n = 0$ which satisfy (3.18) and (3.19).

a) If $r > 0$, $-1 \le j \le r-2$, $1 \le p \le \infty$ and $0 < \alpha < \rho+1+1/p$, then $f \in \mathrm{Lip}(\alpha,r,p)$ if and only if $E_{n,\rho}(f) = O(\overline{\Delta}_n{}^{\alpha})$.

b) For $\alpha = \rho+1+1/p$, $f \in \mathrm{Lip}(\alpha,r,p)$ implies that $E_{n,\rho}(f) = O(\overline{\Delta}_n{}^{\alpha})$ and $E_{n,\rho}(f) = O(\overline{\Delta}_n{}^{\alpha})$ implies that $\omega_r(f,t)_p = O(t^{\alpha}|\log t|)$.

c) For $\alpha > \rho+1+1/p$, $E_{n,\rho}(f) = O(\overline{\Delta}_n{}^{\alpha})$ implies that $\omega_r(f,t)_p = O(t^{\theta})$.

Proof. If $0 < \alpha \le \rho+1+1/p$ then the direct estimate (1.2) gives that for each $f \in \mathrm{Lip}(\alpha,r,p)$, we have $E_{n,\rho}(f) = O(\overline{\Delta}_n{}^{\alpha})$. For the converse direction, we use Theorem 1. Choose a subsequence (Δ_{n_k}) so that

$$C_1 \underline{\Delta}_{n_k} \le \underline{\Delta}_{n_{k+1}} \le 1/2\, \underline{\Delta}_{n_k} \tag{3.20}$$

This is possible because $\lim \underline{\Delta}_n = 0$ and (Δ_n) satisfies (3.19). Let $\Delta_{n_0} = \{0,1\}$. If $0 < \alpha \le \rho+1+1/p$, then (3.17) gives that for any k

$$\omega_r(f,\underline{\Delta}_{n_k})_p \le C\, \underline{\Delta}_{n_k} \sum_{j=0}^{k} \underline{\Delta}_{n_k}^{\theta-\alpha}\, \underline{\Delta}_{n_j}^{\alpha-\theta} \le C\, \underline{\Delta}_{n_k}^{\alpha} \sum_{j=0}^{k} 2^{-j(\theta-\alpha)} \tag{3.21}$$

$$\le \begin{cases} C_\alpha\, \underline{\Delta}_{n_k}^{\alpha}, & \alpha < \rho+1+1/p = \theta \\ C_\alpha\, \underline{\Delta}_{n_k}^{\alpha}\, |\log \underline{\Delta}_{n_k}|, & \alpha = \theta \\ C_\theta\, \underline{\Delta}_{n_k}^{\theta}, & \alpha > \theta \end{cases}$$

where we used (3.18) and the fact that $k \le C|\log \underline{\Delta}_{n_k}|$ because of the lower inequality in (3.20). Using usual properties of the modulus of smoothness this last inequality extends to all values of $t > 0$ because of (3.20). This proves the corollary.

4. Final remarks. We wish to discuss the results of the preceeding section, in particular their sharpness.

Concerning assertion b) of Corollary 1 we construct examples which show that the log term is essential. We do this for $\nu = 2$,

$p = 0$ but this can also be shown in general.

Given $j > 0$ define the hat function $\varphi_j(x) = (2^{-j} - |x|)$ with support $(-2^{-j}, 2^{-j})$ and height 2^{-j}. Then set

$$(4.1) \qquad f_k(x) = \sum_{j=2k+2}^{4k} \varphi_j(x - 2^{-2k-1}) \quad k=1,2,\ldots$$

This f_k is symmetric with respect to 2^{-2k-1}, has support on $[2^{-2k-1} - 2^{-2k-2}, 2^{-2k-1} + 2^{-2k-2}] \subset [2^{-2k-2}, 2^{-2k}]$.
Then for $p = \infty$ our example is to let $\Delta_n = \{k2^{-n}\}_{k=1}^{2^n}$, and

$$f(x) = \sum_{k=1}^{\infty} f_k(x)$$

It is readily seen that $f(x)$ is approximated with order $O(\underline{\Delta}_n)$ by splines from $S_{2,1}(\Delta_n)$. Indeed, setting

$$s_n(x) = \sum_{k=1}^{\infty} \sum_{j=2k+2}^{\min(4k,n)} \varphi_j(x - 2^{-2k-1})$$

we have obviously $s_n \in S_{2,1}(\Delta_n)$ and for $x \in [2^{-2k-2}, 2^{-2k}]$:

$$|f(x) - s_n(x)| \le \sum_{\min(4k,n)+1}^{4k} \|\varphi_j(x)\|_\infty \le \sum_{n+1}^{\infty} 2^{-j} = 2^{-n}.$$

On the other hand, with $t = 2^{-4k}$ and $x = 2^{-2k-1}$ it follows

$$\omega_2(f,t)_\infty \ge |f(x+t) + f(x-t) - 2f(x)| = \sum_{2k+2}^{4k} |\varphi_j(t) + \varphi_j(-t) - 2\varphi_j(0)|$$

$$= 2t(4k - (2k+2)) \ge kt \ge Ct|\log t|$$

For $1 \le p < \infty$ we take $f(x) = \sum_{k=1}^{\infty} k^{-1/p} f_k(x)$ and as the approximating spline function

$$s_n(x) = \sum_{k=1}^{\infty} k^{-1/p} \sum_{2k+2}^{\min(n,4k)} \varphi_j(x - 2^{-2k-1}).$$

On each subinterval $I_k = [2^{-2k-2}, 2^{-2k}]$ $f(x)$ coincides with $k^{-1/p} f_k(x)$ so that

$$\Big\{\int_{I_k}|f(x)-S_n(x)|^p dx\Big\}^{1/p} = \begin{cases} \int_{I_k} 0\,dx = 0\,, & n \ge 4k \\ \le k^{-1/p}\sum_{n+1}^{4k}\Big(\int_{I_k}|\varphi_j(x)|^p dx\Big)^{1/p}, & 2k+2 \le n < 4k \\ \le k^{-1/p}\sum_{2k+2}^{4k}\Big(\int_{I_k}|\varphi_j(x)|^p dx\Big)^{1/p}, & n < 2k+2 \end{cases}$$

$$\le \begin{cases} 0, & n \ge 4k \\ 2^{1/p}k^{-1/p}\sum_{n+1}^{4k} 2^{-j(1+1/p)} \le k^{-1/p}2^{-n(1+1/p)}, & 2k+2 \le n < 4k \\ 2^{1/p}\,k^{-1/p}\,2^{-2k(1+1/p)}, & n > 2k+2 \end{cases}$$

because each φ_j has support on an interval of length 2^{-j+1}. It follows that

$$\int_0^1 |f(x)-S_n(x)|^p dx \le 2\sum_{\frac{n}{4}<k\le\frac{n}{2}} k^{-1}\,2^{-n(p+1)} + \sum_{2k>n-2} k^{-1}2^{-2k(p+1)}$$

$$\le 2\cdot 2^{-n(p+1)} + 2^{-(n-2)(p+1)}$$

$$\le (2+4^{p+1})\,2^{-n(p+1)}\,,$$

so that the approximation order of f is $O(2^{-n(1+1/p)}) = O(\underline{\Delta}_n^{1+1/p})$.

On the other hand, we estimate for $t = 2^{-4n-3}$

$$\int_0^1|f(x+t)+f(x-t)-2f(x)|^p dx \ \ge\ \sum_{k=1}^{n} k^{-1}\int_{I_{k,t}}|f_k(x+t)+f_k(x-t)-2f_k(x)|^p dx$$

where $I_{k,t} = [2^{-2k-1}-t/2,\ 2^{-2k-1}+t/2]$ is contained in the support of $f_k(x\pm t)$ and $f_k(x)$, and even in the support of each $\varphi_j(x\pm t)$, $\varphi_j(x)$ in (4.1). It follows that

$$\int_{I_{k,t}}|f_k(x+t)+f_k(x-t)-2f(x)|^p dx = \int_{-t/2}^{t/2}\Big|\sum_{j=2k+2}^{4k}\big[\varphi_j(x+t)+\varphi_j(x-t)-2\varphi_j(x)\big]\Big|^p dx$$

$$= \int_{-t/2}^{t/2} \Big| \sum_{j=2k+2}^{4k} \Big[-2|x|+|x-t|+|x+t|\Big] \Big|^p dx =$$

$$= (4k-2k-2)^p \int_{-t/2}^{t/2} |2(t-|x|)|^p dx$$

Hence we have

$$\int_0^1 |f(x+t)-f(x-t)-2f(x)|^p dx \geq C \sum_{k=1}^{n} k^{-1} k^p t^{p+1} \geq C\, n^p t^{p+1}$$

and consequently $\omega_2(f,t)_p \geq C\, t^{1+1/p} |\log t|$, with C not depending on t.

This shows that the logarithm term in part b) of the corollary cannot be dropped when $\alpha = \theta = \rho+1+1/p$.

Concerning part c), an approximation order $E_{n,\rho}(f) = O(\underline{\Delta}_n^{\alpha})$, $\alpha > \theta$, cannot imply $\omega_r(t,f) = O(t^{\beta})$ for some $\beta > \theta$ since then e.g. for nested sequences of knot sets

$$\mathcal{S}_{r,\rho}(\Delta_k) \subset \mathrm{Lip}\,(\beta,r,p) \subset C^{\rho+1}\,[a,b]$$

which is a contradiction. However at least in the case of nested sequences there are nontrivial functions (i.e. not splines) for which the approximation order may be $O(\overline{\Delta}_n^{\beta})$ for any $\beta > \theta$. The characterization of these classes of functions is an open problem. The corollary shows that they cannot be described in terms of moduli of continuity. In any case they must be contained in $C^{\rho}[a,b]$ or $W_{\infty}^{\rho+1}(a,b)$ but classical smoothness will not always increase the order of approximation, in particular there is the barrier $W_p^r(a,b)$ for the order $O(\underline{\Delta}_n^{r})$ given by a weak saturation theorem (see [1], [10]).

Finally we wish to discuss the assumption (3.18) on the sequences of knot sets. If we drop it we cannot further conclude assertion a) of the corollary (3.21) and only (3.17) holds. Thus, if $\overline{\lim_{n\to\infty}}\ \overline{\Delta}_n / \underline{\Delta}_n \longrightarrow \infty$, this inequality will give less smoothness (in term of the modulus of continuity). This is in agreement with results on best approximation by splines with optimal knots, where the sequences of partitions may depend on the function to be approximated and only the number of knots $n(1/h \leq \bar{\Delta}/(b-a))$ is prescribed. According to Rice [8], Burchard-Hale [2] even functions with singularities are approximated with order n by piecewise linear splines with optimal knots. In this case, the optimal partitions (meshes) corresponding to the best approximation cannot have uniformly bounded mesh ratio.

On the other hand, by our inverse results we may conclude that for all sequences of partitions with uniformly bounded mesh ratios the order of approximation is the same. Thus in this case the approximation by splines with optimal meshes has no essential advantage over approximation with a priori given sequence of partitions (satisfying (3.18)). According to Burchard-Hale this is e.g. the case when the function to be approximated has a n-th derivative which does not "oscillate too much".

References

1. Ahlberg, J.H., E.N. Nilson and J.L. Walsh: The Theory of Splines and Their Applications, Academic Press, New York 1967.

2. Burchard, H.G. and D.F. Hale: Piecewise polynomial approximation on optimal meshes. J.Approximation Theory 14 (1975), 128-147.

3. Butler, G. and F. Richards: An L_p-saturation theorem for splines, Canad.J.Math 24 (1972), 957-966.

4. De Vore, R.: Degree of Approximation, to appear in the Proceedings of the Symposium on Approximation Theory at the University of Texas, Austin 1976.

5. De Vore, R. and F. Richards: Saturation and inverse theorems for spline approximation. Spline Functions Approx. Theory, Proc.Symp.Univ.Alberta, Edmonton 1972, ISNM21 (1973), 73-82.

6. Johnen, H. and K. Scherer: Direct and inverse theorems for best approximation by Λ - splines. In "Spline Functions", Proc. Symp. Karlsruhe 1975, Springer Lecture Notes in Math. 501, New York 1976, pp. 116-131.

7. Nitsche, J.: Umkehrsätze für Spline Approximation, Compositio Math. 21 (1970), 400-416.

8. Rice, J.R.: On the degree of nonlinear spline approximation. In "Approximation with Special Emphasis on Spline Functions", Academic Press, New York 1969, pp. 349-369.

9. Scherer, K.: Über die beste Approximation von L_p-Funktionen durch Splines. Proc. of Conference on "Constructive Function Theory", Vama 1970, pp. 277-286.

10. Scherer, K.: Some inverse theorems for best approximation by Λ - splines, to appear in the Proceedings of the Symposium on Approximation Theory at the University of Texas, Austin 1976.

11. Timan, A.F.: Theory of Approximation of Functions of a Real Variable. Pergamon Press, New York 1963.

R. De Vore
Department of Mathematics
Oakland University
Rochester, Michigan 48063

K. Scherer
Inst. für Angew. Mathematik
Universität Bonn

ABSCHÄTZUNGEN DURCH STETIGKEITSMODULI BEI FOLGEN VON LINEAREN FUNKTIONALEN

HENNING ESSER (AACHEN)

Es sei $C[a,b]$ $(|a|, |b| < \infty)$ der lineare Raum der auf dem Intervall $[a,b]$ definierten reellwertigen und stetigen Funktionen versehen mit der Max-Norm $\|\cdot\|_C$. $C^m[a,b]$ $(m \geq 1, m \in \mathbb{N})$ bezeichnet den entsprechenden Raum der m mal stetig diffbaren Funktionen auf $[a,b]$ mit der Norm

$$(1) \quad \|g\|_{C^m} = \max\{|g(a)|, |g'(a)|, \ldots, |g^{(m-1)}(a)|, \|g^{(m)}\|_C\}$$

$(g \in C^m)$. C^*, bzw. C^{m*} $(m \geq 1, m \in \mathbb{N})$ sei der duale Raum von C, bzw. C^m mit Norm $\|\cdot\|_{C^*}$, bzw. $\|\cdot\|_{C^{m*}}$. Die Konvergenzgeschwindigkeit von Folgen von Funktionalen beschreiben wir für $f \in C$ durch Stetigkeitsmoduli höherer Ordnung $\omega_m(t,f)$ $(t > 0, m \geq 1, m \in \mathbb{N})$ definiert durch

$$(2) \qquad \omega_m(t,f) = \sup\{|(\Delta_h^m f)(x)|;\quad x, x+mh \in [a,b], |h| \leq t\} \quad ,$$

wobei Δ_h^m die m-te Differenz mit Schrittweite h bezeichnet. - Das Hauptergebnis dieser Note ist enthalten in

<u>Satz 1</u> Sei $\{f_n^*\}_{n=1}^{\infty}$, $f_n^* \in C^*$ $(n = 1,2,\ldots)$, eine Folge von linearen beschränkten Funktionalen, und $f^* \in C^*$ ein gegebenes Funktional. Dann existiert zu jedem $m \geq 1$ $(m \in \mathbb{N})$ eine Konstante $c(m) > 0$, so daß für $f \in C$ und $n = 1,2,\ldots$ gilt

$$(3) \quad |f_n^*(f) - f^*(f)| \leq c(m)\,(1 + \|f_n^*\|_{C^*} + \|f^*\|_{C^*}) \times$$

$$\times \left(\sum_{i=0}^{m-1} |(f_n^* - f^*)(\frac{(\cdot - a)^i}{i!})| \; \|f\|_C + \omega_m(\|f_n^* - f^*\|_{C^{m*}}^{1/m};\; f) \right)$$

<u>Beweis:</u> Wir setzen $R_n = f_n^* - f^*$ $(n = 1,2,\ldots)$. Dann gilt für $f \in C$ und $g \in C^m$ $(m \geq 1)$

(4) $|R_n(f)| \leq \|R_n\|_{C^*} \|f-g\| + \|R_n\|_{C^{m*}} \|g\|_{C^m}$.

Nach dem Riesz (-Sard) Darstellungssatz ([13] S. 139) ist für $g \in C^m$

(5) $$R_n(g) = \sum_{i=0}^{m-1} g^{(i)}(a)\, C_i^{(n)} + \int_a^b g^{(m)}(t)\, d\alpha_n(t) \quad , \quad \text{mit}$$

(6) $$C_i^{(n)} = R_n\left(\frac{(\cdot - a)^i}{i!}\right) \quad (i = 0,1,\ldots m-1) \quad \text{und} \quad \alpha_n \in NBV\,[a,b]^{*)} \quad .$$

$(n = 1,2,\ldots)$

Ferner gilt mit

(7) $$C_m^{(n)} = V_a^b \alpha_n \qquad (n = 1,2,\ldots)$$

die Gleichung

(8) $$\|R_n\|_{C^{m*}} = \sum_{i=0}^{m} |C_i^{(n)}| \quad .$$

Damit erhalten wir aus (4)

(9) $$|R_n(f)| \leq \|R_n\|_{C^*} \|f-g\|_C + \sum_{i=0}^{m-1} |C_i^{(n)}| \max_{0 \leq i \leq m-1} \|g^{(i)}\|_C + |C_m^{(n)}|\, \|g^{(m)}\|_C$$

$(n = 1,2,\ldots)$.

Wegen der Ungleichung $\|g^{(i)}\|_C \leq C_0\{\|g\|_C + \|g^{(m)}\|_C\}$ ($C_0 > 0$ eine Konstante; $0 \leq i \leq m-1$) folgt dann mit einer Konstanten $C_1 > 0$ aus (9)

(10) $$|R_n(f)| \leq C_1\{\|R_n\|_{C^*} \|f-g\|_C + \sum_{i=0}^{m-1} |C_i^{(n)}|\, \|f-g\|_C + \sum_{i=0}^{m-1} |C_i^{(n)}|\, \|f\|_C + \sum_{i=0}^{m} |C_i^{(n)}|\, \|g^{(m)}\|_C\}$$

$$\leq C_2\{\|R_n\|_{C^*}\|f-g\|_C + \|R_n\|_{C^*}\|f-g\|_C + \sum_{i=0}^{m-1} |C_i^{(n)}|\, \|f\|_C + \|R_n\|_{C^{m*}}\, \|g^{(m)}\|_C\}$$

*) NBV = normalisierte Funktionen von beschränkter Variation

wegen (8) und (6). Daher ergibt sich mit einer Konstanten $C_3 > 0$

$$(11)\quad |R_n(f)| \le C_3(1 + \|R_n\|_{C^*}) \left\{ \sum_{i=0}^{m-1} |c_i^{(n)}| \, \|f\|_C + \right.$$

$$\left. + \|f-g\|_C + \|R_n\|_{C^{m*}} \|g^{(m)}\|_C \right\} \quad (n = 1,2,\ldots) \quad .$$

Da $g \in C^m$ in (11) beliebig ist, erhalten wir aus (11) mit dem Peetreschen K-Funktional (z.B. [3])

$$(12)\quad K(t,f;\, C^m,C) = \inf_{g \in C^m} \{\|f-g\|_C + t\|g^{(m)}\|_C\} \qquad (t \ge 0)$$

die Abschätzung

$$(13)\quad |R_n(f)| \le C_3(1 + \|R_n\|_{C^*}) \left\{ \sum_{i=0}^{m-1} |c_i^{(n)}| \, \|f\|_C + \right.$$

$$\left. + K(\|R_n\|_{C^{m*}},\, f;\, C^m,C) \right\} \quad (n = 1,2,\ldots) \quad .$$

Nun ist bekannt, daß das K-Funktional (12) im wesentlichen der m-te Stetigkeitsmodul ist (s. z.B. [9]): Es gibt Konstanten $C_4, C_5 > 0$ mit

$$(14)\quad C_4\omega_m(t,f) \le K(t^m,f;\, C^m,C) \le C_5\omega_m(t,f) \qquad (t > 0) \quad .$$

Setzt man dies in (13) ein, so ergibt sich die Behauptung.

Als eine Anwendung von Satz 1 betrachten wir nun Folgen von positiven Funktionalen. Ein Funktional $f^* \in C^*$ heißt positiv, falls aus $f \in C$, $f(t) \ge 0$ $(t \in [a,b])$ $f^*(f) \ge 0$ folgt. - Offensichtlich ist für $f \in C^m$ $\|f_n^* - f^*\|_{C^{m*}}$ die beste Konstante M in der Abschätzung $|f_n^*(f) - f^*(f)| \le M\|f\|_{C^m}$.

Für positive Funktionale gilt

<u>Satz 2</u> ([6]) Es sei $\{f_n^*\}_{n=1}^{\infty}$, f^* eine Folge von positiven Funktionalen aus C^*. Dann ist

$$(15)\quad \lim_{n\to\infty} f_n^*(f) = f^*(f) \quad (f \in C)$$

genau dann, falls

(16) $$\lim_{n\to\infty} \|f_n^* - f^*\|_{C^{m*}} = 0 \qquad m = 1,2,\ldots .$$

Beweis: Wegen $\|f_n^*\|_{C^*} = |f_n(1)|$ ergibt sich aus (16) leicht (15). Um die Umkehrung zu beweisen, sei $f \in C^1$. Dann gilt mit dem Riesz-Darstellungssatz $f_n^*(f) - f^*(f) = \int_a^b f(t)d(\alpha_n(t) - \alpha(t))$, wobei α_n bzw. $\alpha \in NBV$ die erzeugenden Funktionen von f_n^* bzw. f^* sind. Wegen (15) gilt aber (s. [12] S. 112) $\lim_{n\to\infty} \alpha_n(b) = \alpha(b)$, $\lim_{n\to\infty} \alpha_n(t) = \alpha(t)$ f.ü. in $[a,b]$ und $|\alpha_n(t)| \leq M$ ($t \in [a,b]$, $n = 1,2,\ldots$). Daher folgt durch partielle Integration

$$\begin{aligned}|f_n^*(f) - f^*(f)| &\leq \|f\|_C \, |\alpha_n(b) - \alpha(b)| + \|f'\|_C \int_a^b |\alpha_n(t) - \alpha(t)|dt \\ &\leq C_6(|\alpha_n(b) - \alpha(b)| + \int_a^b |\alpha_n(t) - \alpha(t)|dt)\|f\|_{C^1} ,\end{aligned}$$

woraus (16) für $m=1$ folgt. Wegen (1) folgt dann aber die Aussage für $m > 1$.

Speziell betrachten wir nun Folgen von linearen positiven Operatoren $\{\phi_n\}_{n=1}^\infty$ ($\phi_n \neq I$, $n = 1,2,\ldots$), die $C[a,b]$ in sich abbilden. Ferner seien dann für festes $t \in [a,b]$ positive Funktionale $\phi_{n,t}^*$ ($\in C^*$) definiert durch

(17) $$\phi_{n,t}^*(f) = \phi_n(f;t) \qquad (f \in C;\ n = 1,2,\ldots) ,$$

und die Punktfunktionale mit f_t^* bezeichnet.
Mit $f_i(t) = t^i$ $(i = 0,1,2)$ setzen wir voraus, daß

(18) $$\phi_n(f_i, t) = f_i(t) \qquad (t \in [a,b];\ i = 0,1;\ n = 1,2,\ldots)$$

gilt. Bekanntlich kann ein solcher Approximationsprozeß nicht für f_2 exakt sein (z.B. [4] S. 71). Mit Satz 1 erhalten wir dann punktweise Abschätzungen durch den zweiten Stetigkeitsmodul, und nach Satz 2 ist die Konvergenz für jede stetige Funktion äquivalent mit $\lim_{n\to\infty} \|\phi_{n,t}^* - f_t^*\|_{C^{2*}} = 0$. Wegen (6) und (8) muß dies aber nach dem Satz von Bohmann-Korovkin äquivalent mit der Konvergenz für $f_2(t) = t^2$ sein. Es gilt sogar die folgende Aussage, die man auch aus einer Arbeit von G. Mühlbach erhält ([10]).

Lemma 1 ([6]) Unter der Voraussetzung (18) ist

$$(19)\qquad \|\phi^{*}_{n,t} - f^{*}_{t}\|_{C^{2*}} = \frac{1}{2}\,(\phi_n(f_2,t) - f_2(t)) \qquad (t \in [a,b];\ n = 1,2,\ldots)$$

Damit erhält man nun einfach

Satz 3 ([6]) Sei $\{\phi_n\}_{n=1}^{\infty}$ eine Folge von positiven linearen Abbildungen von $C[a,b]$ in sich, die die Beziehung (18) erfüllen. Dann existiert eine Konstante $C > 0$, so daß für $f \in C[a,b]$ und $n = 1,2,\ldots$ gilt

$$(20)\qquad |\phi_n(f;t) - f(t)| \leq C\,\omega_2\left(\left\{\frac{1}{2}\,(\phi_n(f_2;t) - f_2(t))\right\}^{1/2}; f\right) \qquad t \in [a,b])$$

Entsprechende Abschätzungen durch den ersten Stetigkeitsmodul findet man z.B. zusammengestellt von R. de Vore in [5]. Normabschätzungen der Form (20) mit noch einem Zusatzterm auf der rechten Seite von (20) findet man bei G. Freud ([8]). Für Bernsteinpolynome allerdings haben H. Berens und G. Lorentz ([2]) die Abschätzung (20) bewiesen. Ferner ist noch zu bemerken, daß durch (19) die Saturationsordnung gegeben ist (vergl. hierzu [1], [11]).

LITERATUR

[1] Berens, H.: Pointwise Saturation, in "Spline Functions and Approximation Theory", Proceedings of the Symposium hold at the University of Alberta 1972, 11 - 30, Birkhäuser, Basel 1973.

[2] Berens, H. u. G.G. Lorentz: Inverse theorems for Bernstein polynomials, Indiana Univ. Math. J. 7 (1972), 693 - 708.

[3] Butzer, P.L. u. H. Berens: Semi-Groups of Operators and Approximation, Springer, Berlin 1967.

[4] Cheney, E.W.: Introduction to Approximation Theory, McGraw Hill, New York 1966.

[5] de Vore, R.A.: The Approximation of Continuous Functions by Positive Linear Operators, Lecture Notes in Math. 293, Springer, Berlin 1972.

[6] Esser, H.: Über Konvergenzordnungn diskreter Approximationen, Habilitationsschrift, Aachen 1974.

[7] Esser, H.: On pointwise convergence estimates for positive linear operators on $C[a,b]$, erscheint in Proceedings of the Neth. Acad. of Science.

[8] Freud, G.: On approximation by positive linear methods II, Stud. Sci. Math. Hung. 3 (1968), 365 - 370.

[9] Johnen, H.: Inequalities connected with the moduli of smoothness, Mat. Vesnik 9 (1972), 289 - 303.

[10] Mühlbach, G.: Operatoren vom Bernsteinschen Typ, J. Approx. Theory 3 (1970), 274 - 292.

[11] Mühlbach, G.: Some remarks on pointwise saturation, in "Approximation Theory" (G.G. Lorentz ed.) 433 - 440, Academic Press, New York 1973.

[12] Riesz, F. u. Sz. B. Nagy: Vorlesungen über Funktionalanalysis, Deutscher Verlag der Wissenschaften, Berlin 1956.

[13] Sard, A.: Linear Approximation, Math. Surv. AMS 9 (1963)

Optimale Approximation von linearen Funktionalen auf Klassen periodischer Funktionen

Wilhelm Forst und Mary Mikhail*

1. Einleitung

In diesem Aufsatz betrachten wir Funktionenklassen M des Raumes $C_{2\pi}$ der stetigen reellwertigen 2π-periodischen Funktionen, deren Approximierbarkeit

$$\delta(M,P_n) = \sup_{f\in M} \inf_{g\in P_n} |f-g|$$

durch trigonometrische Polynome vom Grade $\leq n$ aufgrund von Arbeiten von Achieser, Favard u.a. (vgl. [6,Kapitel 8.3 und 8.5]) explizit bekannt ist. Genauer sind dies

$M=H_\beta$, $\beta>0$, die Menge aller $f\in C_{2\pi}$, die sich holomorph in den Streifen $S_\beta:=\{x+iy \mid |y|<\beta\}$ fortsetzen lassen und dort die Ungleichung $|\mathrm{Re}\, f(z)|\leq 1$ erfüllen.

Auf die Klassen

$M=W_r$, $r\geq 1$, die Menge aller $f\in C_{2\pi}$ mit f.ü. existenter r-ter Ableitung, deren L^∞-Norm ≤ 1 ist,

soll an anderer Stelle eingegangen werden.

Gegenstand der vorliegenden Untersuchungen sind Fragen der optimalen Integration und Interpolation bezüglich dieser Klassen H_β und damit zusammenhängende Probleme: Zu vorgegebenen Knoten $x_0<x_1<\ldots<x_{n-1}<x_0+2\pi$ mit den Vielfachheiten $v_0,\ldots,v_{n-1}\geq 1$ und $\varphi\in C_{2\pi}^*$, wobei entweder

$\varphi=\hat{x}$, d.h. $\varphi(f)=f(x)$ für $f\in C_{2\pi}$ $(x\in\mathbb{R}$ fest$)$,

oder

---

* Über die Arbeit wurde vom erstgenannten Autor vorgetragen.

$$\varphi(f)=\frac{1}{2\pi}\int_0^{2\pi} f(x)dx \quad \text{für } f\in C_{2\pi}$$

gilt, sind Gewichte $a=(\alpha_{k,j} | 0\leq k< n,\ 0\leq j< v_k)$ so zu bestimmen, daß für das Fehlerfunktional

$$(1) \qquad R_a := \varphi - \sum_{k=0}^{n-1} \sum_{j=0}^{v_k-1} \alpha_{k,j}\, \hat{x}_k D^j$$

die Norm $|R_a|_\beta := \sup_{f\in H_\beta} R_a(f)$ minimal wird. Diese Optimierungsaufgaben stehen in enger Beziehung zum Ideenkreis von Sard [4] und Schoenberg [5] . Im Gegensatz dazu führt hier die Wahl der Funktionenklassen H_β auf Probleme der L^1-Approximation und wirft damit neue Fragen hinsichtlich der expliziten Bestimmung und Eindeutigkeit des Proximums auf.

Den ersten Anstoß zu den hier angesprochenen Problemen gab eine Arbeit von Schönhage [7] , in der die optimale Integration bei äquidistanten einfachen Knoten $x_k=k\cdot\frac{2\pi}{n}$ behandelt wurde. Die optimale Lagrange-Interpolation hat Mikhail [3] in ihrer Dissertation untersucht. In dieser Arbeit nun werden Teile der Resultate von Mikhail [3] auf die Hermite-Interpolation ausgedehnt. Die von Schönhage [7] bestimmte optimale Quadraturformel läßt sich dann interpolatorisch deuten. Weiter gelingt es mit Hilfe von [2] , den Nachweis der Eindeutigkeit für die L^1-Proxima zu vereinfachen.

Die Untersuchungen ergeben außerdem für n=2N einen interessanten Zusammenhang der optimalen Lagrange-Interpolation mit der Bestimmung von n-Breiten

$$d_n(H_\beta) = \inf \{\delta(H_\beta,U)|\ U\subseteq C_{2\pi}\ ,\ \dim U\leq n\}$$

und extremalen n-dimensionalen Unterräumen. Für diesen Fall erhält man die Optimalität äquidistanter Knoten. Genaueres darüber findet man in [1] ausgeführt.

2. Optimale Interpolation

Wichtig für die folgenden Überlegungen ist die Integraldarstellung

$$(2)\qquad f(z) = \frac{1}{2\pi}\int_0^{2\pi} K_\eta(t-z)\,\mathrm{Re}\, f(t+i\eta)\,dt \qquad (\eta<\beta,\ z\in S_\eta)$$

für Funktionen $f\in H_\beta$ (vgl. [6,p. 197]) ; dabei ist

$$K_\eta(z) = 1+2\sum_{k=1}^{\infty} \frac{\cos(kz)}{\mathrm{ch}(k\eta)} \qquad (z\in S_\eta) \quad .$$

K_η läßt sich fortsetzen zu einer elliptischen Funktion der Ordnung 2 mit den primitiven Perioden 2π und $4i\eta$ (vgl. [3,Satz 1.1]) .

Analog zu [7] bzw. [3,Satz 2.1] erhält man für $\varphi=\hat{x}$ mit Hilfe von (2) den folgenden

Satz 1. Die in (1) definierten Fehlerfunktionale R_a haben die Norm

$$(3)\qquad |R_a|_\beta = \frac{1}{2\pi}\int_0^{2\pi} \Big|K_\beta(t-x)-\sum_{k=0}^{n-1}\sum_{j=0}^{v_k-1} \alpha_{k,j}K_\beta^{(j)}(t-x_k)\Big|\,dt \quad .$$

Satz 1 führt uns auf das L^1-Approximationsproblem, $K_\beta(\cdot-x)$ bezüglich des Unterraumes $U:=\mathrm{span}\{K_\beta^{(j)}(\cdot-x_k)\,|\ 0\le k<n,\ 0\le j<v_k\}$ zu approximieren. Die Existenz eines Proximums ist klar. Bei der Beantwortung der Frage nach der Eindeutigkeit des Proximums kann man ausnutzen, daß K_β ein variationsmindernder Kern ist (vgl. [2] und [3,Satz 3.3]) . Im Falle $\sum_{k=0}^{n-1} v_k = 2N+1$ existiert genau ein Proximum, denn U ist dann ein Haarscher Raum. Nach Überlegungen von Mikhail [3,Kapitel 5] scheint die explizite Bestimmung des Proximums jedoch schwierig zu sein.

Im Falle $\sum_{k=0}^{n-1} v_k = 2N$ ist U kein Haarscher Raum, und wir müssen die Eindeutigkeit auf anderem Wege nachweisen. Wesentliches Hilfsmittel dafür ist

Satz 2. Es existiert ein lineares Funktional $\Psi\in(L^1_{2\pi})^*$ mit $|\Psi|=1$, welches orthogonal zu U ist und eine Darstellung

(4) $$\Psi(f) = \frac{1}{2\pi}\int_0^{2\pi} f(t)\sigma(t)\,dt \qquad \underline{\text{für}}\ f\in L^1_{2\pi}$$

besitzt; dabei ist σ eine Treppenfunktion mit den Werten ± 1 , die pro Periode (dem Maße nach) genau 2N Zeichenwechsel hat. Solch ein lineares Funktional ist bis aufs Vorzeichen eindeutig bestimmt.

Korollar. Für alle $x\in\mathbb{R}$ gilt

$$|\Psi(K_\beta(\cdot - x))| = \inf_{g\in U} \frac{1}{2\pi}\int_0^{2\pi} |K_\beta(t-x)-g(t)|\,dt \quad .$$

Wir wählen zum Beweis von Satz 2 ein festes $x\in\mathbb{R}$, $x\neq x_k \bmod 2\pi$. Ist $g\in U$ ein Proximum zu $K_\beta(\cdot - x)$, so setzen wir $\sigma(t):=\mathrm{sign}(K_\beta(t-x)-g(t))$. Nach [2,Korollar 2] hat σ pro Periode höchstens 2N Zeichenwechsel, und das gemäß (4) definierte Funktional Ψ gehört zu $U^\perp$. Wir zeigen jetzt, daß σ pro Periode genau 2N Zeichenwechsel hat. Dazu definieren wir G per

$$G(z) = \int_0^{2\pi} K_\beta(t-z)\sigma(t)\,dt \qquad (z\in S_\beta) \quad .$$

Für diese Funktion gilt

$$G^{(j)}(x_k)=0 \qquad (0\leq k<n,\ 0\leq j<v_k) \quad ,$$

und so besitzt G pro Periode mindestens 2N Nullstellen im Streifen S_β . Nach [2,Satz 1] hat G keine weitere Nullstelle, denn sonst müßte σ mehr als 2N Zeichenwechsel haben; entsprechend hat σ nicht weniger als 2N Zeichenwechsel, denn sonst könnte G höchstens 2N-2 Nullstellen haben. Also hat σ genau 2N Zeichenwechsel.

Seien nun Ψ_1 und Ψ_2 zwei Funktionale aus $U^\perp$ mit Darstellungen

$$\Psi_j(f) = \frac{1}{2\pi}\int_0^{2\pi} f(t)\sigma_j(t)\,dt \qquad (j=1,2;\ f\in L^1_{2\pi}) \quad ,$$

wobei σ_1 und σ_2 Treppenfunktionen mit den Werten ± 1 sind, die pro Periode genau 2N Zeichenwechsel haben. Wir definieren dann Funktionen G_1 und G_2 durch

$$G_j(z) = \int_0^{2\pi} K_\beta(t-z)\sigma_j(t)\,dt \qquad (j=1,2;\ z\in S_\beta) .$$

G_1 und G_2 haben die gleichen Nullstellen mit gleicher Vielfachheit. Ohne Einschränkung können wir deshalb

$$\operatorname{sign} G_1(t) = \operatorname{sign} G_2(t) \quad \text{für } t\in\mathbb{R}$$

voraussetzen. Im Folgenden zeigen wir $G_1=G_2$, was gleichbedeutend zu $\Psi_1=\Psi_2$ ist. Zum indirekten Beweis nehmen wir an, es existiere $t_o\in\mathbb{R}$ mit $G_2(t_o)>G_1(t_o)>0$. $\lambda:=G_1(t_o)/G_2(t_o)$ erfüllt dann die Ungleichung $0<\lambda<1$, und die Funktion $G:=G_1-\lambda G_2$ hat wegen $G(t_o)=0$ mindestens $2N+2$ Nullstellen. Nach [2,Satz 1] hat folglich $\sigma_1-\lambda\sigma_2$ mindestens $2N+2$ Zeichenwechsel. Andererseits hat $\sigma_1-\lambda\sigma_2$ genauso viele Zeichenwechsel wie σ_1 . Wegen dieses Widerspruchs muß $G_1=G_2$ sein.

Seien $t_o<t_1<\ldots<t_{2N-1}<t_o+2\pi$ die Sprungstellen der Treppenfunktionen σ aus Satz 2 . Ist dann

$$\sum_{k=o}^{n-1}\sum_{j=o}^{v_k-1} \alpha_{k,j}K_\beta^{(j)}(\cdot-x_k)$$

ein L^1-Proximum zu $K_\beta(\cdot-x)$ in U , so verschwindet die Fehlerfunktion an den Stellen $t_o,\ldots,t_{2N-1}$. Folglich ist das lineare Gleichungssystem

$$(5)\qquad \sum_{k=o}^{n-1}\sum_{j=o}^{v_k-1} K_\beta^{(j)}(t_l-x_k)\,\gamma_{k,j} = K_\beta(t_l-x) \qquad (0\le l<2N)$$

mit den Unbekannten $\gamma_{k,j}$ für $x\in\mathbb{R}$ stets lösbar. Für eine Lösung $(\lambda_o,\ldots,\lambda_{2N-1})$ des homogenen Gleichungssystems

$$\sum_{l=o}^{2N-1} K_\beta^{(j)}(t_l-x_k)\,\lambda_l = 0 \qquad (0\le k<n,\ 0\le j<v_k)$$

gilt deshalb die Relation

$$\sum_{l=o}^{2N-1} \lambda_l K_\beta(t_l-x) = 0 \quad \text{für } x\in\mathbb{R} .$$

Daraus folgt $\lambda_o=\ldots=\lambda_{2N-1}=0$ wegen der linearen Unabhängigkeit der Funktionen $K_\beta(t_l-\cdot)$ $(0\le l<2N)$. Mithin ist die Koeffizientendeterminante des Gleichungssystems (5) ungleich Null.

Zusammenfassend gilt somit

Satz 3. Die optimalen Gewichte, die das Integral in (3) minimieren, sind eindeutig bestimmt. Sind $\alpha_{k,j}$ $(0\leq k<n,\ 0\leq j<v_k)$ die optimalen Gewichte für $x\in\mathbb{R}$, so erhalten wir per $w_{k,j}(x):=\alpha_{k,j}$ Funktionen $w_{k,j}$; für den von ihnen aufgespannten Unterraum U' gilt (wegen (5)) $U'=\operatorname{span}\{K_\beta(\cdot-t_j)\mid 0\leq j<2N\}$.

Zum Abschluß dieses Abschnitts betrachten wir zwei Spezialfälle:

1. Es sei $x_k=k\cdot\frac{\pi}{n}$, $v_k=1$ $(0\leq k<2n)$. Dann liefert die Treppenfunktion $\sigma(t):=\operatorname{sign}(\sin nt)$ gemäß (4) ein Funktional $\Psi\in(L^1_{2\pi})^*$, welches Satz 2 genügt. Folglich gilt $t_k=x_k$ $(0\leq k<2n)$. Weiter erhalten wir mit Hilfe des Korollars zu Satz 2 für den Interpolationsfehler

$$\sup_{f\in H_\beta}\left|f(x)-\sum_{k=0}^{2n-1}f(x_k)w_{k,o}(x)\right| = \left|\frac{1}{2\pi}\int_0^{2\pi}K_{n\beta}(t-nx)\operatorname{sign}(\sin t)\,dt\right|$$

$$= \left|\frac{4}{\pi}\sum_{m=0}^{\infty}\frac{\sin((2m+1)nx)}{(2m+1)\operatorname{ch}((2m+1)n\beta)}\right| ;$$

dieser nimmt den Maximalwert

$$\frac{4}{\pi}\sum_{m=0}^{\infty}\frac{(-1)^m}{(2m+1)\operatorname{ch}((2m+1)n\beta)} =: \Delta_{n-1}(\beta)$$

an. Der 2n-dimensionale Unterraum $U'=\operatorname{span}\{w_{k,o}\mid 0\leq k<2n\}$ ist also 2n-extremal, denn nach [1,Satz 1] gilt $d_{2n}(H_\beta)=\Delta_{n-1}(\beta)$. Somit läßt sich in diesem Falle der maximale Interpolationsfehler durch Variation der Knoten nicht mehr verkleinern, d.h. äquidistante Knoten sind optimal.

2. Sei nun $x_k=k\cdot\frac{2\pi}{n}$, $v_k=2$ $(0\leq k<n)$. Wir legen dann $\xi_\gamma\in(0,\frac{\pi}{2})$ eindeutig fest durch die Forderung

$$\int_0^{2\pi}K_\gamma(t)\,\sigma_\gamma(t)\,dt = 0 \quad ,$$

wobei σ_γ durch

$$\sigma_\gamma(t) := \begin{cases} -1 & \text{falls } 0\leq t\leq\xi_\gamma \\ 1 & \text{falls } \xi_\gamma<t<2\pi-\xi_\gamma \\ -1 & \text{falls } 2\pi-\xi_\gamma\leq t\leq 2\pi \end{cases}$$

definiert sei. In diesem Falle liefert die Treppenfunktion $\sigma(t) := \sigma_{n\beta}(nt)$ gemäß (4) ein Funktional $\Psi \in (L^1_{2\pi})^*$, welches Satz 2 genügt. Somit gilt $t_{2k} = x_k + \xi_{n\beta}/n$, $t_{2k+1} = x_{k+1} - \xi_{n\beta}/n$ $(0 \le k < n)$. Für den Interpolationsfehler ergibt sich

$$\sup_{f \in H_\beta} \left| f(x) - \sum_{k=0}^{n-1} (f(x_k) w_{k,0}(x) + f'(x_k) w_{k,1}(x)) \right| = \left| \frac{1}{2\pi} \int_0^{2\pi} K_{n\beta}(t-nx) \sigma_{n\beta}(t) dt \right| ;$$

dieser nimmt z.B. für $x = \frac{\pi}{n}$ den Maximalwert

$$\frac{2}{\pi} \int_{\xi_{n\beta}}^{\pi - \xi_{n\beta}} K_{n\beta}(t)\, dt = 2\, \Delta_{n-1}(\beta) + O(e^{-2n\beta})$$

an.

3. Optimale Integration

Wir approximieren jetzt das Funktional $\varphi \in C^*_{2\pi}$,

$$\varphi(f) = \frac{1}{2\pi} \int_0^{2\pi} f(x) dx \quad \text{für } f \in C_{2\pi} \quad .$$

Analog zu Satz 1 gilt

Satz 4. Die in (1) definierten Fehlerfunktionale R_a haben die Norm

$$|R_a|_\beta = \frac{1}{2\pi} \int_0^{2\pi} \left| 1 - \sum_{k=0}^{n-1} \sum_{j=0}^{v_k - 1} \alpha_{k,j}\, K_\beta^{(j)}(t - x_k) \right| dt \quad .$$

Ziel der weiteren Überlegungen ist die Charakterisierung der L^1-Proxima zu 1 in $U = \text{span}\{K_\beta^{(j)}(\cdot - x_k) \mid 0 \le k < n,\ 0 \le j < v_k\}$. Wir machen dazu die zusätzlichen Annahmen $x_k = k \cdot \frac{2\pi}{n}$ und $v_k = 2m$ $(m \ge 1)$ für $0 \le k < n$. Sei $g_0 \in U$ ein L^1-Proximum zu 1 mit der Darstellung

$$(6) \qquad g_0(t) = \sum_{k=0}^{n-1} \sum_{j=0}^{2m-1} \gamma_{k,j}\, K_\beta^{(j)}(t - x_k) \quad .$$

Dann sind die durch

$$(7)\qquad g_\nu(t) = \sum_{k=0}^{n-1} \sum_{j=0}^{2m-1} \gamma_{k+\nu,j}\, K_\beta^{(j)}(t-x_k) \qquad (\nu=1,\dots,n-1)\ ,$$

$\gamma_{k+n,j} := \gamma_{k,j}$, definierten Funktionen $g_1,\dots,g_{n-1}$ ebenfalls Proxima zu 1 . Folglich ist $g=(g_0+\dots+g_{n-1})/n$ ein L^1-Proximum zu 1 . Wegen $\frac{1}{n}\sum_{k=0}^{n-1} K_\beta(t-x_k)=K_{n\beta}(nt)$ hat g die Darstellung

$$(8)\qquad g(t) = \sum_{j=0}^{2m-1} n^j \gamma_j\, K_{n\beta}^{(j)}(nt)\ ,$$

wobei $\gamma_j=\sum_{k=0}^{n-1}\gamma_{k,j}$ gilt. Mit g(t) ist auch g(-t) Proximum zu 1 , also auch

$$(9)\qquad \overline{g}(t) = (g(t)+g(-t))/2 = \sum_{j=0}^{m-1} \gamma_{2j}\, n^{2j}\, K_{n\beta}^{(2j)}(nt)\ .$$

Bezeichnet σ die Treppenfunktion $\sigma(t):=\operatorname{sign}(1-\overline{g}(t))$, so gilt

$$(10)\qquad \int_0^{2\pi} K_\beta^{(j)}(t-x_k)\,\sigma(t)\,dt = 0 \qquad (0\le k<n,\ 0\le j<2m)\ ,$$

da $\overline{g}\in U$ Proximum zu 1 ist. σ hat genau 2mn Zeichenwechsel pro Periode. Denn einerseits hat σ wegen (10) und [2,Satz 1] mindestens 2mn Zeichenwechsel pro Periode; andererseits folgt aus [2,Korollar 2] und (9) , daß σ pro Periode nicht mehr als 2mn Zeichenwechsel haben kann. Damit genügt σ den Forderungen des Satzes 2 . Sind $t_0<t_1<\dots<t_{2mn-1}<t_0+2\pi$ die Sprungstellen von σ und beachten wir, daß

$$|1-\overline{g}(t)| = \frac{1}{2n}\sum_{\nu=0}^{n-1}(|1-g_\nu(t)|+|1-g_\nu(-t)|) \quad \text{für } t\in\mathbb{R}$$

gilt, so verschwindet $1-g_0$ an den Stellen $t_0,\dots,t_{2mn-1}$. Somit erfüllen die optimalen Gewichte $\gamma_{k,j}$ das Gleichungssystem

$$(11)\qquad \sum_{k=0}^{n-1}\sum_{j=0}^{2m-1} K_\beta^{(j)}(t_l-x_k)\,\gamma_{k,j} = 1 \qquad (0\le l<2mn)\ ,$$

dessen Koeffizientendeterminante aufgrund der Überlegungen in Abschnitt 2 von Null verschieden ist. Die $\gamma_{k,j}$ $(0\le k<n,\ 0\le j<2m)$ sind also eindeutig bestimmt, und es gilt $g_0=\dots=g_{n-1}=g=\overline{g}$. Daraus ergeben sich mit (6), (7), (8) und (9) die Beziehungen

$$\gamma_{2j+1} = 0 \quad (0\le j<m)\ ,$$
$$\gamma_{k,2j+1}=0 \quad (0\le k<n,\ 0\le j<m)$$

sowie

$$\gamma_{k,2j}=\gamma_{2j}/n \quad (0\leq k<n,\ 0\leq j<m) \quad .$$

Zusammenfassend gilt in Verallgemeinerung von [7,Satz 2]

Satz 5. Unter allen Fehlerfunktionalen der Form

$$R_a(f) = \frac{1}{2\pi}\int_0^{2\pi} f(t)dt - \sum_{k=o}^{n-1}\sum_{j=o}^{2m-1}\alpha_{k,j}f^{(j)}(x_k)$$

hat genau dasjenige minimale Norm

$$(12) \qquad |R_a|_\beta = \frac{1}{2\pi}\int_0^{2\pi} |1-\sum_{j=o}^{m-1}\gamma_j\, K_{n\beta}^{(2j)}(t)|dt \qquad (=O(e^{-mn\beta})) \quad ,$$

für das

$$\alpha_{k,2j}=\frac{\gamma_j}{n^{2j+1}} \quad , \quad \alpha_{k,2j+1}=0 \qquad (0\leq k<n,\ 0\leq j<m)$$

gilt; dabei sind $\gamma_o,\ldots,\gamma_{m-1}$ dadurch eindeutig bestimmt, daß sie das Integral (12) minimieren. Überdies ist die optimale Quadraturformel (wegen (11)) interpolatorisch vom Hermite-Typ.

Die vorangehenden Überlegungen wären in befriedigender Weise abgerundet, wenn wir hätten zeigen können, daß äquidistante Knoten als einzige optimal sind. Diese Vermutung scheint aber immer noch ein offenes Problem zu sein.

Literatur

1. Forst, W.: Über die Breite von Klassen holomorpher periodischer Funktionen, eingereicht bei J. Approx. Theory.
2. Forst, W.: Variationsmindernde Eigenschaften eines speziellen Kreinschen Kernes, Math. Z. 148, 67-70 (1976).
3. Mikhail, M.: Optimale Interpolation holomorpher periodischer Funktionen, Dissertation, Tübingen, 1976.
4. Sard, A.: Best approximate integration formulas; best approximation formulas, Amer. J. Math. 71, 80-91 (1949).
5. Schoenberg, I.J.: Spline interpolation and best quadrature formulae, Bull. Amer. Math. Soc. 70, 143-148 (1964).
6. Schönhage, A.: Approximationstheorie, Berlin-New York, 1971.
7. Schönhage, A.: Zur Quadratur holomorpher periodischer Funktionen, J. Approx. Theory 13, 341-347 (1975).

Wilhelm Forst, Mary Mikhail
Fachbereich Mathematik der Universität
Auf der Morgenstelle 10
D-7400 Tübingen

APPROXIMATION DURCH POLYNOME MIT GANZZAHLIGEN KOEFFIZIENTEN

Manfred v.Golitschek

Es sei $C_o[0,1]$ der Banachraum aller stetigen komplexwertigen Funktionen f aufdem abgeschlossenen Intervall $[0,1]$ mit $f(0)=f(1)=0$ und der Norm $\|f\| := \max\{|f(x)|\ ,\ 0 \leq x \leq 1\}$. Sei eine Folge $\Lambda = \{\lambda_k\}_{k=1}^{\infty}$ verschiedener komplexer Zahlen mit positiven Realteilen vorgegeben. Dann untersuchen wir die Frage, ob und wie gut die Funktionen f durch ganzzahlige Λ-Polynome Q approximiert werden können, i.e. durch Funktionen der Form

$$Q(x) = \sum_{j=1}^{s} b_j\, x^{\lambda_j} \quad , \quad b_j \in \mathbf{Z}+i\mathbf{Z} \quad ,$$

wobei **Z** die Menge der ganzen Zahlen bezeichne. Das wesentliche Ziel der vorliegenden Arbeit ist der vollständige Beweis des folgenden allgemeinen Satzes über die Dichtheit der ganzzahligen Λ-Polynome Q .

<u>Satz 1</u>: Es seien δ und M , $\delta > 0$, $0 < M \leq 1$, konstante Zahlen. Die Folge Λ verschiedener komplexer Zahlen erfülle die drei Bedingungen

(i) $|\lambda_k| \geq \delta > 0$, k=1,2,...

(ii) $\operatorname{Re} \lambda_k \geq M|\lambda_k|$, k=1,2,...

(iii) $\sum_{k=1}^{\infty} 1/|\lambda_k| = \infty$.

Dann ist die Menge der ganzzahligen Λ-Polynome Q dicht im Banachraum $C_o[0,1]$.

Ist $\Lambda = \{k\}_{k=1}^{\infty}$ die Folge der natürlichen Zahlen, so sind die Bedingungen (i)-(iii) erfüllt, und Satz 1 besagt, daß die ganzzahligen algebraischen Polynome dicht in $C_o[0,1]$ liegen. Dieses

Ergebnis scheint ursprünglich auf Kakeya [7] zurückzugehen. In den letzten Jahren wurde Satz 1 für wichtige Spezialfälle bewiesen : Ferguson [2] für reelle Folgen Λ im Approximationsintervall $[0,b]$, $0<b<1$; Martirosjan [8] für reelle Folgen Λ unter Zusatzbedingungen im Intervall $[0,1]$. Schließlich zeigten Ferguson-Golitschek [3] für Folgen Λ natürlicher Zahlen, daß die ganzzahligen Λ-Polynome Q dicht in $C_0[0,1]$ sind, wenn die Müntzsche Bedingung (iii) erfüllt ist. Doch ist den beiden letztgenannten Autoren schon seit längerem bekannt, daß die Aussage von Satz 1 zumindest für reelle Exponenten Λ richtig ist. Der nun folgende Beweis von Satz 1 beruht auf der Idee der Arbeit [3] . Doch werden wesentliche Teile sehr vereinfacht und verallgemeinert. Die Grundlage unseres Beweises bilden drei Lemmata, die wir zuerst herleiten werden.

<u>Lemma 1</u>: Für beliebige natürliche Zahlen q und s , $q<s$, gibt es ein Polynom $Q_{qs}(x) = \sum_{i=q+1}^{s} c_{iqs} x^{\lambda_i}$, c_{iqs} komplex , so daß

$$A_{qs} := \| x^{\lambda_q} - Q_{qs}(x) \| \leq 2 \prod_{i=q+1}^{s} \frac{|\lambda_q - \lambda_i|}{|\lambda_q + \bar{\lambda}_i|} ,$$

sowie $Q_{qs}(1) = 1$ erfüllt ist.

<u>Beweis</u>: Für reelle Exponenten λ_i bewies der Autor [4] die Existenz eines Polynoms $\tilde{Q}_{qs}(x) = \sum_{i=q+1}^{s} \tilde{c}_{iqs} x^{\lambda_i}$ mit

$$\| x^{\lambda_q} - \tilde{Q}_{qs}(x) \| \leq \prod_{i=q+1}^{s} \frac{|\lambda_q - \lambda_i|}{|\lambda_q + \bar{\lambda}_i|} .$$

Dieser Beweis ist ohne Schwierigkeiten auf komplexe Exponenten λ_i mit positiven Realteilen übertragbar. Wir setzen nun

$$Q_{qs}(x) := \tilde{Q}_{qs}(x) + (1-\tilde{Q}_{qs}(1)) x^{\lambda_s} .$$

Lemma 2: Seien r und s natürliche Zahlen , $r<s$. Für die komplexen Zahlen d_i , $r+1 \leq i \leq s$, gelte

$$\sum_{i=r+1}^{s} d_i = 0 \quad , \quad \Big|\sum_{i=k}^{s} d_i\Big| < \sqrt{2} \quad \text{für } k = r+2,\ldots,s \quad .$$

Dann ist

$$\Big\|\sum_{i=r+1}^{s} d_i\, x^{\lambda_i}\Big\| \leq 2\, M^{-1} \sum_{k=r+2}^{s} \big(|\lambda_k - \lambda_{k-1}|/|\lambda_k|\big) =: D_{rs} \; .$$

Beweis: Es ist

$$\Big\|\sum_{i=r+1}^{s} d_i\, x^{\lambda_i}\Big\| = \Big\|\sum_{k=r+2}^{s} (x^{\lambda_k} - x^{\lambda_{k-1}})\Big(\sum_{i=k}^{s} d_i\Big)\Big\| \leq \sqrt{2} \sum_{k=r+2}^{s} \|x^{\lambda_k} - x^{\lambda_{k-1}}\|$$

und

$$\|x^{\lambda_k} - x^{\lambda_{k-1}}\| \leq \sqrt{2}\, M^{-1}\, |\lambda_k - \lambda_{k-1}| / |\lambda_k| \quad .$$

Lemma 3: Seien r und s natürliche Zahlen , $r<s$. Zu jeder Funktion $f \in C_0[0,1]$ existieren Zahlen $b_j \in \mathbb{Z}+i\mathbb{Z}$, $1 \leq j \leq s$, so daß

$$\Big\|f(x) - \sum_{j=1}^{s} b_j\, x^{\lambda_j}\Big\| \leq 2\, E_s(f,\Lambda) + \sqrt{2} \sum_{q=1}^{r} A_{qs} + D_{rs} \quad ,$$

wobei A_{qs} in Lemma 1 , D_{rs} in Lemma 2 und $E_s(f,\Lambda)$ durch

$$E_s(f,\Lambda) := \min\Big\{\Big\|f(x) - \sum_{j=1}^{s} a_j\, x^{\lambda_j}\Big\| \;\Big|\; a_j \text{ komplex}\Big\}$$

definiert sind .

Beweis: In [3] wurde dieses Lemma für positive ganze Exponenten λ_j bewiesen. Dieselbe Methode zur Konstruktion der Zahlen b_j ist auch für komplexe Exponenten λ_j möglich, wenn wir (bei Verwendung der Bezeichnungen in [3])

$$b_{q+1} := [\mathrm{Re}\, a_{q+1,q}] + i\,[\mathrm{Im}\, a_{q+1,q}] \quad , \; q=1,\ldots,r \; ,$$

wählen und die Zahlen d_j , $j=r+1,\ldots,s$, nach [3, Formel (27)] getrennt für die Real- und Imaginärteile bestimmen und gemäß [3 , Seite 123 oben] $b_j := a_{jr} - d_j$, $j=r+1,\ldots,s$, setzen.

Wir beweisen nun zuerst den Satz 1 für den Fall, daß die Folge Λ einen endlichen Häufungspunkt λ^* mit $\operatorname{Re}\lambda^* > 0$ besitzt. Wir bezeichnen die Teilfolge von Λ, die gegen λ^* konvergiert, wieder mit $\{\lambda_k\}_{k=1}^{\infty}$. Wegen $\lim \lambda_k = \lambda^*$ für $k \to \infty$ existiert eine Zahl p, $0 < p < 1$, so daß für alle natürlichen Zahlen q und k $|\lambda_q - \lambda_k| / |\lambda_q + \overline{\lambda}_k| \leq 1-p$ erfüllt ist. Daher ist nach Lemma 1

$$A_{qs} \leq 2\,(1-p)^{s-q} \quad \text{für } q=1,\ldots,s-1 .$$

Zu vorgegebener Zahl $\varepsilon > 0$ wählen wir eine natürliche Zahl m so groß, daß $(1-p)^m \leq \varepsilon$ ist, und die natürlichen Zahlen r und $s := r+m$ so groß, daß

$$\sum_{k=r+2}^{s} |\lambda_k - \lambda_{k-1}| \leq \varepsilon \;, \quad |\lambda_k| \geq |\lambda^*|/2 \quad \text{für alle } k=r+2, r+3, \ldots,$$

und $E_s(f,\Lambda) \leq \varepsilon$ ist. Diese letzte Bedingung ist nach Szasz [9] erfüllbar. Aus Lemma 3 folgt dann die Existenz von Zahlen $b_j \in \mathbf{Z}+i\mathbf{Z}$, so daß

$$\Big\| f(x) - \sum_{j=1}^{s} b_j\, x^{\lambda_j} \Big\| \leq (2 + 2^{3/2}\, p^{-1} + 4(M|\lambda^*|)^{-1})\, \varepsilon$$

erfüllt ist. Somit ist Satz 1 vollständig bewiesen, falls Λ einen endlichen Häufungspunkt λ^* besitzt.

Wesentlich aufwendiger wird der Beweis von Satz 1 für allgemeine Folgen Λ, die keinen endlichen Häufungspunkt besitzen. Um auch bösartige (sprich: unregelmäßig wachsende) Folgen Λ einzuschliessen, leiten wir die drei folgenden Lemmata her. Das für uns entscheidende Resultat steht in Lemma 6 und ist eine Erweiterung von [3, Lemma 3], doch ist der hier durchgeführte Beweis wesentlich kürzer und durchsichtiger.

<u>Lemma 4</u>: Es sei $T = \{t_k\}_{k=1}^{\infty}$ eine monoton nicht fallende Folge positiver reeller Zahlen mit den Eigenschaften

(1) $\quad \sum_{k=1}^{\infty} 1/t_k = \infty$ und $t_k \geq k$ für alle $k=1,2,\ldots$.

Dann gibt es beliebig große natürliche Zahlen s , so daß

(2) $\qquad S(q+1,3s) \geq \sqrt{q}/t_q$ für alle $q=1,\ldots,s$

erfüllt ist, wobei die Funktion S durch $S(m,n) := \sum_{k=m}^{n} 1/t_k$ für $m \leq n$ definiert sei.

<u>Beweis</u>: Wir sagen, T genügt der Bedingung $B(m,n)$, $m < n$,falls $S(q+1,n) \geq \sqrt{q}/t_q$ für alle $q=1,\ldots,m$. Wir nehmen an, $B(s,3s)$ sei für alle genügend große natürliche Zahlen s verletzt, etwa für $s \geq s_o$. Dann definieren wir rekursiv für $j=0,1,\ldots$ die Zahl s_{j+1} als die kleinste natürliche Zahl, für die

$$s_{j+1} \geq 3s_j \quad \text{und } B(3s_j, 3s_{j+1}) \quad \text{erfüllt ist .}$$

Es ist $s_{j+1} > 3s_j$, da sonst im Widerspruch zur Annahme $B(s,3s)$ für $s=3s_j$ erfüllt wäre . Nach Definition von s_{j+1} ist folglich $B(3s_j , 3s_{j+1}-3)$ verletzt. Dann existiert eine natürliche Zahl q_j , $1 \leq q_j \leq 3s_j$, so daß

(3) $\qquad S(1+q_j , 3s_{j+1}-3) < \sqrt{q_j}/t_{q_j}$.

Für $j \geq 1$ ist andererseits $B(3s_{j-1} ,3s_j)$ erfüllt und folglich ist $3s_{j-1} < q_j \leq 3s_j$ .Dann ist wegen $q_{j+1} \leq 3s_{j+1}$ und (3)

$$S(1+q_j ,q_{j+1}) \leq S(1+q_j ,3s_{j+1}) < \sqrt{q_j}/t_{q_j} + 3/t_{q_j} < 4/\sqrt{s_{j-1}}$$

und

$$S(1+q_1 ,\infty) < 4 \sum_{j=1}^{\infty} 1/\sqrt{s_{j-1}} < +\infty$$

im Widerspruch zur Müntzschen Bedingung in (1) .

<u>Lemma 5</u>: Die Folge T erfülle (1). Dann gibt es beliebig große natürliche Zahlen u , so daß $t_{3u} \leq 6t_u$ und

(4) $\qquad S(q+1,3u) \geq \sqrt{q}/(7t_q)$ für alle $q=1,\ldots,u$

erfüllt ist.

<u>Beweis</u>: Es gibt beliebig große Zahlen u mit $t_{3u} \leq 6t_u$, da sonst

für genügend großes u

$$S(u+1,\infty) = \sum_{k=o}^{\infty} S(1+u3^k, u3^{k+1}) \leq 2u \sum_{k=o}^{\infty} 3^k/t_{u3^k} \leq 4u/t_u$$

im Widerspruch zur Müntzschen Bedingung in (1) .

Wir wählen nun eine genügend große natürliche Zahl s , für die B(s,3s) erfüllt ist, und nehmen das größte u mit $u \leq s$ und $t_{3u} \leq 6t_u$. Dann ist

$$S(3u+1,3s) \leq 6\, S(u+1,3u) \;,\; S(u+1,3s) \leq 7\, S(u+1,3u) \;,$$

und daher gilt (4) .

<u>Lemma 6</u>: Die Folge T erfülle (1). Dann gibt es zu jeder natürlichen Zahl m beliebig große natürliche Zahlen r und $s:=r+m$, so daß gilt :

$$(5) \qquad (t_s - t_r)/t_r \leq 30\ m/s \quad ,$$

sowie

$$(6) \qquad t_q\, S(q+1,s) \geq \begin{cases} \sqrt{q}/14 & \text{für alle } 1 \leq q < s/2 \\ (s-q)/6 & \text{für alle } s/2 \leq q \leq r \end{cases} \quad .$$

<u>Beweis</u>: Wir wählen ein genügend großes u aus Lemma 5 und bestimmen r so, daß $2u \leq r \leq 3u-m$ und

$$t_{r+m} - t_r = \min\left\{ (t_{i+m} - t_i) \;\middle|\; 2u \leq i \leq 3u-m \right\} .$$

Dann ist

$$t_s - t_r \leq (t_{3u} - t_{2u})/[u/m] \leq 30\ mt_r/s$$

und $S(u+1,3u) \leq 2\, S(u+1,s)$, und folglich für $1 \leq q \leq u$

$$t_q\, S(q+1,s) \geq t_q\, S(q+1,3u)/2 \geq \sqrt{q}/14 \quad .$$

Für $u+1 \leq q \leq r$ ist

$$t_q\, S(q+1,s) \geq t_q\, (s-q)/t_s \geq (s-q)/6 \quad .$$

Für $u+1 \leq q < s/2$ ist aber $(s-q)/6 \geq q/12$, und daher ist Lemma 6 vollständig bewiesen .

Wir beweisen nun Satz 1 für den Fall, daß Λ keinen endlichen Häufungspunkt besitzt. Dann können wir die Elemente von Λ so numerieren, daß

(7) $$0 < |\lambda_1| \leq |\lambda_2| \leq |\lambda_3| \leq \ldots$$

erfüllt ist. Für die Größen A_{qs} in Lemma 1 finden wir folgende Abschätzung:

<u>Lemma 7</u>: Unter der Voraussetzung (7) ist

$$A_{qs} \leq 2 \exp \left\{ -\frac{M^2}{2} |\lambda_q| \sum_{k=q+1}^{s} 1/|\lambda_k| \right\} .$$

<u>Beweis</u>: Sei $\alpha_k := \mathrm{Re}\,\lambda_k$ und $\beta_k := \mathrm{Im}\,\lambda_k$. Dann ist

$$\left|\frac{\lambda_q - \lambda_k}{\lambda_q + \bar{\lambda}_k}\right|^2 \leq \frac{2|\lambda_k|^2 - \alpha_q \alpha_k}{2|\lambda_k|^2 + \alpha_q \alpha_k} \leq \exp\left\{ -M^2 |\lambda_q| / |\lambda_k| \right\} ,$$

wobei wir die Ungleichung $(1-x)/(1+x) \leq e^{-2x}$ für $x = \alpha_q \alpha_k / (2|\lambda_k|^2)$ und die Voraussetzung (ii) des Satzes 1 angewendet haben.

Es sei eine Funktion $f \in C_0[0,1]$ und ein genügend kleines $\varepsilon > 0$ vorgegeben. Es gibt eine Zahl β, $-\frac{\pi}{2} < \beta < \beta + \varepsilon < \frac{\pi}{2}$, so daß die Teilfolge Λ_ε von Λ, die im Winkelraum

$$W_\varepsilon := \left\{ z \in \mathbb{C} \mid \beta \leq \arc z \leq \beta + \varepsilon \right\}$$

des komplexen Zahlkörpers $\mathbb{C}$ liegt, auch den Bedingungen (i)-(iii) des Satzes 1 genügt. Wir bezeichnen die Elemente dieser Teilfolge erneut mit $\Lambda_\varepsilon = \{\lambda_k\}_{k=1}^{\infty}$ und führen alle nun folgenden Betrachtungen ausschließlich für Λ_ε durch. Man beachte, daß auch (7) für Λ_ε gilt. Dann ist für $r < s$ stets

(8) $$\sum_{k=r+2}^{s} \frac{|\lambda_k - \lambda_{k-1}|}{|\lambda_k|} \leq \sum_{k=r+2}^{s} \frac{\varepsilon|\lambda_k| + |\lambda_k| - |\lambda_{k-1}|}{|\lambda_k|}$$

$$\leq \varepsilon(s-r) + (|\lambda_s| - |\lambda_r|)/|\lambda_r| .$$

Wir müssen nun zwei Fälle unterscheiden .

A. Es sei $\liminf_{k\to\infty} |\lambda_k|/k = 0$.

Wir wählen $m := [1/\sqrt{\varepsilon}]$ und die Zahl s so groß, daß

(9) $s \geq 1/\varepsilon$, $E_s(f,\Lambda_\varepsilon) \leq \varepsilon$, $\sum_{q=1}^{m} A_{qs} \leq \varepsilon$

und zusätzlich auch

(10) $|\lambda_s|/s \leq |\lambda_k|/k$ für alle $k=1,2,\ldots,s$

erfüllt ist. Dann ist nach Lemma 7

(11) $A_{qs} \leq 2\exp\left\{-M^2(s-q)|\lambda_q|/(2|\lambda_s|)\right\} \leq 2\exp\left\{-M^2(s-q)q/(2s)\right\}$,

wobei wir in der letzten Ungleichung die Bedingung (10) verwendet haben. Wir setzen $r := s-m$ und gewinnen mit Hilfe von (10)

(12) $(|\lambda_s| - |\lambda_r|)/|\lambda_r| \leq s/r - 1 = m/r$.

Insgesamt folgt jetzt aus Lemma 3 für die oben gewählten Zahlen m, r und s unter Anwendung von (9), (11), (8) und (12) die Existenz von Zahlen $b_j \in Z+iZ$, so daß

(13) $\left\| f(x) - \sum_{j=1}^{s} b_j x^{\lambda_j} \right\| \leq K\sqrt{\varepsilon}$

für eine von ε unabhängigen Konstanten $K > 0$. Damit ist Satz 1 unter der Voraussetzung A vollständig bewiesen .

B. Es sei $\liminf_{k\to\infty} |\lambda_k|/k \neq 0$.

Dann existiert eine Konstante $N > 0$, so daß $|\lambda_k| \geq N\,k$ für alle $k=1,2,\ldots$ gilt . Ohne Beschränkung der Allgemeinheit dürfen wir $N = 1$ annehmen . Dann ist für die Folge $T := \left\{|\lambda_k|\right\}_{k=1}^{\infty}$ die Bedingung (1) in Lemma 4 erfüllt . Wir wählen wieder $m := [1/\sqrt{\varepsilon}]$. Nach Lemma 6 gibt es beliebig große Zahlen r und $s := r+m$, so daß die Ungleichungen (5) und (6) gelten . Wir wählen diese Zahlen r und s so groß, daß auch (9) erfüllt ist.

Ist $\varepsilon > 0$ genügend klein, so ist

$$M^2 \sqrt{q} / 28 \geq 2 \log q \quad \text{für alle} \quad q \geq m ,$$

sowie $M^2 (s-q)/12 \geq 2 \log (s-q)$ für alle $q \leq r$,

und daher folgt aus Lemma 6 und Lemma 7

$$(14) \quad \sum_{q=m+1}^{r} A_{qs} \leq 2 \left(\sum_{m+1 \leq q} \exp\left\{ - \frac{M^2 \sqrt{q}}{28} \right\} + \sum_{q \leq r} \exp\left\{ \frac{-M^2 (s-q)}{12} \right\} \right)$$

$$2 \left(\sum_{m+1 \leq q} q^{-2} + \sum_{q \leq r} (s-q)^{-2} \right) < 6 \sqrt{\varepsilon} .$$

Beachten wir (8) und (5), so folgt

$$(15) \quad \sum_{k=r+2}^{s} (|\lambda_k - \lambda_{k-1}|) / |\lambda_k| \leq \varepsilon m + 30 m/s \leq 31 \sqrt{\varepsilon} .$$

Insgesamt folgt jetzt aus Lemma 3 für die oben gewählten Zahlen m, r und s unter Anwendung von (9), (14) und (15) die Existenz von Zahlen $b_j \in Z + iZ$, so daß (13) erfüllt ist . Nun ist Satz 1 vollständig bewiesen .

Im zweiten Teil dieser Arbeit wollen wir einen kurzen Überblick über die Resultate geben, die die Genauigkeit der Approximation durch ganzzahlige Polynome beschreiben. Ein erstes Ergebnis stammt von Kantorovic [11] :

Satz 2: Zu jeder reellwertigen Funktion $f \in C_o [0,1]$ und jeder natürlichen Zahl n existiert ein ganzzahliges Polynom $Q_n(x) = \sum_{k=o}^{n} b_k x^k$, $b_k \in \mathbb{Z}$, so daß

$$\| f - Q_n \| \leq 2 E_n(f) + O(n^{-1}) ,$$

wobei

$$E_n(f) := \inf \left\{ \| f(x) - \sum_{k=o}^{n} a_k x^k \| , a_k \text{ reell} \right\} .$$

Gelfond [10] und Trigub [12] verschärften dieses Ergebnis. Von Trigub ist der folgende

<u>Satz 3</u>: Sei p eine nichtnegative ganze Zahl und $a < b < a+4$. Die reellwertige Funktion f sei auf $[a,b]$ p-mal stetig differenzierbar. Für f und ihre Ableitungen $f^{(j)}$, $j=1,\dots,p$, sei $f^{(j)}(t) = 0$ für alle $t \in J(a,b)$, wobei die endliche Punktmenge $J(a,b)$ definiert sei durch

$$J(a,b) := \{ t \in [a,b] \mid Q(t)=0 \text{ für alle algebraischen Polynome } Q \text{ mit ganzzahligen Koeffizienten, für die } 0 \leq Q(x) < 1 \text{ für alle } x \in [a,b]\}.$$

Dann existiert zu jeder genügend großen natürlichen Zahl n ein algebraisches Polynom Q_n vom Grade $\leq n$ mit ganzzahligen Koeffizienten, so daß

$$\left| f^{(j)}(x) - Q_n^{(j)}(x) \right| \leq C_p \, (\Delta_n(x))^{p-j} \, w(f^{(p)}, \Delta_n(x))$$

für alle $x \in [a,b]$ und $j=0,\dots,p$. Hierbei ist C_p eine von x und n unabhängige Zahl, $w(f^{(p)}, \cdot)$ bezeichnet den Stetigkeitsmodul von $f^{(p)}$ und $\Delta_n(x) := n^{-2} + n^{-1}\sqrt{(x-a)(b-x)}$.

Um Aussagen über die Approximationsgüte für Λ-Polynome mit ganzzahligen Koeffizienten im Intervall $[0,1]$ zu gewinnen, können wir die Ungleichung in Lemma 3 benützen. Aber im allgemeinen führt dies nicht zu sehr guten Abschätzungen, da die Größe D_{rs} in Lemma 3 meist nicht genügend klein zu machen ist. In der Arbeit [6] konnte ich das Lemma 3 für reelle Folgen Λ verschärfen und so den folgenden Satz beweisen, der obigen Satz 2 von Kantorovic [11] verallgemeinert:

<u>Satz 4</u>: Sei $\Lambda = \{\lambda_k\}_{k=1}^{\infty}$ eine anwachsende Folge positiver reel-

ler Zahlen . Sei $f \in C_o[0,1]$ eine reellwertige Funktion .
Gibt es positive Zahlen B und C und eine natürliche Zahl k_o, so daß

(16) $B \leq \lambda_1$, $\lambda_k \leq B\,k$, $\lambda_{2k} \leq C\,\lambda_k$ für alle $k \geq k_o$,

dann existiert zu jeder natürlichen Zahl s eine ganzzahliges Λ-Polynom $Q_s(x) = \sum_{j=1}^{s} b_j\, x^{\lambda_j}$, $b_j \in \mathbb{Z}$, so daß

(17) $\| f - Q_s \| \leq 2\, E_s(f,\Lambda) + O(s^{-1})$.

Ist hingegen

(18) $\lambda_k \geq B\,k$ und $\lambda_{2k} \leq C\,\lambda_k$ für alle $k \geq 1$,

so gibt eszu jeder natürlichen Zahl s ein ganzzahliges Λ-Polynom Q_s , für das

(19) $\| f - Q_s \| \leq 2\, E_s(f,\Lambda) + O(\varphi(s)^{-B})$

erfüllt ist . Hierbei definieren wir, $\varphi(s) := \exp(\sum_{k=1}^{s} 1/\lambda_k)$.

Verknüpfen wir Satz 4 mit den Jackson-Müntz-Sätzen für Λ-Polynome mit reellen Koeffizienten (siehe Golitschek [5] oder Bak-Newman [1]), so werden wir zu folgendem Korollar geführt:

<u>Korollar</u>: Sei $f \in C_o[0,1]$ eine reellwertige Funktion . Erfüllt Λ die Bedingung (16), so gilt für die ganzzahligen Λ-Polynome Q_s in Satz 4 die Aussage

$$\| f - Q_s \| = O(w(f, s^{-1})) \quad \text{für } s \to \infty .$$

Erfüllt Λ die Bedingung (18) , so ist hingegen

$$\| f - Q_s \| = O(w(f, \varphi(s)^{-B^*})) \quad \text{für } s \to \infty$$

mit $B^* := \min\{B\,;\,2\}$.

Literaturverzeichnis

[1] Bak, J., Newman, D.J.: Müntz-Jackson theorems in $L^p[0,1]$ and $C[0,1]$. Amer.J.Math.94, 437-457 (1972)

[2] Ferguson, L.B.O.: Uniform approximation by polynomials with integral coefficients. Pacific J.Math.27,53-69 (1968), Pacific J.Math.26, 273-281 (1968)

[3] Ferguson, L.B.O., Golitschek, M.: Müntz-Szasz theorem with integral coefficients. Trans.Amer.Math.Soc.213,115-126(1975)

[4] Golitschek, M.: Erweiterung der Approximationssätze von Jackson im Sinne von C.Müntz.J.Approx.Theory 3,72-86(1970)

[5] Golitschek, M.: Jackson-Müntz Sätze in der L_p-Norm. J.Approx.Theory 7,87-106(1973)

[6] Golitschek, M.: The degree of approximation for generalized polynomials with integral coefficients. Trans.Amer.Math. Soc. to appear 1976/77

[7] Kakeya, S.: On approximate polynomials. Tohoku Math.J.6, 182-186 (1914)

[8] Martirosjan, V.A.: Über die gleichmäßige Approximation durch Müntzsche Polynome mit ganzzahligen Koeffizienten. Izvestija Akad.Nauk Armenien SSR VIII, Nr.2,167-175 (1973)

[9] Szasz, O.: Über die Approximation stetiger Funktionen durch lineare Aggregate von Potenzen.Math.Ann.77,482-496 (1916)

[10] Gelfond, A.O.: Über die gleichmäßige Approximation durch Polynome mit ganzzahligen Koeff. Uspehi Mat.Nauk 10,41-65(1955)

[11] Kantorovic, L.: Einige Bemerkungen über die Approximation von Funktionen durch Polynome mit ganzzahligen Koeffizienten. Izvestija Akad.Nauk SSSR, 1163-1168,(1931)

[12] Trigub, R.M.: Approximation von Funktionen durch Polynome mit ganzz.Koeff. Izvestija Akad Nauk SSSR 26,261-280 (1962).

Institut für Angewandte Mathematik und Statistik
D-8700 Würzburg, Bundesrepublik Deutschland

EIN PROBLEM DER BESTAPPROXIMATION IN GEORDNETEN VEKTORRÄUMEN

Gerhard Heindl

I. Einleitung und Problemstellung.

X sei ein reeller geordneter und mit einer monotonen Norm $\|\cdot\|$ versehener Vektorraum. K bezeichne den Ordnungskegel $\{e \in X : e \geqslant 0\}$. Zu gegebenem $i \in X$ und $\emptyset \neq E \subset X$ mit $i \leqslant E$ und $E + K$ konvex, wird das Approximationsproblem betrachtet, Bestapproximierende in E für i, d.h.

$$e \in P_i(E) := \{e' \in E : \|e' - i\| = \inf\{\|e'' - i\| : e'' \in E\}\}$$

zu finden. Probleme dieser Art treten, wie B. Harris in [5] zeigte, (vergl. auch [6]!) u. a. in der Entscheidungstheorie auf. Dabei repräsentiert X den sogenannten Entscheidungsraum, eine Grundmenge von überhaupt möglichen Entscheidungen. "$\leqslant$" definiert eine Präferenzrelation auf X in dem Sinne, daß die Entscheidung $e_1 \in X$ der Entscheidung $e_2 \in X$ vorzuziehen ist, wenn $e_1 < e_2$ gilt. E ist i.a. eine durch vorgeschriebene Restriktionen bestimmte Teilmenge von X, die sogenannte Menge der zulässigen Entscheidungen. i stellt eine Idealentscheidung dar, die man fällen würde, wenn sie zulässig wäre. Da i.a. jedoch i nicht zulässig ist, versucht man i möglichst gut zu approximieren. Nach Festlegung einer geeigneten Norm wird man also versuchen, als Ersatz für i eine Entscheidung $e \in P_i(E)$ zu ermitteln.
Die vorausgesetzte Monotonieeigenschaft der Norm trägt dabei der natürlichen Forderung Rechnung, daß es in E keine Entscheidung geben darf, die jeder Entscheidung aus $P_i(E)$ vorzuziehen ist. Genauer gilt:

<u>1. Lemma:</u> <u>Die Monotonieeigenschaft der Norm ist äquivalent mit der Forderung</u>

(F) $\{e \in E : e < e_o$ <u>für alle</u> $e_o \in P_i(E)\} = \emptyset$ <u>für alle</u> $i \in X$ <u>und</u> $E \subset X$ <u>mit</u> $i \leqslant E$ <u>und</u> $P_i(E) \neq \emptyset$.

<u>Beweis:</u> 1. Ist (F) nicht erfüllt, $i \leqslant E \subset X$, $P_i(E) \neq \emptyset$ und $e \in E$ mit $e < e_o$ für alle $e_o \in P_i(E)$, so ist $e \in E \setminus P_i(E)$, also $\|e - i\| > \|e_o - i\|$ für alle $e_o \in P_i(E)$. Ferner gilt $0 \leqslant e - i < e_o - i$ für alle $e_o \in P_i(E)$. Die Norm kann dann aber nicht monoton sein.

2. Ist (F) erfüllt, so gilt für $e_1, e_2 \in X$ mit $0 \leqslant e_1 \leqslant e_2$: $i := 0 \leqslant E := \{e_1 + t(e_2 - e_1) : t \in [0,1]\} \subset K$ und $P_i(E) \neq \emptyset$, da E kompakt ist. Wegen $e_1 \leqslant E$ ist $e_1 \leqslant P_i(E)$ und wegen $e_1 \in E$ und (F): $e_1 \in P_i(E)$. Damit folgt aber $\|e_1\| \leqslant \|e_2\|$, also die Monotonie der Norm.

Als weiteres Beispiel für ein Problem der beschriebenen Art sei die einseitige Tschebyscheffapproximation genannt (Vergl. etwa [8] S. 62 !)

Zur Lösung der gestellten Aufgabe wurden in [6] aus der Monotonie der Norm und der Voraussetzung $i \leq E$ charakteristische Eigenschaften der Bestapproximierenden abgeleitet, die sich nicht aus den gewöhnlichen Charakterisierungssätzen der Approximationstheorie gewinnen lassen. Im Folgenden soll eine weitere Charakterisierung der Elemente aus $P_i(E)$ angegeben werden. Sie beruht auf einer Verallgemeinerung des von Nikolski eingeführten Begriffes des Fundamentalsystems und enthält Theorem 2 aus [6] als Spezialfall.
Vorbereitend werden einige aus der Monotonie der Norm folgende Eigenschaften des Ordnungskegels, des stetigen Dualraumes X^* von X und des Dualkegels $K^o := \{ l \in X^* : l(e) \geq 0$ für alle $e \in K\}$, sowie verschiedene später zu verwendende Resultate und damit zusammenhängende Bemerkungen zusammengestellt.

II. Vorbereitungen.

1. Zwei einfache Folgerungen aus der Monotonieeigenschaft der Norm sind: a) K <u>ist</u> <u>spitz</u>, <u>d.h.</u>: <u>Aus</u> $e \in K\setminus\{0\}$ <u>folgt</u> $-e \notin K$.

b) <u>Ist</u> $X \neq \{0\}$, <u>so</u> <u>ist</u> $\bar{K} \neq X$ +) <u>und</u> <u>es</u> <u>gibt</u> <u>daher</u> <u>ein</u> $l \in K^o$ <u>mit</u> $\|l\| = 1$.

<u>Beweis:</u> Zu a): Aus $e, -e \in K$ folgt $0 \leq e \leq 0$ und daraus wegen der Monotonie der Norm: $\|e\| \leq \|0\| = 0$.

Zu b): Sei $e \in X\setminus\{0\}$. Nimmt man $\bar{K} = X$ an, so hat man $e, -e \in \bar{K}$. Ist $\{e_k\}$ eine Folge in K, die gegen $-e$ konvergiert, so gilt für alle k: $0 \leq e \leq e_k + e$, also auch $\|e\| \leq \|e_k - (-e)\|$. Da daraus aber $e = 0$ folgt, führt die Annahme $\bar{K} = X$ zu einem Widerspruch.
Die Existenz eines Funktionals $l \in K^o$ mit $\|l\| = 1$ folgt auf bekannte Weise aus der Existenz eines Funktionals in $X^*\setminus\{0\}$, welches eine Kugel in $X\setminus\bar{K}$ von K trennt.

2. Kung-Fu Ng hat in [9] gezeigt, daß die Norm auf X genau dann monoton ist, wenn es zu jedem $l \in X^*$ ein $g \in K^o$ mit $g - l \in K^o$ und $\|g\| \leq \|l\|$ gibt.++) Die Monotonieeigenschaft der Norm drückt sich aber auch in einer Struktureigenschaft der Polaren

$$P := \{ l \in X^* : l(e) \leq \|e\| \text{ für alle } e \in K \} \quad \text{von}$$
$$S_+ := \{ e \in K : \|e\| \leq 1 \}$$

+) Zu $M \subset X$, oder $M \subset X^*$ bezeichne $\bar{M}$ die abgeschlossene Hülle von M bzgl. der Normtopologie.

++) Die in [9] generell vorausgesetzte Vollständigkeit von X wurde zum Beweis dieses Resultates nicht herangezogen!

aus:

2. Lemma: $P = \{l_1 - l_2 : l_1, l_2 \in K^o \text{ und } \|l_1\| \leq 1\} =: \tilde{P}$ ist eine notwendige und hinreichende Bedingung dafür, daß die Norm auf X monoton ist.

Beweis: $\tilde{P} \subset P$ gilt für jede Norm auf X, denn aus $l = l_1 - l_2$ mit $l_1, l_2 \in K^o$ und $\|l_1\| \leq 1$ folgt für alle $e \in K$:

$$l(e) = l_1(e) - l_2(e) \leq l_1(e) \leq \|e\| \ .$$

Die Norm sei nun monoton. Zum Nachweis von $P \subset \tilde{P}$ genügt es zu zeigen, daß es zu jedem $l \in P$ ein $l_1 \in K^o$ mit $l_1 \geq l$ und $\|l_1\| \leq 1$ gibt. Man betrachte dazu zu $l \in P$ die Funktionale

$$p : X \ni e \mapsto \inf\{\|e'\| : e \leq e'\} \in \mathbb{R}$$
$$q : K \ni e \mapsto \sup\{l(e'') : 0 \leq e'' \leq e\} \in \mathbb{R} .$$

p und -q sind sublinear. Ferner folgt für alle $e \in K$ aus $0 \leq e'' \leq e \leq e'$ nacheinander: $l(e'') \leq \|e''\| \leq \|e'\|$, $q(e) \leq \|e'\|$ und $q(e) \leq p(e)$.
Nach einem bekannten Trennungssatz (vergl. etwa [1] !) gibt es ein lineares Funktional l_1 auf X mit

$$l_1(e) \leq p(e) \text{ für alle } e \in X \text{ und}$$
$$l_1(e) \geq q(e) \text{ für alle } e \in K.$$

Wegen $p(e) \leq \|e\|$ für alle $e \in X$ ist $l_1 \in X^*$ mit $\|l_1\| \leq 1$. Aus $0, l(e) \leq q(e)$ für alle $e \in K$ folgt $l_1 \in K^o$ und $l_1 \geq l$.
Ist $P = \tilde{P}$, so gibt es zu jedem $l \in X^* \setminus \{0\}$ wegen $\|l\|^{-1} l \in P$, $l_1, l_2 \in K^o$ mit $\|l_1\| \leq 1$ und $l_1 - l_2 = \|l\|^{-1} l$. Setzt man $g := \|l\| \, l_1$, so ist $g \in K^o$, $g - l = \|l\| \, l_2 \in K^o$ und $\|g\| \leq \|l\|$. Damit folgt aber die Monotonie der Norm auf X aus der (trivialen Richtung der) Charakterisierung monotoner Normen in [9] .

3. Korollar: Aus der Monotonie der Norm auf X folgt

$$\text{Ext } P \subset (\text{Ext } S^*) \cap K^o \ .^{+)}$$

Beweis: l sei ein Extrempunkt von P. Nach Lemma 2 gibt es $l_1, l_2 \in K^o$ mit $\|l_1\| \leq 1$ und $l = l_1 - l_2$. Aus $l + l_2 = l_1 \in S^* \subset P$, $l - l_2 = l_1 - 2l_2 \in P$, $l = 2^{-1}(l + l_2) + 2^{-1}(l - l_2)$ und $l \in \text{Ext } P$ folgt $l_2 = 0$, also $l = l_1$ und $l \in (\text{Ext } S^*) \cap K^o$.

3. Die vorausgesetzte Konvexität von E + K ist, wie man leicht verifiziert, äquivalent mit der Eigenschaft von E:
Für alle $e_1, e_2 \in E$ und $t \in]0,1[$ gibt es ein $e_3 \in E$ mit $e_3 \leq te_1 + (1-t)e_2$.

+) Zu $M \subset X^*$ bezeichne Ext M die Menge der Extrempunkte von M.
$S^* := \{l \in X^* : \|l\| \leq 1\}$.

In Problemen, die aus der Entscheidungstheorie stammen, ist diese Forderung i.a. erfüllt. Häufig ist dabei schon E konvex.

4. Zwischen dem Problem der Approximation von i durch Elemente aus E und dem der Approximation von i durch Elemente aus E + K besteht folgender Zusammenhang:

<u>4. Lemma:</u> $d_i(E) := \inf\{\|e-i\| : e \in E\} = d_i(E+K) := \inf\{\|e-i\| : e \in E + K\}$, <u>und mit</u> $P_i(E + K) := \{e \in E + K : \|e - i\| = d_i(E + K)\}$ <u>ist</u>

$$P_i(E) = E \cap P_i(E + K) .$$

<u>Beweis:</u> Aus $E \subset E + K$ folgt $d_i(E) \geq d_i(E + K)$. Ist $s = e_1 + e_2$ mit $e_1 \in E$ und $e_2 \in K$, so gilt $0 \leq e_1 - i \leq s - i$, also $\|e_1 - i\| \leq \|s - i\|$. Damit folgt aber $d_i(E) \leq \|s - i\|$ für alle $s \in E + K$, d.h. $d_i(E) \leq d_i(E + K)$. Die behauptete Darstellung für $P_i(E)$ ist nun eine Konsequenz von $E \subset E + K$.

Lemma 4 gestattet u.a. die Rückführung eines in [6] direkt bewiesenen Charakterisierungssatzes (Theorem 1) auf einen bekannten Satz der Approximationstheorie.

<u>5. Satz:</u> <u>Ist</u> $X \neq \{0\}$, <u>so gilt</u> $e_o \in P_i(E)$ <u>genau dann, wenn</u> $e_o \in E$ <u>ist und es ein</u> $l \in K^o$ <u>mit</u> $\|l\| = 1$, $l(e_o) = \inf\{l(e) : e \in E\}$ <u>und</u> $l(e_o - i) = \|e_o - i\|$ <u>gibt</u>.

<u>Beweis:</u> 1. Ist $e_o \in E$ und $l \in K^o$ mit $\|l\| = 1$, $l(e_o) = \inf\{l(e) : e \in E\}$ und $l(e_o - i) = \|e_o - i\|$, so gilt für alle $e \in E$:

$$\|e_o - i\| = l(e_o - i) \leq l(e - i) \leq \|e - i\| .$$

2. Ist $e_o \in P_i(E)$, so gilt im Fall $i \in \bar{E}$: $i = e_o$. Aus 1.b) folgt die Existenz eines $l \in K^o$ mit $\|l\| = 1$. Wegen $e_o = i \leq e$ für alle $e \in E$, $e_o \in E$ und $l \in K^o$ ist $l(e_o) = \inf\{l(e) : e \in E\}$.

Im Fall $i \notin \bar{E}$ schließt man aus Lemma 4: $i \notin \overline{E + K}$ und $e_o \in P_i(E + K)$. Da E + K konvex ist, folgt aus dem Satz von Garkavi und Deutsch-Maserick (vergl. [4] Theorem 2.5 !): Es gibt ein $l \in X^*$ mit $\|l\| = 1$, $l(e_o - i) = \|e_o - i\|$ und $l(e_o) = \inf\{l(e) : e \in E + K\}$. Wegen $e_o \in E \subset E + K$ ist dann aber auch $l(e_o) = \inf\{l(e) : e \in E\}$ und $l(e_o + e) \geq l(e_o)$ für alle $e \in K$. Letzteres bedeutet aber $l \in K^o$. Damit ist der Satz bewiesen.

<u>Bemerkung:</u> Ist die auf X gegebene Norm nicht monoton, so lassen sich immer eine einelementige Menge E und ein $i \leq E$ so wählen, daß Satz 5 nicht gilt. Setzt man nämlich zu e_o, $e_1 \in X$ mit $0 \leq e_o \leq e_1$ und $\|e_o\| > \|e_1\|$ $E := \{e_o\}$ und $i := 0$, so ist $i \leq E$ und $e_o \in P_i(E)$. Ein $l \in K^o$ mit $\|l\| = 1$ und $l(e_o - i) = \|e_o - i\|$ kann nicht existieren, weil sonst sowohl $l(e_1) \geq l(e_o)$ als auch $l(e_o) = \|e_o\| > \|e_1\| \geq l(e_1)$ gelten müßte.

III. Ein verallgemeinertes Kolmogoroffsches Kriterium.

Von Nikolski stammt eine auf dem Begriff des Fundamentalsystems beruhende Verallgemeinerung des Kolmogoroffschen Kriteriums auf normierte Räume, die auch zur Charakterisierung der Elemente in $P_i(E)$ herangezogen werden kann. Für das vorliegende Approximationsproblem führt jedoch, wie sich zeigen wird, der Begriff des K-Fundamentalsystems zu einem leistungsfähigeren Kriterium. Dabei soll eine Teilmenge T von S^* <u>K-Fundamentalsystem</u> genannt werden, wenn T in der "schwachen Stern-Topologie" $\sigma(X^*, X)$ abgeschlossen ist und für alle $e_o \in K$

$$T(e_o) := \{l \in T : l(e_o) = \|e_o\|\}$$

nicht leer ist.

(Ein K-Fundamentalsystem ist ein Fundamentalsystem im Sinne von Nikolski, wenn $T(e_o) \neq \emptyset$ für alle $e_o \in X$ gilt. Vergl. [10], [11] und [2] !)

<u>6. Satz:</u> $e_o \in P_i(E)$ <u>gilt genau dann, wenn</u> $e_o \in E$ <u>ist und für ein beliebiges</u> K-<u>Fundamentalsystem</u> T <u>zu jedem</u> $e \in E$ <u>ein</u> $l \in T(e_o - i)$ <u>mit</u> $l(e - e_o) \geq 0$ <u>existiert</u>.

<u>Beweis:</u> +) T sei ein beliebiges K-Fundamentalsystem.

1. Die Bedingung ist hinreichend, denn ist $e_o \in E$, $e \in E$ und $l \in T(e_o - i)$ mit $l(e - e_o) \geq 0$, so folgt

$$\|e - i\| \geq l(e - i) \geq l(e_o - i) = \|e_o - i\| .$$

2. Sei nun $e_o \in P_i(E)$ und $e \in E$.

Die Einschränkung des $\sigma(X^*, X)$-stetigen linearen Funktionals $X^* \ni l' \mapsto l'(e - e_o) \in \mathbb{R}$ auf die $\sigma(X^*, X)$-kompakte Menge $T(e_o - i)$ besitzt ein Maximum. Es gibt also ein $l \in T(e_o - i)$ mit

$$l(e - e_o) = \sup\{l'(e - e_o) : l' \in T(e_o - i)\} =: m .$$

Zu zeigen ist noch $m \geq 0$!

Sei $m < 0$ angenommen.

$H := \{l \in T : l(e - e_o) \geq 0\} \supset T(e - i) \neq \emptyset$, weil $e - i \in K$ und für jedes $l \in T(e - i)$

$$l(e - e_o) = l(e - i) - l(e_o - i) \geq \|e - i\| - \|e_o - i\| \geq 0$$

gilt. Aus $l(e - e_o) \leq m < 0$ für alle $l \in T(e_o - i)$ folgt $H \cap T(e_o - i) = \emptyset$, also $l(e_o - i) < \|e_o - i\|$ für alle $l \in H$. Da H $\sigma(X^*, X)$-kompakt ist, existiert $a := \mathrm{Max}\{l(e_o - i) : l \in H\}$ und es ist $a < \|e_o - i\|$.

Wegen $m < 0$ ist $e - e_o \neq 0$. Es gibt daher ein positives

$$t < \mathrm{Max}\,\{1, \|e - e_o\|^{-1}(\|e_o - i\| - a)\}$$

Aus $e_o + t(e - e_o) - i \in K$, $e_o + t(e - e_o) \in E + K$ und $e_o \in P_i(E + K)$ schließt man: Es gibt ein $l \in T$ mit

$$l(e_o + t(e - e_o) - i) = \|e_o + t(e - e_o) - i\| \geq \|e_o - i\| .$$

+) In Analogie zu den Beweisen von Satz 6 in [2] und der Lemmata 2 und 3 in [3] .

$l \in H$ ist nicht möglich, denn man hätte sonst

$$l(e_o + t(e - e_o) - i) = l(e_o - i) + tl(e - e_o) \leq a + t\|e - e_o\| < \|e_o - i\|.$$

Es kann aber auch nicht $l \in T \setminus H$ sein, weil aus $l(e - e_o) < 0$

$$l(e_o + t(e - e_o) - i) = l(e_o - i) + tl(e - e_o) < \|e_o - i\|$$

folgt. Damit ist die Annahme $m < 0$ zu einem Widerspruch geführt.

Bemerkung: In Satz 6 kann $T(e_o - i)$ durch $T(e_o - i) \cap \mathrm{Ext}\, T$ ersetzt werden. Ist nämlich $e_o \in P_i(E)$ und $e \in E$, so ist $T(e_o - i)$ eine $\sigma(X^*, X)$-kompakte extr. Teilm. von T, $B := \{l \in T(e_o - i) : l(e - e_o) = m\}$ eine $\sigma(X^*, X)$-kompakte extremale Teilmenge von $T(e_o - i)$. B enthält dann aber nach dem Satz von Krein-Milman ein $l \in \mathrm{Ext}\, T$.

Um zu zeigen, daß Satz 6 mehr leistet als die Nikolskische Verallgemeinerung des Kolmogoroffschen Kriteriums, sollen verschiedene K-Fundamentalsysteme angegeben werden, die i.a. keine Fundamentalsysteme sind.

IV. Beispiele für K-Fundamentalsysteme.

1. Da jedes Fundamentalsystem auch ein K-Fundamentalsystem ist, sind z.B. S^* und $\overline{\mathrm{Ext}\, S^*}$ +) K-Fundamentalsysteme. Daneben existieren aber auch K-Fundamentalsysteme, die nur Funktionale aus K^o enthalten:

7. Lemma: $S^*_+ := S^* \cap K^o$ ist ein K-Fundamentalsystem und falls $K \neq \{0\}$, kein Fundamentalsystem.

Beweis: S^*_+ ist $\sigma(X^*, X)$-abgeschlossen. Ist $e_o \in K \setminus \{0\}$, so folgt mit $E := \{e_o\}$ und $i := 0$: $i \leq E$ und $e_o \in P_i(E)$. Nach Satz 5 gibt es daher ein $l \in S^*_+$ mit $l(e_o) = \|e_o\|$. Zusammen mit $S^*_+(0) = S^*_+ \neq \emptyset$ ergibt sich daraus der erste Teil der Behauptung.
Gibt es ein $e_o \in K \setminus \{0\}$, so ist für alle $l \in S^*_+$ $l(e_o) \geq 0$, also $l(-e_o) \leq 0 < \|-e\|$. S^*_+ ist dann aber kein Fundamentalsystem.

Bemerkungen: 1. Aus der Bemerkung zu Satz 6 folgt mit $T = S^*_+$ Theorem 2 in [6].

2. Die Existenz von positiven K-Fundamentalsystemen (d.h. von K-Fundamentalsystemen $T \subset K^o$ ist äquivalent mit der Monotonie der Norm auf X. Gibt es nämlich e_o, $e_1 \in X$ mit $0 \leq e_o \leq e_1$ und $\|e_o\| > \|e_1\|$, so gilt für jedes $l \in T$ mit $l(e_o) = \|e_o\|$: $l(e_1) \leq \|e_1\| < \|e_o\| = l(e_o)$. Wegen $e_1 - e_o \in K$ folgt daraus aber $l \notin K^o$.

2. Jedes K-Fundamentalsystem T enthält ein minimales K-Fundamentalsystem. Man folgert diese Eigenschaft wie die entsprechende für Fundamentalsysteme aus dem Zornschen Lemma. Jede Kette Ω in

$$\gamma := \{T' : T' \subset T \text{ und } T' \text{ K-Fundamentalsystem}\}$$

+) Zu $M \subset X^*$ bezeichne $\overline{M}$ die $\sigma(X^*, X)$-abgeschlossene Hülle von M.

hat nämlich $\hat{T} := \bigcap_{T' \in \Omega} T' \in \gamma$ als untere Schranke.

3. Ist T ein K-Fundamentalsystem, so enthält für jedes $e \in K$ $T(e)$ einen Extrempunkt von T. Daher ist mit T auch Ext T ein K-Fundamentalsystem. Im Fall $K \neq \{0\}$ auch $(\text{Ext } T) \setminus \{0\}$.

4. Von besonderer Bedeutung für die Anwendung von Satz 6 ist die Frage, ob $(\text{Ext } S^*) \cap K^0$ ein K-Fundamentalsystem ist. Diese Frage kann in einigen wichtigen Fällen positiv beantwortet werden. Es gilt

8. Lemma: Ist $K \neq \{0\}$, so ist jede der folgenden drei Bedingungen hinreichend dafür, daß es zu jedem $e_0 \in K \setminus \{0\}$ ein $l \in (\text{Ext } S^*) \cap K^0$ mit $l(e_0) = \| e_0 \|$ gibt: 1. S_+ ist glatt, d.h.: Zu jedem $e \in K$ mit $\|e\| = 1$ gibt es genau ein $l \in S^*$ mit $l(e) = 1$.

2. Die Norm auf X ist additiv auf K.

3. K besitzt innere Punkte relativ K - K.

Beweis: 1. Ist S_+ glatt, so ist für alle $e_0 \in K \setminus \{0\}$ $S^*(e_0)$ eine einelementige extremale Teilmenge von S^*. Wegen $\emptyset \neq S_+^*(e_0) \subset S^*(e_0)$ gibt es dann aber genau ein $l \in (\text{Ext } S^*) \cap K^0$ mit $l(e_0) = \|e_0\|$.

2. Ist die Norm additiv auf K und bezeichnet U den Unterraum K - K von X, so sind die Funktionale

$$p := \|\cdot\| | U \quad \text{und} \quad -q := - \|\cdot\| | K$$

sublinear. Nach dem bereits im Beweis von Lemma 2 benutzten Trennungssatz gibt es ein lineares Funktional l auf U mit den Eigenschaften:

$$l(e) \leq \|e\| \quad \text{für alle } e \in U \text{ und}$$
$$l(e) \geq \|e\| \quad \text{für alle } e \in K.$$

Daraus schließt man $\|l\| = 1$, $l(e) = \|e\|$ für alle $e \in K$ und $l - h \in K^0$ für alle $h \in U^*$ mit $\|h\| \leq 1$. Ferner ist l ein Extrempunkt der Einheitskugel in U^*. Aus der Darstellung $l = 2^{-1}(h_1 + h_2)$ mit $\|h_1\| \leq 1$ und $\|h_2\| \leq 1$ folgt nämlich

$$h := 2^{-1}(h_2 - h_1) = l - h_1 \in K^0 \quad \text{und}$$
$$-h = 2^{-1}(h_1 - h_2) = l - h_2 \in K^0,$$

was aber wegen $U = K - K$ zu $h = 0$, also zu $l = h_1 = h_2$ führt. l kann nach dem Singerschen Erweiterungssatz (vergl. [12] und [13] !) zu einem $\tilde{l} \in \text{Ext } S^*$ erweitert werden. Damit ist über die Behauptung des Lemmas hinausgehend die Existenz eines einelementigen K-Fundamentalsystems $\{\tilde{l}\} \subset (\text{Ext } S^*) \cap K^0$ gezeigt.

3. Nun habe K innere Punkte bzgl. $U = K - K$. Auf Grund des Singerschen Erweiterungssatzes darf ohne Einschränkung $U = X$ angenommen werden. Ferner genügt es zu zeigen, daß P eine konvexe $\sigma(X^*, X)$ - abgeschlossene und $\sigma(X^*, X)$ - lokalkompakte Menge ist, die keine Gerade enthält. Aus einem bekannten Satz von Klee (vergl. [7] 3.3!) folgt dann nämlich, daß für jedes $e_0 \in K$

$$A_{e_o} := \{ l \in P : l(e_o) = \| e_o \| \}$$

als $\sigma(X^*, X)$ - abgeschlossene konvexe und extremale Teilmenge von P einen Punkt aus Ext P enthält. Ext P ist aber nach Korollar 3 eine Teilmenge von $(\text{Ext } S^*) \cap K^o$.

Konvexität und $\sigma(X^*, X)$ - Abgeschlossenheit von P sind klar.

Sei $\hat{e}$ innerer Punkt von K und $r > 0$ so gewählt, daß $e \in K$ gilt für alle $e \in X$ mit $\| e - \hat{e} \| \leq r$.

Für jedes $q \in \mathbb{R}$ ist die Menge $B_q := \{ l \in P : l(\hat{e}) \geq q \}$ beschränkt, weil aus $\| e \| \leq 1$ und $l \in B_q$ folgt:

$\hat{e} \pm re \in K$, $l(\hat{e} \pm re) \leq \| \hat{e} \pm re \| \leq \| \hat{e} \| + r$ und $|l(e)| \leq 1 + r^{-1}(\| \hat{e} \| - q)$.

Da B_q auch $\sigma(X^*, X)$ - abgeschlossen ist, ist B_q $\sigma(X^*, X)$ - kompakt.

Für jedes $\tilde{l} \in P$ ist mit $q := \tilde{l}(\hat{e}) - 1$

$$V := \{ l \in X^* : l(\hat{e}) > q \}$$

eine $\sigma(X^*, X)$ - offene Umgebung von $\tilde{l}$. Aus $P \cap V \subset B_q$ und der $\sigma(X^*, X)$ - Kompaktheit von B_q folgt, daß P $\sigma(X^*, X)$ - lokalkompakt ist.

Falls P eine Gerade enthält, gibt es $h_1, h_2 \in X^*$ mit $h_2 \neq 0$ derart, daß für alle $t \in \mathbb{R}$ $h_1 + th_2 \in P$ gilt. Sei $h \in \{h_2, -h_2\}$. Zu jedem $t > 0$ gibt es dann auf Grund von Lemma 2 ein $l_t \in K^o$ mit $\| h_1 + th + l_t \| \leq 1$. Man gewinnt daraus die Abschätzung $\| h + t^{-1} l_t \| \leq t^{-1}(1 + \| h_1 \|)$, aus der man mit $t \to \infty$ wegen $t^{-1} l_t \in K^o$ $h \in \overline{K^o} = K^o$ erhält. Damit hat man $\pm h_2 \in K^o$, was auf Grund von $K - K = X$ $h_2 = 0$ bedeutet. P kann also keine Gerade enthalten.

Literatur

1. Bonsall, F. F. The decomposition of continuous linear functionals into non-negative components. Proc. Durham Phil. Soc. 13(A) 6-11 (1957)
2. Brosowski, B. Einige Bemerkungen zum verallgemeinerten Kolmogoroffschen Kriterium. Funktionalanalytische Methoden der numerischen Mathematik, ISNM 12, S. 25-34, Birkhäuser - Verlag (1969).
3. Brosowski, B. Nichtlineare Approximation in normierten Vektorräumen. Abstract Spaces and Approximation, ISNM 10, S. 140-159, Birkhäuser - Verlag (1969).
4. Deutsch, F. R. and Maserick, P. H. Applications of the Hahn-Banach Theorem in Approximation Theory. SIAM Rev. 9, 516-530 (1967).
5. Harris, B. Mathematical Models for Statistical Decision Theory. Optimizing Methods in Statistics, 369-389, Ed. J. S. Rustagi, Academic Press, New York (1971).
6. Harris, B. and Heindl, G. The Concept of a Best Approximation as an Optimality Criterion in Statistical Decision Theory. University of Wisconsin - Madison R.N. 1586 (1975).

7. Klee, V. L. Extremal structure of convex sets. Arch. d. Math. 8, 234-240 (1957).
8. Krabs, W. Optimierung und Approximation. Teubner Studienbücher (1975).
9. Kung-Fu Ng. The duality of partially ordered Banach spaces. Proc. London Math. Soc. (3) 19, 269-288 (1969).
10. Nikolski, W. N. Verallgemeinerung eines Satzes von A. N. Kolmogoroff auf Banach-Räume. Untersuchungen moderner Probleme der konstruktiven Funktionentheorie. V.I. Smirnov, Fizmatgiz, 335-337, Moskau (1961) (Russisch).
11. Nikolski, W. N. Ein charakteristisches Kriterium für die am wenigsten abweichenden Elemente aus konvexen Mengen. Untersuchungen moderner Probleme der konstruktiven Funktionentheorie. Verl. d. Akad. d. Wiss. Aserbeidschan, 80-84, Baku (1965) (Russisch).
12. Singer, I. Sur l' extension des fonctionelles linéaires. Rev. Roumaine Math. Pures Appl., 1, 1-8 (1956).
13. Singer, I. On the extension of continuous linear functionals and best approximation in normed linear spaces. Math. Ann., 159, 344-355 (1965).

Gerhard Heindl
Institut für Mathematik
der Technischen Universität München
8 München 2
Arcisstraße 21

A Newton-method for nonlinear Chebyshev approximation

R. Hettich

Department of Applied Mathematics
Twente University of Technology
P.O. Box 217, Enschede, The Netherlands

1. Introduction. In this paper the following approximation problem is considered:

Let $B \subset R^m$ be a compact set, $P \subseteq R^n$ an open set, $f(x)$ and $a(p,x)$ twice continuously differentiable, real-valued functions defined on B and $P \times B$ resp. Then a $p_o \in P$ is to be found, such that

$$\|f - a(p_o,.)\| \leq \|f - a(p,.)\|, \quad p \in U_o \cap P, \tag{1.1}$$

with U_o a neighbourhood of p_o. $\|.\|$ denotes the maximumnorm: $\|g\| = \max_{x \in B} |g(x)|$, g continuous on B. A p_o satisfying (1.1) is called a locally best approximation.

Note that best approximations are defined with respect to the parameter set P and not the function set $\{a(p,.) \mid p \in P\}$. From a practical point of view, this seems more appropriate to us, since otherwise global information about the function set is required, e.g. about points $p^1 \neq p^2$ with $a(p^1,.) = a(p^2,.)$. In practice, apart from special cases, to obtain that sort of information will be very difficult or even impossible.

Define the error function

$$e(p,x) = |f(x) - a(p,x)|. \tag{1.2}$$

If $e(p_o,x)$ has maxima in exactly $r = n + 1$ points $x_o^j \in B$, $j = 1, \ldots, r$, Newton's method is a wellknown and efficient means of computing p_o (cf. [6,8]). If $B = [a,b] \subset R$, $x_o^1 = a$, $x_o^{n+1} = b$, it may be formulated as follows:

p_o, $d_o = \|e(p_o,.)\|$ and x_o^j, $j = 1, \ldots, r$ $(r=n+1!)$, satisfy

$$\begin{aligned} &e(p,x^j) - d = 0, \quad j = 1, \ldots, r, \\ &e_x(p,x^j) = 0, \quad j = 2, \ldots, r-1, \end{aligned} \tag{1.3}$$

a system of 2n - 1 equations for a same number of unknowns. Given initial values

p_1, d_1, x_1^j, (1.3) is solved by Newton's method. Sufficient conditions for local convergence are given in [6].

If $r < n + 1$, there are more unknowns than equations (1.3) and the method is not applicable. In nonlinear programming the same difficulty is encountered if the number of active constraints is less than n. To overcome this difficulty, the equations required by the Kuhn-Tucker condition are added and the dual parameters occuring in this condition are regarded as unknowns too (see e.g. [10]). In Section 4 the same idea is used to formulate a method which is also applicable if $r < n + 1$. In Section 5 conditions for local convergence of the method are given in terms of sufficient conditions of the second order for locally best approximations (cf. Section 4).

Two versions of the method are presented. For linear approximation, the first one is identical to that given (without proof of convergence) in [1] for semi-infinite, linearly constraint programming. We remark that the methods in Section 4 easily can be extended to nonlinear, semi-infinite programming. To prove convergence, conditions given in [9] or [2] may be used instead of those given in Section 3. Using results from [4], approximation subject to constraints (e.g. restricted range approximation) can be treated too.

<u>2. Notation</u>. To facilitate reading we give a list of some essential symbols together with a short description or a reference to the place where they are introduced. We denote by

x	a point in R^m, generally $x \in B$
x^j, x_o^j	maxima of the error function $e(p,x)$, $e(p_o,x)$; $x_o^j \in E_o$
$x^j(t)$	local maxima of $e(p(t),x)$, $x^j(0) = x_o^j$; Theorem 3.1
μ_j	the derivative of $x^j(t)$ in $t = 0$: $\mu_j = x_t^j(0)$
p, p_o	points in $P \subseteq R^n$
$p(t)$	an arc in P, $p(0) = p_o$
ξ	the derivative of $p(t)$ in $t = 0$: $\xi = p_t(0)$
w^{ij}, w_o^{ij}	Lagrangean parameters for extrema of $e(p,x)$, $e(p_o,x)$; (3.11)
w^j, w_o^j	vectors in R^{m_j} with components w^{ij}, w_o^{ij}
$w^j(t)$	parameter vectors for extrema of $e(p(t),x)$, $w^j(0) = w_o^j$; Theorem 3.1
u^j, u_o^j	parameters in first order necessary condition; (3.20), (3.21)
$q(u,\xi)$	for given u a quadratic form in ξ; (3.22)

$f(x)$ the function to be approximated

$a(p,x)$ the approximating function

$e(p,x)$ the error function, $e(p,x) = |f(x) - a(p,x)|$

$g^i(x)$ functions defining B, (3.1)

B a compact region, where f is to be approximated, $B \subset R^m$

P an open set of parameters, $P \subseteq R^n$

E_o the set of x_o^j; (3.3), (3.7)

K a cone; (3.19)

$I, I(x)$ sets of indices; (3.1), (3.2)

G_j, M_j matrices; (3.8), (3.9)

$F_z(z)$ matrix for Newton's method; (4.9)

Derivatives are indicated by lower indices. For instance $p_t(t) = \frac{d}{dt}p(t)$ or

$$e_p(p,x) = \left(\frac{\partial e(p,x)}{\partial p_1}, \ldots, \frac{\partial e(p,x)}{\partial p_n}\right)^T.$$

Lower indices xx, xp, px, pp denote resp. m×m-, m×n-, n×m-, n×n-matrices of second order derivatives. For instance

$$e_{xp}(p,x) = (e_{px}(p,x))^T = \begin{bmatrix} \frac{\partial^2 e(p,x)}{\partial x_1 \partial p_1} \ldots & \frac{\partial^2 e(p,x)}{\partial x_1 \partial p_n} \\ \ldots & \ldots \\ \frac{\partial^2 e(p,x)}{\partial x_m \partial p_1} & \frac{\partial^2 e(p,x)}{\partial x_m \partial p_n} \end{bmatrix}.$$

An upper index j indicates that a function is evaluated for the arguments $p = p_o$, $x = x_o^j$. For instance $e_{xp}^j = e_{xp}(p_o, x_o^j)$ or $g^{ij} = g^i(x_o^j)$.

Finally $C^\nu(A,B)$, $A \subseteq R^k$, $B \subseteq R^\ell$, denotes the set of all functions defined on A with values in B, having continuous derivatives up to order ν. By assumption $f \in C^2(B,R)$, $a \in C^2(P \times B, R)$.

3. Conditions for locally best approximations. In this section some second order conditions for locally best approximations are stated without proof (cf. [3]).

From now the following assumptions are assumed to hold.

Assumption 3.1. The (compact) set B is given by

$$B = \{x \in R^m \mid g^i(x) \leq 0,\ i \in I\}, \tag{3.1}$$

with I a finite set of indices and $g^i \in C^2(R^m,R)$. For $x \in B$ define

$$I(x) = \{i \in I \mid g^i(x) = 0\}. \tag{3.2}$$

Then, for every $x \in B$, $g^i_x(x)$, $i \in I(x)$, are linearly independent.

In the following, $p_o \in P$ is a fixed point such that $\|e(p_o,\cdot)\| > 0$. Define

$$E_o = \{x \in B \mid e(p_o,x) = \|e(p_o,\cdot)\|\}. \tag{3.3}$$

Assumption 3.2. For every $\bar{x} \in E_o$ there are $\bar{w}^i > 0$, $i \in I(\bar{x})$, such that the properties (i) and (ii) hold:

(i) $$e_x(p_o,\bar{x}) - \sum_{i \in I(\bar{x})} \bar{w}^i g^i_x(\bar{x}) = 0. \tag{3.4}$$

(ii) The quadratic form $\mu^T \bar{M} \mu$,

$$\bar{M} = e_{xx}(p_o,\bar{x}) - \sum_{i \in I(\bar{x})} \bar{w}^i g^i_{xx}(p_o,\bar{x}) \tag{3.5}$$

is negative definite on the subspace

$$\bar{T} = \{\mu \in R^m \mid \mu^T g^i_x(p_o,\bar{x}) = 0,\ i \in I(\bar{x})\}. \tag{3.6}$$

Assumption 3.2 implies that E_o is a finite set

$$E_o = \{x^1_o, \ldots, x^r_o\}. \tag{3.7}$$

Thus, for $j = 1, \ldots, r$, there are $w^{ij}_o > 0$, $i \in I(x^j_o)$, such that (i) and (ii) in Assumption 3.2 hold for $\bar{x} = x^j_o$, $\bar{w}^i = w^{ij}_o$.

Let $m_j = \mathrm{card}(I(x^j_o))$ and $w^j_o \in R^{m_j}$ be the vector with components w^{ij}_o, $i \in I(x^j_o)$. Define $m \times m_j$-matrices

$$G_j = (-g^{ij}_x),\quad i \in I(x^j_o). \tag{3.8}$$

Let further

$$M_j = e^j_{xx} - \sum_{i \in I(x^j_o)} w^{ij}_o g^{ij}_{xx} \tag{3.9}$$

and

$$T_j = \{\mu \in R^m \mid \mu^T G_j = 0\}. \tag{3.10}$$

Then, for j = 1, ..., r, we have

$$e^j_x + G_j w^j_o = 0, \tag{3.11}$$

$$\mu^T M_j \mu < 0 \text{ for } \mu \in T_j,\ \mu \neq 0, \tag{3.12}$$

and

$$w^j_o > 0. \tag{3.13}$$

P is an open set. Therefore, for every $\xi \in R^n$ we can find $t^* > 0$ and $p \in C^2([0,t^*],P)$ such that

$$p(0) = p_o,\quad p_t(0) = \xi. \tag{3.14}$$

For every $x^j_o \in E_o$ define $\Phi^j : R^m \times R^{m_j} \times R \to R^{m+m_j}$ by

$$\Phi^j(x,w,t) = \begin{bmatrix} e_x(p(t),x) - \sum_{k \in I(x^j_o)} w_k g^k_x(x) \\ \vdots \\ - g^i(x) \\ \vdots \end{bmatrix},\quad i \in I(x^j_o). \tag{3.15}$$

(3.1) and (3.11) imply

$$\Phi^j(x^j_o, w^j_o, 0) = 0,\quad j = 1, \ldots, r. \tag{3.16}$$

The following theorem is proved in [3].

Theorem 3.1. There are neighbourhoods $U(x^j_o)$, $U(w^j_o)$ of x^j_o, w^j_o, a $t_o > 0$, and functions $x^j \in C^2([0,t_o],U(x^j_o))$, $w^j \in C^2([0,t_o],U(w^j_o))$ such that $x^j(0) = x^j_o$, $w^j(0) = w^j_o$, and

(i) For $(t,x,w) \in [0,t_o] \times U(x^j_o) \times U(w^j_o)$ we have $\Phi^j(x,w,t) = 0$ if and only if $x = x^j(t)$, $w = w^j(t)$.

(ii) For $t \in [0,t_o]$, $e(p(t),x)$ has local maxima in $\bigcup_{j=1}^{r} U(x_o^j)$ in exactly the points $x^1(t), \ldots, x^r(t)$.

(iii) Let G_j and M_j be given by (3.8), (3.9). The derivatives $x_t^j(0)$, $w_t^j(0)$ are uniquely determined by

$$\begin{bmatrix} M_j & G_j \\ G_j^T & 0 \end{bmatrix} \begin{bmatrix} x_t^j(0) \\ w_t^j(0) \end{bmatrix} = \begin{bmatrix} -e_{xp}^j \xi \\ 0 \end{bmatrix}. \tag{3.17}$$

Let

$$\mu_j = x_t^j(0). \tag{3.18}$$

The following conditions are established in [3].

<u>Theorem 3.2</u>. If p_o is a locally best approximation, then, for every $\xi \in K$,

$$K = \{\xi \mid \xi^T e_p^j \le 0,\ j = 1, \ldots, r\}, \tag{3.19}$$

there are real numbers $u_o^j \ge 0$, such that

$$\sum_{j=1}^{r} u_o^j = 1, \tag{3.20}$$

$$\sum_{j=1}^{r} u_o^j e_p^j = 0, \tag{3.21}$$

and

$$q(u_o,\xi) = \xi^T\{\sum_{j=1}^{r} u_o^j e_{pp}^j\}\xi - \sum_{j=1}^{r} u_o^j \mu_j^T M_j \mu_j \ge 0. \tag{3.22}$$

Observe that, by (3.17), we have

$$\mu_j^T M_j \mu_j = \xi^T\{[e_{px}^j \quad 0]\begin{bmatrix} M_j & G_j \\ G_j^T & 0 \end{bmatrix}^{-1}\begin{bmatrix} e_{xp}^j \\ 0 \end{bmatrix}\}\xi. \tag{3.23}$$

Thus, given $u_o = (u_o^1,\ldots,u_o^r)^T$, $q(u_o,\xi)$ is a quadratic form in ξ.

If $q(u_o,\xi) \ge 0$ is replaced by $q(u_o,\xi) > 0$, the condition proves to be sufficient.

Theorem 3.3. If, for every $\xi \in K$, there are $u_o^j \geq 0$, such that (3.20), (3.21) hold and, if $\xi \neq 0$, $q(u_o,\xi) > 0$, then p_o is a locally best approximation in the strict sense, i.e. there is a neighbourhood $U_o \subset P$ of p_o such that $\|e(p,.)\| > \|e(p_o,.)\|$ for $p \in U_o \sim \{p_o\}$.

Obviously, Theorem 3.3 implies:

Theorem 3.4. If (3.20), (3.21) has a unique solution $u_o^j > 0$, $j = 1, \ldots, r$, such that $q(u_o,\xi) > 0$ for $\xi \in K \sim \{0\}$, then p_o is a strict, locally best approximation.

Note that the assumptions in Theorem 3.4 imply that K is a linear subspace: $K = \{\xi \mid \xi^T e_p^j = 0,\ j = 1, \ldots, r\}$. Therefore, the positive definiteness of $q(u_o,\xi)$ on K is sufficient (cf. [9]).

4. The method. Consider the following system of $N = n + r + 1 + mr + \sum_{j=1}^{r} m_j$ equations for the N unknowns $p \in R^n$, $u \in R^r$, $d \in R$, $x^j \in R^m$, $w^j \in R^{m_j}$, $j = 1, \ldots, r$:

$$\sum_{j=1}^{r} u^j e_p(p,x^j) = 0 \tag{4.1}$$

$$e(p,x^j) - d = 0, \quad j = 1, \ldots, r \tag{4.2}$$

$$-\sum_{j=1}^{r} u^j = -1 \tag{4.3}$$

$$e_x(p,x^j) - \sum_{i \in I(x^j)} w^{ij} g_x^i(x^j) = 0, \; j = 1, \ldots, r \tag{4.4}$$

$$g^i(x^j) = 0, \quad i \in I(x^j), \quad j = 1, \ldots, r. \tag{4.5}$$

If p_o is a locally best approximation, the relations (3.21), (3.3), (3.20), (3.4), and (3.2) show that p_o, u_o, $d_o = \|e(p_o,.)\|$, x_o^j, w_o^j solve (4.1) - (4.5).

Method I. Given some approximation $z_1 = (p_1^T, u_1^T, d_1, x_1^{jT}, w_1^{jT})^T \in R^N$ of $z_o = (p_o^T, u_o^T, d_o, x_o^{jT}, w_o^{jT})^T$, system (4.1) - (4.5) is solved by Newton's method. Naturally, convergence is not secure in general (cf. Section 5).

If the system is briefly denoted by

$$F(z) = b, \tag{4.6}$$

approximations z_i, $i = 2, 3, \ldots$, are computed according to

$$z_{i+1} = z_i + \Delta z_i, \tag{4.7}$$

where Δz_i is the solution of the linear system

$$F_z(z_i)\Delta z_i = b - F(z_i). \tag{4.8}$$

If $F_z(z_o)$ is nonsingular and z_1 sufficiently close to z_o, then the z_i are known to converge to z_o, the convergence being superlinear and even quadratic if some additional assumptions hold (cf. [7]). Sufficient conditions for $F_z(z_o)$ to be nonsingular are given in Section 5. $F_z(z)$ may be written as follows

$$F_z(z) = \left[\begin{array}{c|c|c|c|c} A(z) & B(z) & 0 & D(z) & 0 \\ \hline (B(z))^T & 0 & C & S(z) & 0 \\ \hline 0 & C^T & 0 & 0 & 0 \\ \hline D'(z) & 0 & 0 & M(z) & G(z) \\ \hline 0 & 0 & 0 & (G(z))^T & 0 \end{array}\right]. \tag{4.9}$$

Here

$$A(z) = \sum_{j=1}^{r} u^j e_{pp}(p,x^j) \qquad (n\times n\text{-matrix}) \tag{4.10}$$

$$B(z) = (e_p(p_1x^1),\ldots,e_p(p,x^r)) \qquad (n\times r\text{-matrix}) \tag{4.11}$$

$$C^T = (-1,\ldots,-1) \qquad (1\times n\text{-matrix}) \tag{4.12}$$

$$D(z) = (u^1 e_{px}(p,x^1)|\ldots|u^r e_{px}(p,x^r)) \qquad (n\times rm\text{-matrix}) \tag{4.13}$$

$$(D'(z))^T = (e_{px}(p,x^1)|\ldots|e_{px}(p,x^r)) \qquad (n\times rm\text{-matrix}) \tag{4.14}$$

$$S(z) = \begin{bmatrix} (e_x(p,x^1))^T & & 0 \\ & \ddots & \\ 0 & & (e_x(p,x^r))^T \end{bmatrix} \qquad (r\times rm\text{-matrix}) \tag{4.15}$$

$$M(z) = \begin{bmatrix} M_1(z) & & 0 \\ & \ddots & \\ 0 & & M_r(z) \end{bmatrix} \qquad (rm\times rm\text{-matrix}) \tag{4.16}$$

$$M_j(z) = e_{xx}(p,x^j) - \sum_{i\in I(x^j)} w^{ij} g^i_{xx}(x^j), \quad j = 1, \ldots, r \tag{4.17}$$

$$G(z) = \begin{bmatrix} G_1(z) & & 0 \\ & \ddots & \\ 0 & & G_r(z) \end{bmatrix} \qquad (rm\times(\sum_{j=1}^{r} m_j)\text{-matrix}) \tag{4.18}$$

$$G_j(z) = (-g^i_x(x^j)), \quad i \in I(x^j) \qquad (m\times m_j\text{-matrix}). \tag{4.19}$$

Apart from coefficients u^j in $D(z)$ (recall $D'(z)$) and the submatrix $S(z)$, $F_z(z)$ is symmetric. Assuming $u^j > 0$, $j = 1, \ldots, r$ (we need this assumption for our proof of convergence too) a fully symmetric matrix is obtained if in (4.4), (4.5) the resp. j-th group of equations is multiplied by u^j and (4.2) is replaced by $e(p,x^j) - \sum_{i\in I(x^j)} w^{ij} g^i(x^j) - d = 0$. Then, the submatrix $S(z)$ is replaced by a matrix $S^*(z)$ with $S^*(z_0) = 0$ (cf. (3.11)), so that, in practice, this submatrix may be neglected.

A second method based on Theorem 3.1 can be formulated, which, however, will be shown to be essentially equivalent to the first one.

Let z_1 be an approximation of the solution with the property that x^j_1 are exact local maxima of $e(p_1,x)$ and such that (i) and (ii) in Assumption 3.2 hold for $\bar{x} = x^j_1$, $\bar{w} = w^j_1$. Thus, (4.4) and (4.5) hold. Note that Theorem 3.1 is applicable and gives information about the dependence of x^j_1, w^j_1 on p.

<u>Method II.</u> In (4.1), (4.2) x^j are regarded as functions $x^j(p)$ of p. Then (4.1) - (4.3) is a system of n + r + 1 equations for the unknowns p, u, d. Compute p_2, u_2, d_2 by performing one step of Newton's method. Compute x^j_2, w^j_2 such that (4.4), (4.5) are satisfied and start again.

We show that Theorem 3.1 gives us all information about $x^j(p)$ needed to compute p_2, u_2, d_2. We have

$$p_2 = p_1 + \Delta p_1, \qquad u_2 = u_1 + \Delta u_1, \qquad d_2 = d_1 + \Delta d_1, \tag{4.20}$$

where Δp_1, Δu_1, Δd_1 solve the linear system

$$\begin{bmatrix} A'(z_1) & B(z_1) & 0 \\ (B(z_1))^T & 0 & C \\ 0 & C^T & 0 \end{bmatrix} \begin{bmatrix} \Delta p_1 \\ \Delta u_1 \\ \Delta d_1 \end{bmatrix} = b' \tag{4.21}$$

with C given by (4.12), B(z) given by (4.11), b' defined according to (4.6), (4.8), and

$$A'(z_1) = \sum_{j=1}^{r} u_1^j [e_{pp}(p_1,x_1^j) + e_{px}(p_1,x_1^j)x_p^j(p_1)]. \tag{4.22}$$

For this, we have used the relation

$$e_x(p_1,x_1^j)x_p^j(p_1) = 0, \qquad j = 1, \ldots, r, \tag{4.23}$$

which will be proved immediately.

Thus, Method II is fully defined, if $e_{px}(p_1,x_1^j)x_p^j(p_1)$ in (4.22) can be computed. This is possible by means of (3.17) : With $M_j(z_1)$, $G_j(z_1)$ and $e_{xp}(p_1,x_1^j)$ instead of M_j, G_j and e_{xp}^j, for $\xi = e_k \in R^n$, the k-th unit vector, the solution of (3.17) is just the k-th column of the $(m+m_j)\times n$-matrix $\begin{bmatrix} x_p^j(p_1) \\ w_p^j(p_1) \end{bmatrix}$. Thus we have

$$\begin{bmatrix} M_j(z_1) & G_j(z_1) \\ (G_j(z_1))^T & 0 \end{bmatrix} \begin{bmatrix} x_p^j(p_1) \\ w_p^j(p_1) \end{bmatrix} = \begin{bmatrix} -e_{xp}(p_1,x_1^j) \\ 0 \end{bmatrix}. \tag{4.24}$$

Especially $(G_j(z_1))^T x_p^j(p_1) = 0$. Taking account of (3.4), this shows (4.23).

In [3] it is shown that for z_1 in a certain neighbourhood of z_o the matrix on the left of (4.24) is nonsingular. Therefore, we get from (4.24)

$$e_{px}(p_1,x_1^j)x_p^j(p_1) = [e_{px}(p_1,x_1^j),0] \begin{bmatrix} M_j(z_1) & G_j(z_1) \\ (G_j(z_1))^T & 0 \end{bmatrix}^{-1} \begin{bmatrix} -e_{xp}(p_1,x_1^j) \\ 0 \end{bmatrix}. \tag{4.25}$$

By means of (4.25) we can further show that p_2, u_2, d_2 computed according to (4.20) are the same as in Method I for p_1, u_1, d_1, x_1^j, w_1^j satisfying (4.4), (4.5): in this case, Δp, Δx, and Δw satisfy

$$\begin{bmatrix} D'(z_1) \\ 0 \end{bmatrix} \Delta p + \begin{bmatrix} M(z_1) & G(z_1) \\ (G(z_1))^T & 0 \end{bmatrix} \begin{bmatrix} \Delta x \\ \Delta w \end{bmatrix} = 0. \tag{4.26}$$

From (4.26) we get

$$[D(z_1),0]\begin{bmatrix}\Delta x\\ \Delta w\end{bmatrix} = [D(z_1),0]\begin{bmatrix}M(z_1) & G(z_1)\\ (G(z_1))^T & 0\end{bmatrix}^{-1}\begin{bmatrix}D'(z_1)\\ 0\end{bmatrix}\Delta p.$$

Substitution in (4.8) shows that the first $n + r + 1$ equations in (4.8) are identical with (4.21).

Concerning the corrections on x^j, w^j, the methods are different. By Method II x_2^j, w_2^j are computed such that (4.4), (4.5) hold exactly (at least theoretically). Again, this can be done using Newton's method, which is locally convergent due to the nonsingularity of $\begin{bmatrix}M(z_1) & G(z_1)\\ (G(z_1))^T & 0\end{bmatrix}$. By Method I, however, only one step of Newton's method is executed. Taking account of this relationship it is easily shown that local (quadratic) convergence of Method I implies the same for Method II.

If $B = [\alpha,\beta]$(approximation on an interval) $A'(z)$ can easily be computed explicitely. For $x^j \in \{\alpha,\beta\}$ we get $e_{px}(p,x^j)x_p^j(p) = 0$ and for $x^j \in (\alpha,\beta)$

$$e_{px}(p,x^j)x_p^j(p) = -\frac{e_{px}(p,x^j)e_{xp}(p,x^j)}{e_{xx}(p,x^j)} . \tag{4.27}$$

Method II may give rise to the question if we could not proceed more simply as follows: Solve alternately (4.1) - (4.3) with x^j fixed and (4.4), (4.5) with p, u, d fixed. The following example shows that convergence may be destroyed:

Approximate $f(x) = 1 - x^2$ in $[-1,1]$ by $a(p,x) = \frac{1}{2}p^2 - 2px$, $p \in R$. It is easily shown that the sufficient condition given in Theorem 3.4 holds for $p_o = 0$. We have $e(p,x) = |1 - x^2 - \frac{1}{2}p^2 + 2px|$ and $E_o = \{x^1\} = \{0\}$. For $|p| < 1$, the only global maximum of $e(p,x)$ in $[-1,1]$ is $x = p$. (4.1) - (4.5) become:
$u^1(-p+2x^1) = 0$, $1 - (x^1)^2 - \frac{1}{2}p^2 + 2px^1 - d = 0$, $u^1 = 1$, $-2x^1 + 2p = 0$.

Given p_1, $|p_1| < 1$, from the last equation we get $x_1^1 = p_1$ and, therefore, from the first one $p_2 = 2p_1$. Thus for every $p_1 \neq 0$, the method diverges.

On the other hand Method I and II converge: From the last equation we get $x^1 = p$ and, therefore, from the first one $-p + 2p = p = 0$.

<u>5. Convergence</u>. As we have pointed out in the last section, Method I and II converge for a starting point z_1, sufficiently close to a solution z_o of $F(z) = 0$, if $F_z(z_o)$ (see (4.9)) is nonsingular.

Theorem 5.1. If Assumptions 2.1 and 2.2 hold and if the sufficient condition for locally best approximations given in Theorem 3.4 holds in $p = p_o$, then $F_z(z_o)$ is nonsingular.

Proof. Let G_j and M_j be given by (3.8), (3.9). Then we have

$$F_z(z_o)=\left[\begin{array}{c|c|c|c|c}
\sum_{j=1}^{r} u_o^j e_{pp}^j & e_p^1 \;\dots\; e_p^r & 0 & u_o^1 e_{px}^1 \;\dots\; u_o^r e_{px}^r & 0 \\
\hline
\begin{matrix}(e_p^1)^T\\ \vdots \\ (e_p^r)^T\end{matrix} & 0 & \begin{matrix}-1\\ \vdots \\ -1\end{matrix} & 0 & 0 \\
\hline
0 & -1 \;\dots\; -1 & 0 & 0 & 0 \\
\hline
\begin{matrix}e_{xp}^1\\ \vdots \\ e_{xp}^r\end{matrix} & 0 & 0 & \begin{matrix}M_1 & & 0\\ & \ddots & \\ 0 & & M_r\end{matrix} & \begin{matrix}G_1 & & 0\\ & \ddots & \\ 0 & & G_r\end{matrix} \\
\hline
0 & 0 & 0 & \begin{matrix}G_1^T & & 0\\ & \ddots & \\ 0 & & G_r^T\end{matrix} & 0
\end{array}\right] \tag{5.1}$$

We show that the assumption that $F_z(z_o)$ is singular leads to a contradiction.

If $F_z(z_o)$ is singular, then there are $\xi \in R^n$, $\eta \in R^r$, $\delta \in R$, $\xi^1, \dots, \xi^r \in R^m$, and $\omega^j \in R^{m_j}$, $j = 1, \dots, r$, not all equal to zero, such that

$$F_z(z_o)(\xi^T, \eta^T, \delta, \xi^{1^T}, \dots, \xi^{r^T}, \omega^{1^T}, \dots, \omega^{r^T})^T = 0. \tag{5.2}$$

We may assume that

$$\delta \le 0. \tag{5.3}$$

First, we show that $\xi = 0$ implies that all other vectors η, δ, ξ^j, ω^j are zero, contrary to our assumption.

Let $\xi = 0$. Then, for $j = 1, \dots, r$, $G_j^T \xi^j = 0$. Thus $\xi^j \in T_j$ (see (3.10)) and

$$M_j \xi^j + G_j \omega^j = 0. \tag{5.4}$$

Multiplication by $(\xi^j)^T$ shows

$$(\xi^j)^T M_j \xi^j = 0.$$

Therefore, by (3.12), $\xi^j = 0$, $j = 1, \ldots, r$. Hence, by (5.4), we have $G_j\omega^j = 0$. Since the $m\times m_j$-matrix G_j has full rank $m_j \le m$ (see Assumption 3.1), $G_j\omega^j = 0$ is possible if and only if $\omega^j = 0$, $j = 1, \ldots, r$.

Moreover, since $\xi^j = 0$, we have

$$\sum_{j=1}^{r} \eta_j e_p^j = 0, \qquad \sum_{j=1}^{r} \eta_j = 0 \quad (\eta=(\eta_1,\ldots,\eta_r)^T).$$

Hence, by the uniqueness of u_o^j assumed in Theorem 3.4, $\eta = 0$.

Finally, $\xi = 0$ implies $\delta = 0$.

Now assume that $\xi \neq 0$. By (5.3), $\xi^T e_p^j \le 0$, $j = 1, \ldots, r$. Hence, $\xi \in K$ (see (3.19)). From

$$\sum_{j=1}^{r} u_o^j e_{pp}^j \xi + \sum_{j=1}^{r} \eta_j e_p^j + \sum_{j=1}^{r} u_o^j e_{px}^j \xi^j = 0$$

we obtain

$$\xi^T\Big(\sum_{j=1}^{r} u_o^j e_{pp}^j\Big)\xi + \sum_{j=1}^{r} u_o^j \xi^T e_{px}^j \xi^j + \sum_{j=1}^{r} \eta_j \xi^T e_p^j = 0. \tag{5.5}$$

Moreover, the relations

$$e_{xp}^j \xi + M_j \xi^j + G_j \omega^j = 0, \qquad G_j^T \xi = 0 \tag{5.6}$$

hold and show that ξ^j and ω^j are the unique solutions of (3.17) pertinent to ξ. Especially, $\xi^j = \mu_j$, $j = 1, \ldots, r$ (see (3.18)). Finally,

$$\sum_{j=1}^{r} \eta_j \xi^T e_p^j = \delta \sum_{j=1}^{r} \eta_j = \delta 0 = 0. \tag{5.7}$$

Hence, by (5.5), (5.6), (5.7), and $\xi^j = \mu_j$, we obtain

$$q(u_o,\xi) = \xi^T\Big(\sum_{j=1}^{r} u_o^j e_{pp}^j\Big)\xi - \sum_{j=1}^{r} u_o^j \mu_j^T M_j \mu_j = 0,$$

contrary to $q(u_o,\xi) > 0$ for $\xi \in K \sim \{0\}$. This completes the proof.

Numerical example. Consider the problem of approximating $f(x) = \sqrt{x}$ on $[0.25, 1]$ by the H-polynomial (cf [5]) $a(p,x) = \sigma((p_o x+p_1)x+p_2)^2 + p_3$, $\sigma = \pm 1$. In a first step, the discrete problem on $\{x_o, \ldots, x_4\} \subset [0.25, 1]$, $x_i = 0.25 + i(1-0.25)/4$, $i = 0, \ldots, 4$, has been solved by the method described in [5]. The solution is approximately

$$a(p^1,x) = -((0.0874x-0.4956)x+1.118)^2 + 1.501.$$

$e(p^1,x)$ has local maxima in $x = 0.25, 0.387, 0.764, 1.0$ with resp. the values 0.00246, 0.00282, 0.00265, 0.00246.

Carrying out three steps of Method I, we find p^*,

$$p_o^* = 0.088\,090\,539\,351\,, \qquad p_1^* = -\,0.495\,240\,777\,440$$

$$p_2^* = 1.119\,066\,635\,328\,, \qquad p_3^* = 1.504\,174\,868\,404$$

The values of $e(p^*,x)$ in the maxima $x = 0.25, 0.388, 0.760, 1.0$ differ less than 10^{-8}. The maximal error is about 0.00265. It is easily shown that the sufficient condition given in Theorem 3.4 holds.

Approximation by H-polynomials often results in locally best approximations with maximal error in less than $n + 1$ points. For other types of approximation on an interval $B = [\alpha,\beta]$, e.g. rational approximation, this case may be considered degenerate.

The results presented by Wetterling ([11]) indicate that, for $B \subset R^m$, $m > 1$, the case that the number of maxima is less than $n + 1$ is rather a rule than an exception.

References.

1. S.A. Gustafson and K.O. Kortanek, Numerical treatment of a class of semi-infinite programming problems, Nav. Res. Log. Quart., 20 (1973), 477-504.
2. R. Hettich, Extremalkriterien für Optimierungs- und Approximationsaufgaben, Technische Hogeschool Twente, Enschede, Dissertation, 1973.
3. R. Hettich, Kriterien zweiter Ordnung für lokal beste Approximationen, Numer. Math., 22 (1974), 409-417.
4. R. Hettich, Kriterien erster und zweiter Ordnung für lokal beste Approximationen bei Problemen mit Nebenbedingungen, Numer. Math., 25 (1975), 109-122.
5. R. Hettich, Chebyshev approximation by H-polynomials: a numerical method, J. Approximation Theory, 17 (1976).
6. G. Meinardus und D. Schwedt, Nicht-lineare Approximationen, Arch. Rat. Mech. Anal., 17 (1964), 297-326.
7. J.M. Ortega and W.C. Rheinboldt, Iterative solution of nonlinear equations in several variables, Academic Press, New York, 1970.
8. W. Wetterling, Anwendung des Newtonschen Iterationsverfahrens bei der Tschebyscheff-Approximation, insbesondere mit nichtlinear auftretenden Parametern, MTW, 10 (1963), Teil I: 61-63, Teil II: 112-115.
9. W. Wetterling, Definitheitsbedingungen für relative Extrema bei Optimierungs- und Approximationsaufgaben, Numer. Math., 15 (1970), 122-136.

10. W. Wetterling, Über Minimalbedingungen und Newton-Iteration bei nichtlinearen Optimierungsaufgaben, in: Iterationsverfahren, Numerische Mathematik, Approximationstheorie, ISNM, 15 (1970), 93-99, Birkhäuser, Basel-Stuttgart.

11. W. Wetterling, Numerische Anwendung von Kriterien zweiter Ordnung für lokal beste Approximationen, presented at the colloquium underlying these Proceedings.

Approximationen mit Lösungen von Differentialgleichungen

von

K.-H. Hoffmann und A. Klostermair

III. Mathematisches Institut
der Freien Universität Berlin
Arnimallee 2-6
1 Berlin 33

Berlin, September 1976

1. Einleitung und Übersicht

Die folgende klassische Problemstellung der Approximationstheorie ist Gegenstand dieser Arbeit:
Gegeben sei eine Funktion $g \in C[0,1]$ und eine Teilmenge $V \subset C[0,1]$.
Gesucht wird ein Element $v_0 \in V$ mit der Eigenschaft

$$\forall v \in V \quad p(g-v_0) \leq p(g-v) \quad .$$

Hierbei ist p eine stetige Halbnorm auf $C[0,1]$.
Speziell behandeln wir solche Approximationsaufgaben, bei denen die Menge V der Approximierenden durch Lösungen gewisser Differentialgleichungen beschrieben wird.
Für verschiedene Halbnormen p stellen wir Algorithmen zur Berechnung bester Approximierender auf und beweisen deren Konvergenz. Die Algorithmen beruhen auf fortgesetzter Linearisierung der Differentialgleichungen und jeweiliger Lösung des entstehenden linearen Approximationsproblems mit bekannten Verfahren. Wir folgen dabei einer Grundidee von R.BELLMAN (vgl. R.BELLMANN, R.KALABA [1965]), der für den Fall der diskreten Approximation im quadratischen Mittel mit ähnlichen Verfahren eine große Anzahl numerischer Beispiele rechnete.
Im Unterschied zu einem Verfahren von M.R.OSBORNE und G.A. WATSON [1969], das dann von L.CROMME [1975] weiter entwickelt wurde, arbeiten wir nicht mit Linearisierungen einer parameterabhängigen Funktion, die die Approximierenden explizit beschreibt, sondern betrachten die Approximierenden durch Nebenbedingungen implizit gegeben und linearisieren diese. Unser Vorgehen scheint uns bei der praktischen Durchführung gegenüber dem Verfahren von CROMME bei gewissen Anwendungen Vorzüge zu besitzen, weil wir auf die Sonderbehandlung spezieller Ausartungsfälle, wie sie beispielsweise bei der Approximation mit Exponentialsummen auftreten, verzichten können.

Der Anwendungsbereich unserer Verfahren erstreckt sich auf Parameteridentifizierungsprobleme, z.B. die Bestimmung von Reaktionskonstanten in biologischen und chemischen Prozessen, aber auch auf typische Approximationsaufgaben, wie z.B. Approximationen mit Exponentialsummen, die dann als Lösungen von speziellen gewöhnlichen Differentialgleichungen aufgefaßt werden.

2. Ein Parameteridentifizierungsproblem

Ausgangspunkt der vorliegenden Untersuchungen war die Lösung des folgenden von Chemikern gestellten Problems. Eine chemische Reaktion laufe nach dem Schema

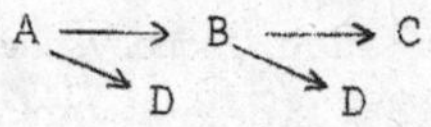

ab. Gemessen wird in einem festen Zeitintervall zu festen Zeiten t_i die Ausbeute an Substanz D_i. Die Reaktion wird durch eine Anfangswertaufgabe beschrieben:

$$(2.1)\quad \begin{cases} \dot{A} = -k_1 A \\ \dot{B} = -k_2 B + k_1 A \\ \dot{C} = k_2 B \\ \dot{D} = k_1 A + k_2 B \end{cases},$$

$A(o) = A_o$, $B(o) = C(o) = D(o) = o$.

k_1 und k_2 sind unbekannte Reaktionskonstanten, die man aus Meßdaten D_i bestimmen möchte.

Man kann dann so vorgehen, daß man die explizite Lösung von (2.1) ausrechnet:

$$(2.2)\quad \begin{cases} A(t) = A_o e^{-k_1 t} \\[1ex] B(t) = \dfrac{k_1 A_o}{k_1 - k_2}\left(e^{-k_2 t} - e^{-k_1 t}\right) \\[1ex] C(t) = \dfrac{A_o}{k_1 - k_2}\left(k_1(1-e^{-k_2 t}) - k_2(1-e^{-k_1 t})\right) \\[1ex] D(t) = A_o\left(2 + \dfrac{2k_2 - k_1}{k_1 - k_2}\, e^{-k_1 t} - \dfrac{k_1}{k_1 - k_2}\, e^{-k_2 t}\right), \end{cases}$$

und durch Variation der Konstanten k_1 und k_2 in (2.2) die Lösung D so bestimmt, daß die Meßdaten bezüglich einer geeigneten Halbnorm p möglichst gut approximiert werden. Für die Halbnorm p kommen sowohl diskrete L_2-Halbnorm als auch Tschebyscheff-Halbnorm in Frage. Es handelt sich hier um eine nichtlineare verkettete Approximation im Sinne von L.COLLATZ (vgl. L.COLLATZ/W.KRABS [1973]).

Für solche Aufgaben stehen im Rahmen der Approximationstheorie wenig Mittel zur Verfügung. Insbesondere sind folgende Fragen weitgehend offen:

a) Existenz von besten Approximierenden,

b) Notwendige Bedingungen,

c) Numerische Algorithmen.

Mit unserem Verfahren berechnen wir eine numerische Lösung dieses Problems ohne Verwendung der expliziten Lösung (2.2).

3. Allgemeine Problemstellung und Bezeichnungen

Es sei $f: [0,1] \times \mathbb{R}^n \times \mathbb{R}^m \to \mathbb{R}^n$, $g \in C[0,1]$ gegeben.

Ferner sei $T: \mathbb{R}^n \to \mathbb{R}$ ein linearer Operator.

Das Parameteridentifizierungsproblem von Abschnitt 2 ist dann ein Spezialfall folgender Approximationsaufgabe, die wir hier behandeln werden:

Gesucht wird eine Lösung $x^*(\cdot|x_o^*,\beta^*) =: x^*$ des Differentialgleichungssystems

$$(3.1) \qquad \dot{x} = f(t,x,\beta) \;, \quad x(o) = x_o \;,$$

so daß

$$(3.2) \qquad p(g-T\circ x^*) = \min$$

unter allen Lösungen $x(\cdot|x_o,\beta)$ des Systems (3.1).

Die Halbnorm p auf C[0,1] wird später genauer spezifiziert.

Durch Umschreiben von (3.1) in das erweiterte Anfangswertproblem

$$(3.1') \qquad \begin{cases} \dot{x} = f(t,x,\beta) \;, & x(o) = x_o \;, \\ \dot{\beta} = 0 \;, & \beta(o) = \beta_o \end{cases}$$

läßt sich die Approximationsaufgabe als ein Problem der Bestimmung eines optimalen Anfangswertes auffassen.

In Zukunft wollen wir die gestellten Probleme als in dieser Form gegeben ansehen.

Dazu sei $\tilde{f}: [0,1] \times \mathbb{R}^n \to \mathbb{R}^n$ eine stetige und stetig partiell nach der zweiten Variablen differenzierbare Abbildung; zur Abkürzung setzen wir $F(x)(t) := \tilde{f}(t,x(t))$. $F: C_{\mathbb{R}^n}[0,1] \to C_{\mathbb{R}^n}[0,1]$ ist dann eine stetig Frechet-differenzierbare Abbildung.

Die Frechetableitung von F an der Stelle $x \in C_{\mathbb{R}^n}[0,1]$ werde mit $F'(x)$ bezeichnet.

Sei $A \subset \mathbb{R}^n$ und $B := \{x \in C^1_{\mathbb{R}^n}[0,1] \mid x(o) \in A\}$.

Problem (P):

Minimiere $p(g-T\circ x)$ unter den Nebenbedingungen

$$x \in B \;, \quad \dot{x} = F(x) \;.$$

Das spezielle Parameteridentifizierungsproblem aus Abschnitt 2 läßt sich in dieser Form schreiben, wenn man setzt:

$$n = 6,$$

$$F(x) = \begin{bmatrix} -x_5x_1 \\ -x_6x_2 + x_5x_1 \\ x_6x_2 \\ x_5x_1 + x_6x_2 \\ 0 \\ 0 \end{bmatrix}, \quad x = (x_1,\dots,x_6) \in C^1_{\mathbb{R}^6}[0,1],$$

$$Tx = x_4, \qquad x = (x_1,\dots,x_6) \in \mathbb{R}^6,$$

$$A = \{x=(x_1,\dots,x_6) \in \mathbb{R}^6 \mid x_1 = A_o,\ x_2=x_3=x_4=0,\ x_5 \geq 0,\ x_6 \geq 0\},$$

$$p(y) = \max\{|y(t_i)| \mid i = 1,\dots,m\}, \quad y \in C[0,1].$$

4. Linearisiertes Problem, exakte Penaltysierbarkeit, Algorithmen

Sei $\tilde{x} \in B$. Wir linearisieren F in $\tilde{x}$ und setzen

$F_{\tilde{x}}(x) := F(\tilde{x}) + F'(\tilde{x})(x-\tilde{x})$. Das im Punkt $\tilde{x}$ linearisierte Problem

$(P_{\tilde{x}})$ lautet damit:

Minimiere $p(g-T\circ x)$ unter den Nebenbedingungen

$$x \in B, \quad \dot{x} = F_{\tilde{x}}(x).$$

Zur Formulierung der Algorithmen und zum Beweis ihrer Konvergenz benötigen wir die folgende Voraussetzung.

(V1) F sei global lipschitzstetig, d.h.

$$\exists L \in \mathbb{R}_o^+ \quad \forall x_1,x_2 \in C[0,1] \quad \|F(x_1)-F(x_2)\| \leq L\| x_1-x_2\| \text{ }^{1)}.$$

---

1) $\|x\| := \max\{|x_i(t)| \mid t \in [0,1], i=1,\dots,n\}$, $x \in C_{\mathbb{R}^n}[0,1]$.

Bekanntlich impliziert (V1) die Existenz einer eindeutig bestimmten Lösung der Anfangswertaufgabe $\dot{x}=F(x)$, $x(o)=x_o$ für jeden Anfangswert $x_o \in \mathbb{R}^n$. Aus (V1) folgt ferner $\| F'(x) \| \leq L$ für jedes $x \in C_{\mathbb{R}^n}[0,1]$ in der induzierten Operatornorm und damit die globale Lipschitzstetigkeit von $F_{\tilde{x}}$ für jedes $\tilde{x}$ mit der gleichen Lipschitzkonstanten L.

4.1 Lemma: Aus (V1) folgt:

(i) $\forall x \in B \;\; \exists \bar{x} \in B \;\; (\dot{\bar{x}}=F(\bar{x}) \wedge \| \bar{x}-x \| \leq e^L \| \dot{x}-F(x) \|)$,

(ii) $\forall \tilde{x} \in B \;\; \forall x \in B \;\; \exists \bar{x} \in B \;\; (\dot{\bar{x}}=F_{\tilde{x}}(\bar{x}) \wedge \| \bar{x}-x \| \leq e^L \| \dot{x}-F_{\tilde{x}}(x) \|)$.

Beweis: Wir zeigen (i) für ein $x \in B$.

Es existiert ein $\bar{x} \in C^1[0,1]$ mit $\dot{\bar{x}}=F(\bar{x})$, $\bar{x}(0) = x(0)$.

Mit der Abschätzung (vgl. F.W.SCHÄFKE,D.SCHMIDT [1973])

$$\| \bar{x}-x \| \leq \int_0^1 |\dot{x}(t)-F(x)(t)| dt \cdot e^L \leq e^L \cdot \| \dot{x}-F(x) \|$$

folgt dann die Behauptung. ∎

Aus diesem Lemma läßt sich auf die exakte Penaltysierbarkeit der Probleme (P) und $(P_{\tilde{x}})$ schließen.

4.2 Korollar: Aus (V1) folgt:

(i) $\forall K \geq e^L C \;\; (\inf\{p(g-T \circ x)+K \| \dot{x}-F(x) \| \;|\; x \in B\} = \inf\{p(g-T \circ x) \;|\; x \in B,\; \dot{x}=F(x)\})$,

(ii) $\forall K > e^L C \;\; \forall \bar{x} \in B$

$(p(g-T \circ \bar{x})+K \| \dot{\bar{x}}-F(\bar{x}) \| = \inf\{p(g-T \circ x)+K \| \dot{x}-F(x) \| \;|\; x \in B\}$

$\Longleftrightarrow \bar{x}$ Minimallösung von (P)).

(iii) Analoge Aussagen gelten für Problem $(P_{\tilde{x}})$, $\tilde{x} \in B$.

Hierbei ist $C:=\sup\{p(T \circ x) \;|\; x \in C_{\mathbb{R}^n}[0,1],\; \| x \| = 1\}$.

Beweis: Wir zeigen "$\geq$" in (i):

Sei $\varepsilon>0$ und $\hat{x}\in B$ mit

$$p(g-T\circ\hat{x})+K\|\dot{\hat{x}}-F(\hat{x})\| \leq \inf\{p(g-T\circ x)+K\|\dot{x}-F(x)\| \,|\, x\in B\}+\varepsilon.$$

Nach Lemma 4.1 existiert ein $\bar{x}\in B$ mit $\dot{\bar{x}}=F(\bar{x})$ und

$\|\bar{x}-\hat{x}\| \leq e^L\|\dot{\hat{x}}-F(\hat{x})\|$. Folglich gilt:

$$\begin{aligned}\inf\{p(g-T\circ x)\,|\,x\in B,\dot{x}=F(x)\} &\leq p(g-T\circ\bar{x}) \leq p(g-T\circ\hat{x})+p(T\circ(\bar{x}-\hat{x}))\\ &\leq p(g-T\circ\hat{x})+C\|\bar{x}-\hat{x}\|\\ &\leq p(g-T\circ\hat{x})+K\|\dot{\hat{x}}-F(\hat{x})\|.\end{aligned}$$

Wir beweisen "$\Rightarrow$" in (ii):

Unter Benutzung von (i) erhält man

$$\begin{aligned}\inf\{p(g-T\circ x)+K\|\dot{x}-F(x)\| \,|\, x\in B\} &= p(g-T\circ\bar{x})+K\|\dot{\bar{x}}-F(\bar{x})\|\\ &= p(g-T\circ\bar{x})+e^L C\|\dot{\bar{x}}-F(\bar{x})\| +(K-e^L C)\|\dot{\bar{x}}-F(\bar{x})\|\\ &\geq \inf\{p(g-T\circ x)+K\|\dot{x}-F(x)\| \,|\, x\in B\}+(K-e^L C)\|\dot{\bar{x}}-F(\bar{x})\|.\end{aligned}$$

Wegen $K-e^L C>0$ folgt $\dot{\bar{x}}=F(\bar{x})$ und daher mit (i) die Behauptung. ∎

<u>4.3 Bemerkung:</u> Im Fall $K=e^L C$ braucht die Äquivalenz in (ii) nicht zu gelten. Betrachte z.B. das Problem:

Minimiere $|x(1)|$ unter der Nebenbedingung $\dot{x}=1$, $x(o)=o$.

Hier gilt $L=o$ und $C=1$. Für $K=1$ hat man mit $\bar{x}(t)\equiv o$ die geforderte Gleichheit in (ii), aber $\bar{x}$ ist nicht Minimallösung von (P).

Die in Korollar 4.2 formulierte exakte Penaltysierbarkeit erweist sich als wesentlich für die Konvergenzbeweise unserer Algorithmen.

<u>4.4 Algorithmus (A1)</u> (Iterationsverfahren ohne λ-Strategie):

Start: Wähle $x^{(1)}\in B$.

Iterationsschritt: $x^{(i)}$ sei bereits konstruiert.

Bestimme $x^{(i+1)}$ als Minimallösung des linearen Approximationsproblems $(P_{x^{(i)}})$.

(Wir nehmen an, daß solche Minimallösungen existieren. Das ist sicher der Fall, wenn A kompakt oder ein linearer Teilraum von $\mathbb{R}^n$ ist).

Von diesem Verfahren wird in Abschnitt 5 gezeigt, daß es unter der Zusatzvoraussetzung der starken Eindeutigkeit (vgl. L.CROMME [1975]) lokal quadratisch konvergiert. Ohne die starke Eindeutigkeit können wir für ein Verfahren mit λ-Strategie, das wir im folgenden beschreiben, noch Konvergenz zeigen. Dazu benötigen wir einige Voraussetzungen und Definitionen.

$$\text{(V2)} \quad \begin{cases} K\in\mathbb{R}^+ \text{ mit } K>e^L\cdot C & \text{(vgl. Korollar 4.2)},\\ 0<\beta<1, & \\ (\lambda_j)_{j\in\mathbb{N}}\subset\mathbb{R}^+,\ \lambda_1=1,\ \alpha:=\inf\limits_{j\in\mathbb{N}}\dfrac{\lambda_{j+1}}{\lambda_j}>0,\ \lim\limits_j \lambda_j=0. & \\ A \text{ konvex, kompakt.} & \end{cases}$$

<u>4.5 Definitionen:</u>

$\hat{x}\in B$ heißt stationärer Punkt von (P):$\Longleftrightarrow$

$\hat{x}$ ist Minimallösung von $(P_{\hat{x}})$.

<u>4.6 Definition</u> (Penaltyfunktion)

$$\varphi: C^1_{\mathbb{R}^n}[0,1] \longrightarrow \mathbb{R}\ , \quad \varphi(x) := p(g-T\circ x)+K\|\dot{x}-F(x)\|\ .$$

<u>4.7 Bemerkung:</u> Sei $x\in B$ und $\bar{x}$ Lösung des Problems (P_x). Dann gilt nach 4.2 und (V2):

$$\begin{aligned} p(g-T\circ\bar{x}) &= p(g-T\circ\bar{x})+K\|\dot{\bar{x}}-F_x(\bar{x})\| \le \\ &\le p(g-T\circ x)+K\|\dot{x}-F(x)\| \\ &= \varphi(x). \end{aligned}$$

4.8 Algorithmus (A2) (Iterationsverfahren mit λ-Strategie):

Start: Wähle $x^{(1)} \in B$.

Iterationsschritt: $x^{(i)}$ sei bereits konstruiert.

Bestimme $\bar{x}^{(i)}$ als Minimallösung des linearen Approximationsproblems $(P_{x^{(i)}})$. Falls $p(g-T\circ\bar{x}^{(i)}) = \varphi(x^{(i)})$, ist $x^{(i)}$ stationärer Punkt von (P) (vgl. Bemerkung 4.7), und die Iteration wird abgebrochen. Anderenfalls ist $\varphi(x^{(i)}) > p(g-T\circ\bar{x}^{(i)})$.

Setze: $h^{(i)} := \bar{x}^{(i)} - x^{(i)}$.

Bestimme den kleinsten Index $j := j^{(i)} \in \mathbb{N}$ mit

(4.9) $$\varphi(x^{(i)}+\lambda_j h^{(i)}) \leq \varphi(x^{(i)}) - \beta\lambda_j(\varphi(x^{(i)}) - p(g-T\circ\bar{x}^{(i)})).$$

Setze: $x^{(i+1)} := x^{(i)}+\lambda_j h^{(i)}$.

5. Konvergenzbeweise

(V1) setzen wir in diesem Abschnitt stets voraus, ohne dies jedesmal zu erwähnen. Für den ersten Konvergenzsatz 5.4 brauchen wir die Existenz einer lokalen Minimallösung x^* von (P) mit folgenden Eigenschaften:

(V3) x^* ist stark eindeutige lokale Minimallösung von (P), d.h.

$$x^* \in B,\ \dot{x}^* = F(x^*) \wedge \exists \delta_1 > 0\ \exists K_1 > 0\ \forall x \in B\ (\| x - x^* \| \leq \delta_1 \wedge \dot{x} = F(x) \Rightarrow$$
$$\Rightarrow p(g-T\circ x) - p(g-T\circ x^*) \geq K_1 \| x - x^* \|).$$

(V4) F' ist lokal lipschitzstetig in einer Umgebung von x^*, d.h.

$$\exists \delta_2 > 0\ \exists L_1 > 0\ \forall x_1, x_2 \in B$$
$$(\| x_1 - x^* \| \leq \delta_2 \wedge \| x_2 - x^* \| \leq \delta_2 \Rightarrow \| F'(x_1) - F'(x_2) \| \leq L_1 \| x_1 - x_2 \|)$$

5.1 Bemerkung: Bekanntlich folgt aus (V4) und B konvex:

$$\forall \bar{x}, x \in B \quad (\| \bar{x}-x^* \| \le \delta_2 \wedge \| x-x^* \| \le \delta_2 \Rightarrow \| F_{\bar{x}}(x)-F(x) \| \le \frac{1}{2} L_1 \| x-\bar{x} \|^2).$$

5.2 Lemma: (V3) $\Rightarrow$

$$\forall x \in B \ (\| x-x^* \| \le \delta_1/2 \Rightarrow p(g-T \circ x)-p(g-T \circ x^*)+K_2 \| \dot{x}-F(x) \| \ge K_1 \| x-x^* \|),$$

wobei $K_2 := (C+K_1)e^L$ gesetzt wurde.

Beweis: Sei $x \in B$ mit $\| x-x^* \| \le \delta_1/2$. Nach Lemma 4.1(i) gibt es $\bar{x} \in B$ mit $\dot{\bar{x}}=F(\bar{x})$ und $\| \bar{x}-x \| \le e^L \| \dot{x}-F(x) \|$. Ohne Einschränkung können wir $\| \bar{x}-x^* \| \le \delta_1$ annehmen. (Andernfalls kann man dann wegen $\| x^*-x \| < \| \bar{x}-x \|$ $\bar{x}$ durch x^* ersetzen.) Damit gilt wegen (V3):

$$K_1 \| x-x^* \| \le K_1(\| x-\bar{x} \| + \| \bar{x}-x^* \|) \le K_1 \| x-\bar{x} \| + p(g-T \circ \bar{x}) - p(g-T \circ x^*)$$
$$\le (C+K_1)e^L \| \dot{x}-F(x) \| + p(g-T \circ x) - p(g-T \circ x^*). \quad \blacksquare$$

5.3 Lemma: (V3) $\wedge$ (V4) $\wedge$ B konvex $\Rightarrow$

$$\forall x \in B \quad p(g-T \circ x) - p(g-T \circ x^*) + K_2 \| \dot{x}-F_{x^*}(x) \| \ge K_1 \| x-x^* \| .$$

(x^* ist also global stark eindeutige Minimallösung von (P_{x^*}).)

Beweis: $\psi : C^1_{\mathbb{R}^n}[0,1] \longrightarrow \mathbb{R}$, definiert durch

$$\psi(x) := p(g-T \circ x) - p(g-T \circ x^*) + K_2 \| \dot{x}-F_{x^*}(x) \| ,$$

ist konvex und es gilt $\psi(x^*) = 0$. Sei $x \in B$. Setze $x_\lambda := \lambda x+(1-\lambda)x^*$ für $0<\lambda<1$. Damit folgt für genügend kleines $\lambda \in \mathbb{R}^+$ wegen B konvex und ψ konvex aus Lemma 5.2 und Bemerkung 5.1:

$$\lambda\psi(x) = \lambda\psi(x) + (1-\lambda)\psi(x^*) \ge \psi(x_\lambda)$$
$$\ge p(g-T \circ x_\lambda) - p(g-T \circ x^*) + K_2(\| \dot{x}_\lambda - F(x_\lambda) \| - \tfrac{1}{2} L \| x_\lambda - x^* \|^2)$$
$$\ge K_1 \lambda \| x-x^* \| - \tfrac{1}{2} L K_2 \lambda^2 \| x-x^* \|^2$$

und hieraus mit $\lambda \longrightarrow 0$ die Behauptung. $\blacksquare$

5.4 Erster Konvergenzsatz für Algorithmus (A1):

Sei A konvex und (V1) erfüllt. x^* sei eine lokale Minimallösung von (P), für die (V3) und (V4) gelten. Dann ist der Algorithmus (A1) (ohne λ-Strategie) lokal und mindestens quadratisch gegen x^* konvergent, d.h.

$$\exists\ \delta>0\quad \exists\ C_1>0\quad \forall x^{(1)}\in B\quad (\|x^{(i)}-x^*\| \leq \delta \rightarrow$$
$$\forall i\in N\quad \|x^{(i+1)}-x^*\| \leq C_1\|x^{(i)}-x^*\|^2).$$

$(x^{(i)})_{i\in N}$ sei dabei die durch (A1) gelieferte Folge, wenn mit $x^{(1)}$ gestartet wird.

Beweis: Wähle δ_o mit $0<\delta_o\leq\delta_2$ und $L_1K_2\delta_o<K_1$. Sei $x\in B$ mit $\|x-x^*\| \leq \delta_o$ und $\bar{x}$ Minimallösung von (P_x). Dann folgt aus Lemma 5.3, mehrfacher Anwendung der Dreiecksungleichung, Korollar 4.2(ii), (V4) und Bemerkung 5.1:

$$\begin{aligned} K_1\|\bar{x}-x^*\| &\leq p(g-T\circ\bar{x})-p(g-T\circ x^*)+K_2\|\dot{\bar{x}}-F_{x^*}(\bar{x})\| \\ &\leq p(g-T\circ\bar{x})-p(g-T\circ x^*)+K_2(\|\dot{\bar{x}}-F_x(\bar{x})\| + \\ &\qquad \|(F'(x^*)-F'(x))(\bar{x}-x^*)\| + \|F(x^*)-F_x(x^*)\|) \\ &\leq 2K_2\ \|F(x^*)-F_x(x^*)\| + L_1K_2\|x^*-x\|\ \|\bar{x}-x^*\| \\ &\leq 2K_2\ \frac{L_1}{2}\ \|x-x^*\|^2 + L_1K_2\delta_o\|\bar{x}-x^*\|, \end{aligned}$$

also $\quad \|\bar{x}-x^*\| \leq C_1\|x-x^*\|^2$ mit $C_1 := \dfrac{K_2L_1}{K_1-L_1K_2\delta_o}$.

Die Behauptung folgt nun für jedes δ mit $0<\delta<\delta_o$ und $C_1\delta<1$. ∎

5.5 Bemerkung: (V3) ist sicher nicht erfüllt, wenn die Halbnorm p an der Stelle $g-T\circ x^*$ Frechet-differenzierbar ist. Falls p eine diskrete Tschebyscheff-Halbnorm ist, gilt "im allgemeinen" (V3). Wir werden in Satz 5.7 zeigen, daß im Falle $p(g-T\circ x^*) = 0$[2] die Voraussetzung (V3) stets gilt, wenn die folgende Bedingung erfüllt ist:

(V5) $\quad \forall x\in B \quad (\dot{x} = F_{x^*}(x) \wedge p(T_\circ(x-x^*)) = 0 \Rightarrow x = x^*)$.

5.6 Lemma: (V5) $\wedge$ B konvex $\Rightarrow$

$\exists K_o>0 \quad \forall x\in B \quad (\dot{x} = F_{x^*}(x) \Rightarrow p(T_\circ(x-x^*)) \geq K_o \| x-x^* \|)$.

Beweis: $B_{x^*} := \{\lambda(x-x^*) | \lambda\in\mathbb{R}, x\in B\}$ ist die lineare Hülle der konvexen Menge $B-x^*$. Auf dem linearen Teilraum

$S_{x^*} := \{x\in B_{x^*} | \dot{x} = F'(x^*)x\} \subset C^1_{\mathbb{R}^n}[0,1]$

ist $p\circ T$ wegen (V5) eine Norm. Auf grund der endlichen Dimension von S_{x^*} sind die Normen $x \longmapsto p(T\circ x)$ und $\|\cdot\|$ auf S_{x^*} äquivalent. Hieraus folgt die Behauptung.

(Beachte: $x\in B_{x^*} + x^* \supset B \wedge \dot{x} = F_{x^*}(x) \Leftrightarrow x-x^*\in S_{x^*}$.) ∎

5.7 Satz: Sei $x^*\in B$ mit $\dot{x}^* = F(x^*)$ und $p(g-T\circ x^*)) = 0$

A sei konvex, (V1) und (V4) seien erfüllt. Dann gilt die Äquivalenz:

(V3) $\Leftrightarrow$ (V5) .

Beweis: "$\Rightarrow$" folgt aus Lemma 5.3. Wir zeigen "$\Leftarrow$";

Sei $x\in B$ mit $\|x-x^*\| \leq \delta_2$, und $\dot{x} = F(x)$. Nach Lemma 4.1(ii) gibt es $\bar{x}\in B$ mit $\dot{\bar{x}} = F_{x^*}(\bar{x})$ und

$\| \bar{x}-x\| \leq e^L \| \dot{x}-F_{x^*}(x)\| = e^L \| F(x)-F_{x^*}(x)\| \leq \frac{1}{2}L_1 e^L \|x-x^*\|^2$,

---

2) Das bedeutet im Identifizierungsproblem von Abschnitt 2, daß das Modell (2.1) den Vorgang genau beschreibt, und die Meßdaten D_i exakt sind.

letzteres wegen Bemerkung 5.1. Nach Lemma 5.6 gilt:

$K_o\|\bar{x}-x^*\| \leq p(T\circ(\bar{x}-x^*)) \leq p(g-T\circ\bar{x}) \leq p(g-T\circ x) + C\|\bar{x}-x\|$, also

$$K_o\|x-x^*\| \leq K_o(\|x-\bar{x}\| + \|\bar{x}-x^*\|) \leq p(g-T\circ x) + (K_o+C)\|\bar{x}-x\|$$

$$\leq p(g-T\circ x) - p(g-T\circ x^*) + \tfrac{1}{2}L_1e^L(K_o+C)\|x-x^*\|^2 .$$

Dann gilt (V3) mit $K_1 = K_o - \frac{1}{2}L_1e^L(K_o+C)\delta_1$ für jedes δ_1 mit $0<\delta_1\leq\delta_2$ und $\frac{1}{2}L_1e^L(K_o+C)\delta_1<K_o$. ∎

5.8 Bemerkung:

In dem Fall einer (diskreten oder kontinuierlichen) L_2-Halbnorm ist (V3) zwar sicher nicht erfüllt, wenn $p(g-T\circ x^*)>0$ gilt. Wir zeigen jedoch für diesen Fall in Satz 5.9 (unter zusätzlichen Voraussetzungen), daß die folgende Abschwächung von (V3):

(V6) $\quad x^*\in B \wedge \dot{x}^* = F(x^*) \wedge \exists\ \delta_1>0\ \exists\ K_1>0\ \forall x\in B$

$$(\|x-x^*\| \leq \delta_1 \wedge \dot{x} = F(x) \Rightarrow p(g-T\circ x)- p(g-T\circ x^*)\geq K_1\|x-x^*\|^2)$$

erfüllt ist. Auf (V6) stützt sich wesentlich der zweite Konvergenzsatz 5.12.

5.9 Satz: Sei B konvex und x^* lokale Minimallösung von (P). (V4)∧(V5) seien erfüllt. Die Halbnorm p werde durch ein positiv semidefinites inneres Produkt $\langle\cdot,\cdot\rangle : C[0,1]\times C[0,1] \to \mathbb{R}$ erzeugt, d.h. es gelte: $\forall y\in C[0,1] \quad p(y)^2 = \langle y,y\rangle$.

Sei $0<\delta_1\leq\delta_2$ (δ_2 wie in (V4)).

Falls $\eta := [Cp(g-T\circ x^*)+K_o(K_o+C)\delta_1]L_1e^L<K_o^2$, gilt (V6) mit

$$K_1 := \frac{K_o^2 - \eta}{2p(g-T\circ x^*)+C\delta_1} \qquad (K_o \text{ wie in Lemma 5.6})\ .$$

Beweis: Sei $x\in B$ mit $\dot{x} = F(x)$ und $\| x-x^*\| \le \delta_1$. Nach Lemma 4.1(ii) gibt es ein $\bar{x}\in B$ mit $\dot{\bar{x}} = F_{x}{*}(\bar{x})$ und

$$\| \bar{x}-x\| \le e^L \| \dot{x}-F_x{*}(x)\| = e^L \| F(x)-F_x{*}(x)\| \le \tfrac{1}{2}L_1 e^L \| x-x^*\|^2 .$$

Da x^* lokale Minimallösung von (P), ist x^* auch globale Minimallösung von $(P_x{*})$ (vgl. Lemma 5.25).

Da $V := \{T\circ\hat{x} \mid \hat{x}\in B \wedge \dot{\hat{x}}=F_x{*}(\hat{x})\}$ konvex, läßt sich das globale Kolmogoroffkriterium (K.-H.HOFFMANN [1975]) anwenden, es liefert wegen $T\circ\bar{x}\in V$: $\langle T\circ\bar{x} - T\circ x^*, g - T\circ x^*\rangle \le 0$. Nach Lemma 5.6 gilt:

$$\begin{aligned}
p(T\circ(x-x^*)) &\ge p(T\circ(\bar{x}-x^*)) - p(T\circ(\bar{x}-x)) \\
&\ge K_o\| \bar{x}-x^*\| - C\| \bar{x}-x\| \\
&\ge K_o\| x-x^*\| -(K_o+C)\tfrac{1}{2}L_1e^L\| x-x^*\|^2 , \quad \text{also} \\
p(T\circ(x-x^*))^2 &\ge K_o^2\| x-x^*\|^2 - 2K_o(K_o+C)\tfrac{1}{2}L_1e^L\| x-x^*\|^3 \\
&\ge (K_o^2 - 2K_o(K_o+C)\tfrac{1}{2}L_1e^L\delta_1)\| x-x^*\|^2 .
\end{aligned}$$

Es folgt:

$$\begin{aligned}
p(g-T\circ x)^2 - p(g-T\circ x^*)^2 &= \langle g-T\circ x, g-T\circ x\rangle - \langle g-T\circ x^*, g-T\circ x^*\rangle \\
&= -2\langle T\circ x - T\circ x^*, g-T\circ x^*\rangle + \langle T\circ x-T\circ x^*, T\circ x - T\circ x^*\rangle \\
&\ge -2\langle T\circ\bar{x}-T\circ x^*, g-T\circ x^*\rangle - 2C\| \bar{x}-x\| p(g-T\circ x^*) + \\
&\quad + p(T\circ(x-x^*))^2 \\
&\ge -2C\| \bar{x}-x\| p(g-T\circ x^*) + (K_o^2 - 2K_o(K_o+C)\tfrac{L_1}{2}e^L\delta_1)\| x-x^*\|^2
\end{aligned}$$

$$\geq (K_o^2 - [Cp(g-T\circ x^*)+K_o(K_o+C)\delta_1]L_1e^L)\| x-x^*\|^2,$$

schließlich

$$p(g-T\circ x)-p(g-T\circ x^*) = \frac{p(g-T\circ x)^2-p(g-T\circ x^*)^2}{p(g-T\circ x)+p(g-T\circ x^*)}$$

$$\geq \frac{K_o^2-\eta}{2p(g-T\circ x^*)+C\| x-x^*\|}\| x-x^*\|^2. \quad \blacksquare$$

Dem zweiten Konvergenzsatz schicken wir zwei Lemmata voraus, die den Lemmata 5.2 bzw. 5.3 entsprechen.

<u>5.10 Lemma:</u> (V6) $\Rightarrow$

$\forall\ 0<\delta\leq\delta_1/2 \quad \forall x\in B$

$$(\| x-x^*\| \leq \delta \Rightarrow p(g-T\circ x)-p(g-T\circ x^*)+K_2(\delta)\| \dot{x}-F(x)\| \geq K_1\| x-x^*\|^2),$$

wobei $K_2(\delta):=[C+K_1(3e^L+1)\delta]e^L$ gesetzt wurde.

Beweis: Sei $x\in B$ mit $\| x-x^*\| \leq \delta$. Wie in Beweis von Lemma 5.2 gibt es $\bar{x}\in B$ mit $\dot{\bar{x}}=F(\bar{x})$, $\| \bar{x}-x\| \leq e^L\| \dot{x}-F(x)\|$ und $\| \bar{x}-x^*\| \leq \delta_1$. Wenn wir $\bar{x}$ wie im Beweis von Lemma 4.1(i) wählen, gilt $\bar{x}(o)=x(o)$ und folglich nach einer bekannten Abschätzung (vgl. F.W.SCHÄFKE,D.SCHMIDT [1973]):

$\| \bar{x}-x^*\| \leq |\bar{x}(o)-x^*(o)|e^L \leq \| x-x^*\| e^L$. Im Falle $\bar{x}=x^*$ ist dies trivial.

Aus (V6) folgt

$$K_1\| x-x^*\|^2 \leq K_1(\| x-\bar{x}\| + \| \bar{x}-x^*\|)^2 = K_1(\| x-\bar{x}\| + 2\| \bar{x}-x^*\|)\| x-\bar{x}\| + K_1\| \bar{x}-x^*\|^2$$

$$\leq K_1(\| x-x^*\| +3\| \bar{x}-x^*\|)\ \| x-\bar{x}\| +p(g-T\circ\bar{x})-p(g-T\circ x^*)$$

$$\leq K_2(\delta)\| \dot{x}-F(x)\| +p(g-T\circ x)-p(g-T\circ x^*). \qquad \blacksquare$$

<u>5.11 Lemma:</u> (V4) $\wedge$ (V6) $\wedge$ B konvex $\Rightarrow$

$\forall$ $o<\delta\leq \min\{\delta_1/2,\delta_2\}$ $\forall x\in B$

$$p(g-T\circ x)-p(g-T\circ x^*)+K_2(\delta)\| \dot{x}-F_x*(x)\| \geq K_1(\delta).\min\{\delta,\| x-x^*\| \}\cdot\| x-x^*\| ,$$

wobei $K_1(\delta) := K_1-\frac{1}{2}L_1K_2(\delta)$ gesetzt wurde.

Beweis: Sei $o<\delta\leq \min\{\delta_1/2,\delta_2\}$. $\psi : C^1_{\mathbb{R}^n}[0,1] \longrightarrow \mathbb{R}$, definiert durch

$$\psi(x) := p(g-T\circ x)-p(g-T\circ x^*)+K_2(\delta)\| \dot{x}-F_x*(x)\| ,$$

ist konvex und es gilt $\psi(x^*) = 0$. Sei $x\in B\setminus\{x^*\}$. [3)]

Mit $\lambda := \min\{\delta/\| x-x^*\| ,1\}$, $x_\lambda := \lambda x+(1-\lambda)x^*$ gilt $o<\lambda\leq 1$, $x\in B$ und $\| x_\lambda-x^*\| \leq\delta$. Aus der Konvexität von ψ, Lemma 5.1o und Bemerkung 5.1 folgt

$$\lambda\psi(x) = \lambda\psi(x)+(1-\lambda)\psi(x^*) \geq \psi(x_\lambda)$$

$$\geq p(g-T\circ x_\lambda)-p(g-T\circ x^*)+K_2(\delta)(\| \dot{x}_\lambda-F(x_\lambda)\| -\tfrac{1}{2}L_1\| x_\lambda-x^*\|^2)$$

$$\geq (K_1-\tfrac{1}{2}L_1K_2(\delta))\| x_\lambda-x^*\|^2 = K_1(\delta)\lambda^2\| x-x^*\|^2 ,$$

also $\psi(x) \geq K_1(\delta)\lambda\| x-x^*\|^2 = K_1(\delta)\min\{\delta,\| x-x^*\| \}\| x-x^*\|$. $\blacksquare$

---

3) Der Fall $x = x^*$ ist trivial.

5.12 Zweiter Konvergenzsatz für Algorithmus (A1)

A sei konvex und x^* sei eine lokale Minimallösung von (P), die (V4) und (V6) erfüllt. Falls

(V7) $K_1 > \frac{5}{2}L_1Ce^L$,

ist der Algorithmus (A1) lokal und mindestens linear gegen x^* konvergent, d.h.

$$\exists\ \delta>0 \quad \exists\ 0\leq C_1<1 \quad \forall x^{(1)}B \quad (\| x^{(1)}-x^*\| \leq \delta \Rightarrow$$
$$\forall i\in\mathbb{N}\ \ \| x^{(i+1)}-x^*\| \leq C_1 \| x^{(i)}-x^*\|).$$

$x^{(i)}$ $i\in\mathbb{N}$ sei dabei die durch (A1) generierte Folge, wenn mit $x^{(1)}$ gestartet wird.

Beweis: Wir modifizieren den Beweis von Satz 5.4.

Sei $0<\delta\leq \min\{\delta_1/2,\delta_2\}$ mit $\delta < \dfrac{K_1-\frac{5}{2}L_1Ce^L}{\frac{5}{2}K_1L_1(3e^L+1)e^L}$

und $x\in B$ mit $\| x-x^*\| < \delta$. $\bar{x}$ sei Minimallösung von (P_x).

Dann folgt aus Lemma 5.11, mehrfacher Anwendung der Dreiecksungleichung, Korollar 4.2(ii), (V4) und Bemerkung 5.1 (vgl. Beweis zu Satz 5.4):

(5.13) $K_1(\delta)\min\{\delta,\| x-x^*\|\}\| \bar{x}-x^*\| \leq$

$$\leq p(g-T\circ\bar{x})-p(g-T\circ x^*)+K_2(\delta)\| \dot{x}-F_{x^*}(\bar{x})\|$$
$$\leq 2K_2(\delta)\cdot\frac{L_1}{2}\| x-x^*\|^2+L_1K_2(\delta)\| x-x^*\|\,\| \bar{x}-x^*\| .$$

Wir werden $q := \frac{\|\bar{x}-x^*\|}{\|x-x^*\|} \leq C_1$ zeigen; dabei sei

$C_1 := \frac{a_1 + (a_1{}^2+4a_1a_2)^{1/2}}{2a_2}$ die positive Lösung von

$a_2C_1^2 - a_1C_1 - a_1 = 0$ mit $a_1 := L_1K_2(\delta)$,

$a_2 := K_1(\delta) = K_1 - \frac{1}{2}a_1$. Wegen $a_1 \geq 0$ und

$K_1 - \frac{5}{2}a_1 = K_1 - \frac{5}{2}L_1Ce^L - \frac{5}{2}L_1K_1(3e^L+1)e^L\ \delta > 0$ gilt

$a_2 > 2a_1 \geq 0$ und deshalb $0 \leq \frac{a_1}{a_2-a_1} \leq C_1 < 1$.

Zum Beweis von $q \leq C_1$ unterscheiden wir zwei Fälle.

1. Fall: $\|\bar{x}-x^*\| \leq \delta$. Aus (5.13) folgt nach Division durch $\|x-x^*\|^2$: $a_2 q^2 - a_1q - a_1 \leq 0$. Folglich gilt $q \leq C_1$

2. Fall: $\|\bar{x}-x^*\| > \delta$. Aus (5.13) folgt nach Division durch $\delta\|x-x^*\|$: $a_2 q \leq a_1\frac{\|x-x^*\|}{\delta}(1+q) \leq a_1(1+q)$.

Also gilt $q \leq \frac{a_1}{a_2-a_1} \leq C_1$. ■

5.14 Bemerkung: Aus Satz 5.9 geht hervor, daß im Falle der L_2-Halbnorm (bei Erfülltsein von (V4) und (V5)) die Voraussetzung (V6) und die Bedingung (V7) in Satz 5.12 erfüllt sind, wenn der lokale Minimalwert $p(g-T\circ x^*)$ genügend klein ist. Weiter läßt sich auf Grund des Beweises von Satz 5.12 vermuten, daß Konvergenzbereich und

Konvergenzgeschwindigkeit um so größer sind, je größer K_1 (im Vergleich zu L_1Ce^L) ist, also (unter Verwendung von Satz 5.9) je kleiner $p(g-T\circ x^*)$ ist.

<u>5.15 Beispiel:</u> Sei $n := 2$,

$$F(x)(t) := \left((1-4t)\frac{x_2(t)^2}{1+x_2(t)^2} + (4-8t)\,x_2(t)\ ,\ 0\right)$$

für $x = (x_1,x_2) \in C^1_{\mathbb{R}^2}[0,1]$,

$Tx := x_1$ für $x = (x_1,x_2) \in \mathbb{R}^2$, $A := \{(x_1,x_2)\in\mathbb{R}^2 \mid x_1=0\}$,

$p(y) := \left(y(\frac{1}{2})^2 + y(1)^2\right)^{\frac{1}{2}}$ für $y \in C[0,1]$,

$g \in C[0,1]$ mit $g(\frac{1}{2}) = 0$ und $a := g(1) \geq 0$ (sonst beliebig).

Die Lösung $x(b)$ des Anfangswertproblems

$\dot{x}(b) = F(x(b))$, $x(b)(0) = (0,b) \in A$ lautet für $b \in \mathbb{R}$:

$$x(b)(t) = \left(t(1-2t)\frac{b^2}{1+b^2} + 4t(1-t)b\ ,\ b\right),\ \text{also}$$

$Tx(b)(\frac{1}{2}) = b$, $Tx(b)(1) = -\dfrac{b^2}{1+b^2}$. Man sieht leicht, daß

$x^* = x(0)$ mit $x^*(t) \equiv 0$ die (einzige lokale) Minimallösung von (P) ist und daß $p(g-T\circ x^*) = a$ ist. Es gilt für $b \in \mathbb{R}$ und $a > 0$:

$$p(g-T\circ x(b)) - p(g-T\circ x^*) = \left(b^2+(a+\frac{b^2}{1+b^2})^2\right)^{\frac{1}{2}} - a$$

$$= \frac{1+2a}{2a} b^2 + O(b^4)$$

$$= \frac{1+2a}{a} \| x(b)-x^* \|^2 (1+O(b^2)).$$

(V6) ist also erfüllt für jedes $K_1 < \frac{1+2a}{2a}$. Außerdem sind (V1) und (V4) erfüllt. Eine Durchrechnung des Algorithmus (A1) zeigt, daß die Folge $\left(x^{(i)}\right)_{i\in\mathbb{N}}$ die Gestalt $x^{(i)} = x(b^{(i)})$, $b^{(i)}\in\mathbb{R}$ für $i>1$ hat mit

$$b^{(i+1)} = -\frac{2(a-d^{(i)}b^{(i)}(1-b^{(i)2}))}{1+4d^{(i)2}} \cdot d^{(i)} = (-2a+O(b^{(i)2}))\cdot b^{(i)} .$$

wobei $d^{(i)} := \frac{b^{(i)}}{(1+b^{(i)2})^2}$ gesetzt wurde.

Algorithmus (A1) ist hier also lokal linear konvergent für $2a < 1$, während er für $2a > 1$ divergiert (x^* ist "abstoßend" für $2a > 1$).

Für den Fall, daß die Halbnorm p durch ein positiv semidefinites inneres Produkt $\langle\cdot,\cdot\rangle$ erzeugt wird, kann man für $x^*\in B$ mit mit $\dot{x}^* = F(x^*)$ und $p(g-T\circ x^*) = 0$ auch ohne Rückgriff auf Satz 5.7 Konvergenz des Algorithmus (A1) mit einem einfachen Fixpunktargument zeigen.

Dazu sei $\bar{x}\in B$ mit $\| x^*-\bar{x}\| \leq \delta_2/2$ und $\Phi[\bar{x}]:[0,1] \longrightarrow \mathbb{R}^{n\times n}$ ein

Fundamentalsystem der homogenen Differentialgleichung $\dot{x} = F'(\bar{x})x$;

d.h. $\dot{\Phi}[\bar{x}] = F'(\bar{x})\Phi[\bar{x}]$, $\Phi[\bar{x}](0) = I$.

Dann gilt für die Lösung des Problems $\dot{x} = F_{\bar{x}}(x)$, $x(0)\in A$ die Darstellung

$$x(t) = \Phi[\bar{x}](t)x(0) + z[\bar{x}](t) \ , \ 0 \leq t \leq 1 \ ,$$

mit $z[\bar{x}](t) := \Phi[\bar{x}](t) \int_0^t \Phi[\bar{x}]^{-1}(s)\left[F(\bar{x})(s)-F'(\bar{x})\bar{x}(s)\right]ds$.

Wir wollen jetzt annehmen, daß A eine lineare Mannigfaltigkeit ist:

$A = \mathrm{span}\,(e_1,\dots,e_k)$ mit einer Basis $e_1,\dots,e_k$ von A.

Das Problem $(P_{\bar{x}})$ lautet dann:

Minimiere

$$\langle g-T\circ(\Phi[\bar{x}](e_1,\dots,e_k)\alpha+z[\bar{x}]) \ , \ g-T\circ(\Phi[\bar{x}](e_1,\dots,e_k)\alpha+z[\bar{x}])\rangle$$

bzgl. $\alpha\in\mathbb{R}^k$.

Notwendig und hinreichend für eine Lösung $\tilde{\alpha}$ von $(P_{\bar{x}})$ ist die Gültigkeit der Normalgleichungen

$$(5.16) \quad \left(\langle T\circ\Phi[\bar{x}]e_i,T\circ\Phi[\bar{x}]e_j\rangle\right)_{i,j=1}^k \begin{bmatrix}\tilde{\alpha}_1\\ \vdots \\ \tilde{\alpha}_k\end{bmatrix} = \left(\langle g-T\circ z[\bar{x}],T\circ\Phi[\bar{x}]e_i\rangle\right)_{i=1}^k ,$$

oder $\tilde{\alpha} = M[\bar{x}]^{-1}\, v[\bar{x}]$, wobei

$$M[\bar{x}] := \left(\langle T\circ\Phi(\bar{x})e_i,T\circ\Phi[\bar{x}]e_j\rangle\right)_{i,j=1}^k \quad \text{und} \quad v(\bar{x}) := \left(\langle g-T\circ z[\bar{x}],T\circ\Phi[\bar{x}]e_i\rangle\right)_{i=1}^k$$

gesetzt und angenommen wurde, daß det $M[\bar{x}] \neq 0$ gilt.

Andererseits gilt:

$$\tilde{x}(t) - \int_0^t F'(\bar{x})(\tilde{x}-\bar{x})(s)ds = \int_0^t F(\bar{x})(s)ds + (e_1,\dots,e_k)M[\bar{x}]^{-1} v(\bar{x}) .$$

Setzt man $\psi(y)(t) := y(t) - \int_0^t F(y)(s)ds$, so gilt:

5.17 Lemma: (V4) $\Rightarrow$

ψ ist stetig differenzierbar, es gilt

$$\psi'(\bar{x})(x)(t) = x(t) - \int_0^t F'(\bar{x})x(s)ds \, , \quad x \in B,$$

und $\psi'(\bar{x})$ besitzt eine beschränkte Inverse.

Beweis: Wir zeigen nur, daß $\psi'(\bar{x})$ eine beschränkte Inverse besitzt:

$$|\psi'(\bar{x})(x)(t)| \geq |x(t)|e^{-2Lt} \cdot e^{2Lt} - \int_0^t \| \tilde{f}_x(s,\bar{x}(s)) \| \cdot |x(s)|e^{-2Ls} e^{2Ls} ds$$

$$\geq |x(t)|e^{-2Lt} \cdot e^{2Lt} - L \cdot \|x\|_w \left(\frac{1}{2L} e^{2Lt} - \frac{1}{2L}\right)$$

$$\geq e^{2Lt} \left(|x(t)|e^{-2Lt} - \frac{1}{2} \|x\|_w\right) .$$

$$\Rightarrow \|\psi'(\bar{x})(x)\|_w \geq \frac{1}{2} \|x\|_w \, , \quad \|x\|_w := \sup_{0 \leq t \leq 1} |x(t)|e^{-2Lt} .$$

Da die Normen $\|x\|_w$ und $\|x\|$ äquivalent sind, folgt

$$\|\psi'(\bar{x})(x)\| \geq \frac{1}{2} \cdot e^{-2L} \|x\|$$

und damit nach dem Satz von Banach (vgl. L.A.LUSTERNIK-V.J.SOBOLEV [1974], p. 1o5) die Existenz einer beschränkten Inversen. ∎

Damit folgt:

$$\psi'(\bar{x})(\tilde{x}-\bar{x}) = -\psi(\bar{x}) + (e_1,\dots,e_k)M[\bar{x}]^{-1}v(\bar{x})$$

und $\tilde{x} = \bar{x} - \psi'(\bar{x})^{-1}\left(\psi(\bar{x})-(e_1,\dots,e_k)M[\bar{x}]^{-1}v(\bar{x})\right)$

oder mit der Abkürzung

$$(5.18)\qquad G(\bar{x}) := \bar{x} - \psi'(\bar{x})^{-1}\left(\psi(\bar{x})-(e_1,\dots,e_k)M[\bar{x}]^{-1}v(x)\right)$$

die Iterationsvorschrift

$$\tilde{x} = G(\bar{x}) .$$

5.19 Lemma: Sei $\det M[x^*] \neq 0$. Dann gilt:

x^* ist Fixpunkt von $G \Leftrightarrow x^*$ ist stationärer Punkt von (P) .

Beweis: $G(x^*) = x^* \Leftrightarrow \psi(x^*) = (e_1,\dots,e_k)M[x^*]^{-1}v(x^*)$.

Damit ist $\psi(x^*)$ eine konstante Funktion: $\psi(x^*) = x^*(0)$.

Dann folgt: $x^* = G(x^*) \Leftrightarrow \dot{x}^* = F(x^*)$ und Normalgleichungen (5.16) sind erfüllt. $\Leftrightarrow x^*$ stationärer Punkt von (P). ∎

Es ist damit die Konvergenz von Algorithmus (A1) gezeigt, wenn, unter der Voraussetzung $\det M[x^{(i)}] \neq 0$ für alle $i\in\mathbb{N}$, die Iterationsfolge $\left(x^{(i)}\right)_{i\in\mathbb{N}}$, $x^{(i+1)} = G\left(x^{(i)}\right)$, konvergiert.

5.2o Dritter Konvergenzsatz für Algorithmus (A1)

Die Halbnorm p werde durch ein positiv semidefinites inneres Produkt $\langle\cdot,\cdot\rangle$ erzeugt. A sei ein linearer Teilraum von $\mathbb{R}^n$.

Es sei $x^*\in B$ mit $\dot{x}^* = F(x^*)$ und $p(g-T\circ x^*) = 0$. Ferner sei $\det M[x^*] \neq 0$, und es gelte (V4) und

(V5') $p(g-T\circ x^*) = 0 \Rightarrow g = T\circ x^*$.

Dann konvergiert der Algorithmus (A1) (ohne λ-Strategie) lokal.

Beweis: Es sei ohne Einschränkung $x^* = \theta$, da man sonst $y := x - x^*$ setzen kann. Mit (V5') folgt dann $g = \theta$. Wir zeigen jetzt, daß $G'(\theta)$ existiert und $G'(\theta) = \theta$ gilt. Dann ist G in einer Null-Umgebung kontrahierend. Wenn dann $M[x]^{-1}$ in dieser Null-Umgebung existiert, so ist die Iteration

$$x^{(i+1)} = G\left(x^{(i)}\right)$$

unbeschränkt durchführbar, und es gilt $\lim\limits_{i\to\infty} x^{(i)} = x^*$.

(i) Die Abbildungen $x \mapsto \Phi[x]$ und $x \mapsto \Phi^{-1}[x]$ sind an der Stelle θ stetig:

Für $\|x\| \leq \delta_2$ gilt mit $\|F'(x)\| \leq \|F'(\theta)\| + L_1\delta_2 =: R$ die Abschätzung:

$$|\dot{\Phi}[x](t)| \leq R\,|\Phi[x](t)|\,, \quad 0\leq t\leq 1\,.$$

Dann folgt:

$$\|\Phi[x]\| \leq e^R \quad \text{für alle } \|x\| \leq \delta_2\,. \tag{5.21}$$

Sei $\|x\| \leq \delta_2$.

$$\Rightarrow \quad |\dot{\Phi}[\theta](t) - \dot{\Phi}[x](t)| = |F'(\theta)\Phi[\theta](t) - F'(x)\Phi[x](t)| \leq$$

$$\leq \|F'(x) - F'(\theta)\| \, |\Phi[x](t)| + \|F'(\theta)\| \, |\Phi[\theta](t) - \Phi[x](t)|$$

$$\leq L_1 \|x\| e^R + R \, |(\Phi[\theta] - \Phi[x])(t)| .$$

$$\Rightarrow \quad |(\Phi[\theta] - \Phi[x])(t)| \leq L_1 \cdot \|x\| e^R \int_0^t \exp\Big(R(t-s)\Big)ds$$

$$\leq \frac{L_1}{R} e^R (e^R - 1) \|x\| \quad \text{für alle } 0 \leq t \leq 1 .$$

Die letzte Abschätzung folgt aus der Gronwallschen Ungleichung (vgl. W.WALTER [1964], p. 13).

Wenn man beachtet, daß $\dot{\overline{\Phi^{-1}}}[x] = - \Phi^{-1}[x]F'(x)$ gilt, folgt mit analogen Überlegungen die Stetigkeit von $x \mapsto \Phi^{-1}[x]$.

(ii) $\exists\ y>0\ \exists\ S>0\ \ (\|x\| < \eta \Rightarrow M[x]^{-1}$ und $\psi'(x)^{-1}$ existieren

und $\max(\|M[x]^{-1}\|, \|\psi'(x)^{-1}\|) < S)$:

Das folgt unmittelbar aus (i) und Lemma 5.17 unter Anwendung des Satzes von Banach (vgl. L.A.LUSTERNIK-V.J.SOBOLEV [1974], p. 1o7).

(iii) $\|v(x)\| \leq o(\|x\|)$ für $\|x\| \to 0$:

Aus $g = \theta$ und $F(\theta) = F(x^*) = \dot{x}^* = \theta$ folgt:

$$\|v(x)\| = \|(\langle -T\circ z(x), T\circ\Phi[x]e_j\rangle)_{j=1}^k\|$$

$$\leq C^2 e^R \|z(x)\|$$

$$= C^2 e^R \, \| \Phi(x)(t) \int_0^t \Phi^{-1}(x)(s)\Big(F(x)(s)-F'(x)x(s)\Big)ds \|$$

$$\leq C^2 e^{3R} \, \| F(x)-F'(x)x \|$$

$$\leq C^2 e^{3R}\Big(\| F(x)-F(\theta)-F'(\theta)x \| + \| (F'(x)-F'(\theta))x \| \Big)$$

$$\leq o\,(\|x\|) \quad \text{für} \quad \|x\| \to 0 \, .$$

(iv) $G'(\theta)$ existiert, und es gilt $G'(\theta) = \theta$:

$$\| G(x)-G(\theta)-\theta x \| = \| G(x) \|$$

$$= \quad \| x-\psi'(x)^{-1}\Big(\psi(x)-(e_1,\ldots,e_k)\, M[x]^{-1} v(x)\Big) \| =$$

$$= \quad \| \psi'(x)^{-1}\Big(\psi'(x)x-\psi(x)-(e_1,\ldots,e_k)\, M[x]^{-1} v(x)\Big) \| \leq$$

$$\leq \quad S\Big[\| \psi(x)-\psi(\theta)-\psi'(\theta)x \| + \| \Big(\psi'(\theta)-\psi'(x)\Big)x \| \Big]$$

$$+ \, S^2 \| v(x) \| \leq o\,(\|x\|) \qquad \text{für} \quad \|x\| \to 0 \, .$$

Damit ist der Beweis vollständig. ∎

5.22 Bemerkung: Wir fassen die bisherigen Ergebnisse zusammen: Der Algorithmus (A1) konvergiert, vorausgesetzt, x^* ist stark eindeutige lokale Minimallösung bzw. (im Fall der L_2-Halbnorm) $p(g-T\circ x^*)$ ist nichtzu groß, lokal quadratisch bzw. linear gegen x^* . Im folgenden werden wir zeigen, daß der Algorithmus (A2) mit λ-Strategie unter wesentlich schwächeren Voraussetzungen und für jede stetige Halbnorm p gegen einen stationären Punkt konvergiert.

5.23 Definition (linearisierte Penaltyfunktion):

Für $\hat{x} \in B$ sei $\varphi_{\hat{x}} : C^1_{\mathbb{R}^n}[0,1] \to \mathbb{R}$, $\varphi_{\hat{x}}(x) := p(g-T\circ x)+K\| \dot{x}-F_{\hat{x}}(x) \|$.

5.24 Lemma: Sei $x \in B$ und $\bar{x} \in B$ Lösung von (P_x) mit $\varphi(x) > p(g - T \circ \bar{x})$.

Dann gilt (β wie in (V2)):

$$\exists\ \delta > 0 \quad \forall\ 0 < \lambda < \delta \quad \varphi(x + \lambda(\bar{x} - x)) \leq \varphi(x) - \beta\lambda(\varphi(x) - p(g - T \circ x)).$$

Beweis: φ besitzt in $x \in B$ eine Richtungsableitung $\varphi'(x)$ und es gilt:

$$\varphi'(x)(\bar{x} - \bar{x}) = \varphi_x'(x)(\bar{x} - x) \leq \varphi_x(\bar{x}) - \varphi(x) = p(g - T \circ \bar{x}) - \varphi(x) < 0 .$$

Hieraus folgt wegen $0 < \beta < 1$ die Behauptung.

5.25 Bemerkung: Aus Lemma 5.24 schließt man:

$$x^* \text{ lokale Minimallösung von (P)} \Rightarrow x^* \text{ stationärer Punkt von (P)}$$
$$\Rightarrow \dot{x}^* = F(x^*).$$

5.26 Konvergenzsatz für Algorithmus (A2)

Sei A konvex, kompakt und sei (V1) und (V2) erfüllt.
Mit einem beliebigen $x^{(1)} \in B$ beginnend, werde das Iterationsverfahren (A2) durchgeführt. Falls das Verfahren beim i-ten Iterationsschritt abbricht, ist $x^{(i)}$ stationärer Punkt von (P). Andernfalls liefert das Verfahren eine beschränkte Folge $\left(x^{(i)}\right)_{i \in \mathbb{N}}$ in $C^1_{\mathbb{R}^n}[0,1]$, die Häufungspunkte besitzt.
Jeder Häufungspunkt ist stationärer Punkt von (P).

Beweis: Das Verfahren breche nicht ab.

(i) Die Folge $\left(x^{(i)}\right)_{i \in \mathbb{N}}$ ist präkompakt in $C^1_{\mathbb{R}^n}[0,1]$:

Nach Konstruktion ist $\left(\varphi(x^{(i)})\right)_{i\in\mathbb{N}}$ monoton fallend und daher gilt für eine Konstante $Q>0$:

$\forall i\in\mathbb{N} \quad \| \dot{x}^{(i)}-F(x^{(i)})\| \leq Q$.

Hieraus folgt

$$\forall i\in\mathbb{N} \quad \forall t\in[0,1] \quad |\dot{x}^{(i)}(t)| \leq Q + |\tilde{f}(t,x^{(i)}(t))| \leq Q + \|F(o)\| + L|x^{(i)}(t)|.$$

Mit Gronwalls Lemma erhält man hieraus die gleichmässige Beschränktheit der Folge $\left(x^{(i)}\right)_{i\in\mathbb{N}}$. Berücksichtigt man das und wendet erneut das Gronwallsche Lemma auf die Ungleichung

$$\begin{aligned}\forall i\in\mathbb{N} \quad \forall t\in[0,1] \quad |\dot{\bar{x}}^{(i)}(t)| &\leq \left|\left(F(x^{(i)})+F'(x^{(i)})(\bar{x}^{(i)}-x^{(i)})\right)(t)\right| \\ &\leq \|F(o)\| + 2L\|x^{(i)}\| + L|\bar{x}^{(i)}(t)|\end{aligned}$$

an, so erhält man auch die gleichmäßige Beschränktheit der Folge $\left(\bar{x}^{(i)}\right)_{i\in\mathbb{N}}$ und dann auch der Folge $\left(\dot{\bar{x}}^{(i)}\right)_{i\in\mathbb{N}}$. Die gleichgradige gleichmäßige Stetigkeit der Folge $\left(\dot{\bar{x}}^{(i)}\right)_{i\in\mathbb{N}}$ folgt aus der Beziehung

$$\begin{aligned}\forall i\in\mathbb{N} \quad \forall s,t\in[0,1] \quad |\dot{\bar{x}}^{(i)}(s)-\dot{\bar{x}}^{(i)}(t)| &\leq |\tilde{f}(s,x^{(i)}(s))-\tilde{f}(t,x^{(i)}(t))| \\ &+ |\tilde{f}_x(s,x^{(i)}(s))(\bar{x}^{(i)}(s)-x^{(i)}(s)) - \\ &- \tilde{f}_x(t,x^{(i)}(t))(\bar{x}^{(i)}(t)-x^{(i)}(t))| \\ &\leq |\tilde{f}(s,x^{(i)}(s))-\tilde{f}(t,x^{(i)}(s))| + L|x^{(i)}(s)-x^{(i)}(t)| \\ &+ |\tilde{f}_x(s,x^{(i)}(s))-\tilde{f}_x(t,x^{(i)}(s))|\cdot|\bar{x}^{(i)}(s)-x^{(i)}(s)| \\ &+ |\tilde{f}_x(t,x^{(i)}(s))-\tilde{f}_x(t,x^{(i)}(t))|\cdot|\bar{x}^{(i)}(s)-x^{(i)}(s)| \\ &+ |\tilde{f}_x(t,x^{(i)}(t))|\cdot\left(|\bar{x}^{(i)}(s)-\bar{x}^{(i)}(t)|+|x^{(i)}(s)-x^{(i)}(t)|\right),\end{aligned}$$

aus der gleichmäßigen Stetigkeit der Abbildungen $\tilde{f}$ und $\tilde{f}_x$ auf jeder kompakten Teilmenge von $[0,1] \times \mathbb{R}^n$ und der gleichmäßigen Beschränktheit und gleichgradigen gleichmäßigen Stetigkeit der Folgen $\left(x^{(i)}\right)_{i\in\mathbb{N}}$ und $\left(\bar{x}^{(i)}\right)_{i\in\mathbb{N}}$. Mit vollständiger Induktion zeigt man schließlich wegen $x^{(i+1)}=(1-\lambda_j)x^{(i)}+\lambda_j\bar{x}^{(i)}$ mit $0\leq\lambda_j\leq 1$, daß auch die Folgen $\left(x^{(i)}\right)_{i\in\mathbb{N}}$ und $\left(\dot{x}^{(i)}\right)_{i\in\mathbb{N}}$ gleichmäßig beschränkt und gleichgradig gleichmäßig stetig sind. Nach Arzela-Ascoli ist die Folge $\left(x^{(i)}\right)_{i\in\mathbb{N}}$ in $C^1_{\mathbb{R}^n}[0,1]$ präkompakt und besitzt folglich Häufungspunkte in $C^1_{\mathbb{R}^n}[0,1]$.

(ii) Jeder Häufungspunkt der Folge $\left(x^{(i)}\right)_{i\in\mathbb{N}}$ in $C^1_{\mathbb{R}^n}[0,1]$ ist stationärer Punkt von (P):

Sei $x^* \in C^1_{\mathbb{R}^n}[0,1]$ Häufungspunkt der Folge $\left(x^{(i)}\right)_{i\in\mathbb{N}}$. Da B abgeschlossen ist, folgt $x^*\in B$. Wir führen den Beweis indirekt und nehmen an, daß x^* kein stationärer Punkt ist, also $\varphi(x^*) > m := \inf\{p(g-T_\circ x) \mid x\in B, \dot{x} = F_{x^*}(x)\}$. Da φ stetig und $\left(\varphi(x^{(i)})\right)_{i\in\mathbb{N}}$ eine monoton fallende Folge ist, gilt:

$\varphi(x^*) = \inf\{\varphi(x^{(i)}) \mid i\in\mathbb{N}\}$. Im Widerspruch hierzu zeigen wir, daß

(5.27) $\varphi(x^{(i+1)}) < \varphi(x^*)$

für ein $i\in\mathbb{N}$ gilt. Dazu benutzen wir die Ungleichung (4.9) und konstruieren ein $i \in \mathbb{N}$, so daß $\varphi(x^{(i)})$ genügend nahe bei $\varphi(x^*)$ liegt und der Term $\beta\lambda_j\left(\varphi(x^{(i)}) - p(g - T\circ\bar{x}^{(i)})\right)$ groß genug ist.

Im weiteren sei $i \in \mathbb{N}$ so gewählt, daß $x^{(i)}$ genügend nahe bei x^* liegt und infolgedessen die nachfolgenden Ungleichungen richtig sind. (Man prüft leicht nach, daß die in diesen Ungleichungen auftretenden Konstanten unabhängig von i sind.)

Sei $\eta := \varphi(x^*) - m > 0$ und $\bar{x} \in B$ mit $\varphi_{x^*}(\bar{x}) = m$. Ferner sei $R > 0$ mit $\forall i \in \mathbb{N} \quad (\| x^{(i)} \| \leq R \wedge \| \bar{x}(i) \| \leq R)$. Da F Frechet-differenzierbar in x^*, gibt es $0 < \delta < 4R$ mit

$$\forall x \in B \quad \left(\| x - x^* \| \leq \delta \Rightarrow \| F(x) - F_{x^*}(x) \| \leq \frac{(1-\beta)\eta}{8R} \| x - x^* \|\right).$$

$\bar{x}^{(i)}$ ist die Minimallösung von $(P_{x^{(i)}})$, folglich gilt

$$\varphi(x^{(i)}) - p(g - T\circ\bar{x}^{(i)}) \geq \varphi(x^*) - \frac{\eta}{4} - \varphi_{x^{(i)}}(\bar{x}) .$$

Auf Grund der Abschätzung (beachte: F stetig Frechet-differenzierbar)

$$\begin{aligned} |\varphi_{x^{(i)}}(\bar{x}) - \varphi_{x^*}(\bar{x})| &\leq K \| F_{x^{(i)}}(\bar{x}) - F_{x^*}(\bar{x}) \| \\ &\leq K \left(\| F(x^{(i)}) - F_{x^*}(x^{(i)}) \| + \| (F'(x^{(i)}) - F'(x^*))(\bar{x} - x^{(i)}) \| \right) \\ &\leq \frac{\eta}{4} \end{aligned}$$

folgt

$$\varphi(x^{(i)}) - p(g - T\circ\bar{x}^{(i)}) \geq \varphi(x^*) - \varphi_{x^*}(\bar{x}) - \frac{\eta}{2} = \frac{\eta}{2} .$$

Wir schätzen $\lambda_j \quad (j = j^{(i)})$ nach unten ab:

Für $0 < \lambda < \frac{\delta}{4R} \leq 1$ gilt wegen

$$\| x^{(i)}+\lambda h^{(i)}-x^{*} \| \leq \| x^{(i)}-x^{*} \| + \lambda \| h^{(i)} \| \leq \frac{\delta}{2} + \frac{\delta}{4R} 2R = \delta$$

die Abschätzung

$$\begin{aligned}
\varphi(x^{(i)}+\lambda h^{(i)}) &\leq \varphi_{x^*}\left(x^{(i)}+\lambda h^{(i)}\right) + \frac{(1-\beta)\eta}{4R}\left(\lambda + \frac{\alpha\delta}{4R}\right) \\
&\leq \varphi_{x^{(i)}}\left(x^{(i)}+\lambda h^{(i)}\right) + \frac{(1-\beta)\eta}{4R}\left(\lambda + \frac{\alpha\delta}{4R}\right) \\
&\leq (1-\lambda)\,\varphi_{x^{(i)}}\left(x^{(i)}\right) + \lambda\,\varphi_{x^{(i)}}\left(\bar{x}^{(i)}\right) + \frac{(1-\beta)\eta}{4R}\left(\lambda + \frac{\alpha\delta}{4R}\right) \\
&\leq \varphi\left(x^{(i)}\right) - \lambda\beta\left(\varphi(x^{(i)}) - p(g-T\circ\bar{x}^{(i)})\right) + \frac{(1-\beta)\eta}{4}\left(\frac{\alpha\delta}{4R} - \lambda\right),
\end{aligned}$$

also $\varphi\left(x^{(i)}+\lambda h^{(i)}\right) \leq \varphi\left(x^{(i)}\right) - \lambda\beta\left(\varphi(x^{(i)}) - p(g-T\circ\bar{x}^{(i)})\right)$,

falls $\frac{\alpha\delta}{4R} \leq \lambda < \frac{\delta}{4R}$. Wegen $\inf\{\frac{\lambda_{j+1}}{\lambda_j} \mid j \in \mathbb{N}\} = \alpha > 0$

gibt es ein $j' \in \mathbb{N}$ mit $\frac{\alpha\delta}{4R} \leq \lambda_{j'} < \frac{\delta}{4R}$. Nach Definition von $j^{(i)}$ im Algorithmus (A2) folgt $j = j^{(i)} \leq j'$ und damit $\lambda_j \geq \lambda_{j'} \geq \frac{\alpha\delta}{4R}$.

Die Ungleichung (5.27) ergibt sich folgendermaßen:

$$\begin{aligned}
\varphi\left(x^{(i+1)}\right) &= \varphi\left(x^{(i)}+\lambda_j h^{(i)}\right) \\
&\leq \varphi\left(x^{(i)}\right) - \lambda_j\beta\left(\varphi(x^{(i)}) - p(g-T\circ\bar{x}^{(i)})\right) \\
&\leq \varphi(x^{*}) + \frac{\alpha\delta\beta\eta}{16R} - \frac{\alpha\delta}{4R}\cdot\beta\cdot\frac{\eta}{2} \\
&< \varphi(x^{*}).
\end{aligned}$$

∎

6. Numerische Resultate

Der Algorithmus (A2) wurde getestet am Parameteridentifizierungsproblem von Abschnitt 2, sowohl für den Fall $p(g-T\circ x^{*}) = 0$ als auch für kleine Werte von $p(g-T\circ x^{*}) \neq 0$, für L_2- und L_∞-Halbnormen. Weiter wurden an mehreren Funktionen (vgl. D.BRAESS [197o]) beste Exponentialapproximationen im L_2- und L_∞-Sinne berechnet. In allen Beispielen erwies es sich als günstig, die Penaltyfaktoren K relativ klein (in der Größenordnung o.o1) zu wählen.

Beim Parameteridentifizierungsproblem erhielten wir auch im Falle kleiner Reaktionskonstanten (Größenordnung o.oo1) noch gute Ergebnisse (vgl. Tabelle 1).
Bei der Exponentialapproximation wurde mit maximal 5 Termen gerechnet (vgl. Tabelle 2).
Die Beispiele wurden auf der Rechenanlage TR 44o des Großrechenzentrums für die Wissenschaft Berlin in der Programmiersprache ALGOL 6o durchgeführt. Herrn W.KOLBE danken wir für die Programmierarbeiten.

Tabelle 1: Parameteridentifizierung (A_o=1)

exakte Werte		Startwerte		berechnete Werte		berechneter Minimalwert	Iterationen
k_1^*	k_2^*	$k_1^{(1)}$	$k_2^{(1)}$	k_1	k_2		
L_2-Approximation mit Minimalwert 0:							
0.11	0.22	0.1	0.2	0.109957	0.259942	$1.380 \cdot 10^{-8}$	3
0.0017	0.00028	0.004	0.0005	0.001700	0.000280	$3.4 \cdot 10^{-10}$	4
L_∞-Approximation mit Minimalwert 0:							
0.11	0.22	0.1	0.2	0.110000	0.260000	$9 \cdot 10^{-11}$	4
0.0017	0.00028	0.004	0.0005	0.001700	0.000280	$6.9 \cdot 10^{-10}$	3
L_∞-Approximation mit Minimalwert 0.0001:							
		0.1	0.2	0.089380	0.234046	$9.5868 \cdot 10^{-5}$	4

Tabelle 2: Exponentialanpassung mit n Termen und Koeffizienten α_i:

	L_2-Approximation			L_∞-Approximation		
	$1/(1+t)$	$t^{\frac{1}{2}}$	$\tan(\frac{\pi}{2}(t-\frac{1}{2}))$	$1/(1+t)$	$t^{\frac{1}{2}}$	$\tan(\frac{\pi}{2}(t-\frac{1}{2}))$
n	4	2	4	5	3	4
λ_1	-10.0000	-4.63694	-8.03480	-9.08935	-96.0903	-7.80937
λ_2	-3.25515	0.59629	-1.96478	-5.00141	-4.16253	-1.93509
λ_3	-1.06512		1.96312	-2.50866	0.56043	1.93509
λ_4	-0.16568		7.98922	-0.97866		7.80938
λ_5				-0.18179		
α_1	0.00151	-0.45982	-0.09881	0.00061	-0.11605	-0.10564
α_2	0.12218	0.55933	-1.04832	0.02099	-0.45651	-1.04536
α_3	0.49186		0.14706	0.15912	0.57829	0.15096
α_4	0.38445		0.00003	0.42839		0.00004
α_5				0.39089		
Min.wert	$1.5\cdot10^{-6}$	$8.2\cdot10^{-3}$	$1.1\cdot10^{-5}$	$1.3\cdot10^{-10}$	$5.7\cdot10^{-3}$	$7.0\cdot10^{-6}$
Iterationen	5	7	9	5	9	4

Literatur:

[1965] Bellmann,R.,R.Kalaba: Quasilinearization and nonlinear boundary value problems,
American Elsevier,New York,1965.

[197o] Braess,D.: Die Konstruktion der Tschebyscheff-Approximierenden bei der Anpassung mit Exponentialsummen,
J. Appr. Theory 3,261-273(197o).

[1973] Collatz,L.,W.Krabs: Approximationstheorie,
Tschebyscheffsche Approximation mit Anwendungen,
Teubner Studienbücher,B.G.Teubner,Stuttgart.

[1975] Cromme,L.: Eine Klasse von Verfahren zur Ermittlung bester nichtlinearer Tschebyscheff-Approximationen,
Numer. Math. 25 ,447-459(1975).

[1975] Hoffmann,K.-H.: Approximationstheorie,
Vorlesungsausarbeitung Universität München.

[1974] Lusternik,L.A.,V.J.Sobolev: Elements of Functional Analysis,
John Wiley & Sons,Inc.-New-York.

[1969] Osborne,M.R.,G.A.Watson: An algorithm for minimax approximation in the nonlinear case,
The Computer J. 12,63-68(1969).

[1973] Schäfke,F.W.,D.Schmidt: Gewöhnliche Differentialgleichungen. Die Grundlagen der Theorie im Reellen und Komplexen,
Heidelberger Taschenbücher,Springer-Verlag Berlin-Heidelberg-New-York.

[1964] Walter,W.: Differential- und Integralungleichungen und ihre Anwendung bei Abschätzungs- und Eindeutigkeitsproblemen,
Springer Tracts in Natural Philosophy Ergebnisse der angewandten Mathematik,Vol.2,Springer-Verlag, Berlin-Göttingen-Heidelberg,New-York.

Galerkin Methods for the Existence and Approximation of Weak Solutions of Nonlinear Dirichlet Problems with Discontinuities

by

Joseph W. Jerome

Department of Mathematics and the
Technological Institute
Northwestern University
Evanston, Illinois 60201

Research supported by National Science
Foundation grant MPS 74-02292 A01

Introduction. Consider the nonlinear Dirichlet problem, in distribution form,

$$(1.1)\quad \begin{array}{ll} \text{(i)} & -\Delta U + H(U) + G(U) = F, \\ \text{(ii)} & U - W \in H_0^1(\Omega), \end{array}$$

where Ω is any bounded open subset of R^N, $N \geq 1$, F is in the dual space $H^{-1}(\Omega)$ of $H_0^1(\Omega)$, W is any prescribed $H^1(\Omega)$ function and H is a monotone increasing function, discontinuous at 0, satisfying,

$$(1.2)\quad \begin{array}{ll} \text{(i)} & H(0-) = 0,\ H(0+) = b > 0 \\ \text{(ii)} & 0 < \inf_{\lambda \neq 0} H'(\lambda) \leq \sup_{\lambda \neq 0} H'(\lambda) < \infty. \end{array}$$

G is assumed to be a Lipschitz continuous function admitting the decomposition,

$$(1.3)\quad \begin{array}{ll} \text{(i)} & G = G_1 + G_2, \\ \text{(ii)} & G_1(u)u \geq 0, \\ \text{(iii)} & \|G_2\|_{Lip} \leq \lambda_1 + \inf_{\lambda \neq 0} H'(\lambda), \end{array}$$

where λ_1 is the smallest eigenvalue of the positive definite, self-adjoint operator $-\Delta$ with domain

$$(1.4)\quad \{u \in H_0^1(\Omega):\ -\Delta u \in L^2(\Omega)\}.$$

Such problems arise, for example, when numerical methods are used to solve the generalized two-phase Stefan problem, given in distribution form by,

$$(1.5)\quad \frac{\partial u}{\partial t} - \nabla \cdot (k(u) \nabla u) + a(u) = f,$$

on a space-time domain $\Omega \times (0, T_0)$ with prescribed initial and boundary conditions and enthalpy discontinuity across the free boundary interface of the time profile of the two phases. The diffusion coefficient k is a positive function with (relatively) compact range, which is discontinuous at 0, and a is an arbitrary Lipschitz body heating function. When the Kirchoff transformation,

$$(1.6) \qquad U = K(u) = \int_0^u k(\lambda)d\lambda,$$

is applied to (1.5), there results the transformed equation,

$$(1.7) \qquad \frac{\partial H(U)}{\partial t} - \Delta U + G(U) = f,$$

where $G(\lambda) = a(K^{-1}(\lambda))$ and where the enthalpy H satisfies (1.2i) and is characterized by its positive slope:

$$(1.8) \qquad H'(\lambda) = \frac{1}{k(K^{-1}(\lambda))}, \ \lambda \neq 0.$$

When a fully implicit Euler scheme is used to approximate $\frac{\partial H}{\partial t}$ by $[H(U_{m+1}) - H(U_m)]/\Delta t$, $m = 0,\ldots,M-1$, one formally obtains from (1.7) M equations of the form (1.1i):

$$(1.9) \qquad -\Delta U_{m+1} + H(U_{m+1})/\Delta t + G(U_{m+1}) = f_{m+1} + H(U_m)/\Delta t.$$

Here, $\Delta t = T_0/M$ and the subscript j denotes evaluation at $j\Delta t$. It can be shown that (1.9) has a solution whenever $\Delta t \leq \|G\|_{Lip}/\inf k$ and an existence and approximation theory can be developed for the generalized Stefan problem (1.5) on the basis of this [8]. Note that in the discretized time dependent problems (1.9), G is an arbitrary Lipschitz function and can be identified with G_2 in the decomposition of (1.3), since Δt is arbitrarily small. The decomposition (1.3), incidentally, is satisfied by every monotone Lipschitz G with the identifications $G_1 = G - G(0)$, $G_2 = G(0)$.

An analysis of (1.1) necessarily involves the use of multivalued mappings T,

$$(1.10) \qquad T: H_0^1(\Omega) \to 2^{H^{-1}(\Omega)},$$

required because $H(U(x))$ is permitted to assume any value on $[0,b]$ if $U(x) = 0$. The solution of (1.1) is then equivalent to determining $V = U - W$ in $H_0^1(\Omega)$ such that $F + \Delta W$ is in the set TV; thus,

$$\text{(1.11)}\quad \begin{array}{ll} \text{(i)} & TV = -\Delta V + H(U) + G(U), \\ \text{(ii)} & TV \ni F + \Delta W. \end{array}$$

A direct existence theory for (1.11) can be constructed by the use of pseudomonotone operators, introduced by Brezis [3]; this was carried out in [7] together with an alternative method involving the smoothing of H, the latter approach being required in the stability analysis for the solutions of the discretized problems (1.9). In this paper, however, in sections two and three, we shall deduce the existence of solutions of (1.1) via convergence of Galerkin solutions, the rate of which is estimated in section four if G is a monotone function. Here convergence results of Ciarlet, Schultz and Varga [5] generalize to non-Lipschitz T. Our major results can now be stated. If $B(\cdot,\cdot)$ denotes the energy inner product on $H_0^1(\Omega)$,

$$B(u,v) = \int_\Omega \nabla u \cdot \nabla v, \tag{1.12}$$

and P the projection in this inner product onto a given finite dimensional space S,

$$B(u,v) = B(Pu,v), \text{ for all } v \in S, \tag{1.13}$$

and if $P^t: H^{-1}(\Omega) \to H^{-1}(\Omega)$ is defined by,

$$\langle P^t F, v\rangle = \langle F, Pv\rangle, \text{ for all } v \in H_0^1(\Omega), \tag{1.14}$$

and

$$F_0 = F - B(W,\cdot) = F + \Delta W, \tag{1.15}$$

then we have the

Theorem 1.1. There is a positive constant $\overline{C}$, depending only on W, T and F, such that the Galerkin "equation",

(1.16) $$P^t TPs \ni P^t F_0,$$

has at least one solution and, indeed, every solution s in the ball,

$$\mathcal{B}_{\overline{C}} = \{v \in H_0^1(\Omega): \|v\|_{H^1} \leq \overline{C}\}.$$

Moreover, if $\{S_n\}$ is a sequence of finite dimensional subspaces of $H_0^1(\Omega)$ such that

(1.17) $$\lim_{n\to\infty} \|v - P_n v\|_{H^1} = 0, \text{ for all } v \in H_0^1(\Omega),$$

then there is a subsequence $\{S_{n_k}\}$ such that the corresponding Galerkin solutions s_{n_k},

(1.18) $$P_{n_k}^t TP_{n_k} s_{n_k} \ni P_{n_k}^t F_0,$$

are convergent in $L^2(\Omega)$ to a solution V of (1.11). The eigenspaces of $-\Delta$, with domain given by (1.4), constitute a sequence satisfying (1.17). Finally, if G_1 is monotone increasing, then (1.11) has a unique solution V and the estimate,

(1.19) $$\|V - s_n\|_{H^1} \leq C_1 \sup\{\|v - P_n v\|_{L^2}: v \in H_0^1(\Omega): \|v\|_{H^1} \leq 1\}$$
$$+ C_2 \|u_0 - P_n u_0\|_{H^1}$$

holds for positive constants C_1 and C_2 for the sequence $\{s_n\}$ of Galerkin solutions; here u_0 is the representer of F_0:

(1.20) $$\langle F_0, v\rangle = B(u_0, v), \text{ for all } v \in H_0^1(\Omega).$$

In particular, if P_n is uniformly convergent to I as a sequence defined on $L^2(\Omega)$, then $s_n \to V$ in $H_0^1(\Omega)$.

Remark. If $\{S_n\}$ is a sequence of finite element spaces with cell diameter h_n, then, under a standard uniformity hypothesis on the basis, together with the existence of a smoothing Green's operator,

$$(-\Delta)^{-1}: L^2(\Omega) \to H^2(\Omega) \cap H_0^1(\Omega), \tag{1.21}$$

we conclude that the first term on the right hand side of (1.19) is of order $O(h_n)$. If $u_0 \epsilon H^2(\Omega)$, then the second term likewise is of order $O(h_n)$. We shall amplify this discussion in section four.

We close the introduction with a brief review of the spaces $H^1(\Omega)$, $H_0^1(\Omega)$ and $H^{-1}(\Omega)$ together with their inner products and/or norms. $H^1(\Omega)$ is the Hilbert space defined by,

$$H^1(\Omega) = \{u \epsilon L^2(\Omega): \frac{\partial u}{\partial x_i} \epsilon L^2(\Omega),\ i = 1,\dots,N\},$$

where differentiation is taken in the sense of distributions and where all functions are real. We have, for the inner product,

$$(u,v)_{H^1} = \sum_{i=1}^{N} \int_\Omega \frac{\partial u}{\partial x_i} \frac{\partial v}{\partial x_i} + \int_\Omega uv, \tag{1.22}$$

and $C^\infty(\Omega) \cap H^1(\Omega)$ is dense in $H^1(\Omega)$ in the norm determined by (1.22). Moreover,

$$H_0^1(\Omega) = \text{closure}[C_0^\infty(\Omega)],$$

in $H^1(\Omega)$. The higher order spaces are defined similarly; cf. e.g., [1]. Also, one frequently writes $H^0(\Omega) = L^2(\Omega)$. Finally,

$$H^{-1}(\Omega) = \text{(topological) dual of } H_0^1(\Omega),$$

and

(1.23)

$$\text{(i)} \quad \|F\|_{H^{-1}(\Omega)} = \sup\{|\langle F,u\rangle|: u \epsilon \mathscr{B}_1\},$$

$$\text{(ii)} \quad \mathscr{B}_1 = \{v \epsilon H_0^1(\Omega): \|v\|_{H^1} \le 1\}.$$

§2. Galerkin Projections and "A Priori' Estimates

Our first result of this section is a coerciveness result which also permits the derivation of 'a priori' estimates for the solution of the Galerkin equations.

Proposition 2.1. Let $D(\cdot,\cdot)$ be a continuous, strongly coercive symmetric bilinear form on $H_0^1(\Omega)$, i.e. there exist positive constants C_0 and C such that,

$$\text{(2.1)}\quad \begin{array}{ll} \text{(i)} & |D(u,v)| \leq C_0 \|u\|_{H^1} \|v\|_{H^1}, \text{ for all } u,v \in H_0^1(\Omega), \\ \text{(ii)} & D(u,u) \geq C\|u\|_{H^1}^2, \text{ for all } u \in H_0^1(\Omega). \end{array}$$

Let H satisfy (1.2) and set

$$\theta_1 = \inf_{\lambda \neq 0} H'(\lambda). \tag{2.2}$$

Let G satisfy (1.3i,ii) and

$$\|G_2\|_{Lip} < C + \theta_1. \tag{2.3}$$

If L denotes the continuous linear mapping of $H_0^1(\Omega)$ onto $H^{-1}(\Omega)$ satisfying,

$$D(u,v) = \langle Lu,v\rangle, \text{ for all } v \in H_0^1(\Omega), \tag{2.4}$$

and the (multivalued) mapping T is given by,

$$TV = LV + H(V + W) + G(V + W), \tag{2.5}$$

then there exist constants c and $\bar{c}$ such that, for $v \in H_0^1(\Omega)$, $v \neq 0$,

$$\frac{\langle Tv,v\rangle}{\|v\|_{H^1}} \geq c\|v\|_{H^1} - \bar{c} \tag{2.6}$$

holds; c is given explicitly by,

$$c = \min(C, C + \theta_1 - \|G_2\|_{Lip}). \tag{2.7}$$

Remark. If λ_1 denotes the smallest positive eigenvalue of the restriction $\tilde{L}$ of L to $L^{-1}(L^2(\Omega))$, then $\lambda_1 \leq C$ and (2.3) is clearly a relaxation of (1.3iii).

Proof: The fact that L is a continuous bijection is a consequence of the Lax-Milgram lemma [2, p. 30] and hence $\tilde{L}$ is compact by the compact injection of $H_0^1(\Omega)$ into $L^2(\Omega)$ [1,p. 99]. In particular, $\tilde{L}$ has a discrete sequence of positive eigenvalues of finite multiplicity converging to $+\infty$ and an associated sequence of eigenfunctions complete in $L^2(\Omega)$ and $H_0^1(\Omega)$.

Now

$$\langle Tv,v\rangle = [D(v,v) + (H(v)-H(0),v)_{L^2} + (G_2(v)-G_2(0),v)_{L^2}]$$
$$+ (G_1(v),v)_{L^2} + (H(v + W)-H(v),v)_{L^2}+(G(v+W)-G(v),v)_{L^2}$$
$$+ (H(0),v)_{L^2} + (G_2(0),v)_{L^2}$$

and, by (2.3), by the monotonicity relation,

$$(H(u) - H(v),u-v)_{L^2} \geq \theta_1(u-v,u-v)_{L^2},$$

and by (1.2ii) we have,

$$\langle Tv,v\rangle \geq c\|v\|^2_{H^1} + (G_1(v),v)_{L^2} - \bar{c}\|v\|_{L^2} \tag{2.8}$$

where,

$$\bar{c} = (\|G\|_{Lip} + \sup_{\lambda\neq 0} H'(\lambda))\|W\|_{L^2} + [\text{meas }\Omega]^{1/2}(b + |G_2(0)|).$$

(2.6) now follows from (1.3ii) upon division of (2.8) by $\|v\|_{H^1}$. This concludes the proof of Proposition 2.1.

Theorem 2.2. Let S be a finite-dimensional subspace of $H_0^1(\Omega)$ and let $P: H_0^1(\Omega) \to S$ be the self-adjoint projection defined by,

(2.9) $$D(u,v) = D(Pu,v), \text{ for all } v\epsilon S.$$

Then the Galerkin equation,

(2.10) $$P^t TPs \ni P^t F_0,$$

has a solution $s\epsilon S$ for a given $F_0 \epsilon H^{-1}(\Omega)$ and

(2.11) $$\|s\|_{H^1} \leq [\|F_0\|_{H^{-1}} + \bar{c}]/c$$

Proof: The mapping T given by (2.5) is a pseudomonotone mapping of $H_0^1(\Omega) \to 2^{H^{-1}(\Omega)}$, i.e., T satisfies

(2.12)

(i) T is bounded;

(ii) $$V_i \to V \text{ in } H_0^1(\Omega) \text{ and } \lim_{i\to\infty} \sup\langle TV_i, V_i - V\rangle \leq 0$$
$$\Longrightarrow \lim_{i\to\infty} \inf\langle TV_i, V_i - Z\rangle \geq \langle TV, V-Z\rangle \forall Z\epsilon H_0^1(\Omega).$$

Indeed, the mapping $V \to H(V + W) = AV$ is a maximal monotone operator [4, p. 25] from $H_0^1(\Omega)$ into $2^{L^2(\Omega)}$, as the subdifferential $A = \partial\varphi$ of the finite-valued, continuous, convex functional,

$$\varphi(V) = \int_\Omega \ell(V + W);$$

here ℓ is the convex primitive of H satisfying $\ell(0) = 0$. The mapping $B = L + G$ is a (single-valued) pseudomonotone mapping; in fact, we have the stronger implication,

$$V_i \to V \text{ in } H_0^1(\Omega) \text{ and } \lim_{i\to\infty} \sup\langle BV_i, V_i - V\rangle \leq 0$$
$$\Longrightarrow V_i \to V \text{ in } H_0^1(\Omega).$$

We omit the details which appear in [7]. Thus, T is pseudomonotone [10, p. 189] as the sum $T = A + B$ of a maximal monotone and pseudomonotone operator. It now follows directly from the definition (2.12) that $P^t TP$ is a pseudomonotone mapping of S into $2^{S'}$; since, by

Proposition 2.1, P^tTP is coercive, i.e.,

$$\frac{\langle P^tTPv,v\rangle}{\|v\|_{H^1}} \to \infty \text{ as } \|v\|_{H^1} \to \infty,\ v\epsilon S,$$

it follows [10, Théorème 2.7, p. 180] that P^tTP is surjective, i.e., the equation (2.10) has a solution $s\epsilon S$.

(2.11) follows by applying (2.10) to (2.6) and the proof is completed.

§3. Convergence of Galerkin Approximations

We shall begin with the estimation of residual approximations.

Proposition 3.1. Let s be a solution of (2.10) and let u_0 satisfy (1.20). Then there exists a selection in Ts and a constant c_1 such that,

$$\begin{aligned}\|Ts-F_0\|_{H^{-1}} \le c_1 \sup\{\|v-Pv\|_{L^2}: v\epsilon\mathscr{B}_1\}\\ + \|u_0-Pu_0\|_{H^1} \sup\{\|v-Pv\|_{H^1}: v\epsilon\mathscr{B}_1\}\end{aligned} \tag{3.1}$$

holds, where c_1 does not depend upon S, and $\mathscr{B}_1$ is given by (1.23ii).

Proof: Since $\langle Ts-F_0,u\rangle = 0$ for $u\epsilon S$, we have, for $v\epsilon H_0^1(\Omega)$,

$$\langle Ts-F_0,v\rangle = \langle Ts-F_0,v-Pv\rangle,$$

and

$$\begin{aligned}\langle Ts-F_0,v\rangle = (H(s+\ W),v-Pv)_{L^2} + (G(s+\ W),v-Pv)_{L^2}\\ - \langle F_0,v-Pv\rangle,\end{aligned} \tag{3.2}$$

since $D(s,v-Pv) = 0$. Thus, from (1.23) and (3.2),

$$\begin{aligned}\|Ts-F_0\|_{H^{-1}} \le \{\|H(s+\ W)\|_{L^2} + \|G(s+\ W)\|_{L^2}\}\times\\ \sup\{\|v-Pv\|_{L^2}: v\epsilon\mathscr{B}_1\}+\end{aligned} \tag{3.3}$$

$$+ \|u_0 - Pu_0\|_{H^1} \sup\{\|v-Pv\|_{H^1} : v \in \mathscr{B}_1\}.$$

Now H and G are bounded operators on $L^2(\Omega)$ and thus we obtain from (2.11) and the elementary estimate $\|s + W\|_{L^2} \le \|s\|_{H^1} + \|W\|_{H^1}/\sqrt{C}$, the existence of a positive constant c_1 such that,

$$(3.4) \qquad \|H(s + W)\|_{L^2} + \|G(s + W)\|_{L^2} \le c_1 .$$

(3.1) is a consequence of (3.3) and (3.4) and the proof of the proposition is completed.

Theorem 3.2. Under the hypotheses of Proposition 2.1, there exists a solution $V \in H_0^1(\Omega)$ of $TV \ni F_0$ for each $F_0 \in H^{-1}(\Omega)$. Moreover, if $\{S_n\}_n$ is a sequence of finite-dimensional subspaces of $H_0^1(\Omega)$ satisfying (1.17), with the associated projections P_n satisfying (2.9), then the residuals $Ts_n - F_0$, determined by

$$(3.5) \qquad P_n^t T P_n s_n = P_n^t F_0, \quad s_n \in S_n, \quad n = 1,2,\dots,$$

are weakly convergent to 0 in $H^{-1}(\Omega)$ and a subsequence of $\{s_n\}$ is convergent in $L^2(\Omega)$ to V, say, $s_{n_k} \to V$.

Proof: A sequence $\{s_n\}$, satisfying (3.5), exists by Theorem 2.2 and

$$(3.6) \qquad \|s_n\|_{H^1} \le \overline{C},$$

where $\overline{C}$ is given by the right hand side of (2.11). In particular, the weak convergence of Ts_n to F_0 in $H^{-1}(\Omega)$ follows from (1.17) and the representations (3.2). Moreover, a subsequence s_{n_k} is weakly convergent, say,

$$(3.7i) \qquad s_{n_k} \rightharpoonup V \text{ (in } H_0^1(\Omega)),$$

and, by the compact injection of $H_0^1(\Omega)$ in $L^2(\Omega)$,

$$(3.7ii) \qquad s_{n_k} \to V \text{ (in } L^2(\Omega)).$$

Without loss of generality, we may assume,

(3.7iii) $$s_{n_k}(x) \to V(x) \text{ a.e. in } \Omega.$$

Since $\{H(s_{n_k} + W)\}$ is bounded in $L^2(\Omega)$, it follows that a subsequence is weakly convergent in $L^2(\Omega)$; for simplicity, we suppose that $\{s_{n_k}\}$ has been chosen so that

(3.7iv) $$H(s_{n_k} + W) \rightharpoonup \chi \epsilon L^2(\Omega).$$

By (3.7ii) and (1.17),

(3.7v) $$G(s_{n_k} + W) \to G(V + W) \text{ (in } L^2(\Omega)).$$

Now L is a continuous linear mapping of $H_0^1(\Omega)$ onto $H^{-1}(\Omega)$ and hence weakly continuous [6, p. 422]. Thus,

(3.8) $$Ls_{n_k} \rightharpoonup LV.$$

It remains to show that $TV \ni F_0$. From (3.7v), (3.8) and the weak residual convergence of Ts_{n_k}, it suffices to show that,

(3.9i) $$\chi\, \Omega_0 \subset [0,b],$$

where

(3.9ii) $$\Omega_0 = \{x \epsilon \Omega: V(x) + W(x) = 0\}.$$

Indeed, by (3.7iii), it follows that $H(V(x) + W(x)) = \chi(x)$ if $V(x) + W(x) \neq 0$; (3.9i) then implies that,

(3.10) $$F_0 = [LV + \chi + G(V + W)] \epsilon TV.$$

To verify (3.9i), suppose, to the contrary, that there exists a set $\Omega_* \subset \Omega$ of positive measure satisfying

(3.11i) $$\chi(x) \geq b + \gamma, \quad x \epsilon \Omega_*,$$

for some $\gamma > 0$, or,

(3.11ii) $$\chi(x) \leq -\gamma, \ x \in \Omega_*.$$

If (3.11i) holds, then, by the weak convergence of $H(s_{n_k} + \quad W)$ to χ in $L^2(\Omega_*)$, and the lower semicontinuity of the norm with respect to weak convergence [6, p. 68] we have,

$$[b + \gamma]^2[\text{measure } \Omega_*] \leq \int_{\Omega_*} [\chi]^2$$

$$\leq \liminf_{k\to\infty} \int_{\Omega_*} [H(s_{n_k} + \quad W)]^2$$

and, by an analogue of the Fatou lemma [6, p. 172], we have,

$$\lim_{k\to\infty} \int_{\Omega_*} [H(s_{n_k} + \quad W)]^2 = \limsup_{k\to\infty} \int_{\Omega_*} [H(s_{n_k} + \quad W)]^2$$

$$\leq \int_{\Omega_*} \limsup_{k\to\infty} [H(s_{n_k} + \quad W)]^2 \leq b^2(\text{measure } \Omega_*]$$

This contradiction establishes that (3.11i) cannot hold. If (3.11ii) holds then,

$$-\gamma[\text{measure } \Omega_*] \geq \int_{\Omega_*} \chi = \lim_{k\to\infty} \int_{\Omega_*} H(s_{n_k} + \quad W).$$

and, by a generalized Fatou lemma [6, p. 172], valid because of (3.4), we have,

$$\lim_{k\to\infty} \int_{\Omega_*} [H(s_{n_k} + \quad W)] \geq \int_{\Omega_*} \liminf_{k\to\infty} [H(s_{n_k} + \quad W)] \geq 0.$$

This contradiction completely establishes (3.9i) and completes the proof of Theorem 3.2.

§4. Rates of Convergence

Throughout this section we make the following standing assumption:

(4.1) G_1 is a monotone increasing Lipschitz function.

<u>Proposition 4.1</u>. Under the hypotheses of Proposition 2.1 and (4.1),

(4.2i) $\langle Tu-Tv,u-v\rangle \geq c(u-v,u-v)_{H^1}$

(4.2ii) $\|u-v\|_{H^1} \leq (c^{-1})\|Tu-Tv\|_{H^{-1}}$

hold for all $u,v\in H_0^1(\Omega)$, where c is given by (2.7). Moreover, there exist positive constants C_1 and C_2 such that if s is a solution of (2.10), then the estimate,

(4.2iii) $\|V-s\|_{H^1} \leq C_1 \sup\{\|v-Pv\|_{L^2}: v\in\mathscr{B}_1\} + C_2\|u_0-Pu_0\|$,

holds in this case with the associated projections P satisfying (2.9).

Proof: By direct calculation, with $Hu = H(u + W)$ etc.,

$$\begin{aligned}\langle Tu-Tv,u-v\rangle &= [D(u-v,u-v) + (Hu-Hv,u-v)_{L^2} \\ &\quad + (G_2u-G_2v,u-v)_{L^2}] + (G_1u-G_1v,u-v)_{L^2} \\ &\geq c(u-v,u-v)_{H^1} + (G_1u-G_1v,u-v)_{L^2}\end{aligned}$$

so that (4.2i) holds if (4.1) holds.

It follows immediately from (4.2i) that

(4.4) $(u-v,u-v)_{H^1} \leq c^{-1}\|Tu-Tv\|_{H^{-1}}\|u-v\|_{H^1}$;

(4.2ii) now follows from (4.4) so that (4.2iii) follows from (4.2ii) and (3.1) with the choices $C_1 = c_1/(c)$ and $C_2 = c^{-1/2}$.

Corollary 4.2. The solution of $TV\ni F_0$ is unique under the hypothesis (4.1).

We state now our approximation results. Recall that $\tilde{L}$ is defined in the remark following (2.7).

Theorem 4.3. Suppose that the operator $\tilde{L}$ satisfies,

(4.4) $\tilde{L}^{-1}: L^2(\Omega) \xrightarrow{\text{continuous}} H^2(\Omega)$, i.e., the Green's operator for $\tilde{L}$ is a smoothing operator, and suppose that $\{S_h\}$ is a net of finite element spaces with the approximation property,

(4.5) $$\|u-P_h u\|_{H^1} \le K_1 h\|u\|_{H^2},$$

for a certain universal constant K_1. Suppose also that $u_0 \epsilon H^2(\Omega)$, where u_0 is given by (1.20). Then the convergence estimate,

(4.6i) $$\|V-s_h\|_{H^1} \le Kh,$$

holds for the unique Galerkin solutions s_h of $P_h^t T P_h s_h \ni P_h^t F_0$; here K does not depend on h and is given explicitly by,

(4.6ii) $$K = C_0 K_1 \tilde{c}\, C^{-1/2} C_1 + K_1 C_2 \|u_0\|_{H^2}.$$

Proof: The argument which we present is an adaptation of Nitsche's trick as presented in Strang and Fix [11, pp. 166, 167]; this adaptation is required because V does not possess increased (i.e., $H^2(\Omega)$) regularity. By (4.2iii), we must estimate,

(4.7) $$\sup\{\|v-P_h v\|_{L^2}: v \epsilon \mathscr{B}_1\},\ \|u_0-P_h u_0\|_{H^1}.$$

To estimate the first quantity in (4.7), we fix v and write,

(4.8) $$\|v-P_h v\|_{L^2} = \sup\{|(g,v-P_h v)_{L^2}|: \|g\|_{L^2} \le 1\}.$$

We then introduce the auxiliary problem, for w,

(4.9) $$D(w,u) = (g,u)_{L^2}, \text{ for all } u \epsilon H_0^1(\Omega),$$

whose solution w, by (4.4), satisfies

(4.10) $$\|w\|_{H^2} \le \tilde{c}\|g\|_{L^2}$$

for some positive constant $\tilde{c}$. With the selection, $u = v-P_h v$ in (4.9) we have, from (2.1i),

$$|(g,v-P_hv)_{L^2}| = |D(w,v-P_hv)| = |D(w-P_hw,v-P_hv)|$$

$$\leq C_0\|w-P_hw\|_{H^1}\|v-P_hv\|_{H^1}$$

$$\leq K_0h\|g\|_{L^2}$$

where $K_0 = C_0K_1\tilde{c}\ \ C^{-1/2}$. If $\|g\|_{L^2} \leq 1$ we thus have,

$$|(g,v-P_hv)| \leq K_0h. \tag{4.11}$$

and (4.11) and (4.8) yield,

$$\sup\{\|v-P_hv\|_{L^2}: v\in\mathscr{B}_1\} \leq K_0h. \tag{4.12}$$

The second term of (4.7) may be estimated directly from the hypothesis (4.5):

$$\|u_0-P_hu_0\| \leq K_1h\|u_0\|_{H^2}. \tag{4.13}$$

Combining (4.2iii),(4.12) and (4.13), we obtain (4.6). This completes the proof of Theorem 4.3.

Remark. Theorem 1.1 is now obtained by selecting $D(\cdot,\cdot) = B(\cdot,\cdot)$ and combining Theorems 2.2, 3.2, Proposition 4.1 and Corollary 4.2. The amplification of the remark following Theorem 1.1 is the content of Theorem 4.3.

It is of interest to note that, when the sequence of eigenspaces of $\tilde{L}$ are used for the approximating spaces S_h, then convergence of s_h to V in $H_0^1(\Omega)$ is guaranteed without the hypothesis (4.4). Specifically, we have,

Theorem 4.4. Under the hypotheses of Proposition 2.1 and (4.1) we have,

$$\|V-s_n\|_{H^1} \leq C_1(C/\lambda_{n+1})^{1/2} + C_2\|u_0-P_nu_0\|_{H^1}, \tag{4.14}$$

if S_n denotes the linear span of the first n eigenfunctions of $\tilde{L}$

and $\{\lambda_\nu\}$ are the nondecreasing eigenvalues. In particular, $s_n \to V$ in $H_0^1(\Omega)$.

The proof of this theorem is immediate from the characterization of the n-width of the set $\{u \in H_0^1(\Omega): D(u,u) \le C\}$ in terms of $(C/\lambda_{n+1})^{1/2}$ (cf. [9]).

Remark. If $\tilde{L}$ is the realization of an elliptic operator, then, under suitable regularity hypotheses on Ω and $\tilde{L}$ (cf. [9] for precise hypotheses), the asymptotic order of $\lambda_{n+1}^{-1/2}$ is $O(n^{-1/N})$; if $u_0 \in H^2(\Omega)$, the n-width of $\{u \in H^2(\Omega): \|u\|_{H^2} \le \|u_0\|_{H^2}\}$ is $O(n^{-1/N})$ and (4.4) yields $O(n^{-1/N})$ convergence for $\|V-s_n\|_{H^1}$.

References

1. S. Agmon, Lectures on Elliptic Boundary Value Problems, Van Nostrand, Princeton, 1965.

2. J. P. Aubin, Approximation of Elliptic Boundary Value Problems, Wiley Interscience, 1972.

3. H. Brezis, Perturbation non linéaire d'opérateurs maximaux monotone, C. R. Acad. Sc. Paris 269(1969), 566-569.

4. ________, Operateurs Maximaux Monotones et semi-groupes de contractions dans les especes de Hilbert, North Holland American Elsevier, Amsterdam and New York, 1973.

5. P. Ciarlet, M. Schultz and R. Varga, Numerical methods of high-order accuracy for nonlinear boundary value problems: V. Monotone operator theory, Numer. Math. 13(1969), 51-77.

6. N. Dunford and J. Schwartz, Linear Operators, Part I, Wiley Interscience, 1957.

7. J. Jerome, Existence and approximation of weak solutions of nonlinear Dirichlet problems with discontinuous coefficients, manuscript.

8. ________, Nonlinear equations of evolution and a generalized Stefan problem, manuscript.

9. ________, Asymptotic estimates of the n-widths in Hilbert space, Proc. Amer. Math. Soc. 33(1972), 367-372.

10. J. Lions, Quelques Méthodes de Résolution des Problèmes aux Limites non Linéaires, Dunod, Paris, 1969.

11. G. Strang and G. Fix, An Analysis of the Finite Element Method, Prentice-Hall, New York, 1973.

NULLSTELLEN VON SPLINES

K. Jetter

This paper is concerned with Budan-Fourier theorems for rather general classes of spline functions. We extend the results known up to now, and at the same time simplify the underlying analysis. Some special cases are added.

Einleitung

Erst in jüngerer Zeit befaßten sich mehrere Autoren mit Nullstellenabschätzungen bzw. Budan-Fourier-Sätzen für Splinefunktionen. Wir greifen diese Fragestellung auf und werden eine Aussage beweisen, die die bisher bekannten Ergebnisse unter einem einheitlichen Gesichtspunkt darstellt und sogar in wesentlichen Teilen verallgemeinert (Satz 2.1).

Leider hat sich bisher in der Literatur keine einheitliche Nullstellenzählung für Splinefunktionen durchgesetzt. Wir lassen uns hier von folgenden Prinzipien leiten: (i) Nullstellenordnungen werden rekursiv definiert. (ii) Vorzeichenwechsel sind genau die Nullstellen ungerader Ordnung. Wie wir sehen werden, führt dies notwendig auf eine verhältnismäßig schwache Nullstellenzählung. Einfache Beispiele zeigen aber, daß sich unsere Ergebnisse auch auf stärkere Nullstellenzählungen übertragen lassen, falls z.B. "Hermite"-Splines betrachtet werden.

Sei (a,b), $a < b$, ein reelles Intervall, $k,n \in \mathbb{N}$, $X = \{x_1,\dots,x_k\}$ mit $a =: x_o < x_1 < \dots < x_k < x_{k+1} := b$ ein Knotenvektor und

$E = (e_{i,j})_{i=1\ j=o}^{k\ \ n}$ eine Inzidenzmatrix (d.h. $e_{i,j} = o$ oder $= 1$).
Das Paar (E,X) definiert einen linearen Raum $S_n(E,X)$ verallgemeinerter Splinefunktionen:

$$s \in S_n(E,X)$$

$$(o.1) \qquad :\Longleftrightarrow \quad \text{(i)}\ s|_{(x_i,x_{i+1})} \in C^n(x_i,x_{i+1}),\ i=o,\dots,k;$$

$$\text{(ii)}\ s^{(n-j)} \text{ unstetig in } x_i \Longrightarrow e_{i,j} = 1.$$

In unseren Überlegungen sei $s \in S_n(E,X)$ ein fest vorgegebener Spline, dessen n-te Ableitung auf (a,b) nicht identisch verschwindet. Bezeichnet $I_q = [a_q,b_q]$, $q=o,\dots,n$, dasjenige kleinste Teilintervall von $[a,b]$, auf dessen Komplement (bezüglich $[a,b]$) die Ableitung $s^{(q)}$ identisch verschwindet, so gilt

$$a \le a_o \le a_1 \le \dots \le a_n < b_n \le b_{n-1} \le \dots \le b_o \le b$$

$$(o.2) \qquad \text{oder}$$

$$\emptyset =: I_{n+1} \subset I_n \subset \dots \subset I_o \subset [a,b],\ I_n \neq \emptyset .$$

Nullstellen von $s^{(q)}$ werden auf geeigneten Äquivalenzklassen von in $[a,b]$ enthaltenen Punkten definiert. Nennt man zwei Punkte z und z' aus $[a,b]$ äquivalent (bezüglich $s^{(q)}$), falls $z = z'$ oder falls $s^{(q)}$ auf (z,z') bzw. auf (z',z) identisch verschwindet, so erhält man als zu $s^{(q)}$ gehörende Äquivalenzklassen die maximalen abgeschlossenen, eventuell entarteten Intervalle $[y_1,y_2]$, in deren Innerem $s^{(q)}$ identisch verschwindet. Nullstellen definieren wir allein auf Äquivalenzklassen $[y_1,y_2]$, die ganz im offenen Intervall (a,b) enthalten sind:

DEFINITION.

(N1) Eine zu $s^{(q)}$, $q < n$, gehörende Äquivalenzklasse $[y_1,y_2] \subset (a,b)$ heißt Nullstelle der Ordnung $\alpha > 1$ für $s^{(q)}$, wenn die Ableitung $s^{(q)}$ in einer Umgebung von $[y_1,y_2]$ stetig ist, auf $[y_1,y_2]$ verschwindet, und wenn $[y_1,y_2]$ für $s^{(q+1)}$ eine Nullstelle der Ordnung $\alpha-1$ ist.

(N2) Eine zu $s^{(q)}$, $q \le n$, gehörende Äquivalenzklasse $[y_1,y_2] \subset (a,b)$

ist genau dann eine Nullstelle ungerader Ordnung für $s^{(q)}$, wenn $s^{(q)}(y_1-\varepsilon)$ und $s^{(q)}(y_2+\varepsilon)$ für alle hinreichend kleinen $\varepsilon > 0$ verschiedenes Vorzeichen haben.

(N3) Eine zu $s^{(n)}$ gehörende Äquivalenzklasse $[y_1,y_2] \subset (a,b)$ ist eine Nullstelle (der Ordnung 1) für $s^{(n)}$, falls $s^{(n)}(y_1-\varepsilon)s^{(n)}(y_2+\varepsilon) < 0$ für alle hinreichend kleinen $\varepsilon > 0$.

Einfache Beispiele zeigen, daß die Stetigkeitsforderung in (N1) die Verträglichkeit von (N1) mit (N2) garantiert. Unstetige Vorzeichenwechsel ("Sprung-Nullstellen") haben stets die Ordnung 1.

(N1) - (N3) liefert die von Schumaker [15,§9] im Fall von "HB ties" verwendete Nullstellenzählung. Identifizieren wir eine Äquivalenzklasse $[y_1,y_2]$ mit y_1 (was wir im folgenden stets annehmen wollen), so erhält man die von Jetter[6] definierte Nullstellenordnung. Der von deBoor und Schoenberg [2] betrachtete Fall einfacher Knoten ist in unseren Betrachtungen enthalten.

§ 1. Der Satz von Budan-Fourier

Es bezeichne $\mathrm{VZW}(s^{(q)};(\alpha,\beta))$ die Anzahl der Vorzeichenwechsel von $s^{(q)}$ im offenen Intervall (α,β), und für $y \in I_r$, $q < r \leq n$, sei $S_q^r(y-)$ bzw. $S_q^r(y+)$ die Anzahl der Vorzeichenwechsel in der Folge $s^{(q)}(y-),s^{(q+1)}(y-),\dots,s^{(r)}(y-)$ bzw. $s^{(q)}(y+),s^{(q+1)}(y+),\dots,s^{(r)}(y+)$. Im Einklang damit, daß eine Nullstelle $[y_1,y_2]$ von $s^{(q)}$ mit y_1 identifiziert wird, werde hierbei im Fall $y_1 < y_2$ formal die folgende Vereinbarung getroffen:

$$\begin{aligned} &\operatorname{sign} s^{(q)}(y_1+) := \operatorname{sign} s^{(q)}(y_2-) := \operatorname{sign} s^{(q)}(y_2+), \\ (1.1)\quad &\text{und für } y \in (y_1,y_2) \\ &\operatorname{sign} s^{(q)}(y-) := \operatorname{sign} s^{(q)}(y+) := \operatorname{sign} s^{(q)}(y_2+). \end{aligned}$$

LEMMA 1.1. *Sei* $s \in S_n(E,X)$, $s^{(n)} \not\equiv o$ *auf* (a,b). *Ist* $q < n$, *und haben* $s^{(q)}$ *und* $s^{(q+1)}$ *auf* (a,b) *nur endlich viele Vorzeichenwechsel, so gilt*

$$(1.2)\quad \begin{aligned} &VZW(s^{(q)};I_{q+1}) - VZW(s^{(q+1)};I_{q+1}) + \sum_{y\in I_{q+1}} \eta_{y,q} = \\ &= S_q^{q+1}(a_{q+1}+) - S_q^{q+1}(b_{q+1}-) + \sum_{y\in I_{q+1}} \gamma_{y,q} . \end{aligned}$$

Hierbei sind $\eta_{y,q}$ *und* $\gamma_{y,q}$ *für* $y \in I_{q+1}$ *definiert durch:*

(1.3)

	$S_q^{q+1}(y-)$	$S_q^{q+1}(y+)$	$VZW(s^{(q)};y)$	$VZW(s^{(q+1)};y)$	$\eta_{y,q}$	$\gamma_{y,q}$
A	o	o	o	o	o	o
B	o	o	1	1	o	o
C	o	1	o	1	o	o
D	o	1	1	o	o	2
E	1	o	o	1	2	o
F	1	o	1	o	o	o
G	1	1	o	o	o	o
H	1	1	1	1	o	o

Der Beweis ist einfach. Sind $z_1,\ldots,z_m$ die Punkte aus I_{q+1}, in denen $s^{(q)}$ oder $s^{(q+1)}$ einen Vorzeichenwechsel hat, und setzt man $z_o := a_{q+1}$ und $z_{m+1} := b_{q+1}$, so folgt $S_q^{q+1}(z_i+) = S_q^{q+1}(z_{i+1}-)$, $i=o,\ldots,m$ oder

$$(1.4)\quad \begin{aligned} &S_q^{q+1}(a_{q+1}+) - S_q^{q+1}(b_{q+1}-) = \\ &= \sum_{i=1}^{m} \{S_q^{q+1}(z_i-) - S_q^{q+1}(z_i+)\} . \end{aligned}$$

Der Rest folgt direkt aus der Tabelle (1.3).

Mit einigen Zusatzüberlegungen läßt sich für die durch (N1)-(N3) definierte Nullstellenzählung eine ähnliche Aussage ableiten. Sei $Z(s^{(q)},s^{(q+1)},\ldots,s^{(q+m)};(a,b))$, $q+m \leq n$, die Anzahl der Nullstellen von $s^{(q)}$, $s^{(q+1)}$, ..., $s^{(q+m)}$ unter Berücksichtigung der Ordnung (aber ohne doppelte Zählung gemeinsamer Nullstellenvielfachheiten) δ_q für $q < n$ die Anzahl der Fälle, in denen $s^{(q+1)}$ eine Nullstelle ungerader Ordnung α hat, die nicht gleichzeitig für $s^{(q)}$ eine Nullstelle der Ordnung $\alpha+1$ ist;

$\gamma_{y,q} = 2$ für $q < n$, falls y in I_{q+1} liegt und durch Fall D der Tabelle (1.3) erfaßt wird,

$= 1$ für $q \leq n$, falls y in $I_q - I_{q+1}$ liegt und $s^{(q)}$ in y eine Sprung-Nullstelle hat,

$= o$ sonst.

SATZ 1.2. *Sei $s \in S_n(E,X)$, $s^{(n)} \not\equiv o$ auf (a,b). Ist $q < n$, und haben $s^{(q)}$ und $s^{(q+1)}$ auf (a,b) nur endlich viele Nullstellen, so gilt*

(1.5)
$$Z(s^{(q)},s^{(q+1)};(a,b)) - \delta_q + 2h_q = \\ = Z(s^{(q+1)};(a,b)) + S_q^{q+1}(a_{q+1}+) - S_q^{q+1}(b_{q+1}-) + \sum_{y\in I_q} \gamma_{y,q}$$

mit einer natürlichen Zahl h_q,

(1.6)
$$o \leq 2h_q \leq \sum_{y\in I_{q+1}} \eta_{y,q} .$$

Beweis. Wegen $Z(s^{(q)},s^{(q+1)};I_q-I_{q+1}) = \sum_{y\in I_q - I_{q+1}} \gamma_{y,q}$ ist allein

$$Z(s^{(q)},s^{(q+1)};I_{q+1}) - \delta_q + 2h_q = \\ = Z(s^{(q+1)};I_{q+1}) + S_q^{q+1}(a_{q+1}+) - S_q^{q+1}(b_{q+1}-) + \sum_{y\in I_{q+1}} \gamma_{y,q}$$

zu zeigen. Diese Identität verifiziert man anhand einer Fallunterscheidung unter Verwendung von Lemma 1.1: Hat z.B. $s^{(q)}$ in $y \in I_{q+1}$ eine Nullstelle gerader Ordnung $\alpha \geq 2$, so ist $s^{(q)}$ auf der zu y gehörenden Äquivalenzklasse stetig. Folglich gilt $S_q^{q+1}(y-) = 1$, $S_q^{q+1}(y+) = o$, und wir haben Fall E der Tabelle (1.3) vorliegen. Damit erhalten wir

$$VZW(s^{(q)};y) - VZW(s^{(q+1)};y) + \eta_{y,q} = 1 = Z(s^{(q)},s^{(q+1)};y) - Z(s^{(q+1)};y)$$

und dieser Fall liefert in δ_q keinen Beitrag.

Die weiteren in Frage kommenden Fälle sind ebenso einfach zu behandeln.

Zusammenfassend setzen wir

(1.7) $\delta := \delta_o+\delta_1+\ldots+\delta_{n-1},\ h := h_o+h_1+\ldots+h_{n-1}.$

Mit

(1.8) $Z(s,s^{(1)},\ldots,s^{(n)};(a,b)) - \delta + 2h = Z(s;(a,b)) + 2H,$

wobei $H \geq h$ eine eindeutig bestimmte natürliche Zahl ist, folgt damit der *Satz von Budan-Fourier für Splines:*

SATZ 1.3. *Sei $s \in S_n(E,X)$, $s^{(n)} \not\equiv o$ auf (a,b). Haben $s, s^{(1)}, \ldots s^{(n)}$ auf (a,b) nur endlich viele Nullstellen, so gilt*

$$Z(s;(a,b)) + 2H = Z(s^{(n)};(a,b)) + \tag{1.9}$$
$$+ \sum_{q=o}^{n-1} \{ S_q^{q+1}(a_{q+1}+) - S_q^{q+1}(b_{q+1}-) + \sum_{y\in I_q} \gamma_{y,q} \} .$$

Ist s speziell ein Polynom vom exakten Grad n, so gilt $a = a_o = \ldots = a_n$, $b = b_o = \ldots = b_n$ und $Z(s^{(n)};(a,b)) = o$. Weiter folgt $\sum_{q=o}^{n-1} \sum_{y\in I_q} \gamma_{y,q} = o$, da Fall D der Tabelle (1.3) nur eintreten kann, wenn y eine Sprung-Nullstelle ist. Damit liefert Satz 1.3 eine starke Form des bekannten Satzes von Budan-Fourier für Polynome; im Fall $(a,b) = (o,\infty)$ erhält man die Vorzeichenregel von Descartes (vgl. z.B.[5]).

§ 2. Abschätzungen unter Verwendung der Struktur von E

Falls $Z_c(s^{(n)};(a,b))$ die Anzahl der Nullstellen von $s^{(n)}$ bezeichnet, in denen $s^{(n)}$ stetig ist, gilt

(2.1) $Z(s^{(n)};(a,b)) = Z_c(s^{(n)};(a,b)) + \sum_{y\in I_n} \gamma_{y,n}.$

Sei α bzw. β die Ordnung von s in einer rechtsseitigen Umgebung von a bzw. einer linksseitigen Umgebung von b, d.h.

$$(2.2)\quad \begin{aligned} \alpha &= n+1, \text{ falls } s^{(n)}(a+\varepsilon) \neq o \text{ für alle hinreichend kleinen } \varepsilon > o, \text{ bzw.} \\ &= \min_{o\leq q\leq n} \{q;\ s^{(q)}(a+\varepsilon) = o \text{ für alle hinreichend kleinen } \varepsilon>o\}; \\ \beta &= n+1, \text{ falls } s^{(n)}(b-\varepsilon) \neq o \text{ für alle hinreichend kleinen } \varepsilon > o, \text{ bzw.} \\ &= \min_{o\leq q\leq n} \{q;\ s^{(q)}(b-\varepsilon) = o \text{ für alle hinreichend kleinen } \varepsilon>o\}. \end{aligned}$$

$|E|$ bezeichne die Anzahl der in E enthaltenen Einsen, und $\gamma(E,\overline{E})$ sei die Anzahl der in E enthaltenen, bezüglich der erweiterten Inzidenzmatrix

$$(2.3)\quad \overline{E} := \begin{bmatrix} o\ldots o1\ldots 1 \\ \hline E \\ \hline o\ldots o1\ldots 1 \end{bmatrix} \begin{matrix} \text{—— Sequenz der Länge } \alpha \\ \\ \text{—— Sequenz der Länge } \beta \end{matrix}$$

gestützten ungeraden Sequenzen (zum Begriff der gestützten ungeraden Sequenz vgl. Lorentz und Zeller [11]).

SATZ 2.1. *Unter den Voraussetzungen von Satz 1.3 gilt*

$$(2.4)\quad \begin{aligned} Z(s;(a,b)) &\leq Z_c(s^{(n)};(a,b)) + S_o^{\alpha-1}(a+) - S_o^{\beta-1}(b-) - n + \\ &\quad + (\beta-1)_+ + |E| + \gamma(E,\overline{E}) - (1-\alpha)_+ - (1-\beta)_+. \end{aligned}$$

(Für $\alpha \leq 1$ bzw. $\beta \leq 1$ setze man hierbei $S_o^{\alpha-1}(.) := o$ bzw. $S_o^{\beta-1}(.) := o$).

<u>Beweis</u>:

Bezeichnet $\tilde{S}_q^{q+1}(b_{q+1}-)$ die Anzahl der Vorzeichenwechsel bei $(-1)^q\, s^{(q)}(b_{q+1}-)$, $(-1)^{q+1}\, s^{(q+1)}(b_{q+1}-)$, so gilt $S_q^{q+1}(b_{q+1}-) = 1 - \tilde{S}_q^{q+1}(b_{q+1}-)$. Wir haben deshalb allein die Abschätzung

$$(2.5)\quad \sum_{q=(\alpha-1)_+}^{n-1} S_q^{q+1}(a_{q+1}+) + \sum_{q=(\beta-1)_+}^{n-1} \tilde{S}_q^{q+1}(b_{q+1}-) + \sum_{q=o}^{n} \sum_{y\in I_q} \gamma_{y,q} \leq |E| + \gamma(E,\overline{E}) - (1-\alpha)_+ - (1-\beta)_+$$

zu zeigen. Hierzu verwenden wir teilweise Ideen, wie sie bei Birkhoff

[1] und in verfeinerter Form bei Lorentz [10] benutzt werden (vgl. auch Jetter [6]).

Ist $\gamma_{y,q} > o$, so ist $y =: y_1$ notwendig Repräsentant einer Sprung-Nullstelle $[y_1,y_2] \subset I_q$. Für $y \in I_q - I_{q+1}$ ist diese Aussage offensichtlich. Für $y \in I_{q+1}$ liegt Fall D der Tabelle (1.3) vor, und $s^{(q)}$ ist weder in y_1 noch in y_2 stetig. Generell gilt $y_1 \in X$, $y_2 \in X$ und $e_{i,n-q} = 1$ für $x_i = y_1$ oder für $x_i = y_2$.

Ist $\gamma_{y,q} = 1$ und $q < n$, so liegt y nicht in I_{q+1}, und folglich erhalten wir $\gamma_{y,q'} = o$ für $q < q' \leq n$.

Ist $\gamma_{y,q} = 2$, also $q < n$ und $y \in I_{q+1}$, so gilt (i) $\gamma_{y,q+1} = o$ und (ii) $\gamma_{y,q-1} = o$, falls $q > o$. Denn gemäß Fall D der Tabelle (1.3) hat die Ableitung $s^{(q+1)}$ in y keinen Vorzeichenwechsel, also keine Sprung-Nullstelle; daraus folgt zunächst $\gamma_{y,q+1} = o$. Im Fall $q > o$ ist deshalb auch $\gamma_{y,q-1} = 2$ nicht möglich. Wäre aber $\gamma_{y,q-1} = 1$, so läge y in $I_{q-1} - I_q$ im Widerspruch zu $y \in I_{q+1} \subset I_q$.

Aus $\gamma_{y,q} = 2$ folgt also insbesondere, daß die Ableitung $s^{(q+1)}$ weder links noch rechts von $[y_1,y_2]$ identisch verschwinden kann. Im Fall $q+1 \geq \alpha$ hat deshalb eine Ableitung $s^{(j)}$, $q+1 \leq j \leq n$, links von y_1 eine Unstetigkeitsstelle, die durch eine in E enthaltene Eins repräsentiert wird. Ebenso hat im Fall $q+1 \geq \beta$ eine Ableitung $s^{(j')}$, $q+1 \leq j' \leq n$, rechts von y_2 eine Unstetigkeitsstelle, die durch eine in E enthaltene Eins repräsentiert wird. Jede zu $\gamma_{y,q} = 2$ gehörende Eins $e_{i,n-q}$ aus E ist demnach in $\bar{E}$ gestützt.

Ist $(\alpha-1)_+ \leq q < n$ und $S_q^{q+1}(a_{q+1}+) = 1$, so sind die folgenden beiden Fälle zu unterscheiden:

1. Liegt a_{q+1} nicht in I_q, ist also $a_q = a_{q+1}$, so kann $s^{(q)}$ in a_q nicht stetig sein (sonst wäre $S_q^{q+1}(a_{q+1}+) = o$). Folglich gilt $a_q =: x_i \in X$ und $e_{i,n-q} = 1$.
2. Liegt a_{q+1} in I_q, ist also $a_q < a_{q+1}$, so kann der Fall eintreten, daß $s^{(q)}$ in a_{q+1} eine Sprung-Nullstelle besitzt, die durch einen Term

$\gamma_{y,q} = 1$ in (2.5) berücksichtigt wird. In diesem Fall existiert aber ein Index j, $j < n-q$, mit $e_{i,j} = 1$ für $x_i := a_{q+1}$ derart, daß $e_{i,j}$ sicher nicht zu einer Sprung-Nullstelle von $s^{(n-j)}$ gehört, also sicher nicht einem eventuell vorkommenden Term $\gamma_{y,n-j} > o$ zuzuordnen ist.

Analoge Aussagen gelten im Fall $(\beta-1)_+ \leq q < n$, $\tilde{S}_q^{q+1}(b_{q+1}-) = 1$.

Aus diesen Überlegungen folgt, daß eine Sequenz der Länge l von E in unserer Abschätzung (2.5) einen Beitrag $\leq$ l+1 liefert, und = l+1 höchstens dann, wenn l ungerade ist und die Sequenz gemäß 2o2...o2 bewichtet wird. In diesem Fall ist die Sequenz in $\overline{E}$ gestützt. Damit ist (2.5) im Fall $\alpha > o$ und $\beta > o$ bewiesen.

Ist aber $\alpha = o$, so existiert eine zu $x_i := a_o$ gehörende Sequenz in E. Im Fall $a_o < a_1$ liefert diese Sequenz keinen Beitrag in (2.5). Ist aber $a_o = a_1$, so wird die vorderste Eins dieser Sequenz nicht zur Abschätzung der Terme $S_q^{q+1}(a_{q+1}+)$ benötigt. Deshalb existiert eine zu a_o gehörende Sequenz der Länge l, die in (2.5) einen Beitrag $\leq$ l-1 liefert. Eine analoge Aussage gilt im Fall $\beta = o$, und im Fall $\alpha = \beta = o$ müssen wegen $s^{(n)} \not\equiv o$ sogar zwei verschiedene unterbewertete Sequenzen in E existieren.

Damit ist der Satz vollständig bewiesen.

<u>Bemerkung</u>: Bezeichnet $S^-(\ldots)$ bzw. $S^+(\ldots)$ die Anzahl der schwachen bzw. starken Vorzeichenwechsel in der Argumentfolge (vgl. Gantmacher und Krein [4]), so folgt

$$
\begin{aligned}
& S_o^{\alpha-1}(a+) - S_o^{\beta-1}(b-) - n + (\beta-1)_+ = \\
&= S^-(s(a+),s^{(1)}(a+),\ldots,s^{(n)}(a+)) + \\
&\qquad - S^+(s(b-),s^{(1)}(b-),\ldots,s^{(n)}(b-)) = \qquad (2.6)\\
&= n - S^+(s(a+),-s^{(1)}(a+),\ldots,(-1)^n s^{(n)}(a+)) + \\
&\qquad - S^+(s(b-),s^{(1)}(b-),\ldots,s^{(n)}(b-)).
\end{aligned}
$$

Wir zeigen nun, wie sich die bisher bekannten Nullstellenaussagen für polynomiale Splines vom Grad n als Spezialfälle aus Satz 2.1 ergeben. In diesen Fällen gilt stets $Z_c(s^{(n)};(a,b)) = o$.

Polynomiale "Hermite"-Splines:

Enthält E nur Sequenzen, die in der ersten Spalte beginnen, so folgt stets $\gamma(E,\bar{E}) = o$ und damit

$$Z(s;(a,b)) \leq S_o^{\alpha-1}(a+) - S_o^{\beta-1}(b-) - n + (\beta-1)_+ +$$

$$(2.7) \qquad + |E| - (1-\alpha)_+ - (1-\beta)_+$$

$$\leq n + |E|.$$

Dies verallgemeinert die Ergebnisse von Johnson [8,Theorem 4], Karlin und Schumaker [9,Lemma 2.1], deBoor und Schoenberg [2,Theorem 1] und Schumaker [15,Theorem 8.1]. Melkman [12,Theorem 2] beweist eine ähnliche Aussage bei anderer Nullstellenzählung.

Monosplines:

Splines vom Typ $M(x) = x^n + s(x)$, s ein polynomialer Spline vom Grad n-1, werden durch Inzidenzmatrizen E erfaßt, die in der ersten Spalte nur Nullen enthalten. Wegen $\alpha = \beta = n+1$ bezeichnet $\gamma(E,\bar{E})$ die Anzahl aller ungeraden Sequenzen in E. Die Abschätzung

$$(2.8) \qquad Z(M;(a,b)) \leq S_o^n(a+) - S_o^n(b-) + |E| + \gamma(E,\bar{E})$$

verallgemeinert die bisher bekannten Nullstellenaussagen für Monosplines, vgl. Johnson [8,Theorem 6], Karlin und Schumaker [9, Lemma 2.2] und Micchelli [13, Theorem 1,erster Teil].

Splines mit kompaktem Träger:

Im Fall $\alpha = \beta = o$ folgt für Splines vom exakten Grad n

$$(2.9) \qquad Z(s;(a,b)) \leq |E| + \gamma(E,E) - n - 2$$

und damit die Aussage von Lorentz [1o,Theorem 1], vgl. auch Jetter [6, Satz 3.4], Birkhoff [1,S. 115,Theorem], Ferguson [3,Theorem 2.1].

§ 3. *Polynomiale "Hermite"-Splines*

Satz 2.1 verschärft die Aussage von Schumaker [15, Theorem 9.3], welche - abgesehen von den oben zitierten Abschätzungen für Splines mit kompaktem Träger - das erste Ergebnis zu sein scheint, das auf Splines mit lakunären Knoten anwendbar ist. Anders verhält es sich im Falle von "Hermite"-Splines, also in dem Sonderfall, daß alle Sequenzen von E in der ersten Spalte beginnen. Dieser Sonderfall läßt sich aufgrund der einfachen Struktur von E besonders leicht behandeln. Sei also $s \in S_n(E,X)$ ein polynomialer Spline (oder allgemeiner ein "Tchebycheffian spline") mit $s^{(n)} \not\equiv o$ auf (a,b), wobei E eine Hermite-Matrix ist, deren i-te Zeile eine Sequenz der Länge k_i enthält.

Zunächst nehmen wir an, daß s auf keinem Teilintervall von (a,b) identisch verschwindet. Auf den Intervallen (x_i, x_{i+1}), $i=o,\ldots,k$, besitze s den exakten Grad n_i (die Ordnung n_i+1). Dann gilt (man vergleiche mit Satz 1.3)

$$Z(s;(x_i,x_{i+1})) + 2H_i = S_o^{n_i}(x_i+) - S_o^{n_i}(x_{i+1}-),\ i=o,\ldots,k \tag{3.1}$$

wobei Z(.) allein durch die übliche Zählung der Nullstellenvielfachheit für Polynome festgelegt ist. Daraus folgt

$$\begin{aligned} Z(s;(a,b)-X) + 2\sum_{i=o}^{k} H_i + \sum_{i=1}^{k} \{S_o^{n_{i-1}}(x_i-) - S_o^{n_i}(x_i+)\} &= \\ = S_o^{n_o}(a+) - S_o^{n_k}(b-) &. \end{aligned} \tag{3.2}$$

Für jede Nullstellenzählung $\tilde{Z}(s;x_i)$ in den Knoten x_i gilt dann die Abschätzung (mit $\tilde{Z}(s;(a,b)) := Z(s;(a,b)-X) + \sum_{i=1}^{k} \tilde{Z}(s;x_i)$)

$$\tilde{Z}(s;(a,b)) \leq S_o^{n_o}(a+) - S_o^{n_k}(b-) + |E| \leq n + |E|, \tag{3.3}$$

falls nur die folgenden Ungleichungen erfüllt sind:

$$\tilde{Z}(s;x_i) - S_o^{n_{i-1}}(x_i-) + S_o^{n_i}(x_i+) - k_i \leq o,\ i=1,\ldots,k\ . \tag{3.4}$$

Verschwindet s auf einem Teilintervall $[x_{i_1},x_{i_2}]$, aber weder in einer rechtsseitigen Umgebung von a noch in einer linksseitigen Umgebung von b, so liefert Induktion über die Anzahl solcher Teilintervalle wiederum die Aussage (3.3), falls für alle (maximalen) Nullintervalle $[x_{i_1},x_{i_2}] \subset (a,b)$ die Ungleichungen

$$\tilde{Z}(s;[x_{i_1},x_{i_2}]) - S_o^{n_{i_1}-1}(x_{i_1}-) + S_o^{n_{i_2}}(x_{i_2}+) - k_{i_1} - k_{i_2} \leq 0 \tag{3.5}$$

gelten; dabei wird (3.4) nur noch für diejenigen Knoten x_i gefordert, für die s weder in einer linksseitigen noch in einer rechtsseitigen Umgebung identisch verschwindet. Wegen $S_o^{n_{i_1}-1}(x_{i_1}-) + k_{i_1} \geq n$ und $k_{i_2} - S_o^{n_{i_2}}(x_{i_2}+) \geq 1$ ist (3.5) sogar erfüllt, wenn die Zählung $\tilde{Z}(s;[x_{i_1},x_{i_2}]) := n+1$ zugrunde gelegt wird.

(3.4) und (3.5) sind insbesondere für stärkere Nullstellenzählungen erfüllt als wir sie in unseren bisherigen Überlegungen verwendet haben (vgl. Melkman [12], Schumaker [15,§7]). Es sei aber betont, daß die dort verwendeten Nullstellenvielfachheiten nicht mehr mit unserer Forderung (N2) verträglich sind.

§ 4. Ergänzungen

Gilt Gleichheit in Satz 2.1 und damit in allen verwendeten Abschätzungen, so liefert unsere Beweisführung einen tiefen Einblick in das Vorzeichenverhalten der Ableitungen von s. Dies wirft die Frage auf, ob die Abschätzung von Satz 2.1 bei fest vorgegebenem Tripel (E,α,β) scharf ist. Für spezielle Tripel (E,α,β) ist diese Frage gelöst (vgl. Fundamentalsatz der Algebra für Monosplines, Micchelli [13], bzw. Aussagen über Interpolationskerne mit maximaler Nullstellenzahl, Jetter [6,7]).

Literatur

[1] BIRKHOFF,G.D., General mean value and remainder theorems with applications to mechanical differentiation and quadrature. Trans. Amer.Math.Soc. 7 (19o6), 1o7-136.

[2] deBOOR,C. and I.J.SCHOENBERG, Cardinal interpolation and spline functions VIII. The Budan-Fourier theorem for splines and applications. In "Spline Functions", K.Böhmer, G.Meinardus and W.Schempp, Eds. Berlin-Heidelberg-New York: Springer-Verlag 1976, 1-79.

[3] FERGUSON,D.R., Sign changes and minimal support properties of Hermite-Birkhoff splines with compact support. SIAM J. Numer. Anal. 11 (1974), 769-779.

[4] GANTMACHER,F.R. und M.G.KREIN, "Oszillationsmatrizen, Oszillationskerne und kleine Schwingungen mechanischer Systeme". Berlin: Akademie-Verlag 196o.

[5] HOUSEHOLDER,M., "The numerical treatment of a single non-linear equation". New York: McGraw Hill, 197o.

[6] JETTER,K., Duale Hermite-Birkhoff-Probleme. Erscheint in J. Approximation Theory.

[7] JETTER,K., Birkhoff interpolation by splines. Erscheint im Tagungsband des Symposiums über Approximationstheorie, Austin 1976.

[8] JOHNSON,R.C., On monosplines of least deviation. Trans.Amer.Math. Soc. 96 (196o), 458-477.

[9] KARLIN,S. and L.L.SCHUMAKER, The fundamental theorem of algebra for Tchebycheffian monosplines. J. d'Anal.Math. 2o (1967), 233--27o.

[1o] LORENTZ,G.G., Zeros of splines and Birkhoff's kernel. Math.Z. 142 (1975), 173-18o.

[11] LORENTZ,G.G. and K.L.ZELLER, Birkhoff interpolation. SIAM J.Numer. Anal. 8 (1971), 43-48.

[12] MELKMAN,A.A., The Budan-Fourier theorem for splines. Israel J. Math. 19 (1974), 256-263.

[13] MICCHELLI,C., The fundamental theorem of algebra for monosplines with multiplicities. Proc.Conf.Oberwolfach 1971, ISNM 2o (1972), 419-43o.

[14] SCHUMAKER,L.L., Zeros of spline functions and applications. Erscheint im J.Approximation Theory.

[15] SCHUMAKER,L.L., Toward a constructive theory of generalized spline functions. In "Spline Functions", K.Böhmer, G.Meinardus and W.Schempp, Eds., Berlin-Heidelberg-New York: Springer-Verlag 1976, 265-331.

Kurt Jetter
Math.Inst.d.Eberhard-Karls-Universität
Auf der Morgenstelle 1o
74oo Tübingen
BRD

LOCAL SPLINE APPROXIMATION METHODS AND OSCULATORY INTERPOLATION FORMULAE

Tom Lyche

Using B-splines, we reformulate in this brief survey most of the osculatory interpolation formulae which can be found in the actuarial litterature. Some new formulae of practical interest are also given.

1. Introduction and discussion.

A large class of local approximation methods for finding a smooth approximation s to a set of data $\langle(x_i, y_i)\rangle$ can be found in the actuarial litterature under the key-word osculary interpolation. I.e. s is a piecewise polynomial of some order $k \geq 2$ (degree $k-1$) and of smoothness C^m for some $1 \leq m \leq k-2$. Moreover s satisfies

$$s^{(r)}(x_i) = A_{ir}y_i \qquad r=0,1,\ldots,m \quad , \text{ all } i \ ,$$

where A_{ir} is a difference operator of the form

$$A_{ir}y_i = \sum_{j=i-p_{ir}}^{i+q_{ir}} c_{irj}y_j \ .$$

Here p_{ir} and q_{ir} are small integers ensuringthe localness of the approximation. A_{ir} should also be consistent with D^r, the r-th derivative operator (at least for small values of r). In the actuarial litterature the datapoints are normally assumed to be equally spaced, and the knots of the spline are either placed at the datapoints(end point formula) or midway between the knots (midpoint formula). In the equally spaced case the operators A_{ir} do not depend on i except near the ends. If $s(x_i) = y_i$ all i, i.e. $A_{io} = I$, the identity, then the formula is called an interpolation formula. Otherwise we will call it an <u>approximation formula</u>. The terms <u>ordinary</u> for interpolation and <u>modified</u> or <u>smoothing</u> for approximation are also used.

With data of unit spacing, $x_i - x_{i-1} = 1$ all i, we have the usual difference operators μ and δ defined by

$$\mu^r z_i = \sum_{j=0}^{r} \binom{r}{j} z_{i-r/2+j}/2^r \qquad r = 0,1,2,\ldots$$

$$\delta^r z_i = \sum_{j=0}^{r} (-1)^{r-j} \binom{r}{j} z_{i-r/2+j} .$$

A classical (1898) example of an interpolation formula is the Karup-King formula. Here s is determined from piecewise cubic Hermite interpolation using $\mu\delta y_i$ instead of y_i'. (Of course a special formula will have to be used at the ends, but this will not be considered). Greville[4] writes the Karup-King formula in the form

$$(1.1) \qquad s(x_i+x) = F(1-x,\delta)y_i + F(x,\delta)y_{i+1} \qquad 0 \le x \le 1$$

where

$$F(x,\delta) = x + x^2(x-1)\delta^2/2 .$$

It is eaily verified that

$$(1.2) \qquad s(x_i) = y_i \quad , \quad s'(x_i) = \mu\delta y_i$$

and that normally $s''(x_i+) \neq s''(x_i-)$. Thus s is a cubic spline of smoothness C^1. (1.1) is a four point formula since in the interval (x_i, x_{i+1}) it involves the four data points $y_{i-1}, y_i, y_{i+1}, y_{i+2}$. We also note that it is an end point formula.

W.A. Jenkins (see [4] for references) has constructed several osculatory interpolation formulae of both the midpoint and the end point type. In [4] his three point quadratic midpoint formula from 1930 is written in the form

$$(1.3) \qquad s(x_{i-\frac{1}{2}}+x) = y_i + G(x,\delta)y_{i+\frac{1}{2}} - G(1-x,\delta)y_{i-\frac{1}{2}} \qquad 0 \le x \le 1$$

where $G(x,\delta) = \frac{1}{2}x^2\delta$. We find $s(x_{i-\frac{1}{2}}) = \mu y_{i-\frac{1}{2}}$, $s'(x_{i-\frac{1}{2}}) = \delta y_{i-\frac{1}{2}}$ all i, so that s is a quadratic spline of smoothness C^1.

We also find

$$(1.4)\qquad s(x_i) = (I+\delta^2/8)y_i \quad , \quad s'(x_i) = \mu\delta\, y_i \ .$$

Formulae like (1.2) and (1.4) are of interest in data smoothing and numerical differentiation, so we shall continue to list the effect of the formulae at the data points.

An important parameter when talking about local approximation methods is the degree of reproduction of the method. A method is said to have <u>degree of reproduction</u> l ([4]) if all polynomilas of degree $\leq l$ are exactly reproduced by the method. Thus it can be shown that the Karup-King formula (1.1) has degree of reproduction two, while in the Jenkins formula (1.3) we have $l=1$.

In 1944 Greville[4] gave a unified treatment of osculatory interpolation listing all classical formulae and giving several new ones. More formulae were obtained by Schoenberg[7] in 1946 using B-splines. Later several local spline approximation methods using B-splines on arbitrary partitions have appeared. We mention here the variation diminishing spline approximation of Schoenberg[8] from 1967 and the quasi-interpolant of deBoor and Fix from 1973 ([3]).

In [6] a general class of local spline approximation methods were considered which turn out to include all the osculatory interpolation formulae which can be found in [4]. To define the methods suppose $N,k \geq 1$ and let $\{t_i\}_{i=1}^{N+k}$ be a nondecreasing sequence satisfying $t_{i+k} > t_i$ $i=1,2,\ldots,N$. Given a linear space of functions $\mathbb{F}$ defined on an interval $[a,b]$ and an $f \in \mathbb{F}$, the methods can be written in the form

$$(1.5)\qquad Qf(x) = \sum_{i=1}^{N} \lambda_i f\, B_{ik}(x)$$

where B_{ik} $i=1,2,\ldots,N$ are the usual B-splines of order k on $t_1,t_2,\ldots,t_{N+k}$, normalized to sum up to one at any point in $[a,b]$ (See [2]). The λ_i's are point evaluator linear functionals which are conveniently written in the form

$$(1.6)\qquad \lambda_i f = \sum_{j=1}^{r} \alpha_{ij} [z_{i1},z_{i2},\ldots,z_{ij}]f \ .$$

Here the brackets denote the usual divided differences. The z_{ij}'s can be chosen among the data points. The α_{ij}'s are constants which can be chosen so that the method has desirable properties. In particular if we insist that the method should have degree of precision l for some $0 \le l \le r-1$ then $\alpha_{i1},\dots,\alpha_{i,l+1}$ are easily calculated ([6] eqn. (3.9)) using Marsdens identity. If $r=k$, $l=k-1$, and $z_{i1}=\dots=z_{ik}$ then we get the quasi-interpolant of deBoor and Fix[3]. If $z_{i1},\dots,z_{ir}$ are distinct and chosen near the support of B_{ik} and the knots are uniformly spaced (but possibly multiple) then we get an osculatory interpolation formula of the form considered in [4]. Moreover since any piecewise polynomial can be written as a linear combination of B-splines on a suitable knot-sequence $\langle t_i \rangle$ it is clear that any osculatory interpolation formula in [4] can be written in the form (1.5) , (1.6).

Writing the formula in terms of B-splines has many advantages:
(i) Given the order k , the smoothness C^m , and the degree of reproduction l, a formula of the form (1.5),(1.6) with these properties can immediatly be written down.
(ii) If $r > l+1$ then the free parameters $\alpha_{i,l+2},\dots,\alpha_{i,r}$ in (1.6) can be used to try to get an approximation Qf of a desired shape. In particular since B-splines are non-negative an obvious sufficient condition for a non-negative approximation is that all the B-spline coefficients are non-negative. On the other hand since ([2])

$$(Qf)' = (k-1)\sum_{i=2}^{N} \frac{\lambda_i f-\lambda_{i-1} f}{t_{i+k-1}-t_i} B_{i,k-1}$$

we get a monotone approximation in an interval $[c,d] \subset [a,b]$ if $\lambda_i f \ge \lambda_{i-1} f$ for all i such that the support of $B_{i,k-1}$ intersects (c,d). Similarly we can get a convex approximation.
(iii) Error bounds can also be derived. We quote a result from [6]: Suppose Q given by (1.5) , (1.6) has degree of reproduction l , $r=l+1$, and f is sufficiently smooth. Then for any n , $0 \le n \le l$

$$\|(f-Qf)^{(j)}\|_{\infty,[t_i,t_{i+1}]} \le K\, h^{n-j}\, \omega(f^{(n)}; h ; I_i) \quad 0 \le j \le n-1$$

Here I_i is an interval slightly larger than $[t_i,t_{i+1}]$, $h = \max_{\nu\,[t_\nu,t_{\nu+1}]\cap I_i \ne \emptyset} (t_{\nu+1}-t_\nu)$, ω is the usual modulus of con-

tinuity, and K is a constant which depends on the placement of the z_{ij}'s and the spacing of the t_{ij}'s in I_i. For certain choices of z_{ij}'s and for small values of j the constant can be made only to depend on k. (See [6] for details).
(iv) By using B-splines it has been shown that Jenkins formula (1.3) is variation diminishing.

The purpose of this paper is to list explicitly all the classical osculatory interpolation formulae in terms of B-splines. In the process we will also get some new formulae of interest when smoothing empirical data. Although one of the main advantages with B-splines is the ease with which nonuniform partitions can be treated we consider here only uniform partitions (with multiple knots). There are several reasons for this: (i) it is important to see what a formula reduces to in the case of uniform partitions. (ii) In practice data are very often uniformly spaced. (iii) If the data are scattered then it is possible to use a two stage process as descibed in [9]. In the first stage we use some form of local least squares producing a nonsmooth approximation. In the second stage uniformly spaced data are taken from this approximation and fed into an osculatory interpolation formula.

The outline for the rest of the paper is as follows: In section 2 and 3 we list for $k\leq 6$ (5) formulae with simple knots $\langle t_i \rangle$ of the end point and midpoint type respectively. Equivalent formulations for $k=3,4$ can be found in [4]. In section 4 we give the C^1 cubic end point formulae of which the Karup-King formula serve as a typical example. Finally in section 5 we consider the other classes of formulae treated in [4]. For each classical formula we give the year of publication. Further references can be found in [4].

2 C^{k-2} end point formulae.

Suppose

$$x_j = t_j = j \quad \text{all } j \text{ ,}$$

and let y_j be data at x_j. Let $f:\mathbb{R}\to\mathbb{R}$ be any function satisfying $f(x_j) = y_j$ all j. Define $t_{j+r}=(1-r)t_j+rt_j$ and $f_{j+r}=f(t_{j+r})$. Given $n\geq 0$ and a difference operator u_k of the form

(2.1) $$u_k = \sum_{j=o}^{n} c_{2j}\delta^{2j}$$

we consider a local spline approximation method Q_k of the form

(2.2) $$Q_k f = \sum_j \lambda_{jk} f\, B_{jk}$$

where

(2.3) $$\lambda_{jk} f = \hat{u}_k f_{j+k/2} \quad , \quad \hat{u}_k = \begin{cases} u_k & k \text{ even} \\ \mu u_k & k \text{ odd} \end{cases} .$$

Now

$$s(t_i) = \sum_{j=i+1-k}^{i-1} \hat{u}_k f_{j+k/2} B_{jk}(t_i) = \hat{u}_k \hat{e}_k f_i$$

where

(2.4) $$\hat{e}_k f_i = \sum_{j=i+1-k}^{i-1} B_{jk}(t_i) f_{j+k/2}.$$

For convenience we list the first few operators $\hat{e}_k$ (the operators m_k are considered in section 3).

k	$\hat{e}_k$	$\hat{m}_k$
2	I	μ
3	μ	$I+\frac{1}{8}\delta^2$
4	$I+\frac{1}{6}\delta^2$	$\mu(I+\frac{1}{24}\delta^2)$
5	$\mu(I+\frac{1}{12}\delta^2)$	$I+\frac{5}{24}\delta^2+\frac{1}{384}\delta^4$
6	$I+\frac{1}{4}\delta^2+\frac{1}{120}\delta^4$	$\mu(I+\frac{1}{8}\delta^2+\frac{1}{1920}\delta^4)$

Table 2.1

Using Schoenbergs formula ([7] p. 73) for differentiating the B-spline series on a uniform partition we can also express the derivatives of s at x_i in terms of y-values. Thus if

(2.5a) $$s(x) = \sum_j f_{j+k/2} B_{jk}(x)$$

then

(2.5b) $$s^{(m)}(x) = \sum_j \delta^m f_{j+\frac{k-m}{2}} B_{j,k-m}(x) \qquad m=0,1,2,\dots,k-1$$

It follows that if $s=Q_kf$ is given by (2.2),(2.3) then by (2.4)

(2.6) $$s^{(m)}(x_i) = \delta^m \hat{u}_k \hat{e}_{k-m} y_i \qquad m=0,1,\dots,k-2$$

where $\hat{u}_k$ is given by (2.1),(2.3) and $\hat{e}_k$ by (2.4). Using the relation $\mu^2 = I+\delta^2/4$ and table 2.1 we get the following table for finding the values $s^{(m)}(x_i)$.

k	$\hat{u}_k$	m=0	m=1	m=2	m=3	m=4
3	$\mu\varphi(0)$	$\varphi(\frac{1}{4})$	$\mu\delta\varphi(0)$			
4	$\varphi(0)$	$\varphi(\frac{1}{6})$	$\mu\delta\varphi(0)$	$\delta^2\varphi(0)$		
5	$\mu\varphi(0)$	$\psi(\frac{1}{3}\ \frac{1}{48})$	$\mu\delta\varphi(\frac{1}{6})$	$\delta^2\varphi(\frac{1}{4})$	$\mu\delta^3\varphi(0)$	
6	$\psi(0)$	$\psi(\frac{1}{4},\frac{1}{120})$	$\mu\delta\varphi(\frac{1}{12})$	$\delta^2\varphi(\frac{1}{6})$	$\mu\delta^3\varphi(0)$	$\delta^4\varphi(0)$

Table 2.2 $\delta^m \hat{u}_k \hat{e}_{k-m}$

Here

(2.7) $$\varphi(\alpha)=c_oI+(\alpha c_o+c_2)\delta^2+\dots+(\alpha c_{2n-2}+c_{2n})\delta^{2n} \quad +\alpha c_{2n}\delta^{2n+2}$$

(2.8) $$\psi(\alpha,\beta)=c_oI+(\alpha c_o+c_2)\delta^2+(\beta c_o+\alpha c_2+c_4)\delta^4+\dots+(\beta c_{2n-4}+\alpha c_{2n-2} +c_{2n})\delta^{2n} \quad +(\beta c_{2n-2}+\alpha c_{2n})\delta^{2n+2}+\beta c_{2n}\delta^{2n+4} .$$

It is customary to require ([4] p.204) symmetry. I.e. it will make no difference in the results of the approximation whether we proceed along the data from left to right or from right to left. This means that values and derivatives of s at x_i of even(odd) order must only contain powers of μ and δ of even (odd) order. It is easily checked that the entries in table 2.2 satisfy the symmetry requirement. This is so because we only consider an even expansion u_k in (2.1).

The degree of reproduction l of the formulae can be determined either from Marsdens identity or making the operators in table 2.2 agree with the expansions of the operators D^m in terms of μ and and powers of δ through terms containing δ^l. Now ([4] p.214)

$$D^0 = I$$
$$D^1 = \mu\delta - \frac{1}{6}\mu\delta^3 + \frac{1}{30}\mu\delta^5 - \frac{1}{140}\mu\delta^7 + \dots$$
$$d^2 = \delta^2 - \frac{1}{12}\delta^4 + \frac{1}{90}\delta^6 - \frac{1}{560}\delta^8 + \dots$$
$$D^3 = \mu\delta^3 - \frac{1}{4}\mu\delta^5 + \frac{7}{120}\mu\delta^7 - \dots$$
$$D^4 = \delta^4 - \frac{1}{6}\delta^6 + \frac{7}{240}\delta^8 - \dots \quad .$$

Thus if $c_o=1$ then all the formulae in table 2.2 will have degree of reproduction at least one. The values of the coefficients c_{2j} which ensures higher degree of reproduction are as follows:

k	l=2	l=3	l=4	l=5
3	$c_2=-\frac{1}{4}$			
4		$c_2=-\frac{1}{6}$		
5		$c_2=-\frac{1}{3}$	$c_4=\frac{13}{144}$	
6		$c_2=-\frac{1}{4}$		$c_4=\frac{13}{240}$

Table 2.3 Degree of reproduction of the end point formula (2.3)

We look closer at the case $k=3$, $n=2$. From table 2.2 we have

(2.9) $$\lambda_{j3}f = \mu(I+c_2\delta^2+c_4\delta^4)f_{j+3/2}$$

(2.10) $$s(x_i)= (I+(\tfrac{1}{4}+c_2)\delta^2+(\tfrac{1}{4}c_2+c_4)\delta^4+\tfrac{1}{4}c_4\delta^6)y_i$$

(2.11) $$s'(x_i)=(\mu\delta+c_2\mu\delta^3+c_4\mu\delta^5)y_i$$

If $c_2=c_4=0$ we have $\lambda_{j3}f = (f_{j+1}+f_{j+2})/2$. This is Jenkins four point, quadratic, approximation formula from 1927. Formulae for other values of c_2 and c_4 are listed in [4] as formula (72) (c_2 arb. , $c_4=0$) , (74) ($c_2=-1/4$, $c_4=0$) , (80) (c_2, c_4 arb.) , (82) ($c_2=-1/4$, c_4 arb.) , and (109) ($c_2=2/3$, $c_4=0$). To arrive at the value $c_2=2/3$ we write $s(x_i) = (I+(1/4+c_2)\delta^2+c_2\delta^4/4)y_i$ in terms of ordinates and choose c_2 to minimize the sum of squares of the coefficients. One can play a similar game with $s'(x_i)$ given by (2.11) and $c_4=0$. This leads to $c_2=2/5$ and the formula

$$(2.12)\qquad s'(x_i) = (-2y_{i-2}-y_{i-1}+y_{i+1}+2y_{i+2})/10 ,$$

which is (for $h=1$) formula (5-8.3) in Lanczos[5]. There it is recommended for numerical differentiation of empirical data. Here (2.12) is part of a smooth approximation formula. We also point out that a reasonable compromise between $c_2=2/3$ and $c_2=2/5$ is $c_2=1/2$. This leads to formula (2.2) with

$$\lambda_{i3}f = (f_i + f_{i+1} + f_{i+2} + f_{i+3})/4 .$$

This has degree of reproduction one for uniformly spaced knots and should be useful for handling empirical data.

We also mention a few formulae with $k=4$. $n=0$ and $c_0=1$ give Schoenberg's cubic variation diminishing spline approximation, while choosing $n=1$ and $c_2=-1/6$ so that according to table 2.3 the degree of reproduction is three we get another Jenkins formula (1927):

$$\lambda_{i4}f = (I-\delta^2/6)f_{i+2} = (-f_{i+1}+8f_{i+2}-f_{i+3})/6$$

This method was also used as a numerical example in [3].

Since $s\in C^2$ when $k=4$ we can use the cubic formulae for numerical determination of second derivatives. Taking $n=3$ and $k=4$ in table 2.2 we have

$$s''(x_i) = (\delta^2+c_2\delta^4+c_4\delta^6+c_6\delta^8)y_i \quad .$$

Now writing this in terms of ordinates and choosing c_2,c_4,c_6 to minimize the sum of squares of the coefficients leads to $c_2=9/7$, $c_4=1/2$, and $c_6= 2/33$. On the other hand with

$$c_2=21/20 \quad , \quad c_4=9/25 \quad , \quad c_6=1/25$$

we get

$$s''(x_i)=\frac{4y_{i-4}+4y_{i-3}+y_{i-2}-4y_{i-1}-10y_i-4y_{i+1}+y_{i+2}+4y_{i+3}+4y_{i+4}}{100}$$

wich is formula (5.10.1) in [5].

3. C^{k-2} midpoint formulae.

Suppose now $t_j=j+\frac{1}{2}$ and $x_j=j$ all j. Let u_k be the operator given by (2.1). We consider a local spline approximation of the form (2.2) where

$$(3.1) \qquad \lambda_{jk}f = \hat{v}_k f_{j+k/2} \quad , \quad \hat{v}_k = \begin{cases} \mu u_k & k \text{ even} \\ u_k & k \text{ odd} . \end{cases}$$

Now with $\hat{m}_k$ given by

$$(3.2) \qquad \hat{m}_k f_{i-\frac{1}{2}} = \sum_{j=i-k}^{i-1} B_{jk}(t_{i-\frac{1}{2}}) f_{j+k/2}$$

and $s=Q_k f$ we find from (2.5)

$$(3.3) \qquad s^{(m)}(x_i) = \delta^m \hat{v}_k \hat{m}_{k-m} y_i \qquad m=0,1,\dots,k-2 .$$

The operators $\hat{m}_k$ are given in table 2.1 for $k \leq 6$. With φ and ψ given by (2.7) and (2.8) the values $s^{(m)}(x_i)$ can be found from the following table

k	$\hat{v}_k$	m=0	m=1	m=2	m=3
3	$\varphi(0)$	$\varphi(\frac{1}{8})$	$\mu\delta\varphi(0)$		
4	$\mu\varphi(0)$	$\psi(\frac{7}{24},\frac{1}{96})$	$\mu\delta\varphi(\frac{1}{8})$	$\delta^2\varphi(\frac{1}{4})$	
5	$\varphi(0)$	$\psi(\frac{5}{24},\frac{1}{384})$	$\mu\delta\varphi(\frac{1}{24})$	$\delta^2\varphi(\frac{1}{8})$	$\mu\delta^3\varphi(0)$

Table 3.1 $\delta^m \hat{v}_k \hat{m}_{k-m}$

The degree of reproduction is one if and only if $c_0=1$. The values of the coefficients c_{2j} which ensure higher degree of reproduction are given in table 3.2.

We shall mainly look at the case $k=3$, $n=2$. Then

$$(3.4) \qquad \lambda_{j3}f = (I+c_2\delta^2+c_4\delta^4)f_{j+3/2}$$

$$(3.5) \qquad s(x_i) = (I+(\tfrac{1}{8}+c_2)\delta^2+(\tfrac{1}{8}c_2+c_4)\delta^4+\tfrac{1}{8}c_4\delta^6)y_i$$

$$(3.6) \qquad s'(x_i) = (\mu\delta + c_2\mu\delta^3 + c_4\mu\delta^5)y_i \;.$$

k	l=2	l=3	l=4
3	$c_2=-1/8$		
4		$c_2=-7/24$	
5		$c_2=-5/24$	$c_4=47/1152$

Table 3.2 Degree of reprocuction of the midpoint formula (2.2), (3.1).

If $c_2=c_4=0$ we have $\lambda_{j3} = f_{j+3/2}$ which defines Schoenbergs quadratic, variation diminishing, spline approximation [8]. Since the values of $s(x_i)$ and $s'(x_i)$ are given by (1.4) this must be identical with Jenkins formula (1.3). (Of course Jenkins did not know that it was varation diminishing.)

Note that the midpoint formula (3.6) and the end point formula (2.11) give the same expression for $s'(x_i)$. Thus again for $c_2=2/5$ $c_4=0$ (3.6) reduces to formula (5-8.3) in [5]. With $c_4=0$ the sum of squares of the coefficients of the ordinates in (3.5) is minimized for $c_2=26/67$. A good choice for handling empirical data should therefore be $c_2=1/3$ which gives

$$(3.7) \qquad \lambda_{j3}f = (f_{j+1/2}+f_{j+3/2}+f_{j+5/2})/3 \;.$$

Before closing this section let us mention the formula

$$\lambda_{j3}f = (-f_{j+1}+4f_{j+3/2}-f_{j+2})/2$$

considered by de Boor ([1]p.234). It uses data both at the knots and midway between them. Such formulae will not be considered here.

4. C^1 cubic end point formulae.

Suppose

$$t_{2j} = t_{2j+1} = x_j = j \qquad \text{all } j$$

and define

(4.1) $$s = Q_4 f = \Sigma \lambda_{j4} f B_{j4}$$

where

$$\lambda_{2j,4} f = u y_{j+1} \quad , \quad \lambda_{2j+1,4} f = v y_{j+1}$$

$$u = I - \mu\delta/3 + c_2\delta^2 - c_3\mu\delta^3 + c_4\delta^4 - c_5\mu\delta^5 + c_6\delta^6$$

$$v = I + \mu\delta/3 + c_2\delta^2 + c_3\mu\delta^3 + c_4\delta^4 + c_5\mu\delta^5 + c_6\delta^6 \ .$$

Thus $Q_4 f$ defines a cubic C^1 end point formula. Since

(4.3) $$s(x_i) = \frac{u+v}{2} y_i = (I + c_2\delta^2 + c_4\delta^4 + c_6\delta^6) y_i$$

(4.4) $$s'(x_i) = \frac{3}{2}(v-u) y_i = (\mu\delta + 3c_3\mu\delta^3 + 3c_5\mu\delta^5) y_i$$

the particular form of u and v ensures symmetry. Moreover the degree of reproduction l is one. If $c_2=0$ then $l=2$, and if in addition $c_3=-1/18$ then $l=3$. For $c_i=0$ $i\geq 2$ we have the Karup-King formula (1.1). Choosing $c_3=-1/18$ and $c_i=0$ otherwise we get another classical formula known as Hendersons formula (1906).

Note that there is a different set of coefficients used for $s(x_i)$ and $s'(x_i)$ in (4.3) and (4.4). Thus they can be chosen independently of each other. Choosing

$$c_2=c_5=c_6=0 \quad , \quad c_3=2/15 \ , \ \text{and} \quad c_4=-3/35$$

(4.3) reduces to the method of smoothing with 4th differences ([5] (5-7.6)), while (4.4) reduces to the optimal formula (2.12).

5. Other formulae.

For completeness we also give the C^1 $k=4$ midpoint formulae, the C^2 $k=5$ end point- and midpoint formulae, and the C^2 $k=6$ end point and midpoint formulae. All these cases were considered by Greville[4] p.261,262.

Cubic C^1 midpoint formulae.

$$t_{2j} = t_{2j+1} = j+\tfrac{1}{2} \quad , \quad x_j = j \ ,$$

$\lambda_{2j,4}f = (\mu-\delta/3+c_2\mu\delta^2+c_3\delta^3+c_4\mu\delta^4+c_5\delta^5)y_{j+1}$

$\lambda_{2j+1,4}f=(\mu+\delta/3+c_2\mu\delta^2-c_3\delta^3+c_4\mu\delta^4-c_5\delta^5)y_{j+1}$

$s(x_i) \quad =(I+(\frac{1}{8}+c_2)\delta^2+(\frac{1}{4}c_2+\frac{3}{8}c_3+c_4)\delta^4+(\frac{1}{4}c_4+\frac{3}{8}c_5)\delta^6)y_i$

$s'(x_i) \quad =\mu\delta+\frac{3}{2}(c_2+c_3)\mu\delta^3+\frac{3}{2}(c_4+c_5)\mu\delta^5$

$$\text{degree of reproduction} = \begin{cases} 3 & c_2=-1/8\ ,\ c_3=1/72 \\ 2 & c_2=-1/8 \\ 1 & \text{otherwise} \end{cases}$$

Quartic C^2 end point formulae (k=5).

$t_{2j} = t_{2j+1} = x_j = j$ all j

$\lambda_{2j,5}f \quad =(I-\delta^2/12+b_4\delta^4+b_6\delta^6)y_{j+1}$

$\lambda_{2j+1,5}f=\mu(I-\delta^2/6+c_4\delta^4)y_{j+3/2}$ ($b_6=c_4=0$, $b_4=1/24$

$s(x_i) \quad =(I+(-\frac{1}{24}+b_4+c_4)\delta^4/2+(\frac{1}{4}c_4+b_6)\delta^6/2)y_i$ gives Jenkins(1926

$s'(x_i) \quad =(\mu\delta-\mu\delta^3/6+c_4\mu\delta^5)y_i$

$s''(x_i) \quad =(\delta^2+6(-\frac{1}{24}b_4+c_4)\delta^4+6(\frac{1}{4}c_4-b_6)\delta^6)y_i$

$$\text{degree of reproduction} = \begin{cases} 4 & b_4=1/144\ ,\ c_4=5/144 \\ 3 & \text{otherwise} \end{cases}$$

Quartic C^2 midpoint formulae (k=5).

$t_{2j} = t_{2j+1} = j+\frac{1}{2}$, $x_j = j$ all j

$\lambda_{2j,5}f \quad =\mu(I-\frac{5}{24}\delta^2+b_4\delta^4)y_{j+3/2}$

$\lambda_{2j+1,5}f=(I-\frac{1}{24}\delta^2 +c_4\delta^4)y_{j+2}$

$s(x_i) \quad =(I+(10b_4+22c_4-13/24)\delta^4/32+(5b_4+c_4)\delta^6/64)y_i$

$s'(x_i) \quad =(\mu\delta-\mu\delta^3/6+(3b_4+c_4)\mu\delta^5/4)y_i$

$s''(x_i) \quad =(\delta^2+3(-1/16+b_4-c_4)\delta^4+3(b_4+c_4)\delta^6/4)y_i$

The degree of reproduction is four if $b_4=47/1152$ and $c_4=7/1152$, and is three otherwise. For $c_4=b_4=0$ we get a Jenkins formula from 1930 ([4] (27)).

Quintic C^2 end point formulae (k=6).

$$t_{3j} = t_{3j+1} = t_{3j+2} = x_j = j \quad \text{all } j$$

$$\lambda_{3j,6}f = (I-2\mu\delta/5+\delta^2/20-a_3\mu\delta^3+a_4\delta^4-a_5\mu\delta^5+a_6\delta^6-a_7\mu\delta^7+a_8\delta^8)y_{j+1}$$

$$\lambda_{3j+1,6}f = (I-\delta^2/20+b_4\delta^4+b_6\delta^6+b_8\delta^8)y_{j+1}$$

$$\lambda_{3j+2,6}f = (I+2\mu\delta/5+\delta^2/20+a_3\mu\delta^3+a_4\delta^4+a_5\mu\delta^5+a_6\delta^6+a_7\mu\delta^7+a_8\delta^8)y_{j+1}$$

$$s(x_i) = (I+(a_4+b_4)\delta^4/2 + (a_6+b_6)\delta^6/2 + (a_8+b_8)\delta^8/2)y_i$$

$$s'(x_i) = (\mu\delta + 5a_3\mu\delta^3/2 + 5a_5\mu\delta^5/2 + 5a_7\mu\delta^7/2)y_i$$

$$s''(x_i) = (\delta^2 + 10(a_4-b_4)\delta^4 + 10(a_6-b_6)\delta^6 + 10(a_8-b_8)\delta^8)y_i$$

The degree of precision is normally two. It is three if $a_3=-1/15$, four if in addition $b_4 = -a_4 = 1/240$, and five if also $a_5=1/75$. when the degree of precision is four and $a_i, b_i=0$ $i>4$ we have Sprague's formula (1880). When $a_3=-1/15$ and $a_i, b_i=0$ $i\geq 4$ we have Buchanans formula (1908).

Quintic C^2 midpoint formulae (k=6).

$$t_{3j} = t_{3j+1} = t_{3j+2} = j+\tfrac{1}{2}, \quad x_j = j$$

$$\lambda_{3j,6}f = (\mu-2\delta/5-3\mu\delta^2/40+a_3\delta^3+a_4\mu\delta^4+a_5\delta^5+a_6\mu\delta^6+a_7\delta^7)y_{j+3/2}$$

$$\lambda_{3j+1,6}f = (\mu- 7\mu\delta^2/40 + b_4\mu\delta^4 + b_6\mu\delta^6)y_{j+3/2}$$

$$\lambda_{3j+2,6}f = (\mu+2\delta/5-3\mu\delta^2/40-a_3\delta^3+a_4\mu\delta^4-a_5\delta^5+a_6\mu\delta^6-a_7\delta^7)y_{j+3/2}$$

$$s(x_i) = (I + (-3+104a_4+24b_4+50a_3)\delta^4/128 + (26a_4+6b_4+50a_5+104a_6 + 24b_6)\delta^6/128 + (26a_6+6b_6+50a_7)\delta^8/128)y_i$$

$$s'(x_i) = (16\mu\delta-(\tfrac{13}{4}35a_3)\mu\delta^3+(35a_5+20a_4+10b_4)\mu\delta^5+(35a_7+20a_6+10b_6\mu\delta^7)\ y_i/16$$

$$s''(x_i) = (\delta^2 + (-\tfrac{1}{8}\tfrac{15}{4}a_3+5(b_4-a_4))\delta^4+(\tfrac{5}{4}(b_4-a_4)+5(b_6-a_6)-\tfrac{15}{4}a_5)\delta^6 + (\tfrac{5}{4}(b_6-a_6)-\tfrac{15}{4}a_7)\delta^8)y_i \ .$$

The degree of precision is normally two. If $a_3=1/60$ then $l=3$. If also $a_4=5/384$ and $b_4=13/384$ then $l=4$, and finally if in addition $a_5=-3/1600$ then $l=5$. If $a_3=3/50$ and $a_i,b_i=0$, $i\geq 4$, then we get our last Jenkins formula (1930).

References.

1. de Boor, C., On uniform approximation by splines, J. Approximation Theory 1(1968), 219-235.
2. de Boor, C., On calculating with B-splines, J. Approximation Theory 6(1972), 50-62.
3. de Boor, C. and G.J. Fix, Spline approximation by quasiinterpolants, J. Approximation Theory 8(1973), 19-45.
4. Greville, T.N.E., The general theory of osculatory interpolation, Trans. Actuarial Soc. America, 45(1944), 202-265.
5. Lanczos, C., Applied Analysis, Prentice Hall, Englewood Cliffs, N.J., 1956.
6. Lyche, T. and L.L. Schumaker, Local spline approximation methods, J. Approximation Theory 15(1975), 294-325.
7. Schoenberg, I.J., Contributions to the problem of approximation of equidistant data by analytic functions, Quart. Appl. Math. 4(1946), 45-99, 112-141.
8. Schoenberg, I.J., On spline functions, in Inequalities (O. Shisha, ed.), Academic Press, New York, 1967, 255-291.
9. Schumaker, L.L., Fitting surfaces to scattered data, to appear in the proceedings of a conference in Approximation Theory held at Austin, Texas, January 1976.

Tom Lyche
Department of Informatics
The University of Oslo
Oslo 3, Norway.

MULTIPLIERS OF STRONG CONVERGENCE

by H.J. Mertens, R.J. Nessel, and G. Wilmes

In this note we would like to extend some classical results concerning multipliers of uniform convergence for one-dimensional trigonometric series to the setting of abstract Fourier expansions in Banach spaces. To this end, let us commence with a brief review of some results typical for the classical situation.

Let $C_{2\pi}$ be the space of 2π-periodic functions f, defined and continuous on the real axis R, with norm $\|f\|_{C_{2\pi}} := \max_{-\pi\le u\le\pi}|f(u)|$. To each $f \in C_{2\pi}$ one may associate its (one-dimensional trigonometric) Fourier series

$$(1) \qquad f(x) \sim \sum_{k=-\infty}^{\infty} f^{\wedge}(k)e^{ikx}, \quad f^{\wedge}(k) := \frac{1}{2\pi}\int_{-\pi}^{\pi} f(u)e^{-iku}du$$

with complex Fourier coefficients $f^{\wedge}(k)$. Since there are functions $f \in C_{2\pi}$ for which the series (1) does not converge uniformly, one is interested in the subspace

$$(2) \qquad (C_{2\pi})_o := \{f \in C_{2\pi}; \quad \lim_{n\to\infty} \| \sum_{k=-n}^{n} f^{\wedge}(k)e^{ikx} - f(x)\|_{C_{2\pi}} = 0\}$$

of those functions having uniformly convergent Fourier series. If $Y \subset C_{2\pi}$ is any subspace, a classical problem then asks for properties an arbitrary sequence $\tau := \{\tau_k\}_{k=-\infty}^{\infty}$ of complex numbers should satisfy such that $f \in Y$ always implies the uniform convergence of the series $\sum_{k=-\infty}^{\infty} \tau_k f^{\wedge}(k)e^{ikx}$, thus defining an element $f^{\tau} \in (C_{2\pi})_o$. Such a factor sequence τ is called a multiplier of uniform or strong convergence, in notation: $\tau \in M(Y,(C_{2\pi})_o)$.

Among the many contributions to this problem we would like to consider the following three in some more detail. The first one, mainly attributed to Karamata [4], is concerned with a necessary and sufficient condition given in terms of the kernel $D_n^{\tau}(u) := \sum_{k=-n}^{n}\tau_k e^{iku}$ of the factor τ.

Theorem A : *The sequence τ of complex numbers belongs to $M(C_{2\pi}, (C_{2\pi})_o)$ if and only if $(1/2\pi)\int_{-\pi}^{\pi} |D_n^\tau(u)|\,du := \|D_n^\tau\|_1 = O(1)$, $n\to\infty$.*

On the basis of this result there is a number of sufficient conditions guaranteeing uniform convergence, provided the function f is already known to possess certain structural properties. For example, let (with some constant $B > 0$)

(3) $\varphi \in C[0,\infty)$, $\varphi(0) = 0$, $\varphi(2u) \leq B\varphi(u)$

be monotonely increasing, $C[0,\infty)$ being the set of functions, continuous on $[0,\infty)$. Let

(4) $(C_{2\pi})_\varphi := \{f \in C_{2\pi};\ E_n(f;C_{2\pi}) = O(\varphi(n^{-1}))\}$,

where $E_n(f;C_{2\pi})$ denotes the best approximation to f by trigonometric polynomials of degree n (cf. (14)). In these terms, Harsiladze [3] formulated the following

Theorem B : *One has $\tau \in M((C_{2\pi})_\varphi, (C_{2\pi})_o)$ if for $n\to\infty$*

$$(i)\ \Big\|\sum_{j=0}^{n} D_j^\tau\Big\|_1 = O(n), \qquad (ii)\ \varphi(n^{-1})\|D_n^\tau\|_1 = o(1).$$

There is a further group of results which give necessary and sufficient criteria in case the sequence τ satisfies certain structural conditions. For example, following Teljakovskii [8], let

(5) $(C_{2\pi})_\omega := \{f \in C_{2\pi};\ \omega(f;\delta) = O(\omega(\delta)),\ \delta \to 0+\}$,

where $\omega(f;\delta) := \sup_{|h|\leq\delta}\|f(u+h) - f(u)\|_{C_{2\pi}}$, and ω is any modulus of continuity. Then

Theorem C : *Let τ be an even and quasi-convex sequence (cf. the case $\alpha=1$ in (11)). Then*

$$\tau \in M((C_{2\pi})_\omega, (C_{2\pi})_o) \Leftrightarrow \tau_n\omega(1/n)\log n = o(1).$$

These and many other results as well as their methods of proof are subject to a long development in which the work of many mathematicians was involved. Apart from the names already given, let us mention those of S.Aljancic, R. Bojanic, R. DeVore, G. Goes, S.A. Husain, S. Kaczmarz -

H. Steinhaus, M. Katayama, S.M. Nikolskii, V.V. Shuk, M. Tomic (cf. [3-6; 8] and the literature cited there). Needless to say that this list is by no means complete.

It is the purpose of this note to discuss the foregoing problem in the frame of abstract Banach spaces, thus to derive some extensions of the above results to a fairly general class of orthogonal expansions.

Let X be an arbitrary (real or complex) Banach space, and [X] the Banach algebra of all bounded linear operators of X into itself. Let $\{P_k\}_{k=0}^{\infty} \subset [X]$ be a given total (i.e. $P_k f = 0$ for all k implies $f = 0$), fundamental (i.e. the linear span of $\bigcup_{k=0}^{\infty} P_k(X)$ is dense in X) sequence of mutually orthogonal projections (i.e. $P_j P_k = \delta_{jk} P_k$). Then with each $f \in X$ one may associate its unique Fourier series expansion

$$(6) \qquad f \sim \sum_{k=0}^{\infty} P_k f.$$

Again one may be interested in the subspace

$$(7) \qquad X_0 := \{f \in X;\ \lim_{n\to\infty} \|S_n f - f\|_X = 0\}$$

of strong convergence, $S_n f := \sum_{k=0}^{n} P_k f$ being the n-th partial sum operator corresponding to (6).

Let s be the set of all sequences $\tau := \{\tau_k\}_{k=0}^{\infty}$ of scalars. For some subspaces $Y, Z \subset X$, a sequence $\tau \in s$ is called a multiplier of type (Y,Z) (corresponding to $\{P_k\}$ and X) if to each $f \in Y$ there exists an element $f^\tau \in Z$ such that $P_k f^\tau = \tau_k P_k f$ for all k. In this terminology multipliers of strong convergence correspond to the particular case $Z = X_0$. If $Y = Z = X$, we abbreviate the notation to $M := M(X,X)$. To each $\tau \in M$ one may associate its multiplier operator T^τ as given via $T^\tau f := f^\tau$, and it is an immediate consequence of the closed graph theorem that $T^\tau \in [X]$. It turns out that M is a Banach algebra with ordinary pointwise operations and norm $\|\tau\|_M := \|T^\tau\|_{[X]}$.

For any $\tau \in s$ we set

$$(8) \qquad \tau(n) := \begin{cases} \tau_k, & 0 \le k \le n \\ 0, & k > n. \end{cases}$$

Obviously, $\tau(n) \in M$ for any n and $\tau \in s$, and one has $T^{\tau(n)} = S_n T^\tau$ if $\tau \in M$. In these terms one has the following extension of Theorem A.

Theorem 1 : A sequence $\tau \in s$ *belongs to* $M(X,X_o)$ *if and only if one has* $\|\tau(n)\|_M = O(1)$ *as* $n\to\infty$.

Proof: As in the classical situation the assertion is a simple consequence of the theorem of Banach - Steinhaus. Indeed if $\tau \in M(X,X_o)$, then in particular $\tau \in M$, and thus $T^\tau \in [X]$. Since by assumption for any $f \in X$ $\lim_{n\to\infty}\|S_n T^\tau f - T^\tau f\|_X = 0$, the uniform boundedness principle implies $\|S_n T^\tau\|_{[X]} = O(1)$, and hence the assertion. Conversely, let $\tau \in s$ be such that $\|\tau(n)\|_M = O(1)$, thus $\|T^{\tau(n)}\|_{[X]} = O(1)$. Obviously, one has for any $f_k \in P_k(X)$

$$\lim_{n\to\infty} \|T^{\tau(n)} f_k - \tau_k f_k\|_X = 0.$$

Since $\{P_k\}$ is fundamental, if follows by the theorem of Banach - Steinhaus that $\tau \in M$ and

$$\lim_{n\to\infty} \|S_n T^\tau f - T^\tau f\|_X = 0$$

for any $f \in X$, thus $\tau \in M(X,X_o)$.

Obviously, Theorem 1 subsumes Theorem A as the particular case

$$(9)\qquad X := C_{2\pi}; \qquad (P_k f)(x) := f\hat{}(-k)e^{-ikx} + f\hat{}(k)e^{ikx},$$

when using the well-known fact that (cf. [1, p. 54])

$$\|D_n^\tau\|_1 = \|T^{\tau(n)}\|_{[C_{2\pi}]} := \|\tau(n)\|_{M(C_{2\pi},C_{2\pi})}.$$

By the way, the above choice of projections already indicates that we shall here restrict ourselves to even or radial multipliers when considering applications to the one - or multi-dimensional trigonometric system, respectively.

In order to generalize Theorem B to the present setting, let us briefly recall a multiplier criterion as developed in our previous papers (cf. [2; 9]): Suppose that for some $\alpha \geq 0$ the Cesàro-(C,α)- means of (6) are uniformly bounded (as usual $A_n^\alpha := \binom{n+\alpha}{n}$):

$$(10)\qquad \Big\|\sum_{k=o}^{n} (A_{n-k}^\alpha / A_n^\alpha)\, P_k f\Big\|_X \leq C_\alpha \|f\|_X \qquad (f \in X).$$

Let l^∞ be the set of bounded sequences and

$$(11) \qquad bv_{\alpha+1} := \{\tau \in l^\infty;\ \sum_{k=0}^{\infty} A_k^\alpha \ |\Delta^{\alpha+1}\tau_k| < \infty\},$$

where $\Delta^{\alpha+1}$ is the (fractional) difference operator of order $\alpha + 1$. Then

$$(12) \qquad bv_{\alpha+1} \subset M.$$

Moreover, if the sequence $\tau := \{\tau_k\}_{k=0}^\infty$ has a sufficiently smooth (α-fold Cossar differentiable) extension $\tau(t)$ to $[0,\infty)$, then its multiplier norm may be estimated by

$$\|\tau\|_M \leq D_\alpha \int_0^\infty t^\alpha |d\tau^{(\alpha)}(t)|,$$

a very convenient fact in applications. In particular, the dilation invariance of the integral immediately delivers

$$(13) \qquad \|\{\tau(k/\rho)\}_{k=0}^\infty\|_M \leq D_\alpha \int_0^\infty \tau^\alpha \ |d\tau^{(\alpha)}(t)|$$

uniformly for $\rho > 0$. For all further details, however, we refer to [2; 9].

For any $f \in X$ let

$$(14) \qquad E_n(f;X) := \inf_{P\in\Pi_n} \|f-P\|_X$$

be the best approximation of f by polynomials of degree n:

$$\Pi_n := \{P \in X;\ \ P := \sum_{k=0}^{n} f_k, \qquad f_k \in P_k(X)\}.$$

With φ as given by (3), let

$$X_\varphi := \{f \in X;\quad E_n(f;X) = O(\varphi(n^{-1})),\ n\to\infty\}.$$

Then one has the following counterpart to Theorem B:

__Theorem 2 :__ Let the system $\{P_k\}$ for the space X be such that condition (10) is satisfied for some $\alpha \geq 0$ and $P_k(X)$ is finite dimensional for each k. Then $\tau \in M(X_\varphi, X_o)$ provided

$$(i)\ \ \tau \in M, \qquad (ii)\ \ \varphi(n^{-1})\ \|\tau(n)\|_M = o(1).$$

Proof : Again the proof follows along classical lines if one uses the following means of de La Vallée-Poussin-type: Let $C_{oo}^{\infty}[0,\infty)$ be the set of arbitrarily often differentiable functions with compact support on $[0,\infty)$, and let

$$\lambda(t) \in C_{oo}^{\infty}[0,\infty), \qquad \lambda(t) := \begin{cases} 1, & 0 \leq t \leq 1 \\ 0, & t \geq 2. \end{cases}$$

Then the means

$$L_n f := \sum_{k=0}^{\infty} \lambda(k/n) P_k f$$

possess the following properties:

(15) (i) $L_n f \in \Pi_{2n-1}$ for all $f \in X$ and n,

(ii) $L_n P = P$ for all $P \in \Pi_n$,

(iii) $\|L_n f\|_X \leq [D_\alpha \int_0^2 t^\alpha |\lambda^{(\alpha+1)}(t)| dt] \|f\|_X$

uniformly for all n and $f \in X$,

(iv) $\|L_n f - f\|_X \leq A\, E_n(f;X)$ for all n and $f \in X$.

Obviously, (i), (ii) are immediate consequences of the definition of λ, whereas the uniform estimate of (iii) follows by (13). In order to show (iv), since $P_k(X)$ is finite dimensional for each k, to each $f \in X$ and n there exists a polynomial $P_n^* \in \Pi_n$ of best approximation, thus

$$E_n(f;X) = \|f - P_n^*\|_X.$$

By (15), (i) - (iii), it then follows that

$$\|L_n f - f\|_X \leq \|L_n f - L_n P_n^*\|_X + \|P_n^* - f\|_X \leq (\|L_n\|_{[X]} + 1)\, E_n(f;X),$$

which implies (iv). [Thus also in this general setting one has the powerful tool of suitable delayed means at one's disposal (cf.[7] and the literature cited there)].

Let $m := [(n+1)/2]$ be the largest integer less than or equal to $(n+1)/2$. Since $\tau \in M$ and $2m-1 \leq n$, it follows by (15) (ii), (iv) that for any $f \in X_\varphi$

$$\|S_n f^\tau - f^\tau\|_X \leqslant \|S_n T^\tau f - L_m T^\tau f\|_X + \|L_m f^\tau - f^\tau\|_X$$

$$\leqslant \|S_n T^\tau [f - L_m f]\|_X + A\, E_m(f^\tau;X)$$

$$\leqslant A\, \|S_n T^\tau\|_{[X]} E_m(f;X) + A\, E_m(f^\tau;X)$$

$$\leqslant A_1(\|\tau(n)\|_M\, \varphi(n^{-1}) + E_m(f^\tau;X)) = o(1).$$

In fact, $\lim_{n\to\infty} E_n(g;X) = 0$ for any $g \in X$ since $\{P_k\}$ is fundamental in X, and $\varphi(m^{-1}) \leqslant B\, \varphi(n^{-1})$ by (3). This completes the proof.

Again Theorem 2 covers Theorem B in the paticular situation (9). Indeed, one may reformulate condition (i) of Theorem B as

$$\|\sum_{k=-n}^{n} (1 - \frac{|k|}{n+1})\tau_k e^{iku}\|_1 = O(1) \qquad (n\to\infty)$$

which then is a familiar necessary and sufficient condition for $\tau \in M(C_{2\pi}, C_{2\pi})$ (cf. [1, pp. 233, 267]).

In connection with possible extensions of Theorem C, our comments here may be very brief. Indeed, in [5] we studied generalizations of (the sufficiency part of) Theorem C to the present frame of abstract Fourier expansions in Banach spaces and considered applications to the convergence of radial partial sums of trigonometric expansions in several variables as well as to expansions into Jacobi polynomials. Further details may also be found in [6] where we will also discuss the matter in terms of multipliers of type (X_1, X_2).

Let us conclude with a brief outline of an application of Theorem 2 to radial partial sums of trigonometric series in Euclidean N-space R^N, $N \geqslant 2$. Let Z^N denote the set of all integer lattice points $j := (j_1, \ldots, j_N)$ in R^N and $Q^N \subset R^N$ the fundamental cube

$$Q^N := \{u \in R^N;\ -\pi \leqslant u_k < \pi, \quad 1 \leqslant k \leqslant N\}.$$

Let $L^1_{2\pi}$, $C_{2\pi}$ be the spaces of functions with period 2π in each variable for which the norm

$$\|f\|_1 := (\frac{1}{2\pi})^N \int_{Q^N} |f(u)|\,du, \qquad \|f\|_{C_{2\pi}} := \max_{u\in Q^N} |f(u)|,$$

respectively, is finite. Then one may define radial projections via

$$(P_k f)(x) := \sum_{|j|^2=k} f^\wedge(j)e^{ijx}, \qquad f^\wedge(j) := (\frac{1}{2\pi})^N \int_{Q^N} f(u)e^{-iju}du$$

(with the trivial interpretation in case for some $k \geq 0$ there does not exist some $j \in Z^N$ such that $|j|^2 = k$). The corresponding partial sum operator (cf. (7)) is then given via

$$(S_n f)(x) := \sum_{|j|^2 \leq n} f^\wedge(j)e^{ijx} = (\frac{1}{2\pi})^N \int_{Q^N} f(x-u)\, D_n(u)du$$

with radial Dirichlet kernel $D_n(u) := \sum_{|j|^2 \leq n} e^{iju}$.

Choosing $X := C_{2\pi}$, it is obvious that $P_k(C_{2\pi})$ is finite dimensional for each k. Moreover, it follows by a classical result of S. Bochner that condition (10) is satisfied for $\alpha > (N-1)/2$. Suppose now that $f \in X$ satisfies some smoothness condition, let's say $f \in Lip_r\beta$ for some $0 < \beta \leq r$, where

$$Lip_r\beta := \{f \in C_{2\pi};\ \sup_{0<|h|\leq\delta} \| \sum_{k=0}^{r} (-1)^{r-k}\binom{r}{k} f(u+kh)\|_{C_{2\pi}} = O(\delta^\beta)\}.$$

Then some classical results in approximation theory imply that $E_n(f;C_{2\pi}) = O(n^{-\beta/2})$. Therefore, if $\tau \in bv_{\alpha+1}$ for some $\alpha > (N-1)/2$, then $\{\tau_{k=|j|^2}\}$ is a radial multiplier of type $(C_{2\pi}, C_{2\pi})$ by (12), so that Theorem 2 finally delivers

Corollary : *Let $f \in Lip_r\beta$ for some $0 \leq \beta \leq r$ and $\tau \in bv_{\alpha+1}$ for some $\alpha > (N-1)/2$. Then there exists $f^\tau \in C_{2\pi}$ such that*

$$\lim_{n\to\infty} \| f^\tau(u) - \sum_{|j|^2 \leq n} \tau_{|j|^2} f^\wedge(j)e^{iju}\|_{C_{2\pi}} = 0$$

provided one has

$$\| \sum_{|j|^2 \leq n} \tau_{|j|^2} e^{iju}\|_1 = o(n^{\beta/2}) \qquad (n\to\infty).$$

Of course, many other applications are now available, particularly, since the crucial condition (10) has by now been studied very intensively for the various classical orthogonal expansions such as those into Bessel, Hermite, Laguerre functions, into Jacobi polynomials, etc. For the details, however, we have to refer to our papers (cf. [5;6]).

The contibutions by H.J. Mertens and G.Wilmes were supported by the " Minister für Wissenschaft und Forschung des Landes Nordrhein-Westfalen", Grant No. II B7 - FA 5844, which is gratefully acknowledged.

Literature

[1] Butzer, P.L. - R.J. Nessel, Fourier Analysis and Approximation, Vol. I, Birkhäuser, Basel and Academic Press, New York 1971.

[2] Butzer, P.L. - R.J. Nessel - W. Trebels, On summation processes of Fourier expansions in Banach spaces, II: Saturation theorems, Tôhoku Math. J. 24 (1972), 551-569.

[3] Harsiladse, F.I., Multipliers of uniform convergence and uniform summability (Russ.),Trudy Tbilissk. mat. Inst. Razmadze 26 (1959), 121-130.

[4] Karamata, J., Suite de fonctionelles linéaires et facteurs de convergence des séries de Fourier, J. Math. Pures Appl. (9) 35 (1956), 87-95.

[5] Mertens, H.J. - R.J. Nessel, Über Multiplikatoren starker Konvergenz für Fourier-Entwicklungen in Banach-Räumen, Math. Nachr. (in print).

[6] Mertens, H.J. - R.J. Nessel - G. Wilmes, Über Multiplikatoren zwischen verschiedenen Banach-Räumen im Zusammenhang mit diskreten Orthogonalentwicklungen, Forschungsberichte des Landes Nordrhein-Westfalen, 1976 (in print).

[7] Nessel, R.J. - G. Wilmes, On Nikolskii-type inequalities for orthogonal expansions, Proc. Approx. Theory Symp., Austin Texas, Jan. 1976 (in print).

[8] Teljakovskii, S.A., Quasiconvex uniform - convergence factors for Fourier series of functions with a given modulus of continuity, Math. Notes 8 (1970), 817-819 (1971) (transl. from Mat. Zametki 5 (1970),619-623).

[9] Trebels, W., Multipliers for (C,α) - Bounded Fourier Expansions in Banach Spaces and Approximation Theory, Lecture Notes in Math. 329, Springer, Berlin 1973.

ZUR LOKALEN KONVERGENZ VON PROJEKTIONEN AUF FINITE ELEMENTE

Joachim Nitsche

Let u respective u_h be the solution of a second order elliptic boundary value problem respective the corresponding Ritz approximations of finite elements S_h. The error $e_h = u-u_h$ can be splitted: $e_h = t_h + E_h$. According to the local regularity of the solution E_h has the corresponding good asymptotic convergence whereas the pollution-term t_h caused by corners of the boundary is an element of a certain linear space T independent of h.

0. Einleitung

Ziel der vorliegenden Arbeit ist die Analyse der Konvergenz der Ritz-Näherungen gegen die Lösung eines elliptischen Randwertproblems. Als Modell-Problem betrachten wir

$$(1)\qquad \begin{aligned} -\Delta u &= f \quad \text{in} \quad \Omega \ , \\ u &= 0 \quad \text{auf} \quad \partial\Omega \ . \end{aligned}$$

Zu einer Folge $\{\overset{o}{S}_h \mid 0 < h < 1\}$ von Approximations-Räumen $\overset{o}{S}_h \subseteq \overset{o}{W}{}_2^1(\Omega)$ sind die Ritz-Näherungen $u_h = R_h u \in \overset{o}{S}_h$ durch

$$(2)\qquad D(u_h,\chi) = (f,\chi) \qquad \text{für} \quad \chi \in \overset{o}{S}_h$$

definiert. Dabei bedeuten $(.,.)$ das L_2-Skalarprodukt und $D(.,.)$ das Dirichlet-Integral.

In $\overset{\circ}{H}_1 = \overset{\circ}{W}{}_2^1(\Omega)$ ist durch

$$\|\nabla v\| := D(v,v)^{1/2}$$

eine Norm erklärt. Der Ritz-Operator R_h läßt sich auch als Orthogonal-Projektor von $\overset{\circ}{H}_1$ auf $\overset{\circ}{S}_h$ bezüglich dieser Norm auffassen:

$$\|\nabla(u-R_h u)\| = \inf_{\chi\in S_h} \|\nabla(u-\chi)\| \quad .$$

In der konstruktiven Funktionen-Theorie beschäftigt man sich mit dem Zusammenhang zwischen dem Konvergenzverhalten wie z.B. $R_h u \to u$ einerseits und gewissen Struktur-Eigenschaften von u andererseits. Diese sind in unserem Fall implizit über die Funktion f gegeben; wir betrachten die Lösungsmenge

$$\mathfrak{L}_k = \{u \mid u \in \overset{\circ}{H}_1 \wedge \Delta u \in W_2^k\} \quad .$$

Bei genügend glattem Rand $\partial\Omega$ ergibt sich $\mathfrak{L}_k = \overset{\circ}{H}_1 \cap W_2^{k+2}$, was jedoch insbesondere bei Vorhandensein von Ecken des Randes nicht mehr gilt.

Der Einfluß einer verminderten Regularität des Randes auf die Konvergenz von Differenzen-Approximationen wurde von verschiedenen Autoren untersucht; wir verweisen auf LAASONEN [16], [17], VEIDINGER [24], [25], [26], VOLKOV [28] und im Fall des entsprechenden Eigenwertproblems auf BRAMBLE-HUBBARD [6], FORSYTHE [10] und REID-WALSH [22]. Im typischen Fall eines L-förmigen Gebietes ergibt sich eine asymptotische Konvergenz gegen die Lösung wie $h^{2/3}$, für die Eigenwerte ergibt sich der Konvergenz-Faktor $h^{4/3}$.

Von prinzipiellem Interesse ist die Art, in welcher ein Rand-Eckpunkt eine Reduktion der Konvergenzgeschwindigkeit bewirkt. Wir werden zeigen, daß sich der Fehler $e_h = u-u_h$ gemäß

(3) $$e_h = t_h + E_h$$

aufspalten läßt. Dabei gilt

(4) $$t_h = \sum_1^I \varkappa_i \; t_i \in T := sp(t_1, \ldots, t_I)$$

mit einem von h bzw. S_h unabhängigem Funktionen-Raum T. Die $\varkappa_i$ lassen sich als lineare Funktionale von f ausdrücken und bewirken globale Beeinflussung des Fehlers; E_h hingegen weist ein gutes Konvergenzverhalten auf.

Der Vollständigkeit halber erwähnen wir die Arbeiten von BABUSKA [1], BABUSKA-KELLOGG [4], BABUSKA-ROSENZWEIG [5], FIX [9] und SCHATZ-WAHLBIN [23], in welchen verschiedene Methoden zur Elimination dieses Pollution-Effektes diskutiert werden. Diese bestehen entweder in einer geeigneten Verfeinerung des Gitters nahe den Ecken oder in der Hinzunahme zusätzlicher "singulärer" Funktionen.

Die Arbeit hängt eng mit [20] zusammen, wo das entsprechende Problem für singuläre Sturm-Liouville-Probleme behandelt wird. Um jedoch Wiederholungen zu vermeiden, wählen wir hier eine modifizierte Beweis-Methode.

1. Regularitätsaussagen

Im Folgenden bezeichne $\Omega \subseteq R^2$ ein beschränktes Gebiet, dessen Rand $\partial\Omega$ mit Ausnahme einer endlichen Anzahl von Ecken hinreichend glatt sei. Für $\Omega' \subseteq \Omega$ bedeute $H_k(\Omega') = W_2^k(\Omega')$ den Sobolev-Raum der Funktionen mit verallgemeinerten, quadratisch integrierbaren Ableitungen bis zur Ordnung k. Wir verwenden die Bezeichnungen

$$\|\nabla^l v\|_{\Omega'} = \left\{ \sum_{|\alpha|=l} \iint_{\Omega'} |D^\alpha v|^2 \, dx \right\}^{1/2} ,$$

(5)

$$\|v\|_{k.\Omega'} = \left\{ \sum_{l\leq k} \|\nabla^l v\|^2_{\Omega'} \right\}^{1/2} .$$

Im Falle $\Omega' = \Omega$ unterdrücken wir den Hinweis auf Ω : $\|\cdot\|_k = \|\cdot\|_{k.\Omega}$.

$\overset{\circ}{H}_k$ ist der Raum, der durch Vervollständigung der Funktionen mit kompaktem, in Ω enthaltenen Träger bezüglich der Norm $\|\cdot\|_k$ entsteht. Sind Ω_1, Ω_2 Teilgebiete von Ω mit $\Omega_2 \subseteq \Omega_1$, so sagen wir, Ω_2 sei echt in Ω_1 enthalten, bzw. schreiben wir $\Omega_2 \subset\subset \Omega_1$, falls

(6) $\operatorname{dist}(\Omega_2, \Omega_1) = \inf\left\{ |x-y| \,\middle|\, x \in \partial\Omega_2 \cap \Omega,\ y \in \partial\Omega_1 \cap \Omega \right\} > 0$

gilt.

Für die Lösung u des Randwertproblems (1) gilt i.a. nicht das "shift-theorem", wonach $f \in H_k$ die Regularität $u \in H_{k+2}$ impliziert. Vielmehr gilt

<u>SATZ 1:</u> a.) <u>Es sei</u> $f \in \overset{\circ}{H}_k (k \geq 0)$. <u>Die Lösung</u> u <u>von (1) gestattet die Aufspaltung</u>

(7) $$u = \sum_{1}^{I(k)} (f, t_i) s_i + \tilde{u}$$

<u>mit</u> $\tilde{u} \in \overset{\circ}{H}_1 \cap H_{k+2}$ <u>und</u>

(8) $$\|\tilde{u}\|_{k+2} \leq c_1 \|f\|_k .$$

b.) <u>Seien die Bedingungen</u>

(9) $$(f, t_i) = 0 \qquad (i = 1, \ldots, I)$$

<u>erfüllt. Weiter seien</u> Ω_1, Ω_2 <u>zwei Gebiete mit</u> $\Omega_2 \subset\subset \Omega_1 \subseteq \Omega$. <u>Aus</u> $f \in L_2(\Omega) \cap H_k(\Omega_1)$

<u>folgt</u> $u \in \overset{\circ}{H}_1 \cap H_{k+2}(\Omega_2)$ <u>und</u>

$$\|u\|_{k+2.\Omega_2} \le c_2\{\|f\|_{k.\Omega_1} + \|f\|\} \quad . \tag{10}$$

Hinsichtlich des Beweises verweisen wir auf GRISVARD [11], [12], KONDRATEV [13], LAASONEN [14], LEHMANN [15], VOLKOV [27] und WASOW [29].

Die Zahl I_k hängt von k und den Ecken von $\partial\Omega$ ab, desgleichen die Konstante c_1 ; in c_2 geht darüber hinaus der Abstand zwischen Ω_1 und Ω_2 ein. t_i sind gewisse in dem Dual-Raum H_{-k} enthaltene harmonische und $s_i \in \overset{\circ}{H}_1$ sogenannte "singuläre" Funktionen. Jedes s_i ist einem Eckpunkt P von $\partial\Omega$ zugeordnet, und nur das Verhalten bei Annäherung an diesen Punkt ist festgelegt.

2. Finite Elemente

Γ_h sei eine Triangulierung von Ω in - verallgemeinerte - Dreiecke; d.h. $\Delta \in \Gamma_h$ ist ein geradliniges Dreieck, falls $\overline{\Delta} \cap \partial\Omega$ aus höchstens einem Punkt besteht, andernfalls liegt eine der Seiten ganz auf $\partial\Omega$. Γ_h heiße $\varkappa$-regulär, wenn für jedes $\Delta \in \Gamma_h$ der In- und Umkreisradius $\underline{r} = \underline{r}(\Delta)$, $\overline{r} = \overline{r}(\Delta)$ den Ungleichungen

$$\varkappa^{-1} h \le \underline{r} \ , \quad \overline{r} \le \varkappa \, h \tag{11}$$

genügt. h gibt ein Maß für die Feinheit der Triangulierung an. Sämtliche im folgenden betrachteten Triangulierungen sollen mit einem festen $\varkappa > 0$ in diesem Sinne regulär sein. Mit $W, W_1, \ldots$ werden wir "Gitter-Gebiete", d.h. Vereinigungen gewisser $\Delta \in \Gamma_h$ bezeichnen; die Menge dieser Gebiete wird durch $\mathfrak{W}$ beschrieben.

Als Approximations-Räume verwenden wir Finite Elemente der Ordnung m : $S_h = S_h(\Gamma_h, m)$ besteht aus allen in Ω stetigen Funktionen, deren Restriktion auf $\Delta \in \Gamma_h$ ein Polynom vom Grad $< m$ ist. In den gekrümmten Elementen wählen wir davon abweichend isoparametrische Modifikationen, vgl. dazu CIARLET-RAVIART [7] und ZLAMAL [30]. $\overset{\circ}{S}_h$ sei der Unterraum $S_h \cap \overset{\circ}{H}_1$.

Es ist nützlich, neben H_k den Raum $H'_k = H'_k(\Gamma_h)$ einzuführen. Er bestehe aus allen Funktionen $v \in L_2(\Omega)$ mit $v_{|\Delta} \in H_k(\Delta)$ für $\Delta \in \Gamma_h$. Neben den Normen und Halb-Normen (5) führen wir für Gittergebiete $W \in \mathfrak{M}$ noch ein

$$\|\nabla^l v\|'_W = \Big\{ \sum_{\Delta \subseteq W} \|\nabla^l v\|^2_\Delta \Big\}^{1/2} \quad ,$$

(12)

$$\|v\|'_{k,W} = \Big\{ \sum_{l \leq k} \|\nabla^l v\|'^2_W \Big\}^{1/2} \quad .$$

Aus der $\varkappa$-Regularität der Triangulierungen und der Tatsache, daß die Elemente aus S_h lokal definierbar sind, folgen unschwer die bekannten, nachfolgend zusammengestellten Eigenschaften:

PROPOSITION 1: Zu jedem $\Omega' \subseteq \Omega$ existiert ein $W \in \mathfrak{M}$ mit $\Omega' \subset W$ und

$$\text{dist}\,(\Omega', W) \leq \varkappa\, h \quad .$$

PROPOSITION 2: Für jedes Gittergebiet W und jedes $\chi \in S_h$ gilt

$$\|\nabla^{k+1} \chi\|'_W \leq c_3 \, \|\nabla^k \chi\|'_W \qquad (k = 0,1,\ldots, m-2) \quad .$$

PROPOSITION 3: Jedes $v \in \overset{\circ}{H}_1 \cap H'_k (k \leq m)$ läßt sich durch ein $\chi \in S_h$ mit

$$\text{dist}\,(\text{supp}\,(v), \text{supp}\,(\chi)) \leq \varkappa\, h$$

approximieren gemäß

$$\|v-\chi\|'_l \leq c_4 \, h^{k-l} \, \|v\|'_k \qquad (0 \leq l < k) \quad .$$

PROPOSITION 4: Sei $\omega \in C^\infty$ fest gewählt und $\Omega_1 := \mathrm{supp}\,(\omega)$. Die Funktion $\omega\chi$ mit $\chi \in S_h$ läßt sich durch ein $\varphi \in S_h$ mit $\mathrm{dist}\,(\mathrm{supp}(\varphi), \mathrm{supp}(\omega)) \le \varkappa\, h$ approximieren gemäß

$$\|\omega\chi-\varphi\|_1' \le c_5\, h^{m-1} \|\chi\|_{m-1.\Omega_1}' \quad .$$

Dabei hängen die Konstanten c_3, c_4 nur von $\varkappa, k$ und m ab, während in c_5 noch das Maximum von ω und seinen Ableitungen bis zur m-ten Ordnung eingeht.

Proposition 1 wurde nur zur Vollständigkeit formuliert. Proposition 2 ist eine direkte Folgerung der Markovschen Ungleichung für Polynome. Proposition 3 ist ausführlich in dem Übersichtsartikel von BABUSKA-AZIZ [3] behandelt. Proposition 5 wurde in [19], [21] eingeführt, vgl. dazu auch DESCLOUX [8]; sie folgt aus Proposition 4 mit $v = \omega\chi$ und $k = m$ unter Berücksichtigung, daß die m-ten Ableitungen von χ verschwinden.

3. Ritz-Verfahren, Fehleranalyse

Wird in der definierenden Beziehung (2) für die Ritz-Näherung u_h rechts f durch $-\Delta u$ ersetzt, so ergibt sich nach partieller Integration für den Fehler $e = e_h = u-u_h$

$$D(e,\chi) = 0 \qquad \text{für } \chi \in \overset{\circ}{S}_h \quad . \tag{13}$$

Bei hinreichend glattem Rand und regulärer Lösung verhält sich der Fehler wie $O(h^k)$ $(k \le m)$ und zwar gleichermaßen global wie lokal. In [19], [21] wurde gezeigt: Seien Ω_o, Ω_1 zwei Gebiete entsprechend

$$\Omega_1 \subset\subset \Omega_o \subseteq \Omega$$

und gelte $u \in \overset{\circ}{H}_1 \cap H_k(\Omega_o)$. Der Fehler e läßt sich dann in Ω_1 gemäß

(14) $$\|e\|_{\Omega_1} + h\|\nabla e\|_{\Omega_1} \leq c\, h^k(\|u\|_{k.\Omega_0} + \|u\|_1)$$

abschätzen. Wir bemerken, daß unabhängig von der Glattheit des Randes bei $\Omega_0 \subset\subset \hat{\Omega} \subseteq \Omega$ stets gilt

(15) $$\|u\|_{k.\Omega_0} + \|u\|_1 \leq c(\|f\|_{k-2.\hat{\Omega}} + \|f\|_{-1}) \quad ,$$

insofern kann die rechte Seite von (14) stets durch die von (15) ersetzt werden.

Wie im folgenden gezeigt werden soll, besitzt e eine Aufspaltung $e = t + E$ (3), wobei nicht e sondern E eine Abschätzung der Gestalt (14) gestattet. In BABUSKA[2] sind Beispiele für die reduzierte Konvergenz-Geschwindigkeit von e angegeben. Der Pollution-Term t liegt dabei genau in dem Raum $T = T_k := sp(t_1,\ldots,t_I)$ mit den Funktionen t_i von Satz 1 .

Es sei nun

(16) $$u \in H_k(\Omega_0) \qquad (k \leq m-1)$$

für ein Gebiet $\Omega_0 \subseteq \Omega$ vorausgesetzt, das in folgendem festgehalten wird. Wir wollen zeigen

SATZ 2: Sei (16) erfüllt. Dann existiert ein $t_e \in T$ aus dem in Satz 1 beschriebenen Teilraum T derart, daß für jedes $\Omega_1 \subseteq \Omega_0$ mit

(17) $$\text{dist}\,(\Omega_1,\Omega_0) \geq c_6 > 0$$

bei $h \leq h_0$ und geeignetem h_0 für $E = e - t_e$ gilt

(18) $$\|E\|_{\Omega_1} + h\,\|E\|_{1.\Omega_1} \leq c_7\, h^k(\|u\|_{k.\Omega_0} + \|u\|_1) \quad .$$

Zum Beweis werden wir nach einigen Vorbereitungen zwei Rekursionsbeziehungen aufstellen (Lemmata 3 und 4) und zunächst zeigen, daß aus diesen die Ungleichung (18) folgt.

Im Abschnitt 4 beweisen wir dann die Lemmata.

Da T endlich dimensional ist und die Funktionen aus T in Ω harmonisch sind, gelten

<u>LEMMA 1:</u> <u>Sei</u> $v \in H_1$ <u>und</u> $\operatorname{supp}(v) \subset\subset \Omega$. <u>Dann gilt</u>

$$D(v,t) = 0 \qquad \text{für } t \in T .$$

<u>LEMMA 2:</u> Sei $\Omega_o \subset\subset \Omega$ <u>fest und</u> Ω', Ω'' <u>in</u> Ω_o <u>enthalten mit</u>

$$d(\Omega'), d(\Omega'') \geq \gamma_1 > 0 .$$

<u>Für beliebige</u> $l', l'' \geq 0$ <u>existiert eine Konstante</u> $\gamma_2 = \gamma_2(\gamma_1, l', l'', k, \Omega_o)$ <u>derart, daß gilt</u>

$$\gamma_2^{-1} \|t\|_{l'.\Omega'} \leq \|t\|_{l''.\Omega''} \leq \gamma_2 \|t\|_{l'.\Omega'} \qquad \underline{\text{für}} \quad t \in T.$$

Dabei ist mit $d(\tilde{\Omega})$ der maximale Radius eines dem Gebiet $\tilde{\Omega}$ einbeschriebenen Kreises bezeichnet.

Lemma 1 ergibt sich unmittelbar durch partielle Integration. Zum Beweis von Lemma 2 beachten wir, daß kein Element $t \in T$ in einem Gebiet $\tilde{\Omega} \subseteq \Omega$ verschwinden kann, ohne in Ω identisch Null zu sein. Daher sind $\|\cdot\|_{l'.\Omega'}$ und $\|\cdot\|_{l''.\Omega''}$ Normen in T. Wegen der endlichen Dimension von T ergibt sich dann durch ein Kompaktheits-Argument die Existenz eines γ_2 mit der angegebenen Abhängigkeit von den Parametern.

Die nachfolgend auftretenden Gebiete sollen alle einen Kreis mit Radius $\geq \gamma_1$ enthalten, wobei $\gamma_1 > 0$ fest gewählt sei. Die Konstanten c_i $(i \geq 7)$ hängen nur von $\varkappa, k, m$ sowie γ_1, Ω_o und schließlich dem in Satz 2 eingeführten c_6 ab, ohne daß darauf hingewiesen wird.

Zum Beweis von Satz 2 wählen wir ein $\Omega_o' \subset\subset \Omega_o$ mit

(19) $$\operatorname{dist}(\Omega_o', \Omega_o) = c_6/2 \quad .$$

Weiter sei $\omega_o \in C^\infty(\overline{\Omega})$ eine Abschneide-Funktion in bezug auf Ω_o', Ω_o, d.h. es gelte $0 \leq \omega_o \leq 1$ und

$$\omega_o(x) = \begin{cases} 1 & \text{für } x \in \Omega_o' \\ 0 & \text{für } x \notin \Omega_o \end{cases} \quad .$$

Das Element t_e wählen wir gemäß

(20) $$(\omega_o(e-t_e), t) = 0 \qquad \text{für } t \in T \quad .$$

Für später schätzen wir t_e ab:

$$\|t_e\|_{\Omega_o'} \leq \|\omega_o^{1/2} t_e\|$$
$$\leq \|\omega_o^{1/2} e\|$$
$$\leq \|e\| \leq \|e\|_1 \leq \|u\|_1$$

Mit Lemma 2 gilt daher bei beliebigem $\tilde{\Omega} \subseteq \Omega_o$ - und $d(\tilde{\Omega}) \geq \gamma_1$ -

(21) $$\|t_e\|_{k.\tilde{\Omega}} \leq c_8 \|u\|_1 \quad .$$

Der wesentliche Schritt beim Beweis von Satz 2 ist eine Rekursionsbeziehung.

<u>LEMMA 3:</u> <u>Es seien</u> Ω', Ω'' <u>zwei Gebiete mit</u> $\Omega'' \subset\subset \Omega' \subseteq \Omega_o'$. <u>Dann gilt</u>

(22) $$\|E\|_{1.\Omega''} \leq c_9(h\|E\|_{1.\Omega'} + h^{k-1}M)$$

<u>mit der Abkürzung</u>

(23) $$M := \|u\|_{k.\Omega_o} + \|u\|_1 \quad .$$

c_9 <u>hängt dabei von</u> $\operatorname{dist}(\Omega'', \Omega')$ <u>ab</u>.

Aus Lemma 3 folgt die Abschätzung (18) von Satz 2 für die 1-Norm von E durch (k-2)-malige Iteration in bezug auf die Gebiete Ω'', Ω'. Entsprechend folgt (18) für die 0-Norm von E aus

LEMMA 4: Seien Ω', Ω'' wie in Lemma 3 gewählt. Dann gilt

(24) $$\|E\|_{\Omega''} \leq c_{10}(h \|E\|_{1.\Omega'} + h^k M) .$$

4. Beweis der Lemmata

Zum Beweis von Lemma 3 schachteln wir zwischen Ω'', Ω' Gebiete $\Omega_3, \Omega_2, \Omega_1$ zwischen gemäß

$$\Omega'' \subset\subset \Omega_3 \subset\subset \Omega_2 \subset\subset \Omega_1 \subset\subset \Omega' .$$

ω sei eine Abschneide-Funktion in bezug auf Ω'', Ω_3. Dann gilt

$$\|E\|_{1.\Omega''} \leq \|\omega E\|_1 .$$

Wie in [20] betrachten wir die Aufspaltung

(25) $$\omega E = (I-R_h)\, \omega(u-t_e) - (I-R_h)\, \omega\, u_h + R_h\, \omega\, E$$

und schätzen die drei Terme getrennt ab:

TERM 1: Unter Berücksichtigung von Proposition 3 ergibt sich unmittelbar mit Lemma 2

$$\begin{aligned}\|(I-R_h)\, \omega(u-t_e)\|_1 &\leq c_{11}\, h^{k-1}\, \|\omega(u-t)\|_k \\ &\leq c_{12}\, h^{k-1}\, M ,\end{aligned}$$

d.h. der erste Term erfüllt die Ungleichung von Lemma 3 .

<u>TERM 2:</u> Mit Proposition 4 erhalten wir

$$\|(I-R_h)\,\omega\, u_h\|_1 \le c_{13} \inf \|\omega\, u_h - \chi\|_1$$

$$\le c_{14}\, h^{m-1}\, \|u_h\|_{m-1.\Omega_3} \quad .$$

Bei $h \le \varkappa^{-1}$ dist (Ω_3, Ω_2) existiert ein Gittergebiet W mit $\Omega_3 \subseteq W \subseteq \Omega_2$. Dann ergibt sich mit Proposition 2

$$h^{m-1}\, \|u_h\|_{m-1.\Omega_3} \le c_{15}\, h^{k-1}\, \|u_h\|_{k-1.W} \quad .$$

Nun sei $\Phi \in \overset{\circ}{S}_h$ eine geeignete Approximation an $u-t_e$. Dann ist insbesondere

$$\|u-t-\Phi\|_{1.W} \le c_{16}\, h^{k-1}\, \|u-t\|_{k.\Omega_1} \quad ,$$

$$\|u-t-\Phi\|_{k-1.W} \le c_{17}\, h\, \|u-t\|_{k.\Omega_1}$$

erfüllbar. Mit Lemma 2 und (23) kann $\|u-t\|_{k.\Omega_1}$ durch M ersetzt werden. Darüber hinaus gilt dann

$$\|\Phi\|_{k-1.W} \le c_{18}\, M \quad .$$

Mit Proposition 2 schließen wir weiter wegen $u_h - \Phi \in \overset{\circ}{S}_h$

$$h^{k-1}\, \|u_h\|_{k-1.W} \le c_{19}(h^{k-1}\, M + h^{k-1}\, \|u_h-\Phi\|_{k-1.W})$$

$$\le c_{19}\, h^{k-1}\, M + c_{20}\, h\, \|u_h-\Phi\|_{1.W}$$

$$\le c_{21}(h^{k-1}\, M + h\, \|E\|_{1.\Omega'}) \quad .$$

Damit erfüllt auch der zweite Term von (25) die Ungleichung (22) von Lemma 3 .

TERM 3: Wegen $R_h(\omega E) \in \overset{\circ}{S}_h$ gilt

$$\|R_h(\omega E)\|_1 \le c_{22} \sup \left\{D(R(\omega E),\chi) \mid \chi \in \overset{\circ}{S}_h \wedge \|\chi\|_1 = 1\right\} .$$

Gemäß der Definition des Ritz-Operators ergibt sich nach partieller Integration

$$\begin{aligned} D(R_h(\omega E),\chi) &= D(\omega E,\chi) \\ &= D(E,\omega\chi) + (E,\Lambda\chi) \end{aligned}$$

mit

$$\Lambda\chi := 2\,\nabla\chi\,\nabla\omega + \chi\,\Delta\omega \quad .$$

Die Funktion $\omega\chi$ läßt sich durch ein $\varphi \in S_h$ mit $\text{supp}\,(\varphi) \subseteq \Omega_2$ - für $h < \varkappa^{-1}$ dist (Ω_3,Ω_2) - approximieren gemäß

$$\begin{aligned} \|\nabla(\omega\chi-\varphi)\| &\le c_{23}\, h^{m-1}\, \|\chi\|_{m-1.\Omega_3} \\ &\le c_{24}\, h\, \|\chi\|_{1.W} \le c_{24}\, h \end{aligned}$$

bei $\|\chi\|_1 \le 1$. Daher ist mit Lemma 1

$$\begin{aligned} |D(E,\omega\chi)| &= |D(E,\omega\chi-\varphi)| \\ &\le c_{24}\, h\, \|E\|_{1.\Omega'} \quad . \end{aligned}$$

Es verbleibt, den Term $(E,\Lambda\chi)$ abzuschätzen. Es gilt bei $\|\chi\|_1 \le 1$

$$\begin{aligned} \text{supp}\,(\Lambda\chi) &\subseteq \Omega_3 \quad , \\ \|\Lambda\chi\| &\le c_{25} \quad . \end{aligned}$$

Jetzt nutzen wir - erstmals - die Wahl von E aus. Mit beliebigem $t \in T$ gilt

$$(E,\Lambda\chi) = (E,\Lambda\chi - \omega_o t) \quad .$$

Wir wählen $t = \hat{t}$ so, daß $\Lambda\chi - \omega_o\hat{t}$ orthogonal zu T ist. Dann gilt

$$\|\hat{t}\|_{\Omega_o'} \leq c_{26} \quad .$$

Nun sei w durch

$$-\Delta w = F := \Lambda\chi - \omega_o\hat{t} \quad \text{in } \Omega \ ,$$

$$w = 0 \qquad \text{auf } \partial\Omega$$

definiert. Nach Satz 1 gilt $w \in H_k$ und

$$\|w\|_{2.\Omega_1} \leq c_{27} \ ,$$

(26)

$$\|w\|_{k.\Omega-\Omega_2} \leq c_{28} \ .$$

Da F orthogonal zu T ist, ergibt sich

$$\begin{aligned}(E,\Lambda\chi) &= (E,F)\\ &= (e,F)\\ &= D(e,w) \quad .\end{aligned}$$

Jetzt sei ρ,σ eine Partition der "1" in bezug auf Ω_1, $\Omega - \Omega_2$, d.h. es gelte

$$\rho + \sigma = 1 \qquad \text{in } \Omega \ ,$$

$$\text{supp}\,(\rho) \subseteq \Omega_1 \ ,$$

$$\text{supp}\,(\sigma) \subseteq \Omega - \Omega_2 \ . \ .$$

Wir schreiben $w_\rho = \rho w$, $w_\sigma = \sigma w$. Nach (26) lassen sich diese Funktionen durch $\varphi_\rho, \varphi_\sigma \in \overset{\circ}{S}_h$ mit

$$\operatorname{supp}(\varphi_\rho) \subseteq \Omega' ,$$

$$\operatorname{supp}(\varphi_\sigma) \subseteq \Omega - \Omega_3$$

approximieren gemäß

$$\|w_\rho - \varphi_\rho\|_1 \le c_{29}\, h ,$$

$$\|w_\sigma - \varphi_\sigma\|_1 \le c_{30}\, h^{k-1} .$$

Indem wir noch - nach Lemma 1 -

$$D(e, w_\rho) = D(E, w_\rho - \varphi_\rho)$$

ausnutzen, erhalten wir so

$$|(E, \Lambda X)| \le c_{31}(h^{k-1} M + h \|E\|_{1.\Omega'}) .$$

Damit ist Lemma 3 bewiesen.

Lemma 3 entspricht Lemma 8 von [20] und Lemma 4 Theorem 3 von [21]. Da der Beweis analog verläuft, unterdrücken wir ihn an dieser Stelle.

Literatur

[1] BABUSKA, I.: Finite element method for domains with corners. Computing 6, 264-273 (1971)

[2] BABUSKA, I.: Solution of problems with interfaces and singularities. Inst. for Fluid Dynamics and Appl. Math., University of Maryland, Techn. Note BN-789 (1974)

[3] BABUSKA, I.: The mathematical foundations of the finite element method with applications to partial differential equations, (Übersichtsartikel in dem gleichnamigen Kongressbericht) K. Aziz and I. Babuska eds., Acad. Press, New York 1972

[4] BABUSKA, I., KELLOGG, R.B.: Numerical solution of the neutron diffusion equation in the presence of corners and interfaces. Inst. for Fluid Dynamics and Appl. Math., University of Maryland, Techn. Note BN-720 (1971)

[5] BABUSKA, I., ROSENZWEIG, M.B.: A finite element scheme for domains with corners. Inst. of Fluid Dynamics and Appl. Math., University of Maryland, Techn. Note BN-720 (1971)

[6] BRAMBLE, J.H., HUBBARD, B.E.: Effects of boundary regularity on the discretization error in the fixed membrane eigenvalue problem. SIAM J. Numer. Anal. 5, 835-863 (1968)

[7] CIARLET, P.G., RAVIART, P.A.: The combined effect of curved boundaries and numerical integration in isoparametric finite element methods. Proc. of Conf. "The mathematical foundations of the finite element method with applications to partial differential equations". Acad. Press, 409-474 (1972)

[8] DESCLOUX, J.: Interior regularity for Galerkin finite element approximations of elliptic partial differential equations (erscheint demnächst)

[9] FIX, G.: Higher-order Rayleigh-Ritz approximations. J. of Math. and Mech. 18, 645-657 (1969)

[10] FORSYTHE, G.E.: Asymptotic lower bounds for the frequencies of certain polygonal membranes. Pacific J. Math. 4, 467-480 (1954)

[11] GRISVARD, P.: Problème de Dirichlet dans un cone. Ricerche Mat. 20, 175-192 (1971)

[12] GRISVARD, P.: Alternative de Fredholm relative au problème de Dirichlet dans un polygone ou un polyèdre. Bollettino U.M.I. 5, 132-164 (1972)

[13] KONDRAT'EV, V.A.: Boundary problems for elliptic equations in domains with conical or angular points. Trans. of the Moscow Math. Soc. 16, 227-313 (1967)

[14] LAASONEN, P.: On the behavior of the solution of the Dirichlet problem at analytic corners. Ann. Acad. Sci. Fennicae, Ser. A, I Mathematica No.241, 3-13 (1957)

[15] LAASONEN, P.: On the degree of convergence of discrete approximations for the solutions of the Dirichlet problem. Ann. Acad. Sci. Fennicae, Ser. A, I Mathematica No.246, 1-19 (1957)

[16] LAASONEN, P.: On the solution of Poisson's difference equation. J. Assoc. Comp. Mach. 5, 370-382 (1958)

[17] LAASONEN, P.: On the discretization error of the Dirichlet problem in a plane region with corners. Ann. Acad. Sci. Fennicae, Ser. A, I Mathematica, 408, 1-16 (1967)

[18] LEHMAN, R.S.: Developments at an analytic corner of the solutions of elliptic partial differential equations. J. Math. Mech. 8, 727-760 (1959)

[19] NITSCHE, J.: Interior error estimates of projection methods. Proceedings Equadiff 3, Czechoslovak conference on differential equations and their applications, 235-239, Brno (1972)

[20] NITSCHE, J.: Der Einfluß von Randsingularitäten beim Ritzschen Verfahren. Numer. Math. 25, 263-278, (1976)

[21] NITSCHE, J., SCHATZ, A.: Interior estimates for Ritz-Galerkin methods. Math. of Comp. 28, 937-958, (1974)

[22] REID, J.K., WALSH, J.E.: An elliptic eigenvalue problem for a reentrant region. J. Soc. Indust. Appl. Math. 13, 837-850 (1965)

[23] SCHATZ, A., WAHLBIN, L.: L_∞-convergence of Galerkin approximations for boundary value problems and domains with corners (erscheint demnächst)

[24] VEIDINGER, L.: Evaluation of the error in finding eigenvalues by the method of finite differences. U.S.S.R. Comp. Math. and Math. Physics 5, 28-42 (1965)

[25] VEIDINGER, L.: On the order of convergence of finite-difference approximations to the solution of the Dirichlet problem in a domain with corners. Studia Sci. Math. Hungar. 3, 337-343 (1968)

[26] VEIDINGER, L.: On the order of convergence of finite-difference approximations to eigenvalues and eigenfunctions. Studia Sci. Math. Hungar. 5, 75-87 (1970)

[27] VOLKOV, E.A.: Differentiability properties of solution of boundary value problems for the Laplace equation on a polygon. Proc. Steklov Inst. Math. 77, 127-159 (1965)

[28] VOLKOV, E.A.: Conditions for the convergence of the method of nets for Laplace's equation at the rate h^2 . Mat. Zametki 6, 669-679 (1969)

[29] WASOW, W.: Asymptotic development of the solution of Dirichlet's problem at analytic corners. Duke Math. J. 24, 47-56 (1957)

[30] ZLAMAL, M.: Curved elements in the finite element method.
Part I : SIAM J. Numer. Anal. 10, 229-240 (1973)
Part II: SIAM J. Numer. Anal. 11, 347-362 (1974).

J.A. Nitsche
Institut für Angewandte Mathematik
Albert-Ludwigs-Universität
Hermann-Herder-Str. 10
7800 Freiburg
Bundesrepublik Deutschland

APPROXIMATION AND PROBABILITY

Arthur Sard

If the stochastic nature of input and desired output are known and if a set of admissible operators is given, one may choose an admissible operator which is efficient or nearly efficient for the approximation of desired output in terms of input; and such an operator may be strongly efficient or nearly so. The object being approximated may be a map.

Arthur Sard
Department of Mathematics
Queens College, City University of New York
Flushing, New York 11367

Current mailing address:
ob dem Hügliacker 16
4102 Binningen
Switzerland

1. Introduction.

Let $G: X \to Y$ be an operator on a linear space X to a space Y, and let $\mathcal{A} = \{A: X \to Y\}$ be a set of operators on X to Y. We shall call the elements of $\mathcal{A}$ admissible approximations of G.

Suppose that we wish to choose one element A of $\mathcal{A}$ in order to approximate Gx by $A(x + \delta x)$, where $x, \delta x \in X$. We may think of A as a machine with contaminated input $x + \delta x$, output $A(x + \delta x)$, and desired output Gx. Thus A is to filter out δx and to approximate G.

We envisage a multiplicity of inputs and desired outputs, perhaps of unequal importance. An effective way to do this is to introduce a trial space Ω and a probability p on Ω. For each $\omega \in \Omega$, with null (p) exceptions, there correspond inputs $x = x_\omega$ and $\delta x = \delta x_\omega$. And if $\Omega_1 \subset \Omega$ is measurable p, then $p\Omega_1$ is the anticipated relative frequency with which trials $\omega \in \Omega_1$ will occur. Thus x and δx are stochastic processes. The nature of these processes will determine our choice of the best admissible approximation.

In Chapter 9 of [2], I consider this problem for the case in which Y is a space of functions and the object being approximated is a function. What is new in the present note is that Y will be a space of maps into a Euclidean space.

The precise problem that we shall consider is in fact more general than the one just described, in that it allows the desired output Gx and the ideal input x to be merely correlated; in other words, it allows G to indicate partial dependence rather than complete dependence.

2. Example.

An industrial process involves a chamber in which turbulent gases interact. Each trial consists of a run. We are interested in the pressure, temperature, and density of the gas as functions of time and position in the chamber. Thus Gx consists of a triple of functions which are interrelated. We are approximating a map of space-time into R^3. Our information $x + \delta x$ about the run consists of observations of some sort, not here specified. Suppose that we know the stochastic character of $x + \delta x$ and Gx. If $\mathcal{A}$ is a set of acceptable approximations, we may choose $A \in \mathcal{A}$ so that $A(x + \delta x)$ will approximate Gx in the best possible way, as we shall see.

Knowledge of the stochastic processes is a demanding hypothesis. If the processes are not known, they may perhaps be estimated. Or, as is sometimes the practice in engineering, the stochastic processes may be assumed tentatively, and such assumptions may be retained as long as they lead to reasonable conclusions.

3. Precise hypotheses and construction.

Let T be a nonempty set and m a measure on T. Let U be a Euclidean space of dimension (real or complex) n. Let Y be the space $L^2(T, m, U)$ of maps on T to U, measurable m, and norm square integrable, two such maps being equivalent if equal almost everywhere m. The inner product in Y is

$$(x, y)_Y = \int_T (x, y)_U \, dm, \qquad x, y \in Y.$$

Assume that Y is separable.

Let X be a separable Hilbert space.

Let $\mathcal{A}$ be a linear set (without topology) of linear, continuous operators on X to Y.

Let Ω be a nonempty set and p a probability on Ω. Let Ψ be the space $L^2(\Omega, p) = L^2(\Omega, p, R)$, assumed separable.

Construct the tensor products $X\Psi$, $Y\Psi$.

Let elements $z \in X\Psi$, $y \in Y\Psi$ be given. Thus z and y are stochastic processes. For each $\omega \in \Omega$, with null (p) exceptions, $z_\omega \in X$ and $y_\omega \in Y$. (Notation: $z = x + \delta x$ and $y = Gx$ of the introduction.) And the error in the approximation of y_ω by Az_ω is

$$e_\omega = Az_\omega - y_\omega.$$

Since $e_\omega \in Y$, e_ω is a map of almost all of T into U: $t \mapsto e_\omega(t)$.

4. Efficient and strongly efficient approximation.

For each $\omega \in \Omega$, with null exceptions, e_ω is the error corresponding to the run ω. The square measure of this error is

$$\|e_\omega\|_Y^2 = \int_T \|e_\omega(t)\|_U^2 \, dm(t),$$

a function on almost all of Ω. This relation shows the role of m in our problem: If dm(t) is large, then $t \in T$ is relatively important, and contrariwise.

The expected value of the above error over all runs is

$$E\|e_\omega\|_Y^2 = \int_T E\|e_\omega(t)\|_U^2 \, dm(t).$$

Here the integrand $E\|e_\omega(t)\|_U^2$ is the expected value of the square error at t, for almost all $t \in T$.

We say that $A^o \in \mathcal{A}$ is <u>efficient</u> if A^o minimizes the expected overall error for all $A \in \mathcal{A}$, that is,

$$E\|e_\omega^o\|_Y^2 \leq E\|e_\omega\|_Y^2, \qquad \text{all } A \in \mathcal{A},$$

where

$$e_\omega^o = A^o z_\omega - y_\omega, \quad e_\omega = Az_\omega - y_\omega;$$

and we say that A° is <u>strongly efficient</u> if for each $t \varepsilon T$, with null exceptions, A° minimizes the expected error at t for all $A \varepsilon \mathcal{A}$; that is,

$$E\|e^\circ_\omega(t)\|^2_U \le E\|e_\omega(t)\|^2_U, \quad \text{almost all } t \varepsilon T; \quad \text{all } A \varepsilon \mathcal{A}.$$

The exceptional set may depend on A.

If A° is strongly efficient, it is efficient.

We say that $A^\circ \varepsilon \mathcal{A}$ is <u>within</u> ρ <u>of</u> <u>efficiency</u>, $\rho > 0$, if

$$E\|e^\circ_\omega\|^2_Y < E\|e_\omega\|^2_Y + \rho^2, \quad \text{all } A \varepsilon \mathcal{A};$$

and we say that A° is <u>within</u> ρ <u>of</u> <u>strong</u> <u>efficiency</u> if a measurable (m) function ζ exists on almost all T such that

$$E\|e^\circ_\omega(t)\|^2_U \le E\|e_\omega(t)\|^2_U + |\zeta(t)|^2, \quad \text{almost all } t \varepsilon T; \quad \text{all } A \varepsilon \mathcal{A};$$

and

$$\int_T |\zeta(t)|^2 \, dm(t) < \rho^2.$$

For any $\rho > 0$, an operator within ρ of efficiency exists. An operator within ρ of strong efficiency may or may not exist. If A is within ρ of strong efficiency, then A is within ρ of efficiency.

We say that <u>efficiency is strong</u> if any efficient operator is strongly efficient and any operator within ρ of efficiency is within ρ of strong efficiency for all $\rho > 0$.

5. Principal results.

For $A \varepsilon \mathcal{A}$, let A' be the tensor product AI, where I is the identity operator on Ψ. Then $A: X \to Y$, $A': X\Psi \to Y\Psi$, and

$$A'(x\psi) = (Ax)\psi, \quad x \varepsilon X, \quad \psi \varepsilon \Psi.$$

In particular ψ may be unity, so that $A'x = Ax$. We shall write $A = A'$; the context will indicate whether A is an operator on X to Y or on $X\Psi$ to $Y\Psi$.

Put $L = \mathcal{A}z$, that is, L consists of all elements Az, $A \varepsilon \mathcal{A}$. Since $\mathcal{A}$ is linear, L is a linear subspace (not necessarily closed) of $Y\Psi$. Put σ equal to the orthogonal projection in $Y\Psi$ of y onto the closure of L. Then $\sigma \varepsilon Y\Psi$. For each $\omega \varepsilon \Omega$, with null exceptions, σ_ω is a map of T into U. For each $t \varepsilon T$, with null exceptions, $\sigma(t)$ is an n-tuple of stochastic variables: $\sigma(t) \varepsilon n\Psi$. Here n is the dimension of U and $n\Psi$ is the direct sum of n replicas of Ψ.

Then $A \varepsilon \mathcal{A}$ is efficient if and only if $Az = \sigma$, and A is within ρ of efficiency, $\rho > 0$, if and only if $E\|Az - \sigma\|^2_Y < \rho^2$.

The space L is separable, since $Y\psi$ is. Hence there is an orthonormal basis

$$\{\lambda^j\}, \quad \lambda^j \varepsilon L, \quad j \varepsilon J,$$

where the index set J is finite or countably infinite. Since $\lambda^j \varepsilon L$, there is an operator $A^j \varepsilon \mathfrak{A}$ such that

$$\lambda^j = A^j z, \quad j \varepsilon J.$$

For each $t \varepsilon T$, with null exceptions, and each $j \varepsilon J$, $\lambda^j(t) \varepsilon n\psi$. Since J is countable, there is a set $T_1 \subset T$ which is almost all of T and is such that

$$\lambda^j(t) \varepsilon n\psi, \quad \text{all } t \varepsilon T_1, \quad \text{all } j \varepsilon J.$$

For each $t \varepsilon T_1$, put

$$M(t) = \text{span}\{\lambda^j(t), \text{ all } j \varepsilon J\} \subset n\psi.$$

Now put $\tau(t)$ equal to the orthogonal projection in $n\psi$ of $y(t)$ onto the closure of $M(t)$, almost all $t \varepsilon T_1$.

An operator $A \varepsilon \mathfrak{A}$ is strongly efficient if and only if $(Az)(t) = \tau(t)$, almost all $t \varepsilon T$; and A is within ρ of strong efficiency, $\rho > 0$, if a function ζ on T exists such that

$$E\|(Az)(t) - \tau(t)\|_U^2 \leq |\zeta(t)|^2, \quad \text{almost all } t \varepsilon T,$$

and

$$\int_T |\zeta(t)|^2 \, dm(t) < \rho^2.$$

Theorem 1. Efficiency is strong if and only if $\sigma(t) = \tau(t)$, almost all $t \varepsilon T$.

Theorem 2. Efficiency is strong if $\mathfrak{A}$ contains all linear continuous operators on X to Y of finite rank.

An operator is of finite rank if its range is finite dimensional.

These theorems are proved as are their counterparts in [2]. Although Theorem 1 is natural enough, the proof that its hypothesis is necessary in the case in which an efficient operator does not exist is somewhat delicate.

Thus the stochastic description of y and z permits us to calculate efficient and nearly efficient operators, and to determine whether efficiency is strong. Furthermore, if $\mathfrak{A}$ is sufficiently large, then efficiency will be strong, whatever $y \varepsilon Y\psi$ and $z \varepsilon X\psi$ may be.

References.

1. A. Sard. Approximation and projection. J. Math. and Phys. 35(1956), 127-144.
2. Linear approximation. American Mathematical Society, Providence, Rhode Island, 1963.

GLOBALE KONVERGENZ VON VERFAHREN ZUR NICHTLINEAREN APPROXIMATION

R. Schaback

This contribution discusses the global convergence behavior of iterative algorithms solving nonlinear approximation problems via a sequence of linear approximation problems. Essentially there are two sufficient conditions for global convergence:

1) The algorithm should have only critical points as accumulation points.

2) The parametrization should imply that the parameters of refined approximations have accumulation points.

For some important nonlinear families of approximating functions it is proved that the above conditions hold, provided that the parametrization is restricted to suitable subfamilies.

1. Einführung

Es sei $A \subset \mathbb{R}^k$ ein offener Parameterbereich, der durch eine Fréchet - differenzierbare Abbildung $F : A \to G = F(A) \subset C(T)$ auf eine (i.a. nicht lineare) Menge G im Raum C(T) der stetigen Funktionen auf einem Kompaktum T abgebildet werde. Die Norm auf C(T) spielt in dieser Arbeit keine besondere Rolle; dadurch wird die gleichzeitige Behandlung von L_p- und Tschebyscheff- Approximation möglich.

Die Approximation einer gegebenen Funktion $f \in C(T)$ durch Funktionen aus $G = F(A) \subset C(T)$ soll durch das folgende, mit (1) bezeichnete Verfahren geschehen (Osborne-Watson [12], Cromme [6-9], Schaback [15], Braess [2], Hettich [10], Hoffmann [11]) :

START: Wähle $a_o \in A$, eine Konstante $K \in (0,\infty]$ und setze $i = 0$.

ITERATION: Gegeben sei $a_i \in A$. Es sei $G_i := F'_{a_i}(\mathbb{R}^k)$ der Tangentialraum zu $G = F(A)$ in $F(a_i)$ (als Bild des $\mathbb{R}^k$ unter der Fréchet - Ableitung F'_{a_i} von F in $a_i \in A$).

Schritt 1: Man berechne $b_i \in \mathbb{R}^k$ als Lösung des (im Falle $K < \infty$ restringierten) linearen Approximationsproblems.

(2) $\quad \| f - F(a_i) - F'_{a_i}(b) \| = \text{Min! über } b \in \mathbb{R}^k \text{ mit } \|b\| \le K$

Schritt 2 : Man bestimme eine Schrittweite $\lambda_i > 0$ mit

(3) $a_{i+1} := a_i + \lambda_i b_i \in A$

(4) $\| f - F(a_{i+1}) \| \leq \| f - F(a_i) \|$.

Dies geschieht i.a. durch näherungsweise Lösung der Minimierungsaufgabe

$$\| f - F(a_i + \lambda b_i) \| = \text{Min!}$$

über $\lambda \geq 0$ mit der Nebenbedingung (3).

Mit (3) kann dann die Iteration wiederholt werden. Die Erfüllbarkeit von (4) folgt sofort aus (2) und der Differenzierbarkeit von F.

Im Gegensatz zu den oben angeführten Arbeiten zu Verfahren dieses Typs soll in dieser Arbeit das globale Konvergenzverhalten im Mittelpunkt stehen. Naturgemäß spielt daher der für die lokale Konvergenz zentrale Begriff der starken Eindeutigkeit keine so große Rolle. Es zeigt sich vielmehr, daß die globalen Konvergenzüberlegungen aufspaltbar sind in zwei grundverschiedene, aber einfach separat zu behandelnde Teile:

Schritt A : Ist $a \in A$ nicht kritisch, d. h. gibt es ein $b \in \mathbb{R}^k$ mit

(5) $\| f - F(a) - F_a'(b) \| < \| f - F(a) \|$,

so kann a kein Häufungspunkt einer durch das Verfahren erzeugten Folge $\{a_i\}$ sein.

Schritt B : Eine Folge $\{a_i\}$ mit (4) besitzt Häufungspunkte in A.

Der Schritt A besteht in einer lokalen Untersuchung des vorgelegten Verfahrens, während Schritt B eine globale Eigenschaft der gegebenen Funktionenfamilie darstellt, die sich unabhängig vom verwendeten Verfahren in einer Reihe von Fällen durch geeignete Parametrisierungen erzwingen läßt. Insofern benötigt man keine globalen Eigenschaften des zugrundeliegenden Verfahrens, sondern lediglich eine geeignete Parametrisierung der gegebenen Funktionenfamilie, um globale Konvergenz zu sichern.

Die Notwendigkeit der Einschränkung auf kritische Punkte und der Schrittweitenoptimierung in Schritt 2 des Verfahrens zeigt

BEISPIEL 1 Es sei $F : \mathbb{R} \to \mathbb{R}^2$ gegeben durch $F(\alpha) = (\cos \alpha, \sin \alpha)$ und $f := (1+\sqrt{3}, 0)$ sei zu approximieren.

Statt des Schrittes 2 des Verfahrens werde α_{i+1} aus

$$(r_{i+1} \cos \alpha_{i+1}, r_{i+1} \sin \alpha_{i+1}) := F(\alpha_i) + F'_{\alpha_i}(\beta_i)$$

mit $|\alpha_i - \alpha_{i+1}| \leq \pi$ bestimmt. Die Iteration führt bei Start mit $\alpha_o \in (-\pi/2, +\pi/2)$, $\alpha_o \neq 0$ zum Iterationskäfig $+\pi/6, -\pi/6, +\pi/6, -\pi/6, \ldots$, wobei für jeden zweiten Schritt (4) verletzt ist.
Dies unterstreicht die Notwendigkeit, (4) durch geeignete Schrittweiten zu erzwingen.

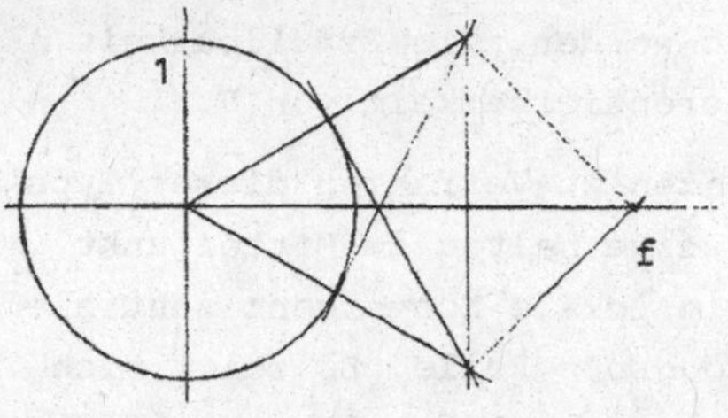

Skizze 1

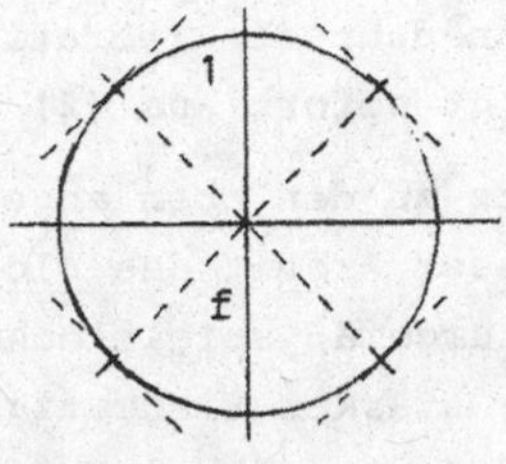

Skizze 2

Approximiert man dagegen (vgl. Skizze 2) den Punkt $f = (0,0)$, so führt der Start des Verfahrens für jedes $\alpha_o \in (k\pi/2, (k+1)\cdot\pi/2), k \in \mathbb{Z}$, zur quadratischen Konvergenz gegen die (stark eindeutige) lokal beste Approximation $k\cdot\pi/2 + \pi/4$. Der Start in den kritischen Punkten $\alpha_o = k\cdot\pi/2$ führt zu keiner Verbesserung, d. h. das Verfahren ist global konvergent gegen kritische Punkte (die, wie man sieht, nicht notwendig lokal beste, geschweige denn global beste Approximationen sind).

2. Nicht-Konvergenz in nicht-kritischen Punkten

Der folgende Satz beinhaltet die Aussage des Schrittes A für das Verfahren (1) :

<u>SATZ 1. Es sei F stetig Fréchet - differenzierbar in einer Umgebung U von</u> $a \in A$ <u>und es gelte</u> (5). <u>Dann existieren</u> $\varepsilon, \delta, \lambda^+ > \lambda^- > 0$, <u>so daß für alle</u>

(6) $\quad b \in K_\delta(a) := \{b \in \mathbb{R}^k \mid \|a-b\| \leq \delta\} \subset U$ <u>und</u>

(7) $\quad c \in \mathbb{R}^k_K := \{c \in \mathbb{R}^k \mid \|c\| \leq K\}, \ 0 < K < \infty$,

mit

(8) $\qquad \| f - F(b) - F_b'(c) \| = \min_{\gamma \in \mathbb{R}_K^k} \| f - F(b) - F_b'(\gamma) \|$

und für alle $\lambda \in (\lambda^-, \lambda^+)$ stets

(9) $\qquad b + \lambda c \in A$ und

(10) $\qquad \| f - F(b + \lambda c) \| \le \| f - F(b) \| - \varepsilon$.

Bei Start in $K_\delta(a)$ liefert ein Schritt des Verfahrens (1) also mindestens eine Verbesserung um ε, auch wenn die Schrittweite λ nicht optimal gewählt wird. Es ist ferner zu bemerken, daß die Lösungen c des Approximationsproblems (8) nicht notwendig eindeutig zu sein oder von b stetig abzuhängen brauchen. Dies macht einen einfachen Beweis des Satzes durch Stetigkeitsüberlegungen unmöglich.

Beweis zu Satz 1: Aus (5) folgt die Existenz von $c_a \in \mathbb{R}_K^k$ und $\delta_1 > 0$ mit

$$\| f - F(a) - F_a'(c_a) \| \le \| f - F(a) \| - \delta_1 .$$

Es gibt dann ein $\delta_2 \in (0, 2\cdot K)$, so daß $K_{\delta_2}(a)$ in A liegt und

$$\phi(b) := \| f - F(b) \| - \| f - F(b) - F_b'(c_a) \| \ge \delta_1 / 2$$

auf $K_{\delta_2}(a)$ gilt. Außerdem sei δ_2 so klein, daß für alle $b \in K_{\delta_2}(a)$, $b + c \in K_{\delta_2}(a)$ stets

$$\| F(b + c) - F(b) - F_b'(c) \| \le \frac{\delta_1}{8K} \| c \|$$

gilt. Für $\delta := \frac{\delta_2}{2}$, $\varepsilon := \frac{\delta_1 \delta_2}{16K}$, $\lambda^+ := \frac{\delta_2}{2K}$, $\lambda^- := \frac{\lambda^+}{2}$, $\lambda \in (\lambda^- \lambda^+) \subset (0,1)$ und eine Lösung c_b von (8) für $b \in K_\delta(a)$ gilt dann (9) wegen

$$\| b + \lambda c_b - a \| \le \delta + \lambda^+ \cdot \| c_b \| \le \frac{\delta_2}{2} + \frac{\delta_2}{2K} \cdot K \le \delta_2$$

und (10) wegen

$$\begin{aligned} \| f - F(b + \lambda c_b) \| &\le \| f - F(b) \| - \lambda (\| f - F(b) \| - \| f - F(b) - F_b'(c_b) \|) \\ &\quad + \| F(b + \lambda c_b) - F(b) - \lambda F_b'(c_b) \| \\ &\le \| f - F(b) \| - \lambda^- (\| f - F(b) \| - \| f - F(b) - F_b'(c_a) \|) \\ &\quad + \frac{\delta_1}{8K} \cdot \lambda^+ \cdot K \\ &\le \| f - F(b) \| - \frac{\delta_2}{4K} \cdot \frac{\delta_1}{2} + \frac{\delta_1}{8K} \cdot \frac{\delta_2}{2K} \cdot K . \end{aligned}$$

Die Einschränkung auf $\mathbb{R}^k_K$ mit $0 < K < \infty$ ist nicht immer notwendig:

SATZ 2. Ist F'_a umkehrbar, so gilt Satz 1 auch für $K = \infty$.

Beweis : Schränkt man die Umgebung U von a so ein, daß $\| F'_a - F'_b \| \leq 1/2 \| F'^{-1}_a \|^{-1}$ für alle $b \in U$ gilt, so folgt für jede Lösung c_b von (8) die Abschätzung

$$\| c_b \| \leq 4 \cdot \| F'^{-1}_a \| \cdot \sup_{b \in U} \| F(b) - f \| =: K \tag{11}$$

aus

$$\begin{aligned} \| c_b \| &\leq \| F'^{-1}_a \| \cdot \| F'_a(c_b) \| \\ &\leq \| F'^{-1}_a \| \cdot \| F'_a(c_b) - F'_b(c_b) + F'_b(c_b) \| \\ &\leq 1/2 \| c_b \| + \| F'_b(c_b) + F(b) - f + f - F(b) \| \cdot \| F'^{-1}_a \| \\ &\leq 1/2 \| c_b \| + 2 \| F(b) - f \| \cdot \| F'^{-1}_a \| . \end{aligned}$$

Das Verfahren läuft also exakt so ab, als hätte man die Konstante K aus (11) von vornherein fixiert. Damit ist der Beweis von Satz 1 übertragbar.

Satz 2 verdeutlicht, daß die Umkehrbarkeit von F'_a ebenso wie die lokale starke Eindeutigkeit der zu berechnenden Approximation nur für die lokalen (Cromme [6]), nicht aber für die globalen Konvergenzaussagen wichtig ist.

3. Globale Konvergenzaussagen.

Der oben angedeutete Schritt B zum Beweis globaler Konvergenzeigenschaften kann ohne jeden Bezug zu irgendeinem numerischen Verfahren axiomatisch formuliert werden:

DEFINITION : Eine Parameterabbildung $F : A \to C(T)$ heißt invers-kompakt bezüglich $f \in C(T)$, wenn jede Folge $\{a_i\} \subset A$ mit (4) mindestens einen Häufungspunkt in A hat.

Dann erhält man zur globalen Konvergenz den einfachen

SATZ 3. Es sei $F : A \to C(T)$ eine Parameterabbildung auf einer offenen Menge $A \subset \mathbb{R}^k$ und $V : A \to A$ beschreibe ein Iterationsverfahren

$$a_i := V(a_{i-1}) = V^i(a_o), \quad a_o \in A. \tag{12}$$

Unter den Voraussetzungen

(13) F ist invers- kompakt bezüglich $f \in C(T)$

(14) $\qquad \| f - F(V(a)) \| \le \| f - F(a) \|$ <u>für alle</u> $a \in A$

(15) $\qquad$ <u>Ist</u> $F(a)$ <u>nicht</u> $\left\{\begin{array}{l}\text{kritischer Punkt} \\ \text{lokal beste Approximation}\end{array}\right\}$ zu f,

<u>so ist a nicht Häufungspunkt einer Folge</u> $\{a_i\} \subset A$ <u>mit</u> (12)

<u>folgt dann für jedes</u> $a_o \in A$ <u>die Existenz eines Häufungspunktes</u> $a \in A$ <u>der Folge</u> (12),

<u>und</u> $F(a)$ <u>ist</u> $\left\{\begin{array}{l}\text{kritischer Punkt} \\ \text{lokal beste Approximation}\end{array}\right\}$ zu f.

<u>Beweis</u> : Aus (14) folgt (4) für die Folge (12). Nach (13) muß (12) mindestens einen Häufungspunkt in A haben, der nach (15) die Behauptung des Satzes erfüllt.

Wie immer muß auch hier eine durch geeignete Axiomatisierung herbeigeführte Vereinfachung der Beweisführung erkauft werden durch den erschwerten Nachweis des Erfülltseins der Axiome. Zur Illustration dient das

<u>BEISPIEL 2.</u> Es sei G eine Funktionenklasse in $C(T)$, $T = [a,b] \subset \mathbb{R}$, und f sei eine Funktion aus $C(T)$ mit einer "Minimalfolge" $\{g_i\} \subset G$, die $\qquad \| g_{i+1} - f \| \le \| g_i - f \|$ und

$$\lim_{i\to\infty} \| g_i - f \| = \inf_{g \in G} \| g - f \|$$

erfüllt, wobei $\{g_i\}$ nicht gleichmäßig in T, sondern nur gleichmäßig auf kompakten Teilmengen von (a,b) gegen eine Grenzfunktion $g \in G$ konvergiert. Diese Situation ist typisch für Ausartungsfälle bei rationalen Funktionen, Exponentialsummen oder allgemeineren γ- Polynomen. Gäbe es eine invers- kompakte globale Parametrisierung $F : A \to G = F(A)$ mit $A \subset \mathbb{R}^k$ und $g_i = F(a_i)$ sowie $g = F(a)$ und $a_i \to a$ für $i \to \infty$, so wäre F nicht stetig in der Normtopologie und a fortiori nicht stetig Fréchet - differenzierbar. Will man größtmögliche Allgemeinheit erzielen, so hat man also zwischen inverser Kompaktheit und Stetigkeit von F zu wählen. Da beide Eigenschaften für numerische Zwecke nicht wesentlich abgeschwächt werden können, hat man durch einschränkende Zusatzstrategien Auswertungen der obrigen Form auszuschalten. Dies wird unten in einigen wichtigen Spezialfällen geschehen.

<u>Bemerkungen:</u> 1) Es ist zu hoffen, daß die Anwendung des Newton--Verfahrens in der von Hettich [10] vorgeschlagenen Form wegen der zusätzlichen Informationen zweiter Ordnung es erlauben wird, Satz 3 in der für lokal beste Approximationen gültigen verschärften Form anzuwenden.

Dazu ist der Nachweis der Verschärfung von (15) durchzuführen.

2) Viele Existenzbeweise für beste Approximationen benutzen Schlüsse, die ähnlich dem im Begriff der inversen Kompaktheit verwendeten ist. Ein wesentlicher Unterschied liegt aber darin, daß die Grenzfunktion Bild eines Parameters aus der offenen Parametermenge sein muß. Wie im Beispiel 2 schon angedeutet wurde, ist dies i. a. nicht ohne Zusatzaufwand erreichbar.

4. Rationale Approximation

Bereits in diesem relativ einfachen Fall können Unannehmlichkeiten analog zum Beispiel 2 auftreten. Diese lassen sich allerdings vermeiden, wenn man wie bei Werner [16] (vgl. auch Collatz [5]) das Walsh - Diagramm bester Approximationen für wachsenden Zähler- bzw. Nennergrad l bzw. r schrittweise aufbaut. Ist etwa für $l, r \geq 1$ eine beste Approximation $g_{l,r}$ in der Klasse $R_{l,r}$ der rationalen Funktionen mit Zählergrad l und Nennergrad r (mit in T positivem Nenner) zu bestimmen, so kann man (vgl. [16]) davon ausgehen, daß $\| g_{l,r} - f\| < \| g_{l-1,r-1} - f \|$ gilt.
Startet man mit der üblichen Parametrisierung F das Verfahren (1) mit der in $R_{l,r}$ eingebetteten Approximation $g_{l-1,r-1} = F(a_o)$, die in $R_{l,r}$ nicht kritisch sein kann, so hat man nach einem Schritt einen Gewinn $\varepsilon > 0$ und das Verfahren läuft auf der offenen Menge A der Parameter a mit $\| f - F(a)\| < \|F(a_o) - f\| - \varepsilon/2$ ab. Nach den schon von Werner [16] gezogenen Schlüssen ist die so eingeschränkte Parametrisierung invers - kompakt und stetig Fréchet - differenzierbar. Außerdem ist die beste Approximation stark eindeutig (Cheney [4], Schaback [14]) und die Fréchet- Ableitungen sind umkehrbar. Somit liefern Satz 2,3 und die Resultate von Cromme [6] den

SATZ 4. Bei der schrittweisen Durchrechnung des Walsh - Diagramms analog zu Werner [16] ist in allen zu bearbeitenden Fällen das Verfahren (1) global quadratisch konvergent; es kann $K = \infty$ und fast immer $\lambda_i = 1$ gesetzt werden.

5. Positive Exponentialsummen

Parametrisiert man die Menge E_n^+ der positiven Exponentialsummen in der üblichen Weise, so kann man völlig analog zum rationalen Fall schrittweise die besten Approximationen in E_1^+, E_2^+, ... berechnen:

START: Gilt $f(t) = -\|f\|$ für ein $t \in T$, so ist 0 die beste Approximation in allen E_n^+. Andernfalls beginne man den Algorithmus (1) in E_1^+ mit der Funktion $g_o = 1/2 \cdot (\|f\| + \min_{t \in T} f(t))$.

ITERATION: In E_n^+ für $n > 1$ beginne man den Algorithmus (1) auf der in üblicher Weise eingebetteten Optimallösung g_{n-1} aus E_{n-1}^+, sofern $g_{n-1} - f$ negativ ist an den äußeren von $2n-1$ aufeinanderfolgenden Alternationspunkten. Andernfalls ist g_{n-1} bereits beste Approximation zu f für alle $N > n$ und das Verfahren kann abgebrochen werden. Hat die Alternante von g_{n-1} eine Länge $\geq 2n+1$, so kann $g_n := g_{n-1}$ gesetzt und ohne Rechnung der nächste Schritt durchgeführt werden.

Man erhält dann den

SATZ 5. Für jeden Schritt ist das Verfahren (1) global quadratisch konvergent und es kann $K = \infty$ sowie fast immer $\lambda_i = 1$ gesetzt werden.

Beweis: durch Verifikation der Voraussetzungen der Sätze 2 und 3 mit Standardschlüssen der Exponentialapproximation, wobei zu bemerken ist, daß der Algorithmus bereits nach einem Schritt in $E_n^+ \setminus E_{n-1}^+$ arbeitet. Die starke Eindeutigkeit nicht ausgearteter Approximationen aus E_n^+ folgt aus [14], womit wieder die Resultate von [6] anwendbar werden.

6. Allgemeine Exponentialsummen

Mit der Funktion

$$\phi(x) = \begin{cases} |x| & |x| \leq 1 \\ 1/2\,(1 + x^2) & |x| \leq 1 \end{cases} \in C^1(\mathbb{R}),$$

und reellen $\lambda_1, \ldots, \lambda_k$, $a_1, \ldots, a_k$ sei das lineare Differentialgleichungssystem

$$u_i' = \lambda_i u_i + u_{i+1} \qquad (1 \leq i \leq k),\ u_{k+1} := 0$$

$$u_i(0) = a_i \prod_{j=1}^{i-1} (1+\phi(\lambda_j)) \qquad (1 \leq i \leq k)$$

vorgegeben. Mit Induktion folgt daraus

$$u_i(x) = \sum_{j=i}^{k} u_j(0)\, \Delta_t^{k-j}(\lambda_j, \ldots, \lambda_k)\, e^{tx} \qquad (1 \leq i \leq k).$$

Auf der Menge

$$\hat{A} = \{(a_1,\ldots, a_n, b_1,\ldots, b_n) \mid -1 \le b_1 \le b_2 \le \ldots \le b_n \le 1\}$$

wird nun folgendermaßen eine Parametrisierung der verallgemeinerten Exponentialsummen konstruiert: ($b_o := -1$, $b_{n+1} := +1$)

Für l, k mit $-1 = b_{l-1} < b_l \le \ldots \le b_k < b_{k+1} = 1$, $1 \le l \le k \le n$

bilde man $\lambda_i := b_i/(1-b_i^2)$ $(l \le i \le k)$

und setze

$$F(a_1,\ldots, a_n, b_1,\ldots, b_n) := \sum_{j=i}^{k} a_j \prod_{i=l}^{j-1} (1+\phi(\lambda_i))\, \Delta_t^{k-j} (\lambda_j,\ldots,\lambda_k)\, e^{tx}$$
$$= u_l(x)$$

Auf $\mathring{A} = \{(a_1,\ldots, a_n; b_1,\ldots, b_n) \mid -1 < b_1 \le \ldots \le b_n < 1\}$ ist F stetig differenzierbar. Nach Standardschlüssen der Exponentialapproximation folgt aus

$$\| F(a)\| \le K,\ a = (a_1,\ldots, a_n, b_1,\ldots, b_n) \in \mathring{A}$$

(d. h. $\| u_1(x)\| \le K$) die Existenz einer von a unabhängigen Konstanten K* mit

$$| u_1^{(j)} (0)| \le K^* \qquad 0 \le j \le n-1,\ K^* \ge K.$$

Mit $| u_{i+1}^{(j)} (0)| \le | u_i^{(j+1)} (0)| + |\lambda_i| \cdot |u_i^{(j)} (0)|$

folgt per Induktion

$$| u_i(0)| \le \prod_{j=1}^{i-1} (1+|\lambda_j|) \cdot K^*$$

$$\le \prod_{j=1}^{i-1} (1+\phi(\lambda_j))\, K^*$$

und somit

$$|a_i| \le K^* \qquad (1 \le i \le n).$$

Dies ermöglicht

<u>SATZ 6. Der Algorithmus erzeugt bei Start in $\mathring{A}$ eine Folge, die entweder</u>

a) <u>eine Folge erzeugt, die einen Häufungspunkt $\hat{a} \in \mathring{A}$ hat, der dann notwendig kritischer Punkt zu f ist</u>

<u>oder</u> b) <u>nur Häufungspunkte in $(A \setminus \mathring{A}) \cap [-K,K]^n \times [-1,+1]^n$ für $K > 0$ hat (d. h. ausgeartete Approximationen sind Limiten).</u>

Der Beweis ist eine einfache Kombination des Satzes 1 mit der Kompaktheit von $A \cap [-K, K]^n \times [-1, +1]^n$ und der Tatsache, daß die Parameter der erzeugten Folge nach der obigen Überlegung in solch einem Kompaktum liegen.

Satz 6 kann naturgemäß nicht als befriedigende Lösung der numerischen Approximation durch Exponentialsummen angesehen werden, denn nur in recht einschränkenden Fällen sind kritische Punkte bei Exponentialsummen auch (eventuell nur lokal) beste Approximationen. Andererseits ist allein aus Informationen über F_a' kein Verfahren denkbar , das bei Start in einen kritischen Punkt diesen verlassen würde. Dazu wären Informationen "zweiter Ordnung" nötig (Hettich [10]). Andererseits besteht Hoffnung (Braess [2]), durch sinnvolle Untergliederung der Exponentialsummen und schrittweise Durchrechnung wie in den beiden vorhergehenden Abschnitten sämtliche kritischen Punkte und damit alle lokal und global besten Approximationen systematisch zu berechnen.

7. Rationale Splinefunktionen

Die vom Verfasser, H. Werner und D. Braess [3, 13, 15, 17] untersuchten rationalen Splinefunktionen lassen sich parametrisieren auf der Menge

$$\hat{A} = \mathbb{R} \times (-1, +1)^{n+2} \times \mathbb{R} \subset \mathbb{R}^{n+4},$$

indem zunächst mit

$$F_1 : \hat{A} \to A = \mathbb{R} \times (0, \infty)^{n+2} \times \mathbb{R},$$

$$F_1(\hat{y}_o, \hat{M}_o, \ldots, \hat{M}_{n+1}) = \left(\hat{y}_o, \frac{1 + \hat{M}_o}{1 - \hat{M}_o}, \ldots, \frac{1 + \hat{M}_{n+1}}{1 - \hat{M}_{n+1}}, \hat{y}_{n+1} \right)$$

die Menge $\hat{A}$ stetig differenzierbar auf die in [15] verwendete Parametermenge abgebildet wird. Man erhält über die zusammengesetzte Parametrisierung F den mit den in [15] geschilderten numerischen Erfahrungen übereinstimmenden

<u>SATZ 7: Der Algorithmus erzeugt bei Start in $\hat{A}$ eine Folge, die entweder</u>

a) <u>gegen eine beste Approximation</u> $F(a)$, $a \in \hat{A}$, <u>zu f konvergiert oder</u>

b) <u>nur Häufungspunkte auf dem Rand von $\hat{A}$ hat.</u>

Zum Beweis ist nur zu bemerken, daß aus den Argumenten in [15] folgt, daß jeder kritische Punkt F(a) mit a ∈ A bereits beste Approximation ist.

Bei den üblichen Spline - Funktionen mit freien Knoten sind noch ungünstigere Resultate zu erwarten als bei den Exponentialsummen. Eine genauere Untersuchung steht noch aus; die bisher durchgeführten numerischen Experimente zeigen stets globale Konvergenz gegen kritische Punkte.

[1] BRAESS, D.: Die Konstruktion der Tschebyscheff - Approximierenden bei der Anpassung mit Exponentialsummen, J. of Approx. Th. 3, 261 - 273 (1970)

[2] BRAESS, D.: Zur numerischen Stabilität des Newton - Verfahrens bei der nichtlinearen Tschebyscheff - Approximation, dieser Band, 53 - 61 (1976)

[3] BRAESS, D. und WERNER, H.: Tschebyscheff - Approximation mit einer Klasse rationaler Spline - Funktionen II, J. of Approx. Th. 10, 379 - 399 (1974)

[4] CHENEY, E. W.: Introduction to Approximation Theory, New York: Mc Graw - Hill 1966

[5] COLLATZ, L.: Tschebyscheffsche Annäherung mit rationalen Funktionen, Abh. Math. Sem. Univ. Hamburg 24, 70 - 78 (1960)

[6] CROMME, L.: Eine Klasse von Verfahren zur Ermittlung bester nichtlinearer Tschebyscheff - Approximationen, Num. Math. 25, 447 - 459 (1976)

[7] CROMME, L.: Bemerkungen zur numerischen Behandlung nichtlinearer Aufgaben der Tschebyscheff - Approximation, in: Numerische Methoden der Approximationstheorie, Bd. III, Basel - Stuttgart: Birkhäuser 1976

[8] CROMME, L.: Zur Tschebyscheff - Approximation bei Ungleichungsnebenbedingungen im Funktionenraum, dieser Band, 143 - 152 (1976)

[9] CROMME, L.: Numerische Methoden zur Behandlung einiger Problemklassen der nichtlinearen Tschebyscheff - Approximation mit Nebenbedingungen, zur Veröffentlichung eingereicht

[10] HETTICH, R.: Ein Newton - Verfahren zur Lösung nichtlinearer Approximationsprobleme, dieser Band, 221 - 235 (1976)

[11] HOFFMANN, K.-H.: Approximationen mit Lösungen von Differentialgleichungen, Vortrag, Bonn (1976)

[12] OSBORNE, M. R., und WATSON, G. A.: An algorithm for minimax approximation in the nonlinear case, Computer Journal 12, 63 - 68 (1969)

[13] SCHABACK, R.: Spezielle rationale Splinefunktionen, J. of Approx. Th. 7, 281 - 292 (1973)

[14] SCHABACK, R.: On Alternation Numbers in Nonlinear Chebyshev Approximation, J. of Approx. Th. (erscheint demnächst)

[15] SCHABACK, R.: Calculation of Best Approximations by Rational Splines, erscheint in den Proceedings des Approximation Theory Symposium, Austin 1976, New York - London: Academic Press

[16] WERNER, H.: Die konstruktive Ermittlung der Tschebyscheff - Approximierenden im Bereich der rationalen Funktionen, Arch. Rat. Mech. Anal. 11, 368 - 384 (1962)

[17] WERNER, H.: Tschebyscheff - Approximation mit einer Klasse rationaler Spline - Funktionen, J. of Approx. Th. 10, 74 - 92 (1974)

Prof. Dr. R. Schaback

Lehrstühle für Numerische
und Angewandte Mathematik
Lotzestraße 16 - 18

D - 3400 - Göttingen

Bundesrepublik Deutschland

Diese Arbeit entstand im Rahmen des von der Deutscher Forschungsgemeinschaft geförderten Sonderforschungsbereichs 72 "Approximation und Optimierung in einer anwendungsbezogenen Mathematik".

Ein Satz vom Jackson-Typ und seine Anwendung auf die Diskretisierung von Kontrollproblemen

E. Schäfer

1. Einleitung

Die Verwendung von Differenzenverfahren zur numerischen Lösung von Kontrollaufgaben wurde z.B. in [1], [2] und [4], [5] untersucht. Solche Verfahren haben den Vorteil, daß man unter schwachen Voraussetzungen Konvergenz zeigen kann. Für Aussagen über die Konvergenzgeschwindigkeit macht man im allgemeinen jedoch hohe Differenzierbarkeitsvoraussetzungen für eine optimale Kontrolle. Derartige Aussagen findet man in [8] und [10] für zwei- beziehungsweise dreimal stetig differenzierbare optimale Steuerungen. Durch Verwendung der L_1-Norm können wir auch die Fälle einer stückweise stetigen bzw. stetigen und stückweise stetig differenzierbaren optimalen Kontrolle quantitativ behandeln. Gerade solche optimalen Steuerungen treten auf Grund des Maximumprinzips sehr häufig bei Aufgaben mit eingeschränktem Kontrollbereich auf.
Im folgenden werden wir für solche optimalen Kontrollen die asymptotische Konvergenzgeschwindigkeit des bei der Diskretisierung auftretenden Fehlers abschätzen. Wesentliches Hilfsmittel dabei ist ein Jacksonsatz für die Polynomapproximation bezüglich der L_1-Norm unter Ungleichungsnebenbedingungen.

2. Problemstellung und Bezeichnungen

Wir behandeln die folgende Kontrollaufgabe für gewöhnliche Differentialgleichungen:

(OC) Minimiere das Kostenfunktional $f(x,u) := k(x(T))$ bezüglich allen auf $I:=[0,T]$ stückweise stetigen Steuerungen $u: I \to \mathbb{R}^m$ unter den Nebenbedingungen

$$\dot{x}(t) = F(t,x(t),u(t)),\ t \in I \text{ fast überall},$$
$$x(0) = x_o,$$
$$g_1(t) \le u(t) \le g_2(t),\ t \in I.$$

Dabei sind $k:\mathbb{R}^n \to \mathbb{R}$, $F:I\times\mathbb{R}^n\times\mathbb{R}^m \to \mathbb{R}^n$, $g_i:I \to \mathbb{R}^m$, $i=1,2$, gegebene Abbildungen. Die Endzeit T und der Anfangszustand x_o sind ebenfalls gegeben.

Bemerkungen zur Form der Aufgabenstellung:

a) Für mehrdimensionale Kontrollen sind die Kontrolleinschränkungen $g_1 \le u \le g_2$ komponentenweise zu verstehen.

b) Die spezielle Gestalt unseres Kostenfunktionals bedeutet keine Einschränkung gegenüber Kostenfunktionalen der Gestalt

$$f(x,u) := k(x(T)) + \int_0^T g(t,x(t),u(t))dt.$$

c) Die im folgenden beschriebene Methode der näherungsweisen Lösung von (OC) läßt sich auf Probleme übertragen, bei denen für die Differentialgleichung eine Randwertaufgabe vorgeschrieben ist (vergleiche jedoch etwa [6], Beispiel 1, für die unter Umständen auftretenden Schwierigkeiten). In dieser Form werden dann auch Probleme mit freier Endzeit erfaßt. Für die asymptotischen Konvergenzaussagen benötigt man eine zu Lemma 8 analoge Aussage für Randwertaufgaben.

Im wesentlichen sind durch obige Formulierung des Problems (OC) also phasenbeschränkte Kontrollaufgaben ausgeschlossen.

Im weiteren setzen wir stets voraus, daß es zu jeder stückweise stetigen Steuerung u genau eine Phase x mit $x(0) = x_o$ und $\dot{x}(t) = F(t,x(t),u(t))$, $t \in I$ fast überall, gibt. Damit ist für obige Anfangswertaufgabe der Lösungsoperator L^{-1} erklärt, $x = L^{-1}u$, und es ist, da keine Phaseneinschränkungen vorliegen,

$$K := \{u:I \to \mathbb{R}^m \mid u \text{ stückweise stetig, } g_1 \le u \le g_2 \text{ auf } I\}$$

die Menge der für (OC) zulässigen Kontrollen.

3. Diskretisierung und allgemeines Konvergenzverhalten

In einem ersten Diskretisierungsschritt lassen wir in der Aufgabe (OC) als Steuerungen nur Polynome vom Höchstgrad r zu (kurz $u \in P_r$), d.h. jede Komponente von u ist ein Polynom vom Höchstgrad r. Dieses Kontrollproblem bezeichnen wir mit $(OC)^r$.

Die Aufgabe $(OC)^r$ diskretisieren wir in einem zweiten Schritt zum Problem $(OC)_h^r$, indem wir die Differentialgleichung auf I durch eine Differenzengleichung auf dem Gitter $I_{h'} \subset I$, $T \in I_{h'}$, maximaler Breite h' ersetzen und die Kontrolleinschränkungen nur auf der endlichen Menge $I_{h''} \subset I$ fordern. Mit $h = (h', h'')$, $|h| := \max(h', h'')$ sei

$(OC)_h^r$ Minimiere das Kostenfunktional

$$f_h(x_h, u_h) := k(x_h(T))$$

bezüglich allen Polynomen $u_h \in P_r$ [1)]

unter den Nebenbedingungen

$$g_1(t) \le u_h(t) \le g_2(t), \quad t \in I_{h''}$$

$$x_h(t+h') - x_h(t) = h' F_{h'}(t, x_h(t), u_h(t)), \quad t \in I_{h'}$$

$$x_h(0) = x_o .$$

Wir fordern wie im kontinuierlichen Fall, daß es zu jedem u_h und x_o genau ein x_h gibt mit $x_h(0) = x_o$, $x_h(t+h') =$

$= x_h(t) + h' F_{h'}(t, x_h(t), u_h(t))$, d.h. es existiert der Lösungsoperator

---

1) Läßt man alle Gitterfunktionen u_h zu, so erhält man das Problem $(OC)_h$

L_h^{-1} mit $x_h = L_h^{-1} u_h$, und $K_h := \{u_h : I_{h,,} \to \mathbb{R}^m |$

$| g_1(t) \leq u_h(t) \leq g_2(t)$ für $t \in I_{h,,}\}$ ist die Menge der für

$(OC)_h$ zulässigen Kontrollen.

Wir untersuchen die Konvergenz der Lösungen der Näherungsaufgaben $(OC)_h^r$ gegen Lösungen der Ausgangsaufgabe (OC).
Dazu bezeichnen wir mit

$$E := \inf \{f(x,u) | (x,u) \text{ zulässig für (OC)}\},$$

$$E^r := \inf \{f(x,u) | (x,u) \text{ zulässig für } (OC)^r\},$$

$$E_h^r := \inf \{f_h(x_h,u_h) | (x_h,u_h) \text{ zulässig für } (OC)_h^r\}$$

die Minimalwerte der einzelnen Aufgaben und mit

$$S := \{u | (x,u) \text{ zulässig für (OC)}, f(x,u) = E\},$$

$$S^r := \{u | (x,u) \text{ zulässig für } (OC)^r, f(x,u) = E^r\},$$

$$S_h^r := \{u_h | (x_h,u_h) \text{ zulässig für } (OC)_h^r, f_h(x_h,u_h) = E_h^r\}$$

die Mengen der zugehörigen optimalen Steuerungen.

Qualitative Konvergenzaussagen der Form

$$E^r \to E \ (r \to \infty), \ E_h^r \to E^r(|h| \to 0),$$

$$E_h^r \to E \ (r \to \infty, \ |h| \to 0)$$

gelten sehr allgemein unter geeigneten Stetigkeitsvoraussetzungen und wurden etwa in [2], [4] und [8] gezeigt. Sind die betrachteten Funktionen lipschitzstetig, kann man quantitative Konvergenzaussagen auf geeignete approximationstheoretische Sätze zurückführen. Dies wird im folgenden gezeigt.

4. Konvergenzgüte

Im folgenden beschränken wir uns auf die Konvergenzuntersuchung der Minimalwerte, die wir in zwei Schritten "$E^r \to E$" und "$E_h^r \to E^r$" durchführen.

Dazu nehmen wir im folgenden immer an, daß $S \neq \emptyset$, $S^r \neq \emptyset$ und $S_h^r \neq \emptyset$ gilt. Weiter setzen wir voraus, daß $k(\cdot)$ und $F(t,\cdot,\cdot)$, gleichmäßig für $t \in I$, auf kompakten Mengen der jeweiligen Argumente gleichmäßig lipschitzstetig sind. Damit ist die Existenz von L^{-1} gesichert und bei Verwendung geeigneter Verfahrensfunktionen F_h, auch die Existenz von L_h^{-1}.

<u>1 Satz</u>: Sei $u^* \in S$. Dann gibt es $C > 0$, so daß für $r \in \mathbb{N}$ gilt:

$$0 \leq E^r - E \leq C \inf \{\|u^*-u\|_1 \mid u \in P_r \cap K\}$$ [1)]

Beweis: Wähle $C_1 > 0$ so, daß

$$\{u \in P_r \cap K \mid \|u^*-u\|_\infty \leq C_1\} \neq \emptyset \text{ für } r \in \mathbb{N} .$$

Dann folgt aus der Lipschitzstetigkeit von k auf kompakten Mengen

$$0 \leq E^r - E \leq$$

$$\leq \inf \{k((L^{-1}u)(T))-k((L^{-1}u^*)(T)) \mid u \in P_r \cap K, \|u^*-u\|_\infty \leq C_1\}$$

$$\leq \text{const} \inf \{\|L^{-1}u(T) - L^{-1}u^*(T)\| \mid u \in P_r \cap K, \|u^*-u\|_\infty \leq C_1\}.$$

Da $L^{-1}: \overset{m}{\underset{1}{X}} L_\infty(I) \to \overset{n}{\underset{1}{X}} C(I)$ beschränkte Mengen in beschränkte Mengen

abbildet, folgt aus der gleichmäßigen Lipschitzstetigkeit von F

---

1) $u : I \to \mathbb{R}^m$ heißt aus $L_p(I)$, falls alle Komponenten aus $L_p(I)$ sind, $1 \leq p \leq \infty$. Für $u = (u_i) \in L_p(I)$ ist $\|u\|_p :=$ $\max \{\|u_i\|_{L_p(I)} \mid 1 \leq i \leq m\}$.

auf Kompakta mit $x := L^{-1}u$, $x^* := L^{-1}u^*$ die Abschätzung

$$\|x(t)-x^*(t)\| \leq \text{const} \int_0^t (\|x(s)-x^*(s)\|+\|u(s)-u^*(s)\|)ds$$

und damit wegen $x(0) = x^*(0)$ nach bekannten Vergleichssätzen (siehe etwa [12], S. 14)

$$0 \leq E^r - E \leq \text{const} \inf \{\|u-u^*\|_1 \mid u \in P_r \cap K\}. \qquad ***$$

Um die rechte Seite der Abschätzung von Satz 1 weiterbehandeln zu können, machen wir geeignete Voraussetzungen an die Schrankenfunktionen g_1 und g_2, die die Anwendung der klassischen Jacksonsätze auf obiges Approximationsproblem unter Nebenbedingungen gestatten.

Für das folgende seien die Schrankenfunktionen g_1 und g_2 stetig und es gelte folgende "Slaterbedingung": $g_1(t) < g_2(t)$, $t \in I$. Bezeichnen wir mit $\omega_p(f,h)$ den L_p-Stetigkeitsmodul von $f \in L_p(I)$ zum Argument h (für vektorwertige Funktionen nehme man das Maximum der komponentenweisen Stetigkeitsmoduli), so gilt der folgende für unsere Zwecke geeignete Satz vom Jacksontyp.

<u>2 Satz:</u> Für $i \in \{0,1\}$ sei $g_1, g_2 \in C^i(I)$ und $g_1 < g_2$. Dann gibt es Dann gibt es $c>0$ und $r_0 \in \mathbb{N}$, so daß $r > r_0$ und $u \in W^{i,1}$ 1) mit $g_1 \leq u \leq g_2$ f.ü. gilt

$$\inf\{\|u-p\|_1 \mid p \in P_r \cap K\} \leq c\, r^{-i}\{\omega_1(u^{(i)},r^{-1})+\omega_\infty(g_1{}^{(i)},g_2{}^{(i)},r^{-1})\}.^{2)}$$

Beweis: Zu u betrachten wir die Konvolution $u_\lambda := u*K_\lambda$ mit dem Steklovkern K_λ (vgl. etwa [11], S.6), wobei wir u über den Rand von I durch Null (für i=0) oder konstant (i=1) auf $\mathbb{R}$ fortsetzen. Dann ist u_λ stetig für $\lambda > 0$ und es gelten (gleichmäßig für $\lambda > 0$ und $\delta > 0$) wegen $\int_{\mathbb{R}} K(s)ds = 1$, $\int_{\mathbb{R}} sK(s)ds = 0$ die Abschätzungen

$$\|u_\lambda - u\|_1 \leq \text{const}\ \lambda^{-i}\omega_1(u^{(i)},\lambda^{-1}) \text{ und } \omega_\infty(u_\lambda{}^{(i)},\delta) \leq \text{const}\ \omega_1(u^{(i)},\delta).$$

---

1) $W^{i,1} := \{v \in C^{i-1}(I) \mid v^{(i-1)}$ absolut stetig$\}$.

2) Abkürzend sei $\omega_\infty(g_1,g_2,h) := \max(\omega_\infty(g_1,h),\omega_\infty(g_2,h))$.

Bilden wir zu g_1 und g_2 die Konvolutionen $g_{1,\lambda}$ und $g_{2,\lambda}$, so folgen aus $g_1 \le u \le g_2$, $t \in I$ f.ü., wegen $K(s) \ge 0$ die Ungleichungen $g_{1,\lambda} \le u_\lambda \le g_{2,\lambda}$, $t \in I$. Beachtet man $\|g_{j,\lambda}(t) - g_j(t)\| \le \text{const}\ \lambda^{-i}\omega_\infty(g_j^{(i)},\lambda^{-1})$, $j=1,2$, so folgt

$$g_1(t) - \text{const}\ \lambda^{-i}\omega_\infty(g_1^{(i)},\lambda^{-1}) \le u_\lambda(t) \le g_2(t) + $$

$$+ \text{const}\ \lambda^{-i}\omega_\infty(g_2^{(i)},\lambda^{-1}) \text{ für } t \in I.$$

Für die beste (uneingeschränkte) Tschebyschewapproximation $p_r^\lambda \in P_r$ für u_λ folgt dann für alle $t \in I$

$$g_1(t) - \text{const}\ \lambda^{-i}\omega_\infty(g_1^{(i)},\lambda^{-1}) - \|u_\lambda - p_r^\lambda\|_\infty \le p_r^\lambda(t) \le$$

$$\le g_2(t) + \text{const}\ \lambda^{-i}\omega_\infty(g_2^{(i)},\lambda^{-1}) + \|u_\lambda - p_r^\lambda\|_\infty.$$

Wie in Lemma 5 von [10] folgt auf Grund der Slaterbedingung die Existenz von r_o aus $\mathbb{N}$, so daß es für $r > r_o$ ein p_r aus P_r gibt mit

$$g_1(t) \le p_r(t) \le g_2(t),\ t \in I$$

und

$$\|p_r - p_r^\lambda\|_\infty \le \text{const}\ (\|u_\lambda - p_r^\lambda\|_\infty + \lambda^{-i}\omega_\infty(g_1^{(i)},g_2^{(i)};\lambda^{-1})).$$

Da nach dem Satz von Jackson $\|u_\lambda - p_r^\lambda\|_\infty \le \text{const}\ r^{-i}\omega_\infty(u_\lambda^{(i)},r^{-1})$ gilt, folgt aus obigen Ungleichungen zusammen mit der Dreiecksungleichung

$$\|u - p_r\|_1 \le \text{const}\ \{\lambda^{-i}\omega_1(u^{(i)},\lambda^{-1}) + r^{-i}\omega_1(u^{(i)},r^{-1}) + $$
$$+ \lambda^{-i}\omega_\infty(g_1^{(i)},g_2^{(i)};\lambda^{-1})\}.$$

Wegen $p_r \in P_r \cap K$ folgt für $\lambda := r$ die Behauptung. **

Da im Beweis von Satz 2 nur die Gültigkeit von Jacksonsätzen ohne Nebenbedingungen ausgenutzt wurde, gilt Satz 2 auch für $S_{2n-1,r}$, den Raum der Polynomsplines vom Grad $2n-1$ mit dem äquidistanten Knotengitter der Breite $1/r$.

<u>3 Bemerkung:</u> Die Aussage von Satz 2 gilt auch für $S_{2n-1,r}$ an Stelle von P_r.

Eine Abschätzung von $|E^r - E|$ erhält man durch die Kombination von Satz 1 und Satz 2. Es gilt

<u>4 Folgerung:</u> Erfüllen g_1 und g_2 die Voraussetzungen wie in Satz 2 und existiert für $i \in \{0,1\}$ ein $u^* \in S \cap W^{i,1}$, dann gibt es $r_o \in \mathbb{N}$ und $C > 0$, so daß für $r > r_o$ gilt:

$$0 \leq E^r - E \leq Cr^{-i}[\omega_1(u^{*(i)}, r^{-1}) + \omega_\infty(g_1^{(i)}, g_2^{(i)}; r^{-1})].$$

Die folgende Bemerkung führt die Abschätzung von $|E^r - E_h^r|$ zurück auf den Fehler bei der Diskretisierung der Differentialgleichung und auf den Zusammenhang zwischen den Gitterwerten eines Polynoms und seinen Werten auf dem ganzen Intervall I.

<u>5 Bemerkung:</u> Für $r \in \mathbb{N}$ und $h > 0$ gilt

$$|E^r - E_h^r| \leq \sup\{|k(L_h^{-1}v(T)) - k(L^{-1}v(T))| \mid v \in P_r \cap K\} +$$
$$+ \sup\{\inf\{|k(L_h^{-1}v(T)) - k(L_h^{-1}v_h(T))| \mid v \in P_r \cap K\} \mid v_h \in P_r \cap K_h\}.$$

Beweis: Wegen $P_r \cap K \subset P_r \cap K_h$ folgt für $u \in S^r$

$$E_h^r - E^r \leq k(L_h^{-1}u(T)) - k(L^{-1}u(T))$$
$$\leq \sup\{|k(L_h^{-1}v(T)) - k(L^{-1}v(T))| \mid v \in P_r \cap K\}.$$

Für $u_h \in S_h^r$ und $v \in P_r \cap K$ gilt

$$E^r - E_h^r \leq k(L^{-1}v(T)) - k(L_h^{-1}u_h(T))$$
$$\leq [k(L_h^{-1}v(T)) - k(L_h^{-1}u_h(T))] + [k(L^{-1}v(T)) - k(L_h^{-1}v(T))]$$
$$\leq |k(L_h^{-1}v(T)) - k(L_h^{-1}u_h(T))| + \sup\{|k(L^{-1}v(T)) - k(L_h^{-1}v(T))| \mid v \in P_r \cap K\}.$$

Da der zweite Summand unabhängig von $v \in P_r \cap K$ ist, kann man im ersten Summanden zum Infimum über $v \in P_r \cap K$ übergehen. Eine anschließende Supremumsbildung über $v_h \in P_r \cap K_h$ vergröbert die Abschätzung höchstens und läßt den zweiten Summanden konstant. ⁂

Die weitere Abschätzung des ersten Summanden in der rechten Seite von Bemerkung 5 hängt vom betrachteten Differenzenverfahren, die des zweiten Summanden von der Wahl des Gitters $I_{h''}$ ab.

Im folgenden sei L_h^{-1} durch ein Runge-Kutta-Verfahren der Konsistenzordnung 2 definiert. Aus der gleichmäßigen Lipschitzstetigkeit von F auf Kompakta folgt die (in h)gleichmäßige Lipschitzstetigkeit von L_h^{-1} auf kompakten Mengen. Zur Abschätzung des zweiten Summanden in Bemerkung 5 benötigen wir also eine Abschätzung von

$$\sup\{\inf\{\|v-v_h\|_{\infty,I_{h''}} \mid v\in P_r \cap K\} \mid v_h\in P_r \cap K_h\},$$

wobei wie üblich $\|f\|_{\infty,I_h} := \max\{\|f(t)\| \mid t\in I_h\}$.

Esser gibt in [8] eine Abschätzung dieses Ausdrucks von der Größenordnung $O(r^{-2})$ für ein äquidistantes Gitter $I_{h''}$, mit $h'' = O(r^{-3})$. Diese Abschätzung wird in [10], Satz 4 und Folgerungen, bezüglich der Gitterbreite wesentlich verbessert unter Verwendung eines "Tschebyschew-Gitters"[1]. Da die Anzahl der Punkte von $I_{h''}$ die Anzahl der Nebenbedingungen in der Optimierungsaufgabe $(OC)_h^r$ bestimmt, ist bei der numerischen Realisierung ein möglichst grobes Gitter wünschenswert.

Zur Abschätzung des ersten Summanden in Bemerkung 5 benötigt man eine Abschätzung der Konvergenzordnung auch für nichtglatte rechte Seiten. Eine für unsere Zwecke geeignete Abschätzung wurde in [3] gezeigt.

Im folgenden zitieren wir die von uns benötigten Aussagen als Hilfssätze, wobei wir die Voraussetzungen für unsere Zwecke geeignet spezialisieren.

---

1) Effekte dieser Art beim Übergang von äquidistanten zu Tschebyschewsgittern sind wohlbekannt, siehe [7].

6 Lemma: Sei $g_1, g_2 \in C(I)$, $g_1 < g_2$. Für $h'' := \frac{1}{r^{1+\alpha}}$

sei $I_{h''} := \{t_i := \frac{T}{2}(1+\cos\frac{2i-1}{2\,[r^{1+\alpha}]}\pi) \mid i=1(1)[r^{1+\alpha}]\}$ 1)

das auf $[0,T]$ transformierte Tschebyschewgitter. Dann existiert $c > 0$ und $r_o \in \mathbb{N}$, so daß für $r > r_o$ gilt:

$$\sup\{\inf\{\|v-v_h\|_\infty \mid v \in P_r \cap K\} \mid v_h \in P_r \cap K_h\}$$

$$\leq c\{r^{-2\alpha} + \inf\{\|g_1-u\|_\infty \mid u \in P_r\} + \inf\{\|g_2-u\|_\infty \mid u \in P_r\}.$$

Beweis: Siehe Satz 4 und Lemma 5 von [10]. ⁂

Für $w \in P_r \cap K_h \supset P_r \cap K$ ist $L_h^{-1}w$ definiert als Lösung x_h der folgenden Differenzengleichungen (vgl. etwa [9] bezüglich der Bezeichnungen für Runge-Kutta-Verfahren):

$$x_h(0)=x_o,\ x_h(t+h')=x_h(t)+h'\sum_{j=1}^{m}\gamma_j k_j(t,x_h(t),w),\ t \in I_{h'};$$

$$k_i(t,x_h(t),w)=F(t+\alpha_i h', x_h(t)+h'\sum_{j=1}^{m}\beta_{ij}k_j(t,x_h(t),w), w(t+\alpha_i h')),$$
$$i=1(1)m.$$

Es seien die folgenden Konsistenzbedingungen (K) erfüllt:

$$(K)\quad \sum_{i=1}^{m}\gamma_i=1,\ \sum_{i=1}^{m}\gamma_i\alpha_i=\frac{1}{2},\ \sum_{j=1}^{m}\beta_{ij}=\alpha_i,\ i=1(1)m.$$

Dann gilt das folgende

7 Lemma: Gelten die Voraussetzungen von Lemma 4 und die Bedingung (K), so existiert ein $c > 0$, ein $h_o > 0$, so daß für alle h mit $|h| < h_o$ und alle $u,v \in P_r \cap K_h$ gilt:

$$\|L_h^{-1}u - L_h^{-1}v\|_{\infty,I_{h'}} \leq c\,\|u-v\|_\infty.$$

---

1) $[r^{1+\alpha}]$ größte ganze Zahl kleiner oder gleich $r^{1+\alpha}$.

Beweis: Aus Satz 4 in [10] folgt, daß die Menge $\{\|u(t)\| \mid t\in I, u\in P_r \cap K_h\}$ beschränkt ist. Da F nach Voraussetzung auf kompakten Mengen gleichmäßig lipschitzstetig ist, folgt Lemma 7 analog Lemma 2.3 in [3]. ⁂

8 Lemma: Sei $F\in C^i$ für $i\in\{1,2\}$ und Bedingung (K) erfüllt. Dann gibt es ein $c > 0$, ein $h_o > 0$ mit

$$\wedge\, r\in\mathbb{N} \;\wedge\; h < h_o \;\wedge\; v\in P_r \cap K$$

$$\|L_h^{-1}v - L^{-1}v\| \le c\, h^i \operatorname{Var}((L^{-1}v)^{(i)}).$$

Dabei ist $\operatorname{Var}((L^{-1}v)^{(i)})$ die Totalvariation der i-ten Ableitung der Lösung $L^{-1}v$ von $\dot{x}=F(t,x,v), x(0)=x_o$.

Beweis: Theoreme 3.4 und 3.5 in [3]. ⁂

Eine Kombination dieser Hilfssätze ermöglicht Aussagen über die asymptotische Konvergenzgüte von $E_h^r \to E$.

9 Satz: In (OC) seien k lokal lipschitzstetig, g_1 und g_2 lipschitzstetig, $g_1 < g_2$ in I, und $F\in C^1$. Ferner existiere eine stückweise lipschitzstetige [1] Lösung u^* von (OC). In $(OC)_h^r$ sei $I_{h'}$ ein Gitter mit $h' = r^{-2}$ und $I_{h''}$ das auf [0,T] transformierte Tschebyschewgitter mit $h'' = r^{-3/2}$. Dann gibt es $c > 0$ und $r_o\in\mathbb{N}$, so daß für $r > r_o$ gilt

$$|E - E_h^r| \le c r^{-1}.$$

Beweis: Für eine stückweise lipschitzstetige Funktion $f : I \to \mathbb{R}$ gilt

$$\omega_1(f,r^{-1}) \le \text{const} \max_{o\le j\le l-1} \{\omega_\infty(f,r^{-1})_{[t_j,t_{j+1}]} + r^{-1}|f(t_j+0)-f(t_j-0)|\},$$

1) Diese hinreichende Voraussetzung kann zu $\omega_1(u^*,h) = O(h)$ abgeschwächt werden (siehe Satz 2).

wobei $\{t_j\}$ die Unstetigkeitspunkte von f sind. Nach Folgerung 4 existiert $r_1 \in \mathbb{N}$, $c_1 > 0$, so daß für $r > r_1$ gilt

$$|E - E^r| \leq c_1\, r^{-1}.$$

Nach Lemma 7 können k und F gleichmäßig lipschitzstetig vorausgesetzt werden. Nach Lemma 6 und 8 folgt aus Bemerkung 5 die Existenz von $r_2 \in \mathbb{N}$ und $c_2 > 0$, so daß für $r > r_2$ gilt:

$$|E^r - E_h^r| \leq c_2[r^{-1} + r^{-2}\sup\{\mathrm{Var}((L^{-1})') \mid v \in P_r \cap K\}].$$

Sei $v \in P_r \cap K$ und $x = L^{-1}v$. Wegen $\|v\|_\infty \leq$ const folgt $\|x\|_\infty, \|x'\|_\infty \leq$ const.
Da $F \in C^1$, folgt

$$\mathrm{Var}(x') \leq \|\tfrac{d}{dt} F(.,x(.),v(.))\|_1$$

$$\leq \text{const } \|v'\|_1$$

$$\leq \text{const } r\, \|v\|_\infty \quad \text{(2. Bernsteinsche Ungleichung).}$$

Insgesamt folgt die Existenz von $c_3 > 0$, so daß für $r > r_o$ gilt

$$|E^r - E_h^r| \leq c_3 r^{-1}. \quad ***$$

Sind die Daten und Lösungen von (OC) glatter, so erhält man eine in r quadratische Konvergenz.

<u>10 Satz:</u> In den Voraussetzungen von Satz 9 sei $F \in C^2$ [1)], g_1 und g_2 habe eine lipschitzstetige Ableitung, u* sei stetig und habe stückweise eine lipschitzstetige Ableitung [2)]. Wählt man $h' = r^{-5/2}$, $h'' = r^{-2}$, so gibt es $c > 0$ und $r_o \in \mathbb{N}$, so daß für $r > r_o$ gilt:

$$|E - E_h^r| \leq c r^{-2}.$$

---

1) Die Voraussetzung $F \in C^2$ bzw. $F \in C^1$ in Satz 9 kann etwas abgeschwächt werden.

2) Wie in Satz 9 reicht aus: $u^* \in W^{1,1}$ mit $\omega_1((u^*)',h) = O(h)$.

Beweis: Wie in Satz 9 gilt auf Grund der Voraussetzungen

$$|E - E^r| \leq \text{const } r^{-2}$$

und

$$|E^r - E_h^r| \leq \text{const}[r^{-2} + r^{-5} \sup\{\text{Var}((L^{-1}v)'') \mid v \in P_r \cap K\}].$$

Wegen $F \in C^2$ folgt wie im Beweis von Satz 9

$$\begin{aligned} \text{Var}((L^{-1}v)'') &\leq \text{const}[r^2 \|v'\|_1 + \|x''\|_1 + \|v''\|_1] \\ &\leq \text{const}[r^3 \|v\|_\infty + r\|v\|_\infty + r^3 \|v\|_\infty] \\ &\leq \text{const } r^3. \end{aligned}$$

⁂

Wir erhalten also dieselbe asymptotische Konvergenzordnung, wie sie in [8], Satz 3.3 für $u \in C^2$ und $h'' = r^{-3}$ gezeigt wurde. Durch die bei uns geforderte geringe Glattheit von u* - stetig und mit stückweise lipschitzstetiger Ableitung - sichert Satz 10 diese Konvergenzordnung auch für praxisnähere Aufgaben.
Bei den in [10] zum numerischen Test dieser Ordnungsabschätzungen gerechneten Beispielen sind die optimalen Kontrollen nicht aus C^1, jedoch die Voraussetzungen von Satz 10 erfüllt. Diese Beispiele zeigen auch numerisch die Konvergenzordnung $O(r^{-2})$.
Das für die numerische Lösung von $(OC)_h^r$ wesentliche, langsamere Wachstum von h''^{-1} wurde auch in [10] erreicht. Dort wurde für ein modifiziertes Problem - zusätzliche Beschränktheit der Ableitungen der Kontrollen - dieselbe Konvergenzordnung für u* aus C^3 gezeigt. Die Voraussetzung u* aus C^3 in diesem modifizierten Problem kann mit unserer Vorgehensweise ebenfalls abgeschwächt werden.

Literatur:

[1] BUDAK, B.M., E.M. BERKOVICH, E.N. SOLOV'EVA: Difference Approximations in Optimal Control Problems. SIAM J. Control 7, 18-31 (1969).

[2] BUDAK, B.M., E.M. BERKOVICH, E.N. SOLOV'EVA: The Convergence of Difference Approximations for Optimal Control Problems. USSR Comput. Math. and math. Phys. 9, 30-65 (1969).

[3] CHARTRES, B.A., R.S. STEPLEMAN: Actual Order of Convergence of Runge-Kutta Methods on Differential Equations with Discontinuities. SIAM J. Numer. Anal. 11, 1193-1206 (1974).

[4] CULLUM, J.: Discrete Approximations to Continous Optimal Control Problems. SIAM J. Control 7, 32-50 (1969).

[5] CULLUM, J.: An Explicite Procedure for Discretizing Continous Optimal Control Problems. JOTA 8, 15-34 (1971).

[6] CULLUM, J.: Finite Dimensional Approximations of State-Constrained Continous Optimal Control Problems. SIAM J. Control 10, 649-670 (1972).

[7] EHLICH, H., K. ZELLER: Schwankung von Polynomen zwischen Gitterpunkten. Math. Z. 86, 41-44 (1964).

[8] ESSER, H.: Zur Diskretisierung von Extremalproblemen. In R. Ansorge, W. Törnig: Numerische, insbesondere approximationstheoretische Behandlung von Funktionalgleichungen. Lecture Notes in Mathematics, 333. Berlin, u.a.: Springer (1973).

[9] GRIGORIEFF, R.D.: Numerik gewöhnlicher Differentialgleichungen, 1. Stuttgart: Teubner (1972).

[10] HOFFMANN, K.-H., E. JÖRN, E. SCHÄFER, H. WEBER: Differenzenverfahren zur Behandlung von Kontrollproblemen. Eingereicht in Numer. Math.

[11] SHAPIRO, H.S.: Smoothing and Approximation of Functions. London: Van Nostrand Reinhold (1969).

[12] WALTER. W.: Differential and Integral Inequalities. Berlin, u.a.: Springer (1970).

Eugen Schäfer
Mathematisches Institut
der Universität
Theresienstraße 39
D 8000 München 2

TWO-STAGE SPLINE METHODS FOR FITTING SURFACES

Larry L. Schumaker

1. Introduction

In this paper we are concerned with numerical methods for handling the following problem:

PROBLEM 1.1. Let D be a domain in the (x,y)-plane, and suppose F is a real-valued function defined on D. Suppose we are given values $F_i = F(x_i,y_i)$ of F at some set of points (x_i,y_i) located in D, $i = 1,2,\ldots,N$. Find a function f defined on D which reasonably approximates F.

This problem arises in a great number of applications, and it is not surprising that a considerable number of papers have been written about numerically usable approximation methods for attacking it. Recently [5], I surveyed some of the available algorithms and compiled a rather extensive bibliography of recent papers on the subject.

Although in preparing [5] I did not have the time to test all of the methods surveyed there, I came away with the distinct impression that users are still not fully satisfied with available algorithms. The purpose of this paper is to describe in detail some new two-stage approximation methods (suggested in [5]) involving piecewise polynomials and splines.

The idea is as follows. In order to construct a method which is applicable to large amounts of genuinely scattered data and which at the same time produces smooth convenient surfaces without excessive computation, we have elected to divide the approximation process into two distinct stages. As a first-stage process we propose certain adaptive local patch approximation schemes which are especially suited to application to scattered data. As a second-stage process we choose direct local spline approximation methods based on B-splines. The details of these two stages are discussed in sections 2 and 3, respectively. In section 4 we consider the properties of the combined two-stage processes, including a brief look at error bounds. To give an idea of how the methods perform, we discuss the results of some numerical tests on real-life data in section 5. A number of remarks are collected in section 6 to close the paper.

2. Stage I. Local Patch Methods

In this section we discuss the construction of a piecewise polynomial patch surface based on data as in Problem 1.1. To describe the methods, suppose first that a rectangle $H = [a,b) \times [c,d)$ is chosen so that $D \subset H$. Let

$$(2.1)\qquad \begin{aligned} a = x_0 < x_1 < \dots < x_{k+1} = b \\ c = y_0 < y_1 < \dots < y_{\ell+1} = d \end{aligned}$$

be partitions of $[a,b]$ and $[c,d]$, respectively. The points define a partition of H into subrectangles given by

$$(2.2)\qquad H = \bigcup_{i=1}^{k} \bigcup_{j=1}^{\ell} H_{ij}, \qquad H_{ij} = [x_i, x_{i+1}) \times [y_j, y_{j+1}).$$

We define the desired patch surface as follows:

$$(2.3)\qquad g(x,y) = \{g_{ij}(x,y), \quad (x,y) \in H_{ij}, \quad \begin{array}{l} i = 0,1,\dots,k \\ j = 0,1,\dots,\ell, \end{array}$$

where each g_{ij} is to be a polynomial of reasonably small degree. To be more specific, suppose that we decide to work with a space of polynomials $\mathcal{P}$ of dimension d spanned by $\{\phi_i\}_1^d$. Then we write

$$(2.4)\qquad g_{ij}(x,y) = \sum_{\nu=1}^{d} c_{ij\nu}\, \phi_\nu(x,y).$$

Since the patch g_{ij} is to represent the surface only in the subrectangle H_{ij}, it is reasonable to try to determine the coefficients of g_{ij} based only on the data in H_{ij}. With scattered data, however, there may be very little or even no data at all in H_{ij}. In this case, the reasonable thing to do would be to construct g_{ij} based on data in a somewhat larger rectangle $\hat{H}_{ij}$ containing both H_{ij} and a sufficient amount of data. The rectangle $\hat{H}_{ij}$ can be chosen adaptively as follows.

Suppose that in order to construct g_{ij} we insist on using a minimum of d_{min} data points. For each set $A \subset H$, let $d(A)$ denote the number of data points lying in A. Then we perform the following iterative process to determine $\hat{H}_{ij}$:

(2.5)
- (a) Set $\hat{H}_{ij} = H_{ij}$;
- (b) If $d(\hat{H}_{ij}) \geq d_{min}$, quit ;
- (c) Replace $\hat{H}_{ij}$ by the union of all subrectangles of H which touch the present $\hat{H}_{ij}$, and return to (b).

To compute the coefficient vector $\{c_{ij\nu}\}_1^d$ based on the data in $\hat{H}_{ij}$, we recommend using either discrete least squares or discrete Tchebycheff approximation with $d_{min} \geq d$. In the case of least squares this can be accomplished by solving the associated system of d normal equations. Since d will be small, this will generally involve a well-conditioned system. In the case of discrete Tchebycheff approximation, the problem of determining the $\{c_{ij\nu}\}_1^d$ can be recast as a linear programming problem and can be attacked by standard linear programming algorithms.

If each of the patches is determined by least-squares fitting with a space $\mathcal{P}$ of polynomials, then the overall method defines a linear operator L_{LSQ} which maps R^N into $\mathcal{PP}$, where

(2.6) $$\mathcal{PP} = \{f : f|_{H_{ij}} \in \mathcal{P}, \quad i = 0,1,\ldots,k; \; j = 0,1,\ldots,\ell\}.$$

Similarly, if the patches are determined by discrete Tchebycheff approximation, then the method defines a linear operator L_{TCH} mapping R^N into $\mathcal{PP}$.

Both methods L_{LSQ} and L_{TCH} involve setting up and solving a total of $k \times \ell$ relatively small approximation problems. It is clear that both methods are local. On the other hand, they both produce a surface which will generally involve jump discontinuities across the partition lines. In some applications the user may be satisfied with a surface of this type. In most cases, however, it will probably be desirable to have a smoother surface, in which case it will be necessary to apply the second-stage process discussed in the following section.

3. Stage II. A Local Spline Approximation Method

To describe this method, we need to introduce a certain class of splines. We begin with the well-known B-splines. Let m and n be positive integers, and choose

$$x_{1-m} \leq \ldots \leq x_0 = a, \quad b = x_{k+1} \leq x_{k+2} \leq \ldots \leq x_{m+k}$$

$$y_{1-n} \leq \ldots \leq y_0 = c, \quad d = y_{\ell+1} \leq y_{\ell+2} \leq \ldots \leq y_{\ell+n}.$$

Let $\{B_i(x)\}_1^{m+k}$ and $\{\tilde{B}_j(y)\}_1^{n+\ell}$ be the B-splines of order m and n associated with the knot sequences $\{x_i\}_{1-m}^{k+m}$ and $\{y_j\}_{1-n}^{\ell+n}$, respectively. For a list of some of the properties of B-splines, see e.g. [1,2,3]. Here we note only that $B_i(x)$, for example, consists piecewise of polynomials of degree m-1 and that it belongs to $C^{m-2}[a,b]$.

Moreover, $B_i(x) > 0$ on (x_{i-m}, x_i) and vanishes outside of $[x_{i-m}, x_i]$. The B-splines can be computed by stable recursion relations (cf. [1,2,3]).

We now define

$$B_{ij}(x,y) = B_i(x)\tilde{B}_j(y), \quad i = 1,2,\dots,k; \quad j = 1,2,\dots,\ell. \tag{3.1}$$

In view of the support properties of the B-splines, it is clear that B_{ij} is a kind of pyramid or hill function with support on $[x_{i-m}, x_i] \times [y_{j-n}, y_j]$.

Let $\mathcal{B} = \text{span } \{B_{ij}\}_{i=1, j=1}^{m+k \;\; n+\ell}$, and let $B(H_D)$ denote the space of all functions defined and bounded on H_D where

$$H_D = \bigcup_{\substack{i=1 \\ H_{ij} \cap D \neq \emptyset}}^{k} \bigcup_{j=1}^{\ell} H_{ij} . \tag{3.2}$$

We now proceed to define a linear operator mapping $B(H_D)$ into $\mathcal{B}$. Let p and q be integers with $1 \le p \le m$ and $1 \le q \le n$. For each $1 \le i \le m+k$ and $1 \le j \le n+\ell$, let

$$\begin{aligned} x_{i-m} &\le \sigma_{i1} < \dots < \sigma_{ip} \le x_i \\ y_{j-n} &\le \tau_{j1} < \dots < \tau_{jq} \le y_j . \end{aligned} \tag{3.3}$$

Let $\{\alpha_{i\nu}\}_{\nu=1}^{p}$ be solutions of the systems of equations

$$\sum_{\nu=1}^{p} \alpha_{i\nu} \, u_r(\sigma_{i\nu}) = \xi_i^{(r)}, \quad r = 1,2,\dots,p, \tag{3.4}$$

where $\{u_r(x) = x^{r-1}\}_{r=1}^{m}$, and

$$\xi_i^{(r)} = \frac{(r-1)!(m-r)!}{(m-1)!} \text{ symm}_{r-1}(x_{i-m+1},\dots,x_{i-1}). \tag{3.5}$$

(Here symm is the usual symmetric function defined by

$$\prod_{i=1}^{d} (x+a_i) = \sum_{\nu=1}^{d} x^{d+1-\nu} \text{ symm}_{\nu-1}(a_1,\dots,a_d).)$$

Similarly, let $\{\beta_{j\mu}\}_{\mu=1}^{q}$ be solutions of the systems

$$\sum_{\mu=1}^{q} \beta_{j\mu} \, u_r(\tau_{j\mu}) = \eta_j^{(r)}, \quad r = 1,2,\dots,q, \tag{3.6}$$

where

$$\eta_j^{(r)} = \frac{(r-1)!(n-r)!}{(n-1)!} \text{ symm}_{r-1}(y_{j-n+1},\dots,y_{j-1}). \tag{3.7}$$

We can now define the operator Q of interest:

(3.8) $$Qg(x,y) = \sum_{\substack{i=1 \\ H_{ij} \cap D \neq \emptyset}}^{m+k} \sum_{j=1}^{n+\ell} \lambda_{ij} g B_{ij}(x,y),$$

where

(3.9) $$\lambda_{ij} g = \sum_{\nu=1}^{p} \sum_{\mu=1}^{q} \alpha_{i\nu} \beta_{j\mu} g(\sigma_{i\nu}, \beta_{j\mu}).$$

It is clear that Q is local. It is also proved in [3] that Q reproduces polynomials of an appropriate order; in particular, Qh = h for all h in

$$\mathcal{P}_p \times \mathcal{P}_q = \left\{ h = \sum_{i=1}^{p} \sum_{j=1}^{q} d_{ij} x^{i-1} y^{j-1} \right\}.$$

Q will not generally be directly applicable to Problem 1.1 with scattered data. It is, however, imminently suited as a second-stage method. To set up the method it is required to solve a total of $(m+k)(n+\ell)$ systems of p linear equations for the α's, and a like number of systems of q equations for the β's. Once these coefficients have been determined, this work need not be repeated if Q is to be reapplied several times. To compute the actual coefficients $\{\lambda_{ij} g\}$ of an approximation to a function g, we need only compute the sums in (3.9). This is what we mean when we say Q is a direct method; at this point no further systems of equations need be solved.

It is easy to write a subroutine to perform the computations needed to set up and apply Q. The locations of the knots and the values of m,n,p, and q provide considerable flexibility in designing an approximation method. It is also easy to prepare a program for use in evaluation of Qf(x,y), or any of its partial derivatives. Because of the local support properties of the B-splines, to evaluate Qf(x,y) it is only necessary to compute a total of $m \times n$ terms in the sum (3.8). The B-splines should be evaluated by the recursions of [1].

4. The Two-stage Methods

We are now in a position to define two linear operators mapping the data vector $\{F_i\}_1^N$ of Problem 1.1 into $\mathcal{S}$. Let

(4.1) $$Q_{LSQ} = QL_{LSQ}$$

and

(4.2) $Q_{TCH} = QL_{TCH}$.

The properties of these two-stage processes follow from the individual properties of each of the stages. It is recommended that the knots for the B-splines be placed at the partition points used in the patch methods. In addition, it is quite natural to choose the space of polynomials used in stage 1 to be $\mathcal{P} = \mathcal{P}_p \times \mathcal{P}_q$. In this case, it is clear that both Q_{LSQ} and Q_{TCH} reproduce $\mathcal{P}$. Thus, for example, if the original surface F defining the data is flat in some area, then the approximation will also be flat in that area.

We devote the remainder of this section to some observations about error bounds for $\|F - Q_{LSQ}F\|$ and $\|F - Q_{TCH}F\|$, assuming that the function F which generates the data is smooth. For convenience we concentrate on Q_{LSQ}, although all of our assertions are also valid for Q_{TCH}. The key to obtaining error bounds is the following simple inequality:

$$\|F - Q_{LSQ}F\| \leq \|F - QF\| + \|\|Q\|\| \, \|F - L_{LSQ}F\|, \tag{4.3}$$

where $\|\|Q\|\|$ is the norm of the operator Q relative to $\|\cdot\|$.

There is no problem with the first term. For example, the following theorem is established in [3]:

<u>THEOREM 4.1</u>. <u>Let</u> m = n, <u>and suppose</u> $s \leq p + q \leq m$. <u>Then for any</u> $F \in C^{s-1}(H_D)$,

$$\|F - QF\|_{L_\infty[H_D]} \leq K\Delta^{s-1}\omega(D^{s-1}; \Delta; H_D),$$

<u>where</u> K <u>is a constant independent of</u> F <u>and</u> Δ,

$$\Delta = \Delta_x + \Delta_y, \quad \Delta_x = \max_{0 \leq i \leq k} (x_{i+1} - x_i), \quad \Delta_y = \max_{0 \leq j \leq \ell} (y_{j+1} - y_j),$$

<u>and</u> ω <u>is the modulus of continuity defined by</u>

$$\omega(D^{s-1}F; \Delta; H_D) = \max_{0 \leq \nu \leq s-1} \omega(f^{(\nu, s-\nu-1)}; \Delta; H_D).$$

Concerning the second term in (4.3), we observe that by following through the arguments in [3], it is possible to establish that $\|\|Q\|\|$ is bounded with respect to the ∞ norm. Error bounds for $\|F - L_{LSQ}\|_\infty$ are, however, considerably more difficult to obtain. The following result shows that if a piecewise constant first-stage surface is used, then the method Q_{LSQ} has an error of order $\omega(F;\Delta)$, provided the data are not too badly scattered.

THEOREM 4.2. Suppose that $\mathcal{P}$ = {constants}. Suppose that the data are sufficiently dense so that for some integer ρ, the height and width of $\hat{H}_{ij}$ is at most ρ times as great as that of H_{ij}, all i, j with $H_{ij} \cap D \neq \emptyset$. Then for every $F \in B(H_D)$,

$$\|F - L_{LSQ}F\|_\infty \leq \frac{\rho}{2}\,\omega(F;\,\Delta;\,H_D), \tag{4.4}$$

where Δ is as in Theorem 4.1. Thus, for some constant K,

$$\|F - Q_{LSQ}F\|_\infty \leq K\omega(F;\,\Delta;\,H_D). \tag{4.5}$$

Proof. Since $L_{LSQ}F$ in each subrectangle is the average of the values of F_{ij} for data in the subrectangle $\hat{H}_{ij}$, it follows that $\|L_{LSQ}F - F\|_{L_\infty[H_{ij}]} \leq \frac{1}{2}\omega(F;\rho\Delta;\hat{H}_{ij})$, and (4.4) follows. Now (4.5) follows from Theorem 4.1 and the above observation that $\|Q\|$ is bounded. □

We may observe that if each H_{ij} contains at least d_{min} data points, then $\rho = 1$. The same result holds for L_{TCH} and Q_{TCH}. If higher-order polynomials are used in the first stage, the analog of Theorem 4.2 will have to depend even more heavily on the distribution of the data points. We shall handle this delicate question in a future paper.

5. A Numerical Example

ALGOL programs to implement each of the methods Q_{LSQ} and Q_{TCH} were written and tested on the Telefunken computer at the University of Munich. These programs have been tested on a number of real-life data sets, as well as on numerous examples where the underlying function F generating the data is known. In this section we shall discuss briefly the results of one such test with real-life data.

The problem involves diagnosing abnormal heart conditions based on the examination of a set of 240 contour maps of the heart potential field as measured at 240 time steps in the course of one heart beat. The data for the construction of these contour maps are produced by fitting the patient with a shirt containing 230 probes. The shape of the shirt is shown in Figure 1. The location of the probes (based on a shirt scaled to be 14 x 27) can be inferred from the entries in Tables 1 and 2 (where the * entries mean no data are available at the associated points).

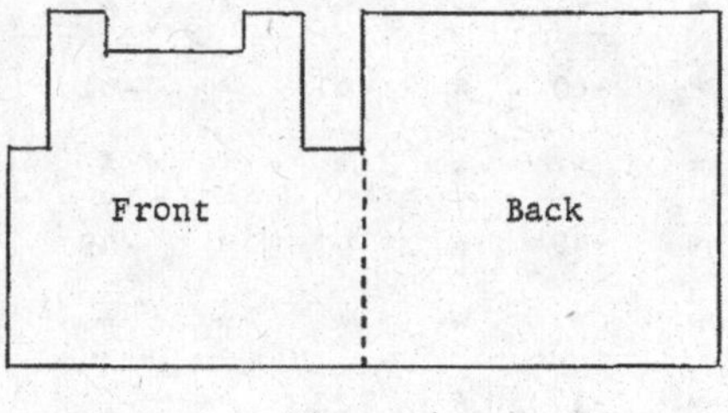

Figure 1

14	*	-43	-49	-62	*	*	*	*	*	*	-72	-66	-64	*	*	*
13	*	-50	-49	-65	*	*	*	*	*	*	-72	-64	-56	*	*	*
12	*	-50	-56	-53	-55	-70	-80	-78	-81	-81	-59	-65	-60	*	*	*
11	*	-43	-49	-48	-42	-59	-57	-65	-62	-57	-46	-56	-57	*	*	*
10	*	-28	-34	-27	-33	-39	-42	-45	-47	-38	-37	-41	-49	*	*	*
9	*	3	-8	-10	-9	-15	-20	-39	-30	-16	-11	-11	-29	*	*	*
8	0	3	7	15	3	4	-11	-27	-9	15	13	95	-3	33	-13	-19
7	20	20	27	34	36	44	37	24	34	61	49	26	19	0	1	-2
6	27	43	47	58	69	78	87	97	95	86	65	55	56	29	23	17
5	43	61	68	82	87	103	97	111	117	110	78	76	72	60	52	41
4	58	66	78	79	102	99	110	107	98	104	100	103	89	75	73	61
3	56	77	88	81	100	95	97	95	96	96	104	100	92	86	76	73
2	70	78	84	90	87	98	90	92	98		100	95	94	86	78	77
1	77	93	85	88	87	82	85	98	99	103	108	104	99	93	78	81
	1	2	3	4	5	6	7	8	9	10	11	12	13	14	15	16

Table 1. The Front

To attack this problem with Q_{LSQ}, we must select the parameters m, n, p, q, d_{min}, and the location of the knots. We have performed tests using various combinations of m, n, p, q with values between 1 and 5. d_{min} was normally chosen to be between the minimum needed for least-squares fitting with $\mathcal{P}_p \times \mathcal{P}_q$ (namely p x q) and about twice this figure. We chose equally spaced knots with spacing between 3 and 5. To get a feel

14	*	*	*	*	*	*	*	*	*	*	*
13	-68	*	-67	*	-67	*	-60	*	-62	*	-48
12	*	*	*	*	*	*	*	*	*	*	*
11	-47	*	-60	*	-61	*	-61	*	-47	*	-48
10	*	*	*	*	*	*	*	*	*	*	*
9	-44	*	-49	*	-51	*	-49	*	-31	*	-32
8	*	*	*	*	*	*	*	*	*	*	*
7	-5	*	-17	*	-22	*∞	-24	*	-17	*	-1
6	*	*	*	*	*	*	*	*	*	*	*
5	29	*	13	*	12	*	13	*	6	*	21
4	*	*	*	*	*	*	*	*	*	*	*
3	69	*	37	*	43	*	43	*	45	*	45
2	*	*	*	*	*	*	*	*	*	*	*
1	84	*	66	*	60	*	66	*	63	*	73
	17	18	19	20	21	22	23	24	25	26	27

Table 2. The Back

for the results, we constructed a rough contour map of the spline surfaces by printing integers between 0 and 9 indicating the height of the surface at points on a grid.

We now give explicit results for a particular choice of parameters. Let H be the rectangle $[1,31] \times [1,16]$. We choose knots 1, 6, 11, 16, 21, 26, 31 in the x-direction, and 1, 6, 11, 16 in the y-direction. Let $m = n = 3$ so that the resulting surface is a bi-quadratic spline (which belongs to $C^1(H)$). We chose $p = q = 2$ so that the patch surface is constructed from bilinear polynomials. We chose $d_{min} = 4$. To construct a contour map of the resulting surface s, we compute the value of $s(x,y)$ at all of the points $\{i, j/4\}_{i=1, j=1}^{31 \quad 64}$. Each of these points was assigned a code height between 0 and 9 as follows:

code height	9	$1 \leq i \leq 8$	0
value	$(-\infty, -60)$	$[20(5-i),\ 20[6-i)]$	$[100, \infty)$

The resulting contour is displayed in Figure 2.

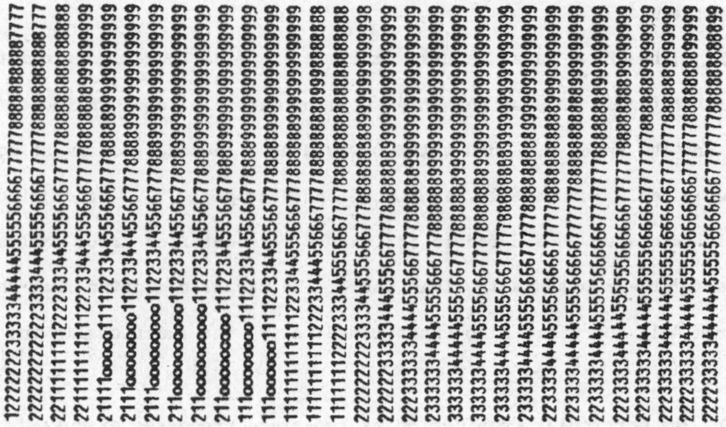

Figure 2. A Biquadratic Spline Fit to Heart Potentials

We found that the use of Q_{TCH} instead of Q_{LSQ} required essentially the same amount of computation and produced similar surfaces. The programs are currently being translated into FORTRAN, and further testing of the methods as well as comparison with other methods is planned.

6. Remarks

(1) In [5] the methods surveyed were divided into four classes: (a) global interpolation, (b) local interpolation, (c) global approximation, and (d) local approximation schemes. The two-stage methods discussed here belong to the last category, which, it seems to me, is probably best suited for fitting fairly large amounts of scattered data. Indeed, when the data are subject to measurement errors it is usually advisable to employ approximation rather than interpolation. Local methods also have the advantage of avoiding the solution of a large (often badly conditioned) linear system. Finally, many physical surfaces are intuitively local in the sense that the behavior of the surface in one area is totally unrelated to its behavior at points a reasonable distance away.

(2) The idea of constructing patch surfaces by solving local discrete least squares or discrete Tchebycheff approximation problems is certainly not new. The methods just haven't been advocated since the resulting surface is usually not smooth

enough for the user.

(3) Although we have discussed the local spline approximation methods only with simple knots, it is also possible to allow multiple knots in the definition of the B-splines and the space $\mathcal{S}$. Such splines would have reduced smoothness across the associated knot lines. For details, see Lyche and Schumaker [3]. Other kinds of linear functionals were also considered in [3], and a very detailed error analysis with specific values for the constants is given there. We may also mention that the error bounds (cf. Theorem 4.1) for Q are also established in [3] for functions in the Sobolev spaces $W_p^s(H_D)$.

(4) The particular combination of two-stage processes discussed here is, of course, only one of many possibilities. Another method which we have tried with some success is to use global continuous least squares as a second stage. This does involve, however, the solution of a large systems of equations, although it is essentially local if one uses B-splines.

(5) The example in section 5 of heart potentials was furnished by Ms. Patrizia Ciarlini of the Instituto per le Applicazioni del Calcolo, Rome.

(6) Although the problem of fitting data in one-dimension is structurally somewhat different from the two-dimensional version studied here, the idea of using a two-stage procedure consisting of a local adaptive process followed by a direct local spline approximation scheme appears very promising. This idea will be fully explored in a separate forthcoming paper [4].

References

1. deBoor, C., On calculating with B-splines, J. Approximation Th. 6 (1972), 50-62.

2. deBoor, C., Splines as linear combinations of B-splines: A survey, in Approximation Theory II, G. G. Lorentz, C. K. Chui, and L. L. Schumaker, eds., Academic Press, New York, 1976, 1-48.

3. Lyche, T., and L. L. Schumaker, Local spline approximation methods, J. Approximation Th. 15 (1975), 294-325.

4. Lyche, T., and L. L. Schumaker, A local adaptive procedure for fitting splines to data in one dimension, to appear.

5. Schumaker, L. L., Fitting surfaces to scattered data, in Approximation Theory II, G. G. Lorentz, C. K. Chui, and L. L. Schumaker, eds., Academic Press, New York, 1976, 203-268.

L. L. Schumaker
Department of Mathematics
and Center for Numerical Analysis
The University of Texas at Austin
Austin, Texas 78712

Supported in part by the Deutsche Forschungsgemeinschaft at the University of Munich and the United States Air Force under grant AFOSR 74-2598B.

ERZEUGUNG UND STRUKTURELLE VERKNÜPFUNGEN VON KERNEN SINGULÄRER FALTUNGSINTEGRALE

E.L. Stark

Herrn Professor Fritz Reutter zum 65. *Geburtstag am* 26.8.1976 *gewidmet.*

1. Motivation. Bei der Approximation von Funktionen durch singuläre Faltungsintegrale werden Kerne, die verschiedenartigsten Ursprungs sind, betrachtet; Zusammenhänge *zwischen* den Kernen werden in den wenigsten Fällen aufgezeigt. Zwei äußerst elementare Hilfsmittel, nämlich *Koppeln* und *Quadrieren* von Kernen decken interessante, bisher verborgene Verbindungen zwischen wohlbekannten Kernen auf.

Es werden (parallel) drei Fälle zu behandeln sein: bei der Approximation 2π-periodischer Funktionen ist zwischen Kernen, die durch ein Polynom bzw. durch eine Fourierreihe gegeben sind, zu unterscheiden. Hinzukommt der Fall der nichtperiodischen Approximation auf der ganzen Zahlengeraden. Um möglichst zahlreiche Beispiele erfassen zu können, werden die Definitionen der Kerne sowie bekannte bzw. die neu gewonnenen Verknüpfungen untereinander in einer Tabelle zusammengefaßt. Zur Illustration wird lediglich das wohl bezeichnendste Beispiel, der Kern von Fejér-Korovkin, etwas ausführlicher diskutiert, u.a. insbesondere, um auch die Querverbindung zwischen dem periodischen und dem nichtperiodischen Fall zu veranschaulichen.

Daß durch "Quadrieren" von Kernen sinnvolle Ergebnisse zu erwarten sind, wird allein durch die Struktur der Kette $\mathcal{D} \to \mathcal{F} \to \mathcal{J}$ ($\mathcal{D}$: Dirichlet-Kern, definiert über die Partialsummen der Fourierreihe; $\mathcal{F}$: Fejér-Kern, definiert als das arithmetische Mittel der Partialsummen; $\mathcal{J}$: Jackson-Kern, historisch erster mit optimaler Ordnung approximierender polynomialer Kern) nahegelegt; man vergleiche die speziell hierauf bezogene Erörterung in [11, p. 41f]. - Die Wirksamkeit von (symmetrischen) Abschnitts-Kopplungen ist nach [17], [18] (vgl. insbesondere [19, p. 128]) durch das Beispiel $\mathcal{D} \to \mathcal{R}$ ($\mathcal{R}$: Rogosinski-Kern, definiert über (1) bzw. (9)) bekannt: zwei Dirichlet-Kerne werden zu einem Polynom

$$R_n^{\alpha}(t) = \frac{1}{2}\{D_n(t-\alpha) + D_n(t+\alpha)\} \qquad (\alpha \in \mathbb{R}) \tag{1}$$

verkoppelt; die Wahl der Verschiebung $\alpha = \alpha(n)$ ist dabei durch die

Forderung eingeschränkt, daß die zugehörigen Lebesgue-Konstanten $L_n(R^\alpha) := \int_0^\pi |R_n^\alpha(t)|dt$ gleichmäßig in n beschränkt bleiben (vgl. auch [23], [24]).

2. Kosinuspolynome. Folgende Klasse normierter Kosinuspolynome vom Grade n wird betrachtet:

$$\text{(2)} \qquad \Pi_n = \{p_n(t) := \frac{1}{2} + \sum_{k=1}^{n} \rho_{k,n}(p)\cos kt;\ \rho_{n,n}(p) \neq 0\} \qquad (n \in \mathbb{N}),$$

$$\text{(3)} \qquad \rho_{k,n}(p) := \frac{2}{\pi}\int_0^\pi p_n(t)\cos kt\,dt,\ 1 \leq k \leq n;\ \rho_{0,n}(p) \equiv 1 \qquad (n \in \mathbb{N});$$

($\Pi_n^+, \Pi_n^\pm$ bezeichnen entsprechend gesondert nichtnegative bzw. oszillierende Polynome). Neben der Polynomdarstellung (2) steht für die meisten der gebräuchlichen Beispiele die geschlossene, d.h., aufsummierte Form des Kerns zur Verfügung; siehe Tab.

Beim Quadrieren erhält man ein positives Polynom vom (doppelten) Grad 2n, wobei die Normierung verlorengeht. Nach Wiederherstellung der Normierung erhält man die formelmäßige Zusammenstellung gemäß

Lemma 1. *Sei*

$$p_n(t) \in \Pi_n,\ \int_0^\pi p_n(t)dt = \frac{\pi}{2} \qquad (n \in \mathbb{N});$$

dann gilt

$$q_{2n}(t) := \frac{\{p_n(t)\}^2}{N_n(p)} \in \Pi_{2n}^+;\ \int_0^\pi q_{2n}(t)dt = \frac{\pi}{2} \qquad (n \in \mathbb{N})$$

$$= \frac{1}{2} + \sum_{k=1}^{2n} \lambda_{k,2n}(p)\cos kt;\ N_n(p) = \frac{1}{2} + \sum_{k=1}^{n} \rho_{k,n}^2(p);$$

$$\text{(4)} \qquad \lambda_{k,2n}(p) = \frac{1}{N_n(p)} \begin{cases} \rho_{k,n} + \sum_{j=1}^{n-k} \rho_{j,n}\rho_{k+j,n} + \frac{1}{2}\sum_{j=1}^{k-1} \rho_{j,n}\rho_{k-j,n}, & 1 \leq k \leq n, \\ \frac{1}{2}\sum_{j=k-n}^{n} \rho_{j,n}\rho_{k-j,n}, & n+1 \leq k \leq 2n. \end{cases}$$

Der *Beweis* erfolgt durch Koeffizientenvergleich nach Ausmultiplizieren zweier (gleicher) Kosinuspolynome unter Verwendung geeigneter trigonometrischer Identitäten.

Bemerkung 1. Es ist erstrebenswert, den verdoppelten Polynomgrad wieder auf den ursprünglichen Grad n zu *reduzieren:* dies ist offensichtlich nur dann möglich, wenn die *abschnittsweise* Definition der Fourierkoeffizienten (= Konvergenzfaktoren) in (4) auf eine einheitliche Darstellung für alle k mit $1 \le k \le 2n$ führt. Von besonderem Vorteil ist in (4) die vereinfachte Berechnung der Konvergenzfaktoren über elementare Summen statt über das definierende Integral (3) mit der oft komplizierten, geschlossenen Kerndarstellung; vgl. z.B. (10).

Lemma 2. *Unter den Voraussetzungen*

$$(5) \qquad 1 \ge \rho_{k,n}(p) > 0 \qquad (1 \le k \le n),$$

$$(6) \qquad \rho_{k,n}(p) \ge \rho_{k+1,n}(p) \qquad (1 \le k \le n-1),$$

$$(7) \qquad p_n(0) = O(n) \qquad (n \to \infty)$$

ist $q_{2n}(t)$ *eine approximierende Identität.*

Von den üblichen Kernen ($\mathcal{D}$) bzw. approximierenden Identitäten (Tab., (2) - (7)) werden diese Voraussetzungen erfüllt.

Beweis. Da $q_{2n}(t) \ge 0$, genügt es zu zeigen, daß $\lim_{n\to\infty} \lambda_{1,2n}(p) = 1$; siehe [4, p. 31, p. 59, Prop. 1.3.10]. Das Wachstumsverhalten des Kerns im Nullpunkt liefert mit (7):

$$\sum_{k=1}^{n} \rho_{k,n}(p) = p_n(0) - \tfrac{1}{2} = O(n) \qquad (n \to \infty);$$

mittels der Cauchy-Schwarz-Ungleichung folgt:

$$\sum_{k=1}^{n} \rho_{k,n}^2 \ge \frac{1}{n} \left(\sum_{k=1}^{n} \rho_{k,n} \right)^2 ;$$

mit (5) und (6) ergibt sich die Abschätzung

$$\sum_{k=1}^{n-1} \rho_{k,n}(\rho_{k,n} - \rho_{k+1,n}) \le \sum_{k=1}^{n-1} (\rho_{k,n} - \rho_{k+1,n}) = \rho_{1,n} - \rho_{n,n}.$$

Damit erhält man schließlich über die Identität

$$1 - \lambda_{1,2n} = \frac{\frac{1}{2} + \rho_{n,n}^2 - \rho_{1,n} + \sum_{k=1}^{n-1} \rho_{k,n}(\rho_{k,n} - \rho_{k+1,n})}{\frac{1}{2} + \sum_{k=1}^{n} \rho_{k,n}^2}$$

$$\leq \frac{\frac{1}{2} + \rho_{n,n}^2 - \rho_{n,n}}{\frac{1}{2} + \frac{1}{n}(p_n(0) - \frac{1}{2})^2} = O(n^{-1}) \qquad (n \to \infty) \blacksquare$$

Bemerkung 2. Das Verfahren gemäß Lemma 1 mit anschließender Reduktion des Polynomgrades liefert exakt den Fejér-Kern $F_n(t)$ als reduziertes Quadrat des Dirichlet-Kerns (der bekanntlich bei der Approximation von $f \in C_{2\pi}$ selbst keine approximierende Identität ist; [4, p. 42]). - Aus $F_{n-1}(t)$ erhält man die übliche Form des Jackson-Kerns $J_{2n-2}(t)$ mit den Konvergenzfaktoren (in der hier günstigsten Gestalt; vgl. [14, p. 42])

$$(8) \qquad \rho_{k,2n-2}(J) = \frac{1}{2n(2n^2+1)} \begin{cases} \frac{(2n-k+1)!}{(2n-k-2)!} - 4\,\frac{(n-k+1)!}{(n-k-2)!}, & 0 \leq k \leq n-2, \\ \frac{(2n-k+1)!}{(2n-k-2)!}, & n-2 \leq k \leq 2n-2; \end{cases}$$

die unmittelbare Reduktion des Polynomgrades ist somit *nicht* möglich. Man vergleiche jedoch die "gewaltsame" Reduktion in [11, p. 41], die durch das unnatürliche Auftreten von Gaußklammern im Parameter n erzwungen wird.

Bemerkung 3. Mit der in (1) zulässigen Verschiebung $\alpha(n) = \pi/2n$ erhält man

$$(9) \qquad R_{n-1}(t) = \frac{1}{2} + \sum_{k=1}^{n-1} \cos\frac{k\pi}{2n} \cos kt = \frac{\sin\frac{\pi}{2n}\cos nt}{2(\cos t - \cos\frac{\pi}{2n})} \in \Pi_{n-1}^{\pm};$$

mit $N_{n-1}(R) = n/2$ folgt

$$\frac{\{R_{n-1}(t)\}^2}{N_{n-1}(R)} = \frac{\sin^2\frac{\pi}{2n}\cos^2 2n\frac{t}{2}}{2n(\cos t - \cos\frac{\pi}{2n})^2} \in \Pi_{2n-2}^{+}.$$

Die Reduktion $2n \to n$ liefert - überraschend - den Kern von Fejér-Korovkin

(10) $$K_{n-2}(t) = \frac{\sin^2 \frac{\pi}{n} \cos^2 n \frac{t}{2}}{n(\cos t - \cos \frac{\pi}{n})^2} \in \Pi^+_{n-2};$$

die Konvergenzfaktoren - in einheitlicher Gestalt - errechnen sich über (4) zu

(11) $$\rho_{k,n-2}(K) = \left(1 - \frac{k}{n}\right) \cos \frac{k\pi}{n} + \frac{1}{n} \cot \frac{\pi}{n} \sin \frac{k\pi}{n} \qquad (0 \le k \le n-2).$$

Es ist damit nachgewiesen, daß dieser in fundamentalen Beziehungen optimale, positive, polynomiale Kern (vgl. z.B. [4], [14]) sich durch das angegebene Verfahren (Koppeln, Quadrieren, Reduzieren) unmittelbar aus dem Dirichlet-Kern herleiten läßt. - Es muß jedoch vermerkt werden, daß dieser Nachweis auf der sehr speziellen Wahl von $\alpha(n) = \pi/2n$ in (1) bzw. (9) beruht: die vollständige Darstellung (9) ÷ sowie ein äußerst kurzer Beweis für die Beschränktheit der zugeordneten Lebesgue-Konstanten ÷ findet sich (allgemein unbemerkt) im *Lehrbuch* von J. FAVARD [10, p. 150 f]; neuerdings wird (9) auch in zahlreichen Arbeiten von V.K. DZJADYK (Kiev) und Schülern herangezogen; vgl. u.a. [7], [8], [6], [22]. Die übliche, nur leicht variierte Verschiebung $\alpha^*(n) = \pi/(2n+1)$ führt zwar auch auf einen Rogosinski-Kern, nunmehr vom Grade n (vgl. u.a. [4], [15]), ist aber für den hier benötigten Zweck völlig ungeeignet. - Entgegen der hier gegebenen *Konstruktion* ist der Kern von Fejér-Korovkin $K_n(t) \in \Pi^+_n$ üblicherweise über mehrere, sehr verschiedenartige Extremalprobleme *definiert:* beispielsweise - am bekanntesten - als eindeutige Lösung des Problems, das normierte, nichtnegative Kosinuspolynom vom Grade n mit maximalem ersten Fourierkoeffizienten, d.h.,

$$\max_{p_n \in \Pi^+_n} \rho_{1,n}(p) \overset{!}{=} \cos \frac{\pi}{n+2} = \rho_{1,n}(K) \qquad (n \in \mathbb{N})$$

zu bestimmen.

Bemerkung 4. Es existiert mindestens ein Kern, der sich gemäß Lemma 1 (Quadrieren und Reduzieren) *selbst reproduziert*, nämlich der Kern von de La Vallée Poussin, s. Tab., (12). Durch diese Erscheinung wird das außergewöhnliche (vgl. etwa [4]) Approximationsverhalten des zugeordneten singulären Integrals aus anderer Sicht erneut beleuchtet. Die Frage der Eindeutigkeit ist noch ungeklärt.

<u>3. Fourierkosinusreihen.</u> Im Fall nichtpolynomialer Kerne läßt sich das Verfahren in vereinfachter Form ebenso durchführen: die Aufspaltung der Konvergenzfaktoren wie in (4) entfällt; (12) nimmt - in gewissem Sinne Grenzfall von (4) - eine einheitliche Form an; eine Reduktion des Polynomgrades erübrigt sich. Der Beweis verläuft analog dem von Lemma 1. Die formelmäßige Verknüpfung wird zusammengefaßt als

<u>Lemma 3.</u> *Sei*

$$p_\xi(t) = \frac{1}{2} + \sum_{k=1}^{\infty} \rho_{k,\xi}(p)\cos kt, \quad \int_0^\pi p_\xi(t)dt = \frac{\pi}{2} \qquad (\xi \in \mathbb{A});$$

dann gilt

$$q_\xi(t) := \frac{\{p_\xi(t)\}^2}{N_\xi(p)}\,; \quad \int_0^\pi q_\xi(t)dt = \frac{\pi}{2}\ (\xi \in \mathbb{A}) \qquad \begin{cases} \xi \to \xi_0, \\ \xi_0 \in \bar{\mathbb{A}}; \end{cases}$$

$$= \frac{1}{2} + \sum_{k=1}^{\infty} \lambda_{k,\xi}(p)\cos kt; \quad N_\xi(p) = \frac{1}{2} + \sum_{k=1}^{\infty} \rho_{k,\xi}^2(p);$$

$$(12) \qquad \lambda_{k,\xi}(p) = \frac{1}{N_\xi(p)}\left\{\rho_{k,\xi} + \sum_{j=1}^{\infty} \rho_{j,\xi}\rho_{k+j,\xi} + \frac{1}{2}\sum_{j=1}^{k-1} \rho_{j,\xi}\rho_{k-j,\xi}\right\}.$$

<u>Bemerkung 5.</u> Unter den gebotenen Modifikationen läßt sich eine Aussage entsprechend der in Lemma 2 beweisen.

<u>Bemerkung 6.</u> Es soll hier nur *ein* repräsentatives Beispiel ausgeführt werden: der Kern von Abel-Poisson (P; Tab., (8)) geht über in den Kern von Ghermanesco (G; Tab., (9); [13, p. 76 ff]). Als Folge von

$$\lim_{r\to 1-} \frac{1-\rho_{k,r}(P)}{1-r} = k, \quad \lim_{r\to 1-} \frac{1-\rho_{k,r}(G)}{(1-r)^2} = \frac{1}{2}k^2$$

weist das G zugeordnete singuläre Integral eine gegenüber dem von Abel-Poisson *verdoppelte* Saturationsordnung auf (vgl. hierzu [20, p. 446, j=5], [21], [3] sowie [12], [16]).

<u>Bemerkung 7.</u> Zusammenhänge zwischen verschiedenen Kernen wurden neben [11] in zwei weiteren Arbeiten diskutiert: in [1] wird ein allgemeiner Kern konstruiert, der sich bei spezieller Parameterwahl auf den

Kern von Fejér bzw. Abel-Poisson reduziert. In [9] wird unter expliziter Angabe sowohl der Konvergenzfaktoren als auch der geschlossenen Gestalt ein Kern behandelt, aus dem sich durch genau bezeichnete Grenzfälle der Parameter gleichzeitig die Kerne von Abel-Poisson, Fejér *und* Fejér-Korovkin ergeben. In dieser - zumindest im Rahmen der Approximationstheorie und insbesondere bei der Untersuchung des singulären Integrals von Fejér-Korovkin unbeachtet gebliebenen - Arbeit wird somit ein erster Zusammenhang zwischen den so wesentlich verschiedenen Kernen F und *K* hergestellt.

4. Approximation auf $(-\infty,\infty)$. Im Fall der Approximation auf $\mathbb{R}$ bietet sich zunächst die Untersuchung singulärer Faltungsintegrale vom Fejér-Typ (FT) mit *geradem* Kern an: für jedes normierte $\Phi \in L^1(\mathbb{R})$, d.h., mit $\int_0^\infty \Phi(t)dt = \pi/2$, definiert

$$\Phi_\xi(t) = \xi\Phi(\xi t) \qquad (t \in \mathbb{R}; \xi \in \mathbb{A} := (0,\infty), \xi_0 = \infty) \tag{13}$$

stets eine (gerade) approximierende Identität auf $\mathbb{R}$; [4, p. 121]. Über

$$\Phi^*_\xi(t) = \sum_{k=-\infty}^{\infty} \xi\Phi(\xi[t+2k\pi]) \tag{14}$$

wird $\Phi(t)$ dann eine *periodische* approximierende Identität zugeordnet; [4, p. 125]. Der Zusammenhang zwischen den Fourierkosinustransformierten (16) von $\Phi(t)$ und den Fourierkoeffizienten (3) von (14) wird (mittels der Poissonschen Summationsformel) hergestellt durch ([4, p. 202])

$$\rho_{k,\xi}(\Phi^*) = \Phi^\wedge\left(\frac{k}{\xi}\right) \qquad (k \in \mathbb{Z}). \tag{15}$$

Die für das Quadrieren von Kernen auf $\mathbb{R}$ (mit anschließender Normierung) relevanten Beziehungen - der Beweis folgt unter Verwendung des Faltungssatzes zur Berechnung von $[\Phi^2]^\wedge(v)$ - werden zusammengestellt als

Lemma 4. *Sei*

$$\Phi(-t) = \Phi(t) \in L^1(\mathbb{R}), \quad \int_0^\infty \Phi(t)dt = \frac{\pi}{2};$$

$$\Phi^\wedge(v) := \frac{2}{\pi}\int_0^\infty \Phi(t)\cos vt\, dt \qquad (v \in \mathbb{R}); \tag{16}$$

dann gilt

$$Q(t) := \frac{\{\Phi(x)\}^2}{N(\Phi)}, \quad \int_0^\infty Q(t)dt = \frac{\pi}{2}; \; N(\Phi) = \int_0^\infty \{\Phi\hat{}(t)\}^2 dt;$$

$$Q\hat{}(v) = \frac{1}{N(\Phi)} \int_0^\infty \Phi\hat{}(v-t)\Phi\hat{}(t)dt \qquad (v \in \mathbb{R}).$$

Neben den periodischen FT-Kernen (14), die über (13) und (15) charakterisiert sind, spielen - durch die Anzahl und Bedeutung der Beispiele bedingt - Kerne vom *gestörten* Fejér-Typ (gFT) eine wichtige Rolle: einem Kern, der nicht FT ist, läßt sich ein FT-Kern zuordnen, so daß - grob gesprochen - beide Kerne bis auf eine Störung höherer Ordnung dasselbe Approximationsverhalten (identischer Saturationsgrenzwert) aufweisen; [21]. Dieser Zusammenhang wird durch die folgenden Beispiele weiter verdeutlicht.

Bemerkung 8. Ausgehend vom $\mathcal{D}$-Kern auf $\mathbb{R}$ erhält man mit Lemma 4 den $\mathcal{F}$-Kern (FT). Durch Quadrieren des $\mathcal{F}$-Kerns erhält man *nicht* - wie in Analogie zum periodischen Fall zu erwarten gewesen wäre - den $\mathcal{J}$-Kern, sondern den Kern von (Jackson-) de La Vallée Poussin; jedoch ist der ursprüngliche $\mathcal{J}$-Kern im Verhältnis zum $\mathcal{JV}$-Kern ein gFT-Kern; [21, p. 357 f].

Bemerkung 9. Ausgehend vom Dirichlet-Kern $\Phi(\mathcal{D};t)$ erhält man die Kopplung

$$\frac{1}{2}\{\Phi(\mathcal{D};t-\alpha) + \Phi(\mathcal{D};t+\alpha)\} = \frac{\alpha.\sin\frac{\alpha}{2}\cos\frac{x}{2} - x\sin\frac{x}{2}\cos\frac{\alpha}{2}}{\alpha^2 - x^2} \qquad (\alpha \in \mathbb{R});$$

mit der sich sofort anbietenden Spezialisierung $\alpha = \pi$ folgt

$$(17) \qquad \Phi(\mathcal{R};x) = \frac{\pi\cos\frac{x}{2}}{\pi^2 - x^2} \in L^1(\mathbb{R});$$

die Fouriertransformierte errechnet sich direkt über (16) zu

$$(18) \qquad \Phi\hat{}(\mathcal{R};v) = \begin{cases} \cos\pi v, & |v| \leq 1/2, \\ 0, & |v| \geq 1/2, \end{cases}$$

woraus über (15) mit $v = k/2n$, $n \in \mathbb{N}$, $1 \leq k \leq n$, sofort die Konvergenzfaktoren $\rho_{k,n-1}(\mathcal{R})$ des periodischen $\mathcal{R}$-Kerns gegeben sind. Man kann also

(17) als Rogosinski-Kern auf $\mathbb{R}$ ansehen. Während dieser *oszillierende* Kern in der Approximationstheorie bisher nicht auftrat, spielt das Paar (17), (18) in den technischen Anwendungen eine gewisse Rolle; siehe z.B. [5, p. 27 (2.54)]. - Quadrieren von (17) führt nun mit Lemma 4 zum Kern von (Bohman-) Zheng Wei-xing (Tab., (7)), dessen periodischer FT-Version der K-Kern als gFT-Kern zur Seite steht; [21, p. 354 f]. Damit dürfte - als durch die Konstruktion begründeter Analogieschluß, im Vergleich mit dem periodischen Fall - die Rolle des Kerns von Fejér-Korovkin auf $\mathbb{R}$ durch den Kern von (Bohman-) Zheng Wei-xing eingenommen werden. Der letztere Kern wird bei der Approximation von Funktionen f mit $f(x)/(1+x^4) \in L^1(\mathbb{R})$ erneut über Extremaleigenschaften definiert; siehe [25, p. 353, p. 361], [14], [26, p. 21 f].

Bemerkung 10. Ein dem vorhergehendem Beispiel ähnlicher Zusammenhang ergibt sich für die Kerne P,G,A der Tab., (8) - (10); siehe auch [1], [20, p. 446, j=5(G), j=6(A)], [21, p. 359 f]. - Auf $\mathbb{R}$ stellt der Weierstrass-Kern (Tab., (12)) ein Beispiel eines sich (im Prinzip: "Reduktion" $x \to x/\sqrt{2}$) selbst reproduzierenden Kerns dar.

Tabelle.

Abkürzungen:

FT	Fejér-Typ
gFT	gestörter Fejér-Typ
K	Kopplung
Q	Quadrieren und Normieren
$\theta_3(x)$	Jacobische Theta-Funktion

Poisson-$\sum$-Formel: Spezifikation des Parameters ξ in (15)

Nr.	Abk.	Kern; Konvergenzfaktoren; $[-\pi,\pi]$	
1	D	Dirichlet $D_n(x) = \frac{\sin(2n+1)\frac{x}{2}}{2\sin(x/2)}$; $\rho_{k,n} = \begin{cases} 1, & 1 \leq k \leq n \\ 0, & k \geq n+1 \end{cases}$	
2	F	Fejér $F_{n-1}(x) = \frac{1}{2n}\left(\frac{\sin n\frac{x}{2}}{\sin\frac{x}{2}}\right)^2$; $\rho_{k,n-1} = \begin{cases} 1-\frac{k}{n}, & 1 \leq k \leq n-1 \\ 0, & k \geq n \end{cases}$	
3	J	Jackson $J_{2n-2}(x) = \frac{3}{2n(2n^2+1)}\left(\frac{\sin(nx/2)}{\sin(x/2)}\right)^4$; $\rho_{k,2n-2}$: s. (8)	
4	JV	Jackson - de La Vallée Poussin $P_{2n-1}(x) = \frac{2+\cos x}{4n^2}\left(\frac{\sin(nx/2)}{\sin(x/2)}\right)^4$; $\rho_{k,2n-1} = \Phi^{\wedge}(\frac{k}{n})$	
5	R	Rogosinski $R_{n-1}(x) = \frac{\sin\frac{\pi}{2n}\cos nx}{2(\cos x - \cos\frac{\pi}{2n})}$; $\rho_{k,n-1} = \begin{cases} \cos\frac{k\pi}{2n}, & 1 \leq k \leq n-1 \\ 0, & k \geq n \end{cases}$	
6	K	Fejér - Korovkin $K_{n-2}(x) = \frac{\sin^2(\pi/n)\cos^2(nx/2)}{n(\cos x - \cos(\pi/n))^2}$; $\rho_{k,n-2}$: s. (11)	
7	Z	Bohman - Zheng Wei-xing $Z_{n-1}(x)$: [21, p. 354]; $\rho_{k,n-1} = \Phi^{\wedge}(\frac{k}{n})$	
8	P	Abel - Poisson $p_r(x) = \frac{1-r^2}{2(1-2r\cos x + r^2)}$; $\rho_{k,r} = r^k$ $(r \to 1-)$	
9	G	Ghermanesco $G_r(x) = \frac{2(1-r^2)}{1+r^2}\,p_r^2(x)$; $\rho_{k,r} = \left\{1 + k\,\frac{1-r^2}{1+r^2}\right\} r^k$	
10	A	Anghelutza $A_r(x)$: [20, p. 446, j = 6]; $\rho_{k,r} = \left\{1 + k \log\frac{1}{r}\right\} r^k$	
11	V	de La Vallée Poussin $V_n(x) = \frac{(2n)!!}{2(2n-1)!!}\cos^{2n}\frac{x}{2}$; $\rho_{k,n} = \begin{cases} \frac{(n!)^2}{(n-k)!(n+k)!}, & 1 \leq k \leq n \\ 0, & k \geq n+1 \end{cases}$	
12	W	Weierstrass $\Theta_3(x) = \frac{1}{2} + \sum_{k=1}^{\infty} e^{-k^2 t}\cos kx$ $(t \to 0+)$	

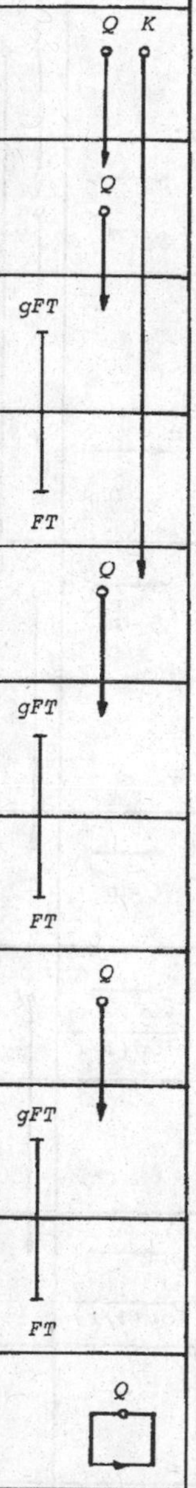

Poisson - $\sum$ - Formel		$\Phi(x);\ \int_0^\infty \Phi(x)dx = \frac{\pi}{2}$	$\Phi^\wedge(v) = \frac{2}{\pi}\int_0^\infty \Phi(x)\cos vx\, dx;$ $\mathbb{R}$	Abk.:
	Q K	$\frac{1}{2}\left(\frac{\sin\frac{x}{2}}{\frac{x}{2}}\right)$	./.	D
← $\xi = n$	Q	$\frac{1}{2}\left(\frac{\sin\frac{x}{2}}{\frac{x}{2}}\right)^2$	$\begin{cases} 1-\lvert v\rvert, & \lvert v\rvert \leq 1 \\ 0, & \lvert v\rvert > 1 \end{cases}$	F
		./.		J
← $\xi = n$		$\frac{3}{4}\left(\frac{\sin\frac{x}{2}}{\frac{x}{2}}\right)^4$	$\begin{cases} 1-\frac{3}{2}v^2+\frac{3}{4}\lvert v\rvert^3, & \lvert v\rvert \leq 1 \\ (2-\lvert v\rvert)^3/4, & 1 \leq \lvert v\rvert \leq 2\ ;\ 0,\ \lvert v\rvert \geq 2 \end{cases}$	JV
← $\xi = n$	Q	$\pi\,\frac{\cos\frac{x}{2}}{\pi^2 - x^2}$	$\begin{cases} \cos \pi v, & \lvert v\rvert \leq 1/2 \\ 0, & \lvert v\rvert \geq 1/2 \end{cases}$	R
		./.		K
← $\xi = n$		$4\pi^2\,\frac{\cos^2\frac{x}{2}}{(\pi^2 - x^2)^2}$	$\begin{cases} (1-\lvert v\rvert)\cos \pi v + \frac{1}{\pi}\sin \pi\lvert v\rvert, & \lvert v\rvert \leq 1 \\ 0, & \lvert v\rvert \geq 1 \end{cases}$	Z
← $\xi = \frac{1}{\log(1/r)}$	Q	$\frac{1}{1+x^2}$	$e^{-\lvert v\rvert}$	P
		./.		G
← $\xi = \frac{1}{\log(1/r)}$		$\frac{2}{(1+x^2)^2}$	$(1+\lvert v\rvert)e^{-\lvert v\rvert}$	A
		./.		V
← $\xi = \frac{1}{\sqrt{t}}$	Q	$\frac{\sqrt{\pi}}{2}\, e^{-x^2/4}$	e^{-v^2}	W

5. Literatur.

[1] ANGELESCO, M., Sur un procédé de sommation des séries trigonométriques. C.R. Acad. Sci. Paris 165 (1917) 419-422.

[2] ANGHELUTZA, T., Une remarque sur l'intégrale de Poisson. Bull. Sci. Math. Bibliothèque École Hautes Études = Darboux Bull. (2) 48 (1924) 138-140.

[3] BASKAKOV, V.A., On operators which, in the best way, approximate twice differentiable functions (Russ.). In: Application of Functional Analysis in Approximation Theory, III (Russ.). (Ed. A.L. Garkavi, A.V. Efimov, V.N. Nikol'skiĭ, A.M. Rubinov) Kalinin. Gos. Univ., Kalinin 1974, 145 pp.; pp. 30-32.

[4] BUTZER, P.L. - R.J. NESSEL, Fourier Analysis and Approximation, I. New York - London - Basel - Stuttgart 1971, xvi + 553 pp.

[5] CHAMPENEY, D.C., Fourier Transforms and Their Physical Applications. London - New York 1973, x + 256 pp.

[6] DZJADYK, V.K., On the application of linear methods to the approximation by polynomials of functions which are solutions of Fredholm integral equations of the second kind, I (Russ.). Ukrain. Mat. Ž. 22 (1970) 461-480 = Transl. Ukrain. Math. J. 22 (1970) 394-410.

[7] DZJADYK, V.K. - V.T. GAVRILJUK - A.I. STEPANECK, On approximation of Hölder class functions by Rogozinsky's polynomials (Ukrain.; Engl. sum.). Dopovīdī Akad. Nauk Ukrain. RSR 1969, no. 3 (1969) 203-206.

[8] DZJADYK, V.K. - V.T. GAVRILJUK - A.I. STEPANECK, Exact upper bound for approximations on classes of differential periodic functions using Rogosinski polynomials (Russ.). Ukrain. Mat. Ž. 22 (1970) 481-493 = Transl. Ukrain. Math. J. 22 (1970) 411-421.

[9] EGERVÁRY, E., Verschärfung eines Harnackschen Satzes und anderer Abschätzungen für nichtnegative harmonische Polynome. Math. Z. 34 (1932) 741-757.

[10] FAVARD, J., Cours D'Analyse de L'École Polytechnique, II. Paris 1960, 578 pp.

[11] FEINERMAN, R.P. - D.J. NEWMAN, Polynomial Approximation. Baltimore 1972, viii + 148 pp.

[12] GALBRAITH, A.S. - J.W. GREEN, A note on the mean value of the Poisson kernel. Bull. Amer. Math. Soc. 53 (1947) 314-320.

[13] GHERMANESCO, M., Sur l'intégrale de Poisson. Bull. Sci. Math. École Polytechnique de Timişoara 4, fasc. 3-4 (1932) 159-184.

[14] GÖRLICH, E. - E.L. STARK, Über beste Konstanten und asymptotische Entwicklungen positiver Faltungsintegrale und deren Zusammenhang mit dem Saturationsproblem. Jber. Deutsch. Math.-Verein. 72 (1970) 18-61.

[15] NATANSON, I.P., Konstruktive Funktionentheorie. Berlin 1955, xiv + 515 pp.

[16] PINNEY, E., On a note of Galbraith and Green. Bull. Amer. Math. Soc. 54 (1948) 527.

[17] ROGOSINSKI, W., Über die Abschnitte trigonometrischer Reihen. Math. Ann. 95 (1925) 110-134.

[18] ROGOSINSKI, W., Reihensummierung durch Abschnittskoppelungen, I. Math. Z. 25 (1926) 132-149.

[19] ROGOSINSKI, W., Fourier Series. New York 1959, vi + 176 pp.

[20] STARK, E.L., On a generalization of Abel-Poisson's sigular integral having kernels of finite oscillation. Studia Sci. Math. Hungar. 7 (1972) 437-449.

[21] STARK, E.L., Nikolskiĭ constants for positive singular integrals of perturbed Fejér-type. In: Linear Operators and Approximation (Proc. Conf. Math. Res. Inst. Oberwolfach, Black Forest, 14.-22.8.1971; Ed. P.L. Butzer - J.P. Kahane - B. Sz.-Nagy; ISNM 20). Basel-Stuttgart 1972, 506 pp.; pp. 348-363.

[22] STEPANECK, A.I., Approximation of continuous periodic functions by means of one linear method (Ukrain.; Engl. sum.). Dopovīdī Akad. Nauk Ukrain. RSR 1974, no. 5 (1974) 408-412.

[23] TIMAN, A.F., On the Lebesgue constants for certain methods of summation (Russ.). Dokl. Akad. Nauk (N.S.) 61 (1948) 989-992.

[24] TIMAN, A.F. - M.M. GANZBURG, On the convergence of certain processes for the summation of Fourier series (Russ.). Dokl. Akad. Nauk. (N.S.) 63 (1948) 619-622.

[25] ZHENG WEI-XING, On the extreme property of the operator $B_\sigma(f;x)$ (Chin.). Acta Math. Sinica 16 (1966) 54-62 = Transl. Chinese Math. 6 (1965) 353-362.

[26] ŽUK, V.V. - G.I. NATANSON, On the accuracy of the approximation of periodic functions by linear methods (Russ.; Engl. sum.). Vestnik Leningrad. Univ. no. 13 Mat. Meh. Astronom. Vyp. 3 (1975) 19-24.

Dr. Eberhard L. Stark
Lehrstuhl A für Mathematik
Rheinisch-Westfälische Technische Hochschule
Templergraben 55

D 5100 Aachen
Bundesrepublik Deutschland

CHARAKTERISIERUNG DER BESTEN ALGEBRAISCHEN APPROXIMATION DURCH LOKALE LIPSCHITZBEDINGUNGEN

R.L. STENS

1. Einleitung

Will man den Fundamentalsatz der besten Approximation 2π-periodischer stetiger Funktionen durch trigonometrische Polynome, nämlich die Äquivalenz der Sätze von Jackson und Bernstein, auf die Approximation von Funktionen, die auf einem endlichen Intervall, etwa $[-1,1]$, definiert sind, durch algebraische Polynome übertragen, so stellt sich heraus, daß der Bernstein-Satz nicht mehr gilt. In [2,3] wurde gezeigt, daß man durch Modifikation der Definition von "Ableitung" und "Stetigkeitsmodul" dennoch zu einem Satz gelangen kann, der dem trigonometrischen entspricht. Natürlich möchte man dabei diese neuen Begriffe durch die entsprechenden klassischen ausdrücken.

Man definiert dazu für $f \in C[-1,1]$ einen Translationsoperator τ_h und einen Differenzenoperator $\overline{\Delta}_h^\alpha$ durch

$$(\tau_h f)(x) := \tfrac{1}{2}\{f(xh+\sqrt{(1-x^2)(1-h^2)}) + f(xh-\sqrt{(1-x^2)(1-h^2)})\} \quad (x,h\in[-1,1]),$$

$$(\overline{\Delta}_h^\alpha f)(x) := (-1)^{[\alpha]} \sum_{j=0}^{\infty} (-1)^j \binom{\alpha}{j} (\tau_h^j f)(x) \quad (x,h\in[-1,1]).$$

Dabei ist α eine beliebige positive Zahl und $[\alpha]$ die größte ganze Zahl kleiner oder gleich α. Existiert nun eine Funktion $g \in C[-1,1]$, für die gilt

$$\lim_{h\to 1-} \left\| \frac{(\overline{\Delta}_h^\alpha f)(x)}{(1-h)^\alpha} - g(x) \right\|_C = 0,$$

dann heißt g die Chebyshev-Ableitung der Ordnung $\alpha>0$ und wird mit $D^\alpha f$ bezeichnet.

Entsprechend definiert man einen Chebyshev-Stetigkeitsmodul durch

$$\omega_\alpha^T(f;\eta;C) := \sup_{\eta\leq h\leq 1} \|\overline{\Delta}_h^\alpha f\|_C \quad (0\leq\eta\leq 1)$$

und die zugehörige Lipschitzklasse der Ordnung $\sigma>0$ mit Index $\alpha>0$ als

$$\mathrm{Lip}_\alpha^T(\sigma;C) := \{f\in C[-1,1];\ \omega_\alpha^T(f;\eta;C) = O((1-\eta)^\sigma), \eta\to 1-\}.$$

Bezeichnet man die Menge der algebraischen Polynome vom Grade $\leq n$ mit P_n, dann erhält man als Spezialfall eines allgemeineren Ergebnisses (vgl. [3])

<u>*Satz 1.*</u> *Ist* $f\in C[-1,1]$, $r\in \mathbb{P}=\{0,1,\ldots\}$, $0<\sigma<1$, *dann sind äquivalent:*

(i) $$E_n(f;C) := \inf_{p_n\in P_n} \|f-p_n\|_C = O(n^{-r-\sigma}) \qquad (n\to\infty)$$

(ii) $$D^{r/2}f\in \mathrm{Lip}_1^T(\sigma/2;C).$$

Unser Ziel ist es nun, die Bedingung (ii) mittels der üblichen Ableitungen und durch Lipschitzbedingungen, die über die gewöhnliche Translation definiert sind, zu charakterisieren. Wir benutzen dabei einen lokalen Stetigkeitsmodul, der wahrscheinlich zum ersten Mal von A.L. Fuksman [4] benutzt wurde, und dessen Ergebnis im wesentlichen in unserem enthalten ist. Darüber hinaus können wir aber aufgrund unserer gegenüber [4] modifizierten Beweistechnik zusätzlich eine Aussage vom Stečkin-Typ beweisen.

<u>2. Der lokale Stetigkeitsmodul</u>

<u>*Definition 1.*</u> *Sei* $f\in C[-1,1]$, $0<\delta<1$, *dann heißt*

(2.1) $$\Omega(f;\delta;x;C) := \sup\{|f(x+h)-f(x)|;\ h\in[-\delta,\delta]\cap[-1-x,1-x]\}$$

der lokale Stetigkeitsmodul von f *an der Stelle* $x\in[-1,1]$.

Im Gegensatz zum üblichen Stetigkeitmodul

$$\omega(f;\delta;C) := \sup\{|f(x+h)-f(x)|;\ x, x+h\in[-1,1],\ |h|\leq\delta\}$$

wird das Supremum in (2.1) bei festem $x\in[-1,1]$ über die verschiedenen Werte von h gebildet, die den Bedingungen $|h|\leq\delta$ und $x+h\in[-1,1]$ genügen. Es gilt

$$\sup_{x\in[-1,1]} \Omega(f;\delta;x;C) = \omega(f;\delta;C) \qquad (0<\delta<1).$$

Über den Stetigkeitsmodul Ω kann man jetzt eine lokale Lipschitzklasse definieren. Mit $\Lambda(x;\delta) := \mathrm{Min}\{\delta^2/(1-x^2),\delta\}$, $x \in [-1,1]$, $0<\delta<1$, setzen wir

$$(2.2)\quad \mathrm{P-Lip}_1(\sigma;C) := \{f \in C[-1,1];\ \Omega(f;\delta;x;C) = O([\Lambda(x;\delta)]^\sigma),\ \delta\to 0+\}.$$

Fuksman [4] benutzt anstelle von Λ die Funktion $\Lambda_1(x;\delta) := \delta/[(1-x^2)^{1/2} + \delta^{1/2}]$. Das folgende Lemma zeigt neben einer Ungleichung, die wir später benötigen, daß man in (2.2) Λ durch Λ_1^2 ersetzen kann.

Lemma 1. *Für* $0<\delta<1$ *gilt*

(i) $$[\Lambda_1(x;\delta)]^2 \leq \Lambda(x;\delta) \leq 4[\Lambda_1(x;\delta)]^2 \qquad (x \in [-1,1]);$$

(ii) $$\mathrm{Min}\left\{\frac{\delta^2}{(1-\delta)^2 - x^2}, \delta\right\} \leq 36\,\Lambda(x;\delta) \qquad (|x| \leq 1-\delta).$$

Beweis. Die linke Ungleichung in (i) gilt wegen

$$\left(\frac{\delta}{(1-x^2)^{1/2} + \delta^{1/2}}\right)^2 \leq \frac{\delta^2}{(1-x^2) + \delta} \leq \mathrm{Min}\left\{\frac{\delta^2}{1-x^2}, \delta\right\},$$

und die rechte Seite folgt aus

$$\left(\frac{\delta}{(1-x^2)^{1/2} + \delta^{1/2}}\right)^2 \geq \frac{\delta^2}{(2\,\mathrm{Max}\{(1-x^2)^{1/2}, \delta^{1/2}\})^2} = \frac{1}{4}\,\mathrm{Min}\left\{\frac{\delta^2}{1-x^2}, \delta\right\}.$$

Um (ii) zu zeigen, betrachte man die Funktion

$$g_\delta(x) = \frac{\delta}{[(1-\delta)^2 - x^2]^{1/2} + \delta^{1/2}}\,[\Lambda_1(x;\delta)]^{-1} \qquad (|x| \leq 1-\delta).$$

g_δ ist monoton steigend auf $[0,1-\delta]$, und wegen $g_\delta(x) = g_\delta(-x)$ gilt $0 \leq g_\delta(x) \leq g_\delta(1-\delta) = (2-\delta)^{1/2} + 1 \leq 3$ für $|x| \leq 1-\delta$, d.h.

$$(2.3)\quad \frac{\delta}{[(1-\delta)^2 - x^2]^{1/2} + \delta^{1/2}} \leq 3\,\Lambda_1(x;\delta) \qquad (|x| \leq 1-\delta).$$

Ersetzt man auf der rechten Seite von (i) jetzt x durch $(x^2 + 2\delta - \delta^2)^{1/2}$, benutzt (2.3) und die linke Seite von (i), so folgt (ii).

Neben den Stetigkeitsmoduln für Funktionen, definiert auf $[-1,1]$, benötigen wir noch solche für periodische Funktionen. Sei $F \in C_{2\pi}$, der Menge aller auf $\mathbb{R}$ stetigen 2π-periodischen Funktionen, dann setzen wir

$$\omega_1(F;\delta;C_{2\pi}) = \sup_{|\varphi|\leq\delta} \|F(\cdot+\varphi) - F(\cdot)\|_{C_{2\pi}}$$

$$\omega_2(F;\delta;C_{2\pi}) = \sup_{|\varphi|\leq\delta} \|F(\cdot+\varphi) + F(\cdot-\varphi) - 2F(\cdot)\|_{C_{2\pi}}$$

$$\mathrm{Lip}_r(\sigma;C_{2\pi}) = \{F \in C_{2\pi};\ \omega_r(F;\delta;C_{2\pi}) = O(\delta^\sigma),\ \delta\to 0+\} \qquad (r=1,2).$$

Die Wahl der Funktion Λ in (2.2) erscheint natürlich zunächst sehr willkürlich. Wir wollen jetzt zeigen, daß aufgrund dessen $\mathrm{Lip}_1^T(\sigma;C) = P-\mathrm{Lip}_1(\sigma;C)$ für $0<\sigma<1/2$ gilt. Den ersten Schritt dazu liefert

Lemma 2. *Sei* $f \in P-\mathrm{Lip}_1(\sigma;C)$ *für ein* $\sigma>0$, *dann folgt* $f\circ\cos \in \mathrm{Lip}_1(2\sigma;C_{2\pi})$.

Beweis. Aus der Taylorformel erhält man $|\cos(\theta+\varphi) - \cos\theta| \leq |\varphi\sin\theta| + \varphi^2$ und daraus für $\theta \in \mathbb{R}$ und $0<|\varphi|<1/2$

$$\begin{aligned}|f(\cos(\theta+\varphi)) - f(\cos\theta)| &\leq \Omega(f;|\cos(\theta+\varphi)-\cos\theta|;\cos\theta;C)\\ &\leq \Omega(f;|\varphi\sin\theta| + \varphi^2;\cos\theta;C)\\ &\leq M\left(\mathrm{Min}\left\{\frac{(|\varphi\sin\theta| + \varphi^2)^2}{\sin^2\theta},\ |\varphi\sin\theta| + \varphi^2\right\}\right)^\sigma.\end{aligned}$$

Für $|\sin\theta| \geq |\varphi|$ gilt $(|\varphi\sin\theta| + \varphi^2)^2/\sin^2\theta \leq (|\varphi|+|\varphi|)^2 = 4\varphi^2$, und für $|\sin\theta| \leq |\varphi|$ gilt $|\varphi\sin\theta| + \varphi^2 \leq 2\varphi^2$, so daß für alle $\theta \in \mathbb{R}$ und $0<|\varphi|<1/2$ folgt

$$|f(\cos(\theta+\varphi)) - f(\cos\theta)| \leq 4M|\varphi|^{2\sigma},$$

womit die Behauptung bewiesen ist.

Bevor wir unser erstes Hauptergebnis beweisen, benötigen wir noch eine verallgemeinerte Form der Bernstein- und Markov-Ungleichung (vgl. [4]).

Lemma 3. *Sei* $p_n \in P_n$, *dann gilt für* $m,j \in \mathbb{P}$, $0\leq m\leq j$, *mit einer Konstanten* $M(j)$, *die nur von j abhängt.*

(2.4) $$|p_n^{(j)}(x)| \leq M(j)n^{(j+m)}(1-x^2)^{(m-j)/2}\|p_n\|_C \qquad (x\in[-1,1]).$$

Beweis. Wir benutzen zwei Ungleichungen, die man unmittelbar aus [5, S. 227 (51)] erhält, und zwar

(2.5) $$|p_n^{(k)}(x)| \leq M(k)\left(\frac{n}{(1-x^2)^{1/2}}\right)^k \|p_n\|_C \qquad (x\in[-1,1];k\in\mathbb{P}),$$

(2.6) $$\|p_n^{(k)}\|_C \leq M(k)n^{2k}\|p_n\|_C \qquad (k\in\mathbb{P}).$$

Wendet man zunächst (2.5) auf $p_n^{(m)}$ mit $k = j-m$ an und dann (2.6) mit $k = m$, so folgt unmittelbar (2.4).

Satz 2. *(a)* $P-\mathrm{Lip}_1(\sigma;C)\subset \mathrm{Lip}_1^T(\sigma;C)$ $(\sigma>0)$.

(b) $P-\mathrm{Lip}_1(\sigma;C) = \mathrm{Lip}_1^T(\sigma;C)$ $(0<\sigma<1/2)$.

Beweis. (a) Aus $f\in P-\mathrm{Lip}_1(\sigma;C)$ folgt nach La. 2 $f\circ\cos\in \mathrm{Lip}_1(2\sigma;C_{2\pi})$ und somit auch $f\circ\cos\in\mathrm{Lip}_2(2\sigma;C_{2\pi})$, was zu $f\in\mathrm{Lip}_1^T(\sigma;C)$ äquivalent ist (vgl. [2, La. 3]).

(b) Es bleibt noch $\mathrm{Lip}_1^T(\sigma;C)\subset P-\mathrm{Lip}_1(\sigma;C)$ für $0<\sigma<1/2$ zu zeigen. Ist also $f\in\mathrm{Lip}_1^T(\sigma;C)$, dann folgt nach Satz 1 mit $r=0$ die Existenz einer Folge von Polynomen $p_n\in\mathcal{P}_n$, $n\in\mathbb{P}$, mit

(2.7) $$\|f-p_n\|_C = O(n^{-2\sigma}) \qquad (n\to\infty).$$

Setzt man $Q_0 = p_1$ und $Q_\nu = p_{2^\nu} - p_{2^{\nu-1}}$, $\nu\in\mathbb{N}$, dann gilt $\sum_{\nu=0}^\infty Q_\nu(x) = f(x)$ gleichmäßig auf $[-1,1]$, und aus (2.7) folgt die Abschätzung

(2.8) $$\|Q_\nu\| \leq M\,2^{-2\nu\sigma} \qquad (\nu\in\mathbb{P}).$$

Mit La. 3 und (2.8) erhält man somit für beliebiges $\mu\in\mathbb{N}$ mit einem geeigneten $\theta\in(0,1)$ für $m=0,1$ sowie $x,x+h\in[-1,1]$

$$|f(x+h)-f(x)| = \left|\sum_{\nu=0}^{\mu}(Q_\nu(x+h)-Q_\nu(x)) + \sum_{\nu=\mu+1}^{\infty}(Q_\nu(x+h)-Q_\nu(x))\right|$$

$$\leq |h|\sum_{\nu=0}^{\mu}|Q_\nu'(x+\theta h)| + 2\sum_{\nu=\mu+1}^{\infty}\|Q_\nu\|_C \leq$$

$$\leq M_1\Big(|h|(1-(x+\theta h)^2)^{(m-1)/2} \sum_{\nu=0}^{\mu} 2^{\nu(1+m-2\sigma)}$$

$$+ \sum_{\nu=\mu+1}^{\infty} 2^{-2\nu\sigma}\Big).$$

Mit Hilfe der Ungleichungen $\sum_{\nu=0}^{\mu} 2^{\nu(1+m-2\sigma)} \leq M_2\, 2^{\mu(1+m-2\sigma)}$ und $\sum_{\nu=\mu+1}^{\infty} 2^{-2\nu\sigma} \leq M_3\, 2^{-2\mu\sigma}$ folgt daraus für m=0 bzw. m=1

$$(2.9) \quad |f(x+h)-f(x)| \leq M_4\{|h|(1-(x+\theta h)^2)^{-1/2}\, 2^{\mu(1-2\sigma)} + 2^{-2\mu\sigma}\}$$

$$(2.10) \quad |f(x+h)-f(x)| \leq M_5\{|h| 2^{2\mu(1-\sigma)} + 2^{-2\mu\sigma}\}.$$

Sei nun $0<\delta<1$, $|x|<1-\delta$, und wählt man $\mu \in \mathbb{N}$, so daß

$$2^{\mu-1} < [(1-\delta)^2 - x^2]^{1/2}/\delta \leq 2^{\mu}$$

gilt, dann liefert (2.9) für $|h|\leq\delta$ wegen $1-(x+\theta h)^2 \geq (1-\delta)^2 - x^2$

$$(2.11) \quad \Omega(f;\delta;x;C) \leq M_6\left(\frac{\delta^2}{(1-\delta)^2 - x^2}\right)^{\sigma} \qquad (|x|\leq 1-\delta;0<\delta<1).$$

Wählt man dagegen $2^{\mu-1} < \delta^{-1/2} \leq 2^{\mu}$, dann folgt aus (2.10)

$$(2.12) \quad \Omega(f;\delta;x;C) \leq M_7\, \delta^{\sigma} \qquad (x\in[-1,1];0<\delta<1).$$

Kombiniert man (2.11) und (2.12) dann folgt mit La. 1 (ii)

$$(2.13) \quad \Omega(f;\delta;x;C) \leq M_8[\Lambda(x;\delta)]^{\sigma} \qquad (|x|\leq 1-\delta;0<\delta<1).$$

Für $1-\delta\leq|x|\leq 1$ läßt sich (2.12) ebenfalls in der Form (2.13) schreiben, d.h. (2.13) gilt sogar für alle $x\in[-1,1]$, womit der Satz vollständig bewiesen ist.

3. Der Fundamentalsatz der besten algebraischen Approximation

In Abschnitt 2 haben wir die Klasse $\mathrm{Lip}_1^T(\sigma;C)$ mit Hilfe des lokalen Stetigkeitsmoduls Ω beschrieben. Dieses Ergebnis liefert zwar eine gute Charakterisierung von $f\in \mathrm{Lip}_1^T(\sigma;C)$, ist aber für Aussagen vom Typ $D^{r/2}f\in \mathrm{Lip}_1^T(\sigma;C)$, $r\in\mathbb{N}$, noch unbefriedigend. Diese Fälle wollen wir im folgenden näher untersuchen.

Sei $f \in C[-1,1]$, $r,j \in \mathbb{P}$, wir setzen

$$\psi_{j,r}(f;x) := \begin{cases} f^{(j)}(x) & , \quad 0 \leq j \leq r/2 \\ f^{(j)}(x)(1-x^2)^{j-r/2} & , \quad r/2 < j \leq r, \end{cases}$$

für die $x \in (-1,1)$, für die die rechte Seite sinnvoll ist. Weiter setzen wir

$$\left.\begin{aligned} \psi_{j,r}(f;1) &:= \lim_{x\to 1-} \psi_{j,r}(f;x) \\ \psi_{j,r}(f;-1) &:= \lim_{x\to(-1)+} \psi_{j,r}(f;x) \end{aligned}\right\} \qquad (0 \leq j \leq r),$$

sofern diese Grenzwerte existieren.

Mittels dieser Funktionen läßt sich die Bedingung $(f \circ \cos)^{(r)} \in \mathrm{Lip}_1(2\sigma;C_{2\pi})$, die wir für $r=0$ schon im Beweis von Satz 2 betrachtet haben, durch die Klasse $P-\mathrm{Lip}_1(\sigma;C)$ beschreiben.

<u>*Lemma 4.*</u> *(a) Ist* $f \in C[-1,1]$ *im Punkt* $x \in (-1,1)$ *r-mal differenzierbar, dann ist* $f \circ \cos$ *in jedem Punkte* $\theta \in \mathbb{R}$ *mit* $\cos\theta = x$ *r-mal differenzierbar, und es gilt*

$$(3.1)\qquad (f\circ\cos)^{(r)}(\theta) = \sum_{j=0}^{[r/2]} t_{j,r}(\theta) f^{(j)}(\cos\theta) + \sum_{j=[r/2]+1}^{r} t_{j,r}(\theta) f^{(j)}(\cos\theta)\sin^{2j-r}\theta$$

mit irgendwelchen trigonometrischen Polynomen $t_{j,r}$.

(b) Ist $f \in C[-1,1]$ *und erfüllen die Funktionen* $\psi_{j,r}$ *die Bedingungen*

$$(3.2)\qquad \psi_{j,r}(f;\cdot) \in C[-1,1] \qquad (0 \leq j \leq r),$$

$$(3.3)\qquad \psi_{j,r}(f;1) = \psi_{j,r}(f;-1) = 0 \qquad (r/2 < j \leq r),$$

$$(3.4)\qquad \psi_{j,r}(f;\cdot) \in P-\mathrm{Lip}_1(\sigma;C) \qquad (0 \leq j \leq r),$$

für ein $\sigma \in (0,1/2)$, *dann ist* $f \circ \cos$ *auf* $\mathbb{R}$ *r-mal differenzierbar, und es gilt*

$$(f \circ \cos)^{(r)} \in \mathrm{Lip}_1(2\sigma; C_{2\pi}). \tag{3.5}$$

Beweis. (a) erhält man unmittelbar aus der Kettenregel sowie durch vollständige Induktion über r. Sei jetzt (3.2) erfüllt, dann ist f r-mal stetig differenzierbar auf (-1,1) und mit (a) folgt, daß $f \circ \cos$ auf jedem Intervall $(\nu\pi, (\nu+1)\pi)$, $\nu \in \mathbb{Z} = \{0, \pm 1, \pm 2, \ldots\}$, stetig differenzierbar ist. Setzt man jetzt $s_j(\theta) := \sin^{2j-r}\theta$, dann ist noch die Existenz folgender Grenzwerte zu zeigen:

$$\lim_{\theta \to \nu\pi} (f \circ \cos)^{(r)}(\theta) = \lim_{\theta \to \nu\pi} \sum_{j=0}^{[r/2]} t_{j,r}(\theta) f^{(j)}(\cos\theta) + \sum_{j=[r/2]+1}^{r} t_{j,r}(\theta) f^{(j)}(\cos\theta) s_j(\theta) \qquad (\nu \in \mathbb{Z}). \tag{3.6}$$

Dies folgt aber aus der Stetigkeit der $\psi_{j,r}(f;\cdot)$ in den Punkten ± 1, wobei man im Falle ungerader r noch (3.3) benötigt.

Zum Beweis der Lipschitzbedingung erhalten wir aus (3.4) und La. 2

$$\psi_{j,r}(f;\cos\cdot) = f^{(j)}(\cos\cdot) \in \mathrm{Lip}_1(2\sigma; C_{2\pi}) \qquad (0 \leq j \leq r/2), \tag{3.7}$$

$$\psi_{j,r}(f;\cos\cdot) = f^{(j)}(\cos\cdot)|s_j(\cdot)| \in \mathrm{Lip}_1(2\sigma; C_{2\pi}) \qquad (r/2 < j \leq r). \tag{3.8}$$

Man zeigt leicht, daß aus (3.8) auch

$$f^{(j)}(\cos\cdot)s_j(\cdot) \in \mathrm{Lip}_1(2\sigma; C_{2\pi}) \qquad (r/2 < j \leq r)$$

folgt, und Darstellung (3.1) für $(f \circ \cos)^{(r)}$ liefert dann (3.5).

Ist f nun eine Funktion, die den Bedingungen (3.2), (3.3) und (3.4) genügt, dann folgt nach dem gerade Bewiesenen und dem Jackson-Satz für die beste trigonometrische Approximation (vgl. [1, Thm. 2.2.3]) die Existenz einer Folge gerader trigonometrischer Polynome t_n vom Grade $\leq n$ mit

$$\| f \circ \cos - t_n \|_{C_{2\pi}} = O(n^{-r-2\sigma}) \qquad (n \to \infty).$$

Die Substitution $\theta = \arccos x$ liefert dann wegen $t_n \circ \arccos \in P_n$

$$E_n(f;C) \leq \| f - t_n \circ \arccos \|_C = \| f \circ \cos - t_n \|_{C_{2\pi}} = O(n^{-r-2\sigma}) \qquad (n \to \infty).$$

Mit Satz 1 erhält man daraus

Folgerung 1. *Unter den Voraussetzungen von La. 4 (b) an die Funktionen* $\psi_{j,r}$ *gilt* $D^{r/2} f \in \mathrm{Lip}_1^T(\sigma;C)$.

Die Umkehrung zu diesem Ergebnis beweisen wir in zwei Schritten. Zunächst eine Aussage vom Stečkin-Typ.

Lemma 5. *Sei* $D^{r/2} f \in \mathrm{Lip}_1^T(\sigma;C)$ *für ein* $r \in \mathbb{P}$ *und ein* $\sigma \in (0,1/2)$, *dann erfüllen die Funktionen* $\psi_{j,r}$ *die Bedingungen (3.2), (3.3) und für die Polynome bester Approximation zu* $f \in C[-1,1]$ *aus* P_n, $p_n^* = p_n^*(f)$, *gilt*

$$\text{(3.9)} \quad \| \psi_{j,r}(f;x) - (p_n^*)^{(j)}(x) \|_C = O(n^{2j-r-2\sigma}) \qquad (n \to \infty; 0 \leq j \leq r/2)$$

$$\text{(3.10)} \quad \| \psi_{j,r}(f;x) - (p_n^*)^{(j)}(x)(1-x^2)^{j-r/2} \|_C = O(n^{-2\sigma}) \qquad (n \to \infty; r/2 < j \leq r).$$

Beweis. Mit $P_{n,\nu}(x) = p^*_{n2^\nu}(x)$, $\nu \in \mathbb{P}$, erhalten wir

$$\text{(3.11)} \qquad \sum_{\nu=0}^{k} (P_{n,\nu}(x) - P_{n,\nu+1}(x)) = p_n^*(x) - p^*_{n2^{k+1}}(x) \qquad (k \in \mathbb{P}).$$

Wendet man auf die einzelnen Summanden La. 3 an, so folgt

$$\text{(3.12)} \qquad \left| \frac{d^j}{dx^j} [P_{n,\nu}(x) - P_{n,\nu+1}(x)] \right|$$

$$\leq M(j) n^{j+m} 2^{\nu(j+m)} (1-x^2)^{(m-j)/2} \{ \| P_{n,\nu} - f \|_C + \| P_{n,\nu+1} - f \|_C \}.$$

Nach Voraussetzung gilt mit Satz 1 $\| P_{n,\nu} - f \|_C = O((n2^\nu)^{-r-2\sigma})$, $n \to \infty$, und aus (3.12) folgt

$$\text{(3.13)} \qquad \left| (1-x^2)^{(j-m)/2} \frac{d^j}{dx^j} [P_{n,\nu}(x) - P_{n,\nu+1}(x)] \right|$$

$$\leq M_1(j) n^{j+m-r-2\sigma} 2^{-\nu(r+2\sigma-j-m)} \qquad (x \in [-1,1]; j \in \mathbb{P}; 0 \leq m \leq j).$$

Sei jetzt $j+m \leq r$, $m \leq j$, dann folgt aus (3.13), daß die Reihe

$$\sum_{\nu=0}^{\infty} \frac{d^j}{dx^j} [P_{n,\nu}(x) - P_{n,\nu+1}(x)]$$

gleichmäßig auf jedem kompakten Teilintervall von (-1,1) konvergiert. Wegen der gleichmäßigen Konvergenz von (3.11) gegen $p_n^*(x)-f(x)$ für $k\to\infty$, liefert dies die Existenz von $f^{(j)}(x)$ auf (-1,1) und

$$(3.14)\qquad \sum_{\nu=0}^{\infty} (1-x^2)^{(j-m)/2} \frac{d^j}{dx^j} [P_{n,\nu}(x) - P_{n,\nu+1}(x)]$$

$$= (1-x^2)^{(j-m)/2} [(p_n^*)^{(j)}(x) - f^{(j)}(x)] \qquad (x \in (-1,1)).$$

Weiter folgt aus der gleichmäßigen Konvergenz der linken Seite von (3.14)

$$(3.15)\qquad \lim_{|x|\to 1-} (1-x^2)^{(j-m)/2} f^{(j)}(x)$$

$$= \lim_{|x|\to 1-} (1-x^2)^{(j-m)/2} (p_n^*)^{(j)}(x)$$

$$- \sum_{\nu=0}^{\infty} \lim_{|x|\to 1-} (1-x^2)^{(j-m)/2} \frac{d^j}{dx^j} [P_{n,\nu}(x) - P_{n,\nu+1}(x)].$$

Aus (3.13) und (3.14) erhält man noch

$$(3.16)\qquad \| (1-x^2)^{(j-m)/2} [(p_n^*)^{(j)}(x) - f^{(j)}(x)] \|_C$$

$$\leq M_1(j) n^{j+m-r-2\sigma} \sum_{\nu=0}^{\infty} 2^{-\nu(r+2\sigma-j-m)} = O(n^{j+m-r-2\sigma}) \qquad (n\to\infty).$$

Aus den Gleichungen (3.14), (3.15), (3.16), die alle nur unter den Nebenbedingungen $j+m\leq r$, $m\leq j$ gelten, können wir durch geeignete Wahl von m die Behauptung ableiten. Wählen wir zunächst $m=j$, dann folgt für $j\leq [r/2]$ aus (3.14) die Existenz und Stetigkeit von $f^{(j)}(x)$ auf (-1,1), aus (3.15) die Stetigkeit auf [-1,1], d.h. es gilt (3.2) für $0\leq j\leq r/2$, und (3.16) liefert (3.9). Mit $m=r-j$, $r/2<j\leq r$, folgt aus (3.14) die Stetigkeit der $f^{(j)}$ auf (-1,1) und aus (3.15) folgen die Randbedingungen

$$\lim_{|x|\to 1-} (1-x^2) f^{(j)}(x) = \psi_{j,r}(f;\pm 1) = 0,$$

womit (3.2) für $r/2<j\leq r$ und (3.3) gezeigt sind. Aus (3.16) erhält man schließlich noch (3.10).

Um die Umkehrung zu Folgerung 1 zu vervollständigen, werden wir im nächsten Satz aus (3.9) und (3.10) auf (3.4) schließen.

<u>*Lemma 6.*</u> *Sei* $f\in C[-1,1]$, *so daß die* $\psi_{j,r}(f;\cdot)$ *die Bedingungen (3.2) und (3.3) erfüllen. Existiert eine Folge von Polynomen* $p_n\in P_n$, $n\in \mathbb{P}$, *und ein* $\sigma\in(0,1/2)$ *mit,*

$$(3.17)\qquad \|\psi_{j,r}(f;x)-p_n^{(j)}(x)\|_C=O(n^{2j-r-2\sigma})\qquad (n\to\infty;0\leq j\leq r/2),$$

$$(3.18)\qquad \|\psi_{j,r}(f;x)-p_n^{(j)}(x)(1-x^2)^{j-r/2}\|_C=O(n^{-2\sigma})\quad (n\to\infty;r/2<j\leq r),$$

dann gilt $\psi_{j,r}(f;\cdot)\in P-\mathrm{Lip}_1(\sigma;C)$, $0\leq j\leq r$.

<u>Beweis.</u> Aus (3.17) folgt

$$E_n(\psi_{j,r}(f;\cdot);C)\leq\|\psi_{j,r}(f;\cdot)-p_n^{(j)}(\cdot)\|_C=O(n^{-2\sigma})\qquad(n\to\infty;0\leq j\leq r/2),$$

und Satz 1 liefert zusammen mit Satz 2(b) $\psi_{j,r}\in P-\mathrm{Lip}_1(\sigma;C)$, $0\leq j\leq r/2$.

Für $r/2<j\leq r$ setzen wir $x=\cos\theta$ in (3.18), und mit $s_j(\theta)=\sin^{2j-r}\theta$ ergibt dies

$$(3.19)\qquad \|f^{(j)}(\cos\theta)|s_j(\theta)|-p_n^{(j)}(\cos\theta)|s_j(\theta)|\|_{C_{2\pi}}=O(n^{-2\sigma})$$
$$(n\to\infty;r/2<j\leq r).$$

In (3.19) kann man aber $|s_j(\theta)|$ durch $s_j(\theta)$ ersetzen, d.h.

$$\|f^{(j)}(\cos\theta)s_j(\theta)-p_n^{(j)}(\cos\theta)s_j(\theta)\|_{C_{2\pi}}=O(n^{-2\sigma})\qquad(n\to\infty;r/2<j\leq r).$$

Daraus folgt zum einen die Stetigkeit von $f^{(j)}(\cos\theta)s_j(\theta)$, und da $p_n^{(j)}(\cos\theta)s_j(\theta)$ ein trigonometrisches Polynom vom Grade $\leq n$ ist, folgt mit dem Bernstein-Satz für die beste trigonometrische Approximation (vgl. [1, Thm. 2.3.4])

$$f^{(j)}(\cos\theta)s_j(\theta)\in \mathrm{Lip}_1(2\sigma;C_{2\pi})\qquad(r/2<j\leq r).$$

Damit gilt aber auch

$$f^{(j)}(\cos\theta)|s_j(\theta)| = f^{(j)}(\cos\theta)(1-\cos^2\theta)^{j-r/2} \in \mathrm{Lip}_1(2\sigma;C_{2\pi}) \qquad (r/2<j\leq r),$$

was wiederum nach [2, La. 3] und Satz 2(b) zu

$$f^{(j)}(x)(1-x^2)^{j-r/2} = \psi_{j,r}(f;x) \in P-\mathrm{Lip}_1(\sigma;C) \qquad (r/2<j\leq r)$$

äquivalent ist.

Mit diesem Ergebnis läßt sich nun Satz 1 folgendermaßen erweitern.

<u>*Satz 3.*</u> *Sei* $f \in C[-1,1]$, $r \in \mathbb{P}$, $0<\sigma<1$, *dann sind folgende Aussagen äquivalent:*

(i) $D^{r/2}f \in \mathrm{Lip}_1^T(\sigma/2;C)$

(ii) *die Funktionen* $\psi_{j,r}(f;\cdot)$ *erfüllen folgende Bedingungen:*

$$\psi_{j,r}(f;\cdot) \in C[-1,1], \qquad (0\leq j\leq r),$$

$$\psi_{j,r}(f;\pm 1) = 0, \qquad (r/2<j\leq r),$$

$$\psi_{j,r}(f;\cdot) \in P-\mathrm{Lip}_1(\sigma/2;C), \qquad (0\leq j\leq r).$$

(iii) *die Funktionen* $\psi_{j,r}(f;\cdot)$ *und die Polynome bester Approximation,* $p_n^* = p_n^*(f)$, *erfüllen folgende Bedingungen:*

$$\psi_{j,r}(f;\cdot) \in C[-1,1], \qquad (0\leq j\leq r),$$

$$\psi_{j,r}(f;\pm 1) = 0, \qquad (r/2<j\leq r),$$

$$\|\psi_{j,r}(f;x) - (p_n^*)^{(j)}(x)\|_C = O(n^{2j-r-\sigma}), \qquad (n\to\infty; 0\leq j\leq r/2),$$

$$\|\psi_{j,r}(f;x) - (p_n^*)^{(j)}(x)(1-x^2)^{j-r/2}\|_C = O(n^{-\sigma}), \quad (n\to\infty; r/2<j\leq r).$$

<u>Beweis.</u> Die Aussage (i) ⇒ (iii) folgt aus La. 5, (iii) ⇒ (ii) ist durch La. 6 bewiesen, und Folgerung 1 liefert (ii) ⇒ (i).

Bemerkung. In (ii) kann man die zweite Bedingung für gerade r fallenlassen, und in der dritten Bedingung kann man sich auf $(r/2\leq j\leq r)$ beschränken.

4. Literatur

[1] BUTZER, P.L. - R.J. NESSEL, Fourier Analysis and Approximation, Vol. 1. Birkhäuser Verlag, Basel; Academic Press, New York, 1971.

[2] BUTZER, P.L. - R.L. STENS, Chebyshev transform methods in the theory of best algebraic approximation. Abh. Math. Sem. Univ. Hamburg 45 (1976), 165-190.

[3] BUTZER, P.L. - R.L. STENS, Fractional Chebyshev operational calculus and best algebraic approximation. In: Proc. Symposium on Approximation Theory, Jan. 1976, Austin, Texas. (Im Druck)

[4] FUKSMAN, A.L., Structural characteristic of functions such that $E_n(f;-1,1)\leq Mn^{-(k+\alpha)}$. (Russ.) Uspehi Mat. Nauk 20 (1965), 187-190.

[5] TIMAN, A.F., Theory of Approximation of Functions of a Real Variable. Pergamon Press, Oxford-London-New York-Paris, 1963.

R.L. Stens
Lehrstuhl A für Mathematik
Rheinisch-Westfälische Technische Hochschule

D 5100 Aachen
Bundesrepublik Deutschland

Approximative properties of splines

Ju. N. Subbotin (Sverdlovsk)

Let $\tilde{C}[a,b]$ denote the space of continuous functions $f(x)$ with period $(b-a)$ and norm $\|f\| = \max_{a\le x\le b} |f(x)|$, $\tilde{C}^{(k)}[a,b]$ the space of all continuous and along with k-th derivative $(b-a)$-periodic functions, put $\tilde{C}^{(o)}[a,b] = \tilde{C}[a,b]$. We consider the approximative properties of $(b-a)$-periodic interpolation splines $S_{2n-1,\tau}(f;x)$ of degree $2n-1$ and deficiency 1 with knots

$$a = x_{o,\tau} < x_{1,\tau} < \ldots < x_{i,\tau} < \ldots < x_{i_2,\tau} < \ldots < x_{\tau,\tau} = b \ , \qquad (1)$$

where

$$x_{i,\tau} - x_{i-1,\tau} = h_{i,\tau} = \begin{cases} h_{1,\tau} (i=1,2,\ldots,i_1) \\ h_{2,\tau} (i=i_1+1,\ldots,i_2) \\ h_{1,\tau} (i=i_2+1,\ldots,\tau) \ , \end{cases} \qquad (2)$$

$h_{1,\tau} \cdot h_{2,\tau}^{-1}$ equals 1 or 2 $\tau \in \mathbb{N}$, $\mathbb{N}$ the set of all natural numbers.

The next result holds.

Theorem 1. Let $f \in \tilde{C}[a,b]$ and the $(b-a)$-periodic spline $S_{2n-1,\tau}(f;x)$ of degree $2n-1$ and deficiency 1 interpolate $f(x)$ in nodes (1), then

$$\|S_{2n-1,\tau}(f;x)\| \le K_n \|f\| \ (\tau \in \mathbb{N}) \qquad (3)$$

with K_n independent of both f and τ *.

In the proof of the theorem we use a representation of splines $S_{2n-1,\tau}(f;x)$ in a form of $(b-a)$-periodic B-splines which are defined as follows (see [1] - [4]).

---

* In the sequel K_n, $K_{n,p}$ are constants, different in general, dependent only on what indexed.

Set $\tilde{B}_{i,2n-1,\tau}(x)=(\bar{x}_{i+2n-1}-\bar{x}_{i-1,\tau})[\bar{x}_{i-1,\tau},\ldots,\bar{x}_{i+2n-1,\tau};(t-x)_+^{2n-1}]$,

where $[\bar{x}_0,\bar{x}_1,\ldots,\bar{x}_{2n};(t-x)_+^{2n-1}]$ is a divided difference of order 2n on the set $\bar{x}_0,\bar{x}_1,\ldots,\bar{x}_{2n}$, with respect to t for a function $(t-x)_+^{2n-1} = \max[0,(t-x)^{2n-1}]$, and

$$\bar{x}_{i+\tau s,\tau} = x_{i,\tau}+(b-a)s \quad (i=0,1,\ldots,\tau;\ s=0,\pm 1,\ldots). \tag{4}$$

Let $B_{i,2n-1,\tau}(x)$ be (b-a)-periodic extension of a function $\alpha_i\tilde{B}_{i,2n-1,\tau}(x)$ where $\alpha_0=1$ and α_i are such that $\alpha_i B_{i,2n-1,\tau}(x_{i-1+n})=\tilde{B}_{0,2n-1,\tau}(x_{n-1})\ (i=1,2,\ldots,\tau-1)$.

For $B_{i,2n-1,\tau}(x)$ there will be used a notation $B_i(x)$ as well. Then any (b-a)-periodic splines $S_{2n-1,\tau}(x)$ of degree 2n-1 and deficiency 1 with knots (1) may be put in a form

$$S_{2n-1,\tau}(x) = \sum_{i=n+1}^{\tau-n} c_i B_i(x) \tag{5}$$

in unique manner, with $c_{-p} = c_{\tau-p}\ (p=1,2,\ldots,n-1)$.

Lemma 1. For any $\tau\in\mathbb{N}$ the following inequality holds:

$$\max_{0\le i\le\tau-1} |C_i^*| \le K_n \max_{0\le i\le\tau-1} |d_i|, \tag{6}$$

where K_n is a constant dependent only on n, $\{C_i^*\}_{i=0}^{\tau-1}$ being a solution of the linear system

$$\sum_{i=0}^{\tau-1} C_i^* B_i(x_k) = d_k\ (k=0,1,\ldots,\tau-1). \tag{7}$$

Validity of theorem 1 follows directly from Lemma 1 and from the fact that at any x only 2n-1 functions $B_i(x)$ don't vanish.

Theorem 2. Let $f\in\hat{C}^{(k)}[a,b]\ (0\le k\le 2n+2)$ then

$$\|f^{(p)}(x)-S^{(p)}_{2n+1,\tau}(f;x)\| \le K_{p,n}\omega_{2n+2-p}(f^{(p)},\delta_\tau)\ ; \tag{8}$$

in particular

$$\|f^{(p)}(x)-S^{(p)}_{2n+1,\tau}(x)\| \le 2^{k-p}K_{p,n}h^{k-p}\omega_{2n+2-k}(f^{(k)},\delta_\tau) \tag{9}$$

$(p=0,1,\dots,k)$ where $\delta_\tau=\max(h_{1,\tau},h_{2,\tau})$ and $\omega_k(f,\delta)$ is a smothness modulus of order k in the uniform metric.

Proof. When $i=0$ the inequality (8) follows from Lebesgue inequality $\|f(x)-S_{2n+1,\tau}(f;x)\| \le \|f(x)-S_{2n+1,\tau}(x)\|$ $(\|S_{2n+1,\tau}(f)\|_{\tilde{C}}^{\hat{C}}+1)$ and from well known results [5]-[7] on the orders of approximation by the best splines. The proof in case of $1\le p\le k$ leans on the following statement.

Theorem 3. Under assumptions of Theorem 2

$$\|S^{(p)}_{2n-1,\tau}(f;x)\| \le K_{p,n}\|f^{(p)}\| \quad (p=1,2,\dots,k). \tag{10}$$

Let, indeed, be $n_p=p+2(2n-p)(p-1)$, $\bar{h}_{ip}=h_{i+1,\tau}n_p^{-1}$ and $x'_{i,j}=x_{i,\tau}+j\bar{h}_{i,p}$ $(j=0,1,\dots,n_p)$ then using (9) we obtain

$$|\bar{h}_{i,p}^{-p}\Delta^p_{\bar{h}_{i,p}} S_{2n-1,\tau}(f,x_{i,2jp})|=|S^{(p)}_{2n-1,\tau}(f,\xi_{2jp})| \le$$

$$\le |\bar{h}_{i,p}^{-p}\Delta^p_{\bar{h}_{i,p}} [S_{2n-1,\tau}(f,x_{i,2jp})-f(x_{i,2jp})]| +$$

$$+ |\bar{h}_{i,p}^{-p}\Delta^p_{\bar{h}_{i,p}} f(x_{i,2jp})| \le K_{p,n}\|f^{(\tau)}(x)\|,$$

where $\Delta^p_h f(x) = \sum_{s=0}^{p}(-1)^{p-s}C^s_p f(x+sh)$ and $x_{i,2jp}<\xi_{2jp}<x_{i(2j+1)p}$ $(j=0,1,2,\dots,2n-1)$. Noting $|\xi_{2j+2,p}-\xi_{2j,p}|\ge\bar{h}_{i,p}$, we then construct an interpolation Lagrange polynomial $S^{(p)}_{2n-1,\tau}(f,x)$ $(x_{ip}<x<x_{i+1,\tau})$ of degree $2n-1-p$ with nodes ξ_{2jp}, and arrive to (10).

The use of both (10) and the result on orders of approximation by the best splines mentioned above gives (8) in case $i=1,2,\dots,k$.

Theorem 4. Let $f(x)$ be from $\tilde{C}[a,b]$ and $S_{n,\tau}$ be a set of $(b-a)$-periodic splines of degree $2n-1$ and deficiency 1 with knots (1). If $S_{n,\tau}(f,x)_2 \epsilon S_{n,\tau}$ is the best approximant to $f(x)$ in $L_2[a,b]$ then

$$\|S_{n,\tau}(f;x)\| \leq K_n \|f\| . \tag{11}$$

Obviously the proof of (11) may be based on considering only functions with mean value equal to zero.

Let $F(x)$ be the $(n+1)$-fold periodic integral of $f(x)$ with $\int_a^b f(x)dx=o$ and $S_{2n+1,\tau}(F,x)$ be a spline from $S_{2n+1,\tau}$ interpolating $F(x)$ in (1). Since in case of periodic functions and periodic interpolation -splines of odd degree and deficiency 1 the first integral relation, and therefore, the best approximation property holds [8] then from this and from $S^{(n+1)}_{2n+1,\tau}(F,x)\epsilon S_{n,\tau}$ and validity of (10) $i=k=n+1$ substituting F for f one arrives to (11).

Theorem 4 for the case of nonperiodic functions and nonperiodic splines was proven in [9],[10].

Theorems 2 and 4 make possible to expand Z. Ciesielski's and J. Domsta's results [10] on the existence of orthogonal normed bases in a space of nonperiodic k-times differentiable functions to the space of periodic functions. Let $\hat{C}(-\pi,\pi)$ be a subspace of even functions in $\tilde{C}(-\pi,\pi)$; then in particular the following statement holds.

<u>Theorem 5.</u> For any $m \epsilon \mathbb{N}$ there exists in $\hat{C}(-\pi,\pi)$ an orthonormal basis which consists of functions $\phi_{i,m}(x)=\phi_i(x)\epsilon\hat{C}(-\pi,\pi)$ and is such that

$$\|f(x)-S_\tau(f,x)\| \leq K_m \omega_{m+1}(f,\tfrac{1}{\tau}), \tag{12}$$

$S_\tau(f,x)$ the partial sum of the expansion $f(x) = \sum_{i=o}^{\infty} c_i(f)\phi_i(x)$.

In the proof of theorem 5 we use a method of construction of bases realised in [10]. As imbedded τ-dimensional subspaces, $\tau=1,2,\ldots$, we pick up subspaces $\hat{S}_{2n-1,\tau}$ of all, with respect to $x=o$, even splines of degree $2n-1$ and deficiency 1 with knots

$$x_{i,\tau} = \begin{cases} \pi_i 2^{-\nu-1} & , i=0,\pm1,\pm2,\ldots,\pm2\mu \\ \pi_i 2^{-\nu} & , i=\pm(\mu+1),\pm(\mu+2),\ldots,\pm2^{\nu}, \end{cases}$$

where $\tau=2^{\nu}+\mu$, $0\le\mu<2^{\nu}$; $\mu,\nu\in\mathbb{N}$.

The best approximation-properties of interpolation splines in L_2 (see the proof of theorem 4) and theorem 2 yield (12) in the case $m=2n-1$. Inequality $\omega_{m+1}(f,\frac{1}{\tau}) \le 2\omega_m(f,\frac{1}{\tau})$ leads to (12) in case of even m.

In what follows, bases with good properties of convergence for analytic-function-spaces will be constructed. The basis existence problem for separable Banach spaces goes back to S. Banach [11]. It turns out (see [12]), that in general the the problem has a negative solution.

Let A^k denote the space of functions $f(z)$ analytic in unit circle with k-th derivative continuous in closed circle $|z|\le1$. This space is normed in the usual way

$$\|f(z)\|_k = \max_{o\le s\le k} \max_{|z|=1} |f^{(s)}(z)|.$$

Theorem 6. For any integers $k\ge o$, $m\ge o$ in A^k there exists a basis $\{g_s(z)\}_{s=o}^{\infty}$ and

$$\|f(z)-S_p(f,z)\|_k \le K_m[\omega_{m+1}(u,\tfrac{1}{p})+\omega_{m+1}(v,\tfrac{1}{p})] \qquad (13)$$

where $u(t)=\mathrm{Re}f^{(k)}(e^{it})$, $v(t)=\mathrm{Im}f^{(k)}(e^{it})$ and $S_p(f,z)$ is the partial sum of order p of decomposition f in the basis.

S. V. Bockarjov [13] has proven theorem 6 if k=m=o. Let $\phi_p(t) = \phi_{p,2n-1}(t)$ (p=o,1,...) be the orthonormal basis from theorem 5, $\phi_o(t) = (2\tau)^{1/2}$, $\phi_{p,2n-1}(t) \subset \hat{S}_{2n-1,p}$. Put

$g_0(t) = \frac{1+i}{2\sqrt{\pi}}$, $i=\sqrt{-1}$ and $g_s(t) = \frac{1}{\sqrt{2}}[\phi_s(t)+i\tilde{\phi}_s(t)]$ where $\tilde{\phi}_s(t)$ is conjugate to $\phi_s(t)$. Since $\phi_s(t)$ is continuously differentiable then $\tilde{\phi}_s(t) \in \tilde{C}(-\pi,\pi)$. Set

$$g_s(z)=g_{s,2n-1}(z)=\frac{1}{2\pi}\int_{-\pi}^{\pi} g_s(t)P(\tau,x-t)dt \quad (s=o,1,\ldots) \tag{14}$$

where $P(\tau,x)$ is the Poisson kernel, $z = \tau e^{ix}$.

We show that the set of functions (14) constitutes a basis in A^0 and the estimate (13) holds with k=o. The set (14) consists of orthogonal functions with respect to the inner product defined as

$$(g_s,g_q) = \int_{-\pi}^{\pi} g_s(e^{ix})\bar{g}_q(e^{ix})dx .$$

Let $f(z) \in A^0$; then $f(e^{ix})=u(x)+iv(x)+\mathrm{Im}f(o)$ where both $u(x)$ and $v(x)$ are real valued continuous 2π-periodic functions and $v(x) = \tilde{u}(x)$. If

$$S_p(f,e^{ix}) = \sum_{s=o}^{p} c_s g_s(e^{ix})$$

where

$$c_s = c_s(f) = \int_{-\pi}^{\pi} f(e^{ix})\bar{g}_s(e^{ix})dx,$$

then analogously [13] we have

$$S_p(f,e^{ix})=\frac{1}{2\pi}\int_{\pi}^{\pi} u_1(x)dx + \sum_{s=1}^{p}(u_1,\phi_s)\phi_s(x)-\sum_{s=1}^{p}(v_1,\phi_s)\tilde{\phi}_s(x) +$$

$$+ i\mathrm{Im}f(o)+i\left[\sum_{s=1}^{p}(v_1,\phi_s)\phi_s(x)+\sum_{s=1}^{p}(u_1,\phi_s)\tilde{\phi}_s(x)\right] \tag{15}$$

where $u_1(x)=\frac{1}{2}[u(x)+u(-x)|$, $v_1(x)=\frac{1}{2}[v(x)+v(-x)]$,

$u_2(x)=u(x)-u_1(x), v_2(x)=v(x)-v_1(x), (u,\phi_s)=\int_{-\pi}^{\pi} u(x)\phi_s(x)dx.$

Since $u_1(x)$ is even, then due to Theorem 5

$$\|u_1(x)-\frac{1}{2}\int_{-\pi}^{\pi} u_1(x)dx-\sum_{s=1}^{p}(u_1,\phi_s)\phi_s(x)\| \leq K_n\omega_{2n}(u,\frac{1}{p}) \tag{16}$$

Set $S_p(v,x) = \sum_{s=o}^{p}(v_1,\phi_s)\phi_s(x)$.

We show in the manner of [13] that

$$\|\tilde{S}_p(v_1,x)\| \leq K_n[\|v_1\|+\|\tilde{v}_1\|] \quad (p\in\mathbb{N}). \tag{17}$$

Let $p=2^\nu+\mu$, $o\leq\mu<2^\nu$, $\nu=o,1,2,\ldots$; since the Dirichlet kernel

$$K_p(x,y) = \sum_{s=o}^{p}\phi_s(x)\phi_s(y)$$

for the set of functions $\{\phi_s(x)\}$ is a spline of degree 2n-1 in each variable with the distance between knots not less than $(2p)^{-1}$, the application of Markov's inequality

$$\|P'_{2n-1}(x)\|_{C[a,b]} \leq K_n(b-a)^{-1}\int_b^a |P_{2n-1}(x)|dx$$

to algebraic polynomials of degree 2n-1 leads to

$$\int_o^{1/p} t^{-1}|K_p(x+t,y)-K_p(x-t,y)|dt \leq \int_o^{1/p} t^{-1}\int_{-t}^{t}|K'_p(x+u,y)|du\, dt \leq$$

$$\leq K_n\cdot p\int_{5/p}^{5/p}|K_p(x+t,y)|dt. \tag{18}$$

We prove the theorem for the case m=2n-1. The proof in case of even m follows that of Theorem 5. Let $\phi_s(t) = \phi_{s,2n-1}(t)$ be formed in the same way as in Theorem 5. Since $\{\phi_s(t)\}$ constitutes a basis in $\tilde{C}(-\pi,\pi)$, then

$$\max_{-\pi\leq x\leq\pi}\int_{-\pi}^{\pi}|K_p(x,y)|dy \leq K_n \quad \text{for any } p. \quad \text{Next,}$$

$$\int_{-\pi}^{\pi}\Big|\int_{-\pi}^{\pi}K_p(x,y)g(y)dy\Big|dx\le \int_{-\pi}^{\pi}|g(y)|\int_{-\pi}^{\pi}|K_p(x,y)|dy\,dx\le K_n\int_{-\pi}^{\pi}|g(y)|dy.$$

Therefore S_p is bounded operator from $L[-\pi,\pi]$ in to $L[-\pi,\pi]$.

Using the known results [5]-[7] on orders of approximation by the best splines valid for periodic functions as well, for any $g(y)$ from $L(-\pi,\pi)$ we have

$$\int_{-\pi}^{\pi}|g(y)-S_p(g,y)|dy\le K_n\omega(g,\tfrac{1}{p})_1 , \qquad (19)$$

where $S_p(g,y)$ is the partial sum of order p of decomposition of a function g with respect to the orthogonal basis

$\{\phi_{s,2n-1}(t)\}$ and $\omega(g,\delta)_1 = \sup\limits_{|t|\le\delta}\int_{-\pi}^{\pi}|g(x+t)-g(x)|dx.$

From (19), in analogy of [13], we obtain (16).

Let us denote $\tau_p(\nu_1,x)$ the Vallée-Poussin mean,

$$\tau_p(\nu_1,x)=\frac{1}{p}[\hat{S}_p(\nu_1,x)+\hat{S}_{p+1}(\nu_1,x)+\ldots+\hat{S}_{2p-1}(\nu_1,x)],$$

where

$$\hat{S}_p(\nu_1,x)=\sum_{s=1}^{p}a_s\cos s\,x,\quad a_s=\frac{1}{\pi}\int_{-\pi}^{\pi}\nu_1(x)\cos s\,x\,dx.$$

It is known [14],[15], that

$$\|\nu_1(x)-\tau_p(\nu_1,x)\|\le 4E_p(\nu_1)\le K_n\omega_{2n}(\nu_1,\tfrac{1}{p}).$$

Moreover

$$\|\tilde{\tau}_p(\nu_1,x)-\tilde{S}_p(\tau_p,x)\|\le\Big\|\frac{1}{\pi}\int_0^{1/p}\frac{1}{2tg\ t/2}\int_{-t}^{t}|R''(x+u)|du\,dt\Big\| +$$

$$+\Big\|\frac{1}{\pi}\int_{1/p}^{\pi}\frac{R'(x+t)-R'(x-t)}{2tg\ t/2}dt\Big\| = I_1+I_2,$$

where $R'(x)=\tau_p(\nu_1,x)-S_p(\tau_p,x)$.

Let $F(x)$ be equal to the $2n+1$-fold periodic integral of $\tau_p'(\nu_1,x)$. Note that $S_p(\tau_p,x)$ coincides with $2n$-th derivative of spline $S_{4n-1,p}(F,x)$ interpolating $F(x)$. From above inequality and Theorem 2 we obtain

$$I_1 \le \frac{2}{\pi p} \| R''(x) \| \le \frac{K_n}{p} \omega_{2n-1}(\tau'_p, \frac{1}{p}).$$

Integrating by parts and using Theorem 2 we have

$$I_2 \le \| \frac{R(x+t)-R(x-t)}{2\mathrm{tg}\,\frac{t}{2}} \Big|_{t=1/p}^{\pi} \| + \|R(x)\| K \int_{1/p}^{\pi} \frac{dt}{t^2} \le K_n \omega_{2n}(\tau_p, \frac{1}{p}).$$

With the help of these inequalities we have

$$|\tilde{\nu}_1(x) - \tilde{S}_p(\nu_1, x)| = |\tilde{S}_p(\nu_1 - \tau_p, x) + \tilde{S}_p(\tau_p, x) - \tilde{\tau}_p(\nu_1, x) +$$

$$+ \tau_p(\tilde{\nu}_1, x) - \tilde{\nu}_1(x)| \le K_n[\omega_{2n}(\nu_1, \frac{1}{p}) + \omega_{2n}(\tilde{\nu}_1, \frac{1}{p}) +$$

$$+ \frac{1}{p}\omega_{2n-1}(\tau'_p, \frac{1}{p}) + \omega_{2n}(\tau_p, \frac{1}{p})]. \qquad (20)$$

It easily follows from the result [16], [17] generalizing Bernstein inequalities that

$$\frac{1}{p}\omega_{2n-1}(\tau'_p, \frac{1}{p}) \le K_n \omega_{2n}(\tau_p, \frac{1}{p}).$$

Next,

$$\omega_{2n}(\tau_p, \frac{1}{p}) \le \omega_{2n}(\nu_1, \frac{1}{p}) + 2^{2n} \| \nu_1 - \tau_p(\nu_1, x) \| \le K_n \omega_{2n}(\nu_1, \frac{1}{p}).$$

Thus (19) and $\tilde{\nu}_1(x) = -u_2(x)$ give

$$\| u_2(x) + \tilde{S}_p(\nu_1, x) \| \le K_n[\omega_{2n}(\nu, \frac{1}{p}) + \omega_{2n}(u, \frac{1}{p})]. \qquad (21)$$

It follows from (15),(16) and (21) that

$$\| \mathrm{Re} f(e^{ix}) - \mathrm{Re} S_p(f, e^{ix}) \| \le K_n[\omega_{2n}(u, \frac{1}{p}) + \omega_{2n}(\nu, \frac{1}{p})].$$

In an analogue fashion a similar inequality can be proved for

$$\| \mathrm{Im} f(e^{ix}) - \mathrm{Im} S_p(f, e^{ix}) \|.$$

Thus (13) is proved for k=o. Uniqueness of decomposition follows from orthogonality of $\{g_s(z)\}$. So Theorem 6 is valid for k=o.

Let k=1. Set $Tf(z) = \int_o^z f(w)\,dw$ where the integration is on the straight line between o and z.

Set

$$\theta_s(z) = \begin{cases} z^s/s!\ , & s=o,1,\dots,k-1 \\ T^k g_{s-k}(z)\ , & s=k,k+1,\dots, \end{cases} \qquad (22)$$

where $g_s(z)$ is defined in (14).

Then expanding $f(z) \in A^k$ into a series $\sum_{s=o}^{\infty} c_s \theta_s(z)$ with respect to the inner product

$$(\theta_p, \theta_q) = \sum_{s=o}^{k-1} \theta_p^{(s)}(o)\bar{\theta}_q^{(s)}(o) + \int_{-\pi}^{\pi} \theta_p^{(k)}(e^{ix})\bar{\theta}_q^{(k)}(e^{ix})dx$$

and using the prooven case k=o we easily derive that the theorem is true for any k.

Note that the proof of lemma 1 is based on results of [18], [2] and mainly of [4].

References

1. H.B.Curry, I.J.Schoenberg, On Pólya functions IV: The fundamental spline functions and their limits, J.d'Analyse Math. 17(1966), 71-107.

2. I.J.Schoenberg,Cardinal interpolation and spline functions, J.Approximation Theory, 2(1969). 167-206.

3. J.R.Rice, the approximation of functions, Vol. II, Addison-Wesley, Reading, 1969.

4. J.Domsta, A theorem on B-splines, Studia Math., T. XLI (1972), 291-314.

5. Ju.N.Subbotin, Extremal functional interpolation and spline approximation, Math. zametki, 16, N 5 (1974), 843-854.

6. Ju.N.Subbotin, Extremal functional interpolation and spline approximation, Thesis, Sverdlovsk, 1973.

7. K.Scherer, On Bernstein's inequalities in Banach space, Math.zametki, 17, N 6 (1975), 925-937.

8. J.H.Ahlberg, E.N.Nilson, J.L.Walsh, The theory of splines and their applications, Academic Press, New York and London, 1967.

9. Z.Ciesielski, Properties of the orthonormal Franklin system, II, Studia Math. 27 (1966), 289-323.

10. Z.Ciesielski, J. Domsta, Construction of an orthonormal basis in $C^m(I^d)$ and $W_p^m(I^d)$, Studia Math., T. XLI (1972) 211-224.

11. S.Banach, Théorie des opérations linéaires, Warszawa, 1932.

12. P.Enflo, A conterexample to the approximation problem in Banach spaces, Acta math., Vol. 130, N 3-4 (1973), 309-317.

13. S.V.Bockarjov, Existence of basis in space of analytic in a circle functions and some properties of Franklin's system

(Russian), Math. Sbornik, V.95 (137), 1 (1974), 3-18.

14. I.P.Natanson, Constructive theory of function. M-L., 1949.

15. S.B.Steckin, On the order of best approximation of continuous functions, Isvestia AN USSR (Ser.Math.) 15 (1951), 219-242.

16. S.M.Nikolskij, Generalization of one Bernstein's inequality, Dokl. AN USSR, 60, N 9 (1948), 1507-1510.

17. S.B.Steckin, Generalization of some Bernstein's inequalities, Dokl. AN USSR, 60 (1948), 1511-1514.

18. Ju.N.Subbotin, On the relations between finite differences and the corresponding derivatives, Proc. Steklov Inst. Math., 78 (1965), pp. 24-42.

ON THE APPROXIMATION BEHAVIOR OF THE RIESZ MEANS IN $L^p(R^n)$

Walter Trebels

Starting point of the following considerations is a series of Comptes Rendues Notes and a monograph due to M. Zamansky [15, 16]. Among other things Zamansky carries out the following idea: Given a summation method of a general Cesàro-summable series, describe its approximation behavior by that of an appropriate typical mean as accurate as possible, i.e., "gage" (with respect to the approximation behavior) the given summation method by an appropriate typical mean*); then only the latter ones have to be discussed in detail.

Here we indicate how this idea can be carried over to approximation processes on $L^p(R^n)$, $1 \le p \le \infty$, (where in abuse of notation we interpret $L^\infty = C_0$) which are of convolution type. In particular we are interested in the Riesz means

$$(1) \qquad R_{\alpha,\gamma}(f;\rho) = \int_{R^n} \rho^n r_{\alpha,\gamma}(\rho y) f(x-y)\,dy \quad , \qquad f \in L^p \quad ,$$

where $r_{\alpha,\gamma}$ is the Fourier transform of $r^\wedge_{\alpha,\gamma}(v) = (1-|v|^\gamma)^\alpha_+$.
In the following we always assume that α is chosen sufficiently large (e.g. $\alpha > (n-1)|1/p - 1/2|$) so that $R_{\alpha,\gamma}$ is a bounded operator on L^p : $\|R_{\alpha,\gamma}(f;\rho)\|_p \le C\|f\|_p$.

Now some comparison theorems between different approximation methods are known [8], [2], [13], [4], [12]; e.g. on L^p the approxima-

---

*) M.F. Timan [10], inspired by results on the comparison of approximation processes due to Shapiro and Boman [8], [2], develops a closely related concept by estimating the approximation behavior of a given approximation process in $L^p_{2\pi}$ from above and below by combinations of the best approximation. An advantage of Zamansky's approach is the fact that it is also good at the saturation phenomenon.

tion behavior of the generalized Weierstrass integral

$$W_\gamma(f;\rho) = \int_{R^n} \rho^n w_\gamma(\rho y) f(x-y)\,dy \quad , \qquad w_\gamma^\wedge(v) = e^{-|v|^\gamma} \quad ,$$

is completely equivalent**) to that of the corresponding Riesz means [12; p. 83, p. 58]:

$$(2) \qquad \|W_\gamma(f;\rho) - f\|_p \approx \|R_{\alpha,\gamma}(f;\rho) - f\|_p \; .$$

In view of these comparison theorems we restrict ourselves only to a discussion of the Riesz means $R_{\alpha,\gamma}$ in $L^p(R^n)$, which we will carry out by a combination of Fourier multiplier methods and a modification of Zamansky's methods. Some of the following results can be deduced from general approximation theorems due to Butzer-Scherer [5]. The advantage of the deduction given here is that it is quite elementary and illustrates the power of partial integration and Fourier transformation as methods of proof.
The author is grateful to H. Berens for various useful suggestions.

First we observe that (2) shows that for fixed γ all Riesz means (1) (as long as they are bounded) are equivalent with respect to their approximation behavior; so without loss of generality we may assume α integer.

<u>*Theorem 1.*</u> *a) If $\gamma \geq \delta > 0$, then*

$$\|R_{\alpha,\gamma}(f;\rho)-f\|_p \leq C\|R_{\alpha,\delta}(f;\rho)-f\|_p \; , \quad 1 \leq p \leq \infty \; .$$

b) If $\gamma,\delta>0$, $\Omega(\rho)$ is monotone decreasing in ρ, and

$$\|R_{\alpha,\delta}(f;\rho)-f\|_p = O(\Omega(\rho)) \; , \qquad 1 \leq p \leq \infty \; ,$$

then

$$\|R_{\alpha,\gamma}(f;\rho)-f\|_p \leq \frac{C}{\rho^\gamma}\int_0^\rho t^{\gamma-1}\Omega(t)\,dt \; , \quad 1 \leq p \leq \infty \; .$$

**) If two non-negative functions a(t,f), b(t,f) can be estimated by a(t,f) ≤ Cb(t,f) ≤ C'a(t,f), C,C' being independent of t and f, we write a(t,f) ≈ b(t,f).

Proof. a) is proved by multiplier methods (cf. [12; p. 61]). To verify b) define first $e(t) = (1-t^{\gamma/\delta})_+^{\alpha}$ and observe that by partial integration one has [12; p. 25]

$$e(t^\delta) = C \int_{t^\delta}^{\infty} (s-t^\delta)^\alpha de^{(\alpha)}(s) \ , \quad e(0) = C \int_0^\infty s^\alpha de^{(\alpha)}(s) = 1 \ .$$

Then, by an interchange of integration order, one obtains for $\phi \in S$ (the set of infinitely differentiable functions on R^n, vanishing rapidly at infinity)

$$R_{\alpha,\gamma}(\phi;\rho) - \phi = \int_{R^n} (1-(\tfrac{|v|}{\rho})^\gamma)_+^\alpha \phi^\wedge(v) e^{ix\cdot v} dv - \phi(x)$$

$$= C \int_0^\infty \int_{R^n} \{(1-(\tfrac{|v|}{\rho})^\delta \tfrac{1}{s})_+^\alpha - 1\} \phi^\wedge(v) e^{ix\cdot v} dv \ s^\alpha de^{(\alpha)}(s)$$

$$= C \int_0^\infty \{R_{\alpha,\delta}(\phi;\rho s^{1/\delta}) - \phi\} s^\alpha de^{(\alpha)}(s) \ .$$

Since S is dense in the considered L^p-spaces, there is equality of the left and the right hand side for all $f \in L^p$. So Minkowski's integral inequality, the hypothesis, and an evaluation of the variation of $e^{(\alpha)}(s)$ give finally

$$\|R_{\alpha,\gamma}(f;\rho) - f\|_p \le |C| \int_0^{1+} s^\alpha \Omega(\rho s^{1/\delta}) |de^{(\alpha)}(s)|$$

$$\le C' \frac{1}{\rho^\gamma} \int_0^\rho t^{\gamma-1} \Omega(t) dt \ .$$

In case $\gamma \ge \delta > 0$ part a) of the theorem is clearly stronger than part b). There naturally arises the question: How much stronger? To examine it consider the σ-modulus of continuity due to H.S. Shapiro [8; p. 219] in case of the Riesz means:

$$\|R_{\alpha,\delta}(f;\rho) - f\|_p^* = \sup_{\zeta \ge \rho} \|R_{\alpha,\delta}(f;\zeta) - f\|_p \ .$$

Obviously, this σ-modulus is monotone decreasing and satisfies the relation

$$(3) \quad \|R_{\alpha,\delta}(f;t) - f\|_p^* \le C(1+(\tfrac{\rho}{t})^\delta) \|R_{\alpha,\delta}(f;\rho) - f\|_p^* \ , \quad t > 0$$

which is easy to verify for the generalized Weierstrass integral and, therefore by (2), is also true for the Riesz means. Thus, choosing $\Omega(\zeta) = \|R_{\alpha,\delta}(f;\zeta)-f\|_p^*$, one is led to

$$\|R_{\alpha,\gamma}(f;\zeta)-f\|_p \le C\|R_{\alpha,\delta}(f;\rho)-f\|_p^* \frac{1}{\zeta^\gamma}\int_0^\zeta t^{\gamma-1}(1+(\tfrac{\rho}{t})^\delta)dt$$

which converges uniformly in $\zeta \ge \rho$ in case $0 \le \delta < \gamma$. Thus we obtain a result slightly weaker than Theorem 1a without multiplier methods (which, however, in our deduction are hidden in (3)):

Corollary. *If $\gamma > \delta > 0$, then*

$$\|R_{\alpha,\gamma}(f;\rho)-f\|_p^* \le C\|R_{\alpha,\delta}(f;\rho)-f\|_p^* .$$

Also by the method of Theorem 1a Zamansky-type theorem and its inverse can be deduced.

Theorem 2. *a) Let $\gamma>0$, $1 \le p \le \infty$, and $\phi\in S$, then*

$$(4)\qquad \Big\|\int_{R^n} |v|^\gamma(1-\frac{|v|}{\rho})_+^\alpha \phi^\wedge(v)e^{ix\cdot v}dv\Big\|_p \le C\rho^\gamma\|R_{\alpha,\gamma}(\phi;\rho)-\phi\|_p ,$$

C being independent of $\phi\in S$.

b) Conversely, if the left side of (4) behaves like $O(\Omega(\rho))$, Ω being as in Theorem 1, and $\int_\rho^\infty \Omega(t)dt/t < \infty$, then

$$(5)\qquad \|R_{\alpha,\gamma}(f;\rho)-f\|_p \le C\Big\{\int_\rho^\infty \frac{\Omega(t)}{t}\,dt + \frac{1}{\rho^\gamma}\int_0^\rho t^{\gamma-1}\Omega(t)dt\Big\} .$$

(4) is proved by showing that $|v|^\gamma(1-|v|)_+^\alpha/(1-(1-|v|^\gamma)_+^\alpha)$ belongs to M_p (the set of bounded multipliers for L^p; [9; p. 94]), e.g. with the aid of [12; p. 83]. To prove (5) first establish by partial integration for $m>\alpha$

$$R_{2m,\gamma}(\phi;\rho)-\phi = C\int_0^\infty t^{2m+\gamma}\int_{R^n}(\frac{|v|}{\rho t})^\gamma(1-\frac{|v|}{\rho t})_+^{2m}\phi^\wedge(v)e^{ix\cdot v}dv\,de^{(2m)}(t) ,$$

where $e(t) = t^{-\gamma}\{1-(1-t^\gamma)_+^{2m}\}$. Since trivially $(1-|v|)_+^{2m-\alpha}\in M_p$ and the approximation behavior of $R_{\alpha,\gamma}$ for different parameters α is the same, one arrives at

$$\|R_{\alpha,\gamma}(\phi;\rho)-\phi\|_p \le C'\int_0^\infty t^{2m+\gamma}\Big\|\int_{R^n}(\frac{|v|}{\rho t})^\gamma(1-\frac{|v|}{\rho t})_+^\alpha\phi^\wedge(v)e^{ix\cdot v}dv\Big\|_p|de^{(2m)}(t)| ,$$

from which the assertion follows if one uses the hypothesis $\|\ldots\|_p = O(\Omega(\rho t))$.
We conclude this discussion with a slight generalization of (4), proved analogously, namely $(\gamma,\delta>0)$

$$(4') \qquad \Big\|\int_{R^n} |v|^{\gamma}\Big(1-\Big(\frac{|v|}{\rho}\Big)^{\delta}\Big)_+^{\alpha}\phi^{\wedge}(v)e^{ix\cdot v}dv\Big\|_p \le C\rho^{\gamma}\|R_{\alpha,\gamma}(\phi;\rho)-\phi\|_p \; .$$

More interesting are estimates of $\|R_{\alpha,\gamma}(f;\rho)-f\|_p^*$ from above and below by the standard modulus of continuity:

$$w_{m,p}(f;t) = \sup_{|h|\le t}\|\Delta_h^m f\|_p \; , \qquad \Delta_h^m f(x) = \sum_{k=o}^{m}(-1)^k\binom{m}{k}f(x+kh)$$

To this end, introduce $B_{\rho,p}$ being the set of all entire functions on C^n of exponential type ρ, whose restrictions to R^n belong to L^p. We recall the well-known fact [7; p. 185-186] that to each $k\ge 1$ and each $f\in L^p$ there exists a family $\{F_\rho\}_{\rho>0}$, $F_\rho\in B_{\rho,p}$, such that

$$(6) \qquad \|f-F_\rho\|_p \le C\, w_k(f;\tfrac{1}{\rho}) \quad .$$

Further, if one considers a family of entire functions $\{F_\rho\}_{\rho>0}$ with $\|F_\rho\|_p = O(1)$ then, analogous to [16; p. 53], the classical Bernstein proof gives

$$(7) \qquad \|D^j F_\rho\|_p \le C\int_o^{\rho} t^{|j|-1}\|F_t-f\|_p^* dt \quad ,$$

where $j = (j_1,\ldots,j_n)$ is a multi-index with non negative integer coefficients j_k, $|j| = j_1+\ldots+j_n$, and $D^j = \partial^{|j|}/\partial x_1^{j_1}\ldots\partial x_n^{j_n}$.

Theorem 3. *Let m,m' be natural numbers, $1\le p\le\infty$, and $\gamma>0$. Then*

a) $$w_{m,p}(f;\tfrac{1}{\rho}) \le \frac{C}{\rho^m}\int_o^{\rho} t^{m-1}\|R_{\alpha,\gamma}(f;t)-f\|_p^* \, dt \quad ;$$

b) $$\|R_{\alpha,\gamma}(f;\rho)-f\|_p \le \frac{C}{\rho^{\gamma}}\int_o^{\rho} t^{\gamma-1}w_{m',p}(f;\tfrac{1}{t})dt \quad ;$$

c) $$w_{m,p}(f;\tfrac{1}{\rho}) \le C\|R_{\alpha,\gamma}(f;\rho)-f\|_p^* \le C' w_{m',p}(f;\tfrac{1}{\rho}) \; , \quad m' < \gamma < m \; .$$

Proof. a) (Compare [16; p. 61]) Since

$$\Delta_h^m f = \Delta_h^m (f-R_{\alpha,\gamma}(f;\rho)) + \Delta_h^m R_{\alpha,\gamma}(f;\rho) \quad ,$$

one obtains by Taylor's formula and Minkowski's integral inequality

$$\|\Delta_h^m f\|_p \le 2^m \|R_{\alpha,\gamma}(f;\rho)-f\|_p^* + C \sum_{|j|=m} |h|^m \|D^j R_{\alpha,\gamma}(f;\rho)\|_p$$

which by (7) and the monotonicity of the σ-modulus gives assertion a).

b) (Compare [16; p. 62]) First consider the case $\gamma = 2m$. By (6) it follows that

$$\|R_{\alpha,2m}(f;\rho)-f\|_p \le \|R_{\alpha,2m}(f-F_\rho;\rho)\|_p + \|R_{\alpha,2m}(F_\rho;\rho)-f\|_p$$

$$\le C\, w_{m',p}(f;\tfrac{1}{\rho}) + \sum_{k=1}^{\alpha} \|\Delta^{mk} F_\rho\|_p \, \rho^{-2mk} \quad ,$$

where $\Delta = \sum_{k=1}^n \partial^2/\partial x_k^2$ is the Laplacian and $\Delta^i = \Delta\Delta^{i-1}$. Now apply (7), observe that $w_{m'}$ is monotone, and trivially

$$\frac{1}{\rho^\kappa}\int_0^\rho t^{\kappa-1}\Omega(t)dt \le \frac{1}{\rho^\lambda}\int_0^\rho t^{\lambda-1}\Omega(t)dt \ , \qquad 0 < \lambda < \kappa \ ,$$

for monotone decreasing Ω, so b) follows in case $\gamma = 2m$. In case of arbitrary γ, $0 < \gamma < 2m$, Theorem 1b with $\Omega(t) = \|R_{\alpha,2m}(f;t)-f\|_p^*$ and the above result give the assertion.

c) By (3) and the well-known property

(8) $$w_{k,p}(f;\tfrac{1}{t}) \le (1+\tfrac{\rho}{t})^k \, w_k(f;\tfrac{1}{\rho}) \ , \qquad k\in N \ ,$$

the assertion c) follows.

Theorem 3c suggests to consider moduli of continuity with fractional index $\kappa>0$ (in order to improve Theorem 3c to fractional m,m' arbitrary close to γ):

$$w_{\kappa,p}(f;t) = \sup_{|h|\le t} \|\Delta_h^\kappa f\|_p \quad ,$$

where the fractional difference operator is defined by

$$\Delta_h^{\kappa} f(x) = \sum_{i=0}^{\infty} A_i^{-\kappa-1} f(x+ih) \quad , \quad A_i^{-\kappa-1} = \frac{\Gamma(i-\kappa)}{\Gamma(i+1)\Gamma(-\kappa)} \quad .$$

<u>*Theorem 4.*</u> *I. Let $n=1$, $\gamma>0$, and $1 \leq p \leq \infty$. Then*

a) $$\omega_{\gamma,p}(f;\tfrac{1}{\rho}) \approx \|R_{\alpha,\gamma}(f;\rho)-f\|_p^* \quad , \qquad 1 < p < \infty ,$$

which is also true in case $p=1$ and ∞ provided $\gamma = 2m$, $m\in N$.

b) Otherwise there holds

$$\omega_{\kappa,p}(f;\tfrac{1}{\rho}) \leq C\|R_{\alpha,\gamma}(f;\rho)-f\|_p^* \leq C'\omega_{\lambda,p}(f;\tfrac{1}{\rho}) \quad , \qquad \lambda < \gamma < \kappa .$$

II. Let $n\geq 2$ and $\gamma>0$. Then

a) $$\omega_{m,p}(f;\tfrac{1}{\rho}) \approx \|R_{\alpha,m}(f;\rho)-f\|_p^* \quad , \quad m\in N , \qquad 1 < p < \infty .$$

b) $$\omega_{\gamma,p}(f;\tfrac{1}{\rho}) \leq C\|R_{\alpha,\gamma}(f;\rho)-f\|_p^* \quad , \qquad 1 < p < \infty .$$

c) $$\sup_{|t|\leq 1/\rho} \Big\|\sum_{k=1}^{n} \Delta_{te^k}^2 f\Big\|_p \leq C\|R_{\alpha,2}(f;\rho)-f\|_p^* \leq C'\omega_{2,p}(f;\tfrac{1}{\rho})$$

for $1 \leq p \leq \infty$, where $t\in R$ and e^k is the unit vector along the k-th axis. Analogous results hold for $\gamma = 2m$, $m\in N$.

Proof. Following Boman [1] consider for $\gamma>0$, $1 \leq p \leq \infty$,

$$H_\gamma^p = \{f\in L^p;\ I_\gamma * f\in L^p,\ I_\gamma^\wedge(v) = |v|^\gamma\} \quad ,$$

supplied with the semi-norm $|f|_{p,\gamma} = \|I_\gamma * f\|_p$. It is convenient to introduce the (modified) Peetre K-functional

$$K(t^\gamma,f;L^p,H_\gamma^p) = \inf_{f=f_1+f_2} \{\|f_1\|_p + t^\gamma |f_2|_{p,\gamma}\} .$$

Then, with the aid of (4') ($\delta=\gamma$) and the standard saturation argument,

(9) $$\|R_{\alpha,\gamma}(f;\rho)-f\|_p + \rho^{-\gamma}|R_{\alpha,\gamma}(f;\rho)|_{p,\gamma}$$

$$\leq C\|R_{\alpha,\gamma}(f;\rho)-f\|_p \leq C'\{\|f-g\|_p + \rho^{-\gamma}|g|_{p,\gamma}\} \quad .$$

Taking the infimum over all $g\in H^p_\gamma$ leads to

$$(9') \qquad K(\rho^{-\gamma},f;L^p,H^p_\gamma) \simeq \|f-R_{\alpha,\gamma}(f;\rho)\|^*_p \ , \qquad 1 \le p \le \infty \ .$$

I (n=1). On account of the boundedness of the Hilbert transform on L^p, $1 < p < \infty$, H^p_γ coincides for all $\gamma>0$ with

$$^\#H^p_\gamma = \{f\in L^p;\ ^\#I_\gamma * f\in L^p,\ ^\#I^\wedge_\gamma(v) = (iv)^\gamma\} \ ;$$

trivially $^\#H^p_{2m} = H^p_{2m}$, $1 \le p \le \infty$. Now using results in [14] one can show analogously to [3] that

$$(10) \qquad w_{\gamma,p}(f;t) \approx K(t^\gamma,f;L^p,{}^\#H^p_\gamma) \ , \qquad 1 \le p \le \infty \ .$$

By (9') this implies Ia). Ib) is established as soon as we can show

$$(11) \quad K(t^\kappa,f;L^p,{}^\#H^p_\kappa) \le C\, K(t^\gamma,f;L^p,H^p_\gamma) \le C' K(t^\lambda,f;L^p,{}^\#H^p_\lambda)$$

(for a closely related result see [5; p. 339]). Now, on the one hand

$$K(t^\kappa,f;L^p,{}^\#H^p_\kappa) \le \|f-R_{2n,\gamma}(f;\rho)\|_p + \rho^{-\kappa}\|{}^\#I_\kappa * R_{2n,\gamma}(f;\rho)\|_p \qquad (\rho t=1)$$

$$\le C\{\|f-R_{n,\gamma}(f;\rho)\|_p + \rho^{-\gamma}|R_{n,\gamma}(f;\rho)|_{p,\gamma}\}$$

since $(iv)^\kappa|v|^{-\gamma}(1-|v|^\gamma)^n_+\in M_1$ for $\kappa>\gamma$ (apply e.g. Lemma 1 in [1]). Thus, by (9) and (9') the left hand side of (11) holds. On the other hand one obtains by direct estimate for arbitrary $g\in {}^\#H^p_\lambda$

$$\|f-R_{\alpha,\gamma}(f;\rho)\|_p \le C\{\|f-g\|_p + \rho^{-\lambda}\|{}^\#I_\lambda * g\|_p\} \ ,$$

since $(iv)^{-\lambda}\{(1-(1-|v|^\gamma)^\alpha_+\}\in M_1$ for $\gamma>\lambda$. Thus, (11) is true by (9').

II (n≥2). Note that by the Marcinkiewicz-multiplier criterion [9; p. 1o9] there holds $(e^{ih\cdot v}-1)^\gamma|v|^{-\gamma}|h|^{-\gamma}\in M_p$, $1 < p < \infty$; thus b) (and a priori one part in a)) is true. The left inequality in c) follows by results in [11].

Hence, in a) and c) only $\|R_{\alpha,\gamma}(f;\rho)-f\|^*_p$ has to be estimated. Now choose in (9) $g = F_\rho$ such that (6) holds with k=m and observe that

$$|F_\rho|_{p,m} \leq C \sum_{|j|=m} \|D^j F_\rho\|_p$$

in case $1 \leq p \leq \infty$ if m is an even integer and in case $1 < p < \infty$ if m is a natural number, since the Riesz transform is bounded on L^p. So we are finished as soon as we have shown

<u>*Lemma*</u>. $\|f-F_\rho\|_p \leq C\, w_{m,p}(f;\frac{1}{\rho})$, $1 \leq p \leq \infty$, $m \in N$, *implies*

$$\|D^j F_\rho\|_p \leq C\, \rho^m\, w_{m,p}(f;\tfrac{1}{\rho}) \quad , \qquad |j| = m \quad .$$

Proof. (Compare [16; p. 55]) Defining

$$\Delta^j_{1/\rho} F_\rho = \Delta^{j_1}_{e^1/\rho} \cdots \Delta^{j_n}_{e^n/\rho} F_\rho = \Delta^j_{1/\rho}(F_\rho - f) + \Delta^j_{1/\rho} f \quad ,$$

it follows by Taylor's formula (applied upon $\Delta^j_{1/\rho} F_\rho$) and relations (6), (7) that

$$\|D^j F_\rho\|_p \leq C\, \rho^m w_{m,p}(f;\tfrac{1}{\rho}) + \rho^m \|\Delta^j_{1/\rho} f\|_p \quad .$$

Thus it remains to estimate the last term. Let D^* be the power set of $\{1,\ldots,m\}$ and $T(h)$ be the translation operator: $T(h)f(x) = f(x+h)$; then, for $h^1,\ldots,h^m \in R^n$, (see [6])

$$\Delta_{h^1} \cdots \Delta_{h^m} f = \sum_{D \in D^*} (-1)^{m-|D|}\, T(\sum_{k \notin D} h_k)\, \Delta^m_{\sum_{k \in D} h^k/k} f \quad .$$

Choosing $\Delta_{h^1} \cdots \Delta_{h^m} = \Delta^j_{1/\rho} f$, clearly $|h^k| \leq 1/\rho$, and therefore by (8), by the boundedness of the translation operator, and by the monotonicity of the modulus of continuity

$$\|\Delta^j_{1/\rho} f\|_p \leq C\, w_{m,p}(f;\tfrac{1}{\rho}) \quad ,$$

which completes the proof of the Lemma.

<u>Remark</u>. i) We mention that with the aid of multiplier methods Theorem 4, II, c) can be improved to (cf. Theorem 3c)

$$w_{3,p}(f;\tfrac{1}{\rho}) + \sup_{|t| \leq 1/\rho} \Big\|\sum_{k=1}^{n} \Delta^2_{te^k} f\Big\|_p \approx \|R_{\alpha,2}(f;\rho) - f\|^*_p$$

(clearly, the left side is $\leq w_{2,p}(f;\frac{1}{\rho})$).

ii) Via the K-functional a very simple and neat description (9') of the approximation behavior of the Riesz kernel is given. Difficulties arise when one tries to express the smoothness properties of f (hidden in the K-functional) by the standard modulus of continuity. In the one-dimensional case, Theorem 4,I gives a satisfactory answer to this problem, whereas in the more-dimensional case there are essential gaps and it would be interesting to remove them.

iii) Via the K-functional (see (9')) comparison theorems may be derived.

REFERENCES

[1] BOMAN, J., Saturation problems and distribution theory, in: Lecture Notes in Math. Vol. 187, pp. 249-266, Springer, Berlin 1971.

[2] BOMAN, J. - H.S. SHAPIRO, Comparison theorems for a generalized modulus of continuity, Ark. Mat. 9 (1971), 91-116.

[3] BUTZER, P.L. - H. DYCKHOFF - E. GÖRLICH - R.L. STENS, Best trigonometric approximation, fractional order derivatives and Lipschitz classes (to appear).

[4] BUTZER, P.L. - R.J. NESSEL - W. TREBELS, On summation processes of Fourier expansions in Banach spaces I; II, Tôhoku Math. J. 24 (1972), 127-14o; 551-569.

[5] BUTZER, P.L. - K. SCHERER, Jackson and Bernstein-type inequalities for families of commutative operators in Banach spaces, J. Approximation Theory 5 (1972), 3o8-342.

[6] JOHNEN, H., Sätze vom Jackson-Typ auf Darstellungsräumen, kompakter, zusammenhängender Liegruppen, in: Linear Operators and Approximation, pp. 254-272, ISNM 2o, Birkhäuser, Basel 1972.

[7] NIKOLSKIĬ, S.M., Approximation of Functions of Several Variables and Imbedding Theorems, Springer, Berlin 1975.

[8] SHAPIRO, H.S., Topics in Approximation Theory, Lecture Notes in Math. Vol. 187, Springer, Berlin 1971.

[9] STEIN, E.M., Singular Integrals and Differentiability Properties of Functions, Princeton Univ. Press, Princeton N.J. 197o.

[1o] TIMAN, M.F., The best approximation of periodic functions by trigonometric polynomials, and transforms of convolution type, Soviet Math. Dokl. 12 (1971), 897-9oo.

[11] TREBELS, W., Generalized Lischitz conditions and Riesz derivatives on the space of Bessel potentials L^p_α, Applicable Anal. 1 (1971), 75-99.

[12] TREBELS, W., Multipliers of (C,α)-Bounded Fourier Expansions in Banach Spaces and Approximation Theory, Lecture Notes in Math. Vol. 329, Springer, Berlin 1973.

[13] TRIGUB, R.M., Linear summation methods and absolute convergence of Fourier series (Russian), Izv. Akad. Nauk, Sér. Mat. 32 (1968), 24-49.

[14] WESTPHAL, U., An approach to fractional powers of operators via fractional differences, Proc. London Math. Soc. (3) 29 (1974), 557-576.

[15] ZAMANSKY, M., Théorie de l'approximation, C.R. Acad. Sci. Paris, Sér. A, 279 (1974), 855-858; 28o (1975), 5o1-5o4; 933-936; 1431-1434; 281 (1975), 1o89-1o9o.

[16] ZAMANSKY, M., Approximation. Théorèmes généraux pour les procédés d'approximation dans un espace normé. Application: approximation des fonctions périodiques, Faculté des Sciences, Univ. Paris VI, Paris 1975.

Tschebyscheff-Approximation by Regular Splines with Free Knots.

by

Helmut Werner and Henry Loeb

In this and in a subsequent paper we will discuss the phenomena which may arise if regular splines are used as approximating functions and the knots are allowed to move freely in the domain of definition. We consider Tschebyscheff-Approximation and show existence in the closure under compact convergence of regular splines. Furthermore, we give some sufficient criteria for a spline to be a Tschebyscheff-Approximation.

This generalizes the results given earlier in Werner [10,11]. On the other hand the axiom of positivity introduced below will exclude some of the phenomena observed with polynomial splines as described by Schumaker [8,9]. Of course, our results are close to those given by Schomberg [7] in his dissertation, where special rational splines were considered. An announcement of our results was made at the Symposium on Approximation Theory held at Austin 1976 [12]. The axioms basically originate from a paper by Schaback [6].

1. Definitions and Axioms

Let α, β be reals and $I = [\alpha,\beta]$. Denote by m and k natural numbers. We consider partitions π of I by sets of knots x_j, j=0,...,m, where $x_{j-1} < x_j$, $x_0=\alpha$, $x_m=\beta$, and we introduce $I_j = [x_{j-1}, x_j]$.

To define the spline in each subinterval I_j we use classes of real-valued functions

$$\mathcal{T}_j:\ t_j(x\ ;\ I_j,\ m_{j-1}, m_j) \text{ for } x \in I_j\ .$$

satisfying the following axioms

(I) Each class shall be <u>smooth</u>, i.e. of class $C^k(I_j)$, and continuous with respect to the parameters x_{j-1}, x_j (which define I_j) and m_{j-1}, m_j.

Let $\mathcal{P}_{k-1}$ be the class of polynomials of degree less than k.

Nonlinear Splines shall be of the form

$$S_{m,k} := \{s \mid s \in C^k(I), \exists \pi\ \forall_j : s|_{I_j} = p_j + t_j,\ p_j \in \mathcal{P}_{k-1},\ t_j \in \mathcal{T}_j\}.$$

(II) The class $S_{m,k}$ shall be regular, i.e. for each j and every pair $x^*, x^{**} \in I_j$ and every pair $t^*, t \in \mathcal{T}_j$ we have the implication

$$D^k(t-t^*) = 0 \text{ for } x^* \text{ and } x^{**} \Rightarrow t = t^* \text{ on } [x^{**}, x^*].$$

This property is used to parametrize the functions t_j by means of their kth order derivatives evaluated at the end points x_{j-1} and x_j of I_j. Throughout this paper we assume that this has been done. Consequently, something should be stated about the range of $D^k t$. For the sake of simplicity suppose that all families $\mathcal{T}_j$ coincide, $\mathcal{T}_j \equiv \mathcal{T}$.

(III) Positivity of $S_{m,k}$: For every x and every $t \in \mathcal{T}$ we have $D^k t(x) > 0$.

(IV) Solvability: For every quadruple x^{**}, x^*, $m^{**} > 0$, $m^* > 0$ there is a $t \in \mathcal{T}$ such that

$$D^k t(x^{**}) = m^{**},\ D^k t(x^*) = m^*. \text{ Let } J = [x^{**}, x^*].$$

From the definition of $S_{m,k}$ it is obvious that $S_{m,k}$ is invariant under the change of classes $\mathcal{T}$ to $\tilde{\mathcal{T}}$ such that

$$\tilde{\mathcal{T}}:\ \tilde{t}(x;\ J, m^{**}, m^*) = t(x;\ J, m^{**}, m^*) + p(x;\ J, m^{**}, m^*)$$

where $p(x,\ J, m^{**}, m^*) \in \mathcal{P}_{k-1}$ continuously depends upon the parameters.
This fact may be used for the following

(V) Normalization: Choose $p(x;\ J, m^{**}, m^*)$ such that

$$\tilde{t}(x;\ J, m^{**}, m^*) = t(x;\ J, m^{**}, m^*) + p(x;\ J, m^{**}, m^*)$$

satisfies $D^v\tilde{t}(x^*; \ldots) = 0$ for $v = 0, \ldots, k-1$.
Similarly define $\tilde{\tilde{t}}(x;\ J, m^{**}, m^*)$ to satisfy

$$D^v\tilde{\tilde{t}}(x^{**}; \ldots) = 0 \text{ for } v = 0, \ldots, k-1.$$

Before giving the last axiom we illustrate the assumptions made so far by the example of <u>special rational splines</u>.

Let $J := [-1,0]$ and $z = \frac{x - x_j}{x_j - x_{j-1}}$, $x_j \neq x_{j-1}$, then define

$$\tilde{t}(z,c,d) = c \cdot \frac{z^k}{d-z} = c \cdot \left(\frac{d^k}{d-z} - z^{k-1} - z^{k-2} \cdot d - \ldots - d^{k-1}\right) \text{ for } z \in J,$$

constituting a polynomial added to a simple pole.

Apparently $D^v\tilde{t}(0) = 0$ for $v=0,\ldots,k-1$. (c and d are omitted.)

To express c and d in terms of $m(0) = D^k\tilde{t}(0)$ and $m(-1) = D^k\tilde{t}(-1)$, we have only to solve

$$m(0) = \frac{k!\cdot c}{d}, \quad m(-1) = \frac{k!\cdot c\cdot d^k}{(1+d)^{k+1}},$$

to obtain

$$\frac{1}{d} = \sqrt[k+1]{\frac{m(0)}{m(-1)}} - 1, \quad c = \frac{m(0)\cdot d}{k!}.$$

For positive $m(0)$, $m(-1)$ it is apparent that d and c have equal signs.

For convenience we retain the parameters c, d to discuss what happens if $m(0)$ or $m(-1)$ approaches 0 or ∞.

For fixed $c \in \mathbb{R}_+$ and $d \to 0+$ we have the following limit values

	$D^k\tilde{t}(z)$	$D^{k-1}\tilde{t}(z)$	$D^v\tilde{t}(z)$, $v<k$
$z = 0$	∞	0	0
$z < 0$	0	$-c\cdot(k-1)!$	$-cD^vz^{k-1}$

These cases correspond to $m(0) \to +\infty$ (if $c > 0$), i.e. $\tilde{t}$ converges to a polynomial in $[-1,0]$, its (k-1)st order derivative has a jump at 0 and its kth order derivative tends to infinity at this point. The kth order derivative $m(-1)$ approaches zero monotonically.

Under the assumption $m(0) > m(-1) > 0$ we have the following cases.

$\lim \frac{m(0)}{m(-1)}$	∞	in $(1,\infty)$	1
$\lim\ d$	0	$(0,\infty)$	∞

and $c = m(0)\cdot d/k!$ may approach any value in $[0,\infty]$.

It is apparent that $0 < \frac{m(0)}{m(-1)} < 1$ corresponds to the cases $-1 > d > -\infty$.

Similar situations are observed if $\frac{c}{d-z}$ is replaced by an exponential or logarithmic term.

To formalize the properties exhibited above we formulate the axiom of <u>steepness</u>.

(VI) Denote $m = D^k t(x^{**})$ and $m^* = D^k t(x^*)$. Then the families $\mathfrak{T}$ shall satisfy the following properties

(i) If $m^* \geq \delta > 0$ and $m \to \infty$ then $\int^{x^*} t(x;\ J, m, m^*)dx \to \infty$ for fixed $x^{**} < x^*$. If $\Big|_{x^{**}}$ $|t(x;\ J,\ m,\ m^*)|$ is uniformly bounded and $m \to \infty$, then $D^k t \Rightarrow 0$ on every closed subinterval of $(x^{**}, x^*]$.

(ii) For every constant $a > 0$ there is a function $m^*(m)$ such that $m \to \infty$ implies $m^*(m) \to 0$ and

$\lim_{m\to\infty} D^v\tilde{t}(x;J,\ m, m^*(m)) = a\cdot D^v (x-x_{j-1}^{**})^{k-1}$ for $x \in (x_{j-1}^{**}, x_j^*]$, $v=0,\ldots,k-1$.

Similarly there is $m(m^*)$ with $m(m^*) \to 0$ for $m^* \to \infty$ and

$\lim_{m^*\to\infty} D^v\tilde{t}(x;\ J, m(m^*), m^*) = -a\cdot D^v (x-x^*)^{k-1}$ for every $x \in [x^{**}, x^*)$, v as before.

(iii) If m^* is bounded and $m \to 0$ then $D^v\tilde{t}(x,\ J, m, m^*) \to 0$ on $[x^{**}\ ,\ x^*)$, uniformly on every closed subinterval, for $v=0,\ldots,k$.

It is apparent that (VIii) weakens and replaces the multiplicativity of the parameter c in the above example.

2. Sequences of Regular Splines

We have to consider the limits of compactly converging sequences on I, i.e. sequences uniformly converging on every closed subinterval of (x_o, x_m).

Consider any sequence of uniformly bounded functions $\{s^v\}_{v=1,2,\ldots} \subset S_{m,k}$. Generalizing "Hilfssatz 3.2" of Werner [10] we have

Lemma 2.1: Let $I(\delta) = (\alpha+\delta, \beta-\delta)$ and suppose $s \in S_{m,k}$ and $|s(x)| \leq K(\delta)$ for every $x \in I(\delta)$, then there exists a constant c_1 such that for $\delta^* > \delta > 0$

$$|s^v(x)| \leq c_1 \cdot K(\delta) \cdot (\delta^* - \delta)^{-k-1} \text{ for every } x \in I(\delta^*)$$
$$\text{and } v=1,\ldots,k-1.$$

The proof uses the positivity (III) of regular splines, from which monotonicity of $D^{k-1}s$ follows. If $I(\delta^*)$ is not empty, the (k-1)st difference quotients over the equidistant points $\alpha+\delta$, $\alpha+\delta+\Delta,\ldots,$ $\alpha+\delta^*$ and $\beta-\delta^*$, $\beta-\delta^*+\Delta,\ldots,\beta-\delta$ with $\Delta = (\delta^*-\delta)/(k-1)$ provide upper and lower bounds for $D^{k-1}s$ in $I(\delta^*)$ of the type given in the statement of the lemma. Integration and another use of difference quotients proves the inequality inductively for lower order derivatives.

This lemma implies

Corrollary 2.2: The functions satisfying the hypotheses of Lemma 2.1 are equicontinuous in $I(\delta^*)$ together with their derivatives up to order k-2.

Given any infinite sequence of bounded functions in $S_{m,k}$ it is possible to pass to a subsequence that is (together with its derivatives up to order k-2) compactly convergent on I, utilising only the standard diagonal method.

In fact, by continuing the selection of subsequences we may assume that the partitions π^v associated with the functions s^v and the values of $D^{k-1}s^v$ and $D^k s^v$ considered at the points of π^v converge, possibly to infinity.

Hence we may assume in the following, that the sequences considered possess all of these properties:

$$\left.\begin{array}{ll} \pi^v = (x_o^v, x_1^v, \ldots, x_m^v), & x_j^v \to x_j \\ & D^{k-1}s^v(x_j) \to s^{(k-1)}(x_j) \\ m_j^v := D^k s^v(x_j) & \to m_j \end{array}\right\} \begin{array}{l} j=0,\ldots,m , \\ \text{for } v \to \infty . \end{array}$$

We investigate the functions $s(x)$ arising in this way. To get the local properties at x_j we may assume that the classes $\tilde{\mathcal{F}}$ and $\tilde{\tilde{\mathcal{F}}}$ are normalized so that all derivatives of the families t_j and t_{j-1} up to order k-1 vanish at the point x_j^v. Convergence of the lower order derivatives at x_j implies convergence, and therefore boundedness, of the polynomial parts of the restrictions of s^v to I_{j-1}^v and I_j^v; consequently $t^v = s^v - p^v$ is bounded too.

3. Local Properties of Limits of Regular Splines.

To analyse the functions $s(x)$ described in the previous section we introduce the following two notations.

Def. 3.1: A knot x_j is n-fold with respect to $s(x)$ if n sequences of knots converge to x_j, i.e. $\lim_{v\to\infty} x_i^v =: x_i = x_j$ for $i=j,j+1,\ldots,j+n-1$.

Def. 3.2: A knot x_j of s is of the nth kind, if $s \in C^{n-1}$ locally and the nth order derivative exists in a deleted neighborhood of x_j and has at most a jump discontinuity at x_j.

Under the assumptions made in the previous section the limit function $s(x)$ of the sequence $s^v(x)$ is at least of class C^{k-2} in (x_o, x_m). Therefore only knots of kth and (k-1)st kind may appear. Observe that the functions $s^v(x)$ are uniformly bounded. We consider all possible cases arising from multiple knots and reduction of differentiability. Consider a knot x_j of $s(x)$, $x_j \neq \alpha, \beta$. We will see that the limit function may reduce to a polynomial in some subinterval of I. We will say, in this case, that s is polynomial, otherwise we say s is regular in that subinterval.

<u>1-fold knot x_j</u>

Since x_j is 1-fold there is a fixed neighborhood $U(x_j)$ containing none of the knots $x^v_{j\pm1}$ of the approximating functions $s^v(x)$. Taking difference quotient of kth order, a reasoning similar to one used before furnishes points $x^{*v}_{j\pm1}$ at which the derivatives $D^k s^v$ are uniformly bounded and these points and derivatives may (after selection) be assumed to converge, $\lim x^{*v}_{j\pm1} = x^*_{\pm} \neq x_j$, $D^k t^v(x^{**}_{j\pm1}) \to m^*_{\pm}$.

(i) Suppose $m_j = \lim m^v_j = 0$.

By (VIiii) we can conclude that $s(x)$ has a vanishing kth order derivative in (x^*_-, x^*_+) and the left- and right-hand limits of the lower order derivatives are equal. In other words $s(x)$ is reduced to a polynomial of degree less than k in the neighborhood $U(x_j)$.

(ii) Suppose $m_j \in (0,\infty)$.

Since $\mathfrak{T}_{j-1}$ and $\mathfrak{T}_j$ consist of functions continuous in the parameters, the limit functions t_{j-1} and t_j are determined by the values m^*_-, m_j and m_j, m^*_+ respectively, continuity of $D^k s(x)$ prevails in (x_j, x^*_+) and (x^*_-, x_j). Assume $m^*_+ > 0$. If in addition $m^*_- > 0$ then we find continuity of $D^k s$ throughout (x^*_-, x^*_+).
If $m^*_- = 0$, then s reduces to a polynomial of degree k-1 in (x^*_-, x_j), by (VIiii). In this case a knot of the k-th kind has arisen at x_j and s is polynomial on one side, regular on the other side of x_j.

(iii) Let $m_j = \infty$.

Since the sequence $\{s^v\}$ is uniformly bounded, (VIi) shows that $m^*_{\pm}$ are zero, and in (x^*_-, x_j) and (x_j, x^*_+) the function s reduces to a polynomial of degree less than k.
Since $m^v_j \to \infty$, however, it is possible that

$$D^{k-1}s(x^*_+) - D^{k-1}s(x^*_-) \geq \underline{\lim} \int_{x^*_-}^{x^*_+} D^k s^v(x)\,dx > 0,$$

so that there is a jump of the (k-1)st derivative at x_i,

i.e. we find a knot of the (k-1)st kind.

These three cases are also typical for limits of subclasses $S_{m,k}(\pi)$, for the partition π fixed, hence no coalescing of knots is possible. For variable knots there are additional cases.

<u>2-fold knot $x_j = x_{j+1}$</u>

It is obvious that all cases mentioned before may appear again for multiple knots, e.g. if $m_j = m_{j+1}$. We remark that there is symmetry with respect to x_j and x_{j+1}, hence it suffices to consider the following additional situations. As before construct $x_-^* < x_j = x_{j+1} < x_+^*$.

(iv) $0 < m_j < m_{j+1}$

This means that the sequences m_j^v and m_{j+1}^v have a uniform finite upper and positive lower bound.
Since the k-th order derivatives of s^v in x_j^v, x_{j+1}^v depend continuously on the interval and the parameters m_j^v and m_{j+1}^v, they are uniformly bounded. Hence

$$D^{k-1}s(x)\Big|_{x_j^v}^{x_{j+1}^v} = \int_{x_j^v}^{x_{j+1}^v} D^k t(x; I_{j+1}^v, m_j^v, m_{j+1}^v)dx = O(x_{j+1}^v - x_j^v) \to 0 \text{ for } v \to \infty,$$

i.e. the (k-1)st derivative is continuous at x_j.
Contrary to case (ii) the function s may be regular on both sides of x_j but $D^k s$ may have a jump at this point, i.e. we encounter a knot of the kth kind.

(v) $0 \leq m_j < m_{j+1} = \infty$

By boundedness of s we conclude from (VIi) that s will be polynomial on the right-hand side of $x_{j+1} = x_j$.
Since $\frac{m_{j+1}^v}{m_j^v} \to \infty$ the expression

$$D^{k-1}s(x)\Big|_{x_j^v}^{x_{j+1}^v} = \int_{x_j^v}^{x_{j+1}^v} D^k t(x; I_{j+1}, m_j, m_{j+1})dx$$

may have a non zero limit for $v \to \infty$ in spite of the coalescence of the two limits x_j and x_{j+1}, hence $D^{k-1}s(x)$

might have a jump at x_j.

Since m_j may be positive, however, it is possible that s is regular on the left-hand side of x_j.

Looking at the behavior of s in the neighborhood of x_j we see that cases (i) - (v) exhaust all possible cases compatible with 1-fold or 2-fold knots.

So far we have not found that s regular on both sides of x_j and $D^{k-1}s$ discontinuous at x_j is feasible. We discover this in case of a

<u>3-fold knot $x_{j-1} = x_j = x_{j+1}$.</u>

To show that the said case may occur let

(vi) $0 < m_{j-1}, m_{j+1} < m_j = \infty$.

Since $m_{j\pm1}$ are finite the function s may be regular on the right- and left-hand side of x_j, just as in previous cases.

To compare $\lim D^{k-1}s^v(x^v_{j+1})$ and $\lim D^{k-1}s^v(x_{j-1})$ (the existence of the limits is guaranteed by the presupposed selection) we consider the expression

$$D^{k-1}s^v(x)\Big|_{x^v_{j-1}}^{x^v_{j+1}} = \int_{x^v_{j-1}}^{x^v_j} D^k s^v(t)dt + \int_{x^v_j}^{x^v_{j+1}} D^k s^v(t)dt.$$

Since $D^k s^v(x^v_j) \to \infty$ it is feasible due to (VIi) that the right-hand sum tends to a positive value in spite of the vanishing of $\lim (x^v_{j+1}-x^v_{j-1})$.

This shows that a knot of the (k-1)st kind can indeed arise.

Since $D^{k-2}s$ is continuous by construction of s there can not be any (k-2)nd kind knots. Coalescence of more knots can not give anything new.

In all cases where $m_j=\infty$ and $x_j < x_{j+1}$ (or $x_{j-1} < x_j$) we saw that s would be polynomial between x_j and $x_{j\pm1}$. Therefore $m_{j\pm1}$ was necessarily equal to zero and, by case (i), must be polynomial between $x_{j\pm1}$ and $x_{j\pm2}$ too (if $j\pm2$ does not exceed α or β).

Properly interpreted this also holds if x_{j+1} and x_{j+2} coalesce. It looks as if every interior polynomial segment of s "absorbs" one limit knot x_j.

Finally, we remark that a transition between regular and polynomial sections of a limit function s can take place only at knots, for between them the convergence of the polynomials and the elements t_j^V is uniform. For the same reason it is possible to define s on $[\alpha,\beta]$ as the continuous extension of s (defined on (α,β)).

We summarize all of these oberservations in

Theorem 1:

If $S_{m,k}$ denotes the class of regular splines defined above (for a given family $\mathcal{F}$) then the limit functions of compactly converging sequences of $S_{m,k}$ are piecewise regular or polynomial. Transitions may only occur at knots, and are bound to the following rules that indicate the "worst" cases that may cccur (obviously 3-fold includes 2-fold includes 1-fold):

multiplicity \ kind of knot	k - 1	k	else
1	polynomial, polynomial	polynomial, regular	regular, regular
2	" , regular	regular , regular	
3	regular , regular		

Counting multiplicities at the knots (including α and β) and assigning to each interior polynomial piece multiplicity 1, one can say that the sum of all multiplicities cannot exceed m+1, the number of knots admitted in $S_{m,k}$.

This way of counting multiplicities was introduced in the dissertation of Schomberg [7].

4. Existence of Tschebyscheff-Approximation

Since the class $S_{m,k}$ is not closed there is no chance to prove an existence theorem. On the other hand such a theorem is almost trivial if one uses the closure $\tilde{S}_{m,k}$ of $S_{m,k}$ under compact convergence.

Theorem 2: For every $f \in C(I)$ there exists an element s^* of $\tilde{S}_{m,k}$ best approximating f, i.e.

$$\eta := \| f-s^* \| \leq \| f-s \| \quad \text{for every } s \in \tilde{S}_{m,k} .$$

($\| \cdot \|$ denotes the Tschebyscheff-Norm on I.)

Proof: Take a minimizing sequence $\{s^\nu\} \subset S_{m,k}$. The elements s^ν have uniformly bounded norms. Hence we may take a subsequence compactly converging to an element $s^* \in \tilde{S}_{m,k}$.

For every $\delta > 0$ we have

$$x \in I(\delta) \text{ implies } |s^*(x) - f(x)| \leq \lim \| s^\nu - f \| = \eta .$$

The statement of the theorem follows from the continuity of $s^*(x)$ and $f(x)$ in I.

The existence theorem is also true if one uses only a subclass of regular splines with respect to a fixed partition π. We turn to the more important point of characterizing best approximations.

5. Zeros of Differences of Regular Splines.

The regularity axiom states that a kth order derivative of the difference of any two regular splines has at most one (separated) zero in each subinterval $[x_{j-1}, x_j]$. Therefore, by Rolle's theorem, the difference itself has at most m+k zeros. Here we may even count multiplicities up to order k+1.

This statement carries over to the limit functions $s \in \tilde{S}_{m,k}$, if we introduce $m^*(s)+1$ as sum of the multiplicities of knots and polynomial segments as defined above. Obviously $m^*(s) = m(s)$ if s is a (proper) regular spline, i.e. an element of $S_{m,k}$.

Theorem 3: Let $s_1, s_2 \in \tilde{S}_{m,k}$, then the number of separated zeros of $s_1 - s_2$ is at most

$$m^*(s_1) + m^*(s_2) + k - 1 .$$

Remark: This bound could be sharpened, for example by taking into account common knots.

Proof: Write $r = s_1 - s_2$ and assume $r \neq 0$. Denote its (separated) zeros by $\tilde{z}_1, \ldots, \tilde{z}_{N+k-2}$ (N an integer). $D^{k-2}r$ is continuous and by Rolle's theorem it has N zeros, $z_1, \ldots, z_N$, say.

In each interval (z_{j-1}, z_j) select a point y_i and add z_j to this set if it is a double zero of $D^{k-2}r$. We add an extra point in the intervals $[\alpha, z_1]$ and $[z_N, \beta]$ different from z_1 (resp. z_N) if this point is a double zero of $D^{k-2}r$, so that the differences of the values $D^{k-2}r$ evaluated at successive points y_j alternate in sign.

By our assumption about s_1, s_2 there are two elements $u_i \in S_{m_i,k}$, $m_i = m^*(s_i)$, such that $D^{k-2}(u_1 - u_2)$ also have differences, evaluated at successive points y_j, alternating in sign. Since u_1 and $u_2 \in C^k(I)$, the last statement is equivalent to the fact that $D^{k-1}(u_1 - u_2)$ changes sign N-1 times, i.e. has N-1 zeros. A last application of Rolle's theorem shows that $D^k(u_1 - u_2)$ has at least N-2 zeros and this, by regularity, is less than or equal to M, the number of intervals generated by the knots of u_1 and u_2. Hence the number of zeros of r itself is less than or equal to k+M.

Now $M \leq m(u_1) + m(u_2) - 1 \leq m^+(s_1) + m^+(s_2) - 1$.

Corollary: For a given partition π consider the regular splines $S_{m,k}(\pi)$. Then for any two functions $s_1, s_2 \in S_{m,k}(\pi)$ the difference $s_1 - s_2$ has at most m+k zeros.

The proof is the same as before if one observes that M = m in the above proof, since s_1, s_2 respectively u_1 and u_2 have the same m+1 knots.

An immediate application of theorem 3 is the

Characterization theorem (Variable knots.)

The function $s^* \in \tilde{S}_{m,k}$ is a Tschebyscheff-Approximation to a function f(x) if there is an interval $J \subset I$ containing an alternant of the error function $f(x)-s^*(x)$ of length $m+m^*(s^*,J) + k+1$, where $m^*(s,J)$ counts knots with multiplicities and interior polynomial sections of s^* in J as defined before.

The proof follows standard patterns. Analoguously, we could formulate a sufficient criterion for the fixed-knot approximation case, compare Werner [11]. If one takes special structural properties into account then it is possible to get stronger results as in the example of rational splines, Braess-Werner [5].

References

1 Arndt, H., Interpolation mit regulären Spline-Funktionen, Dissertation Münster 1974.

2 Barrar, B.R. and H.L. Loeb, Existence of Best Spline Approximations with Free Knots, J. Math. Anal. & Appl. 31 (1970), 383-390.

3 Barrar, B.R. and H.L. Loeb, Spline Functions with Free Knots as the Limit of Varisolvent Families, J. Approximation Theory 12 (1974), 70-77.

4 Braess, D., Chebyshev Approximation by Spline Functions with Free Knots, Numer. Math. 17 (1971), 357-366.

5 Braess, D. and H. Werner, Tschebyscheff-Approximation mit einer Klasse rationaler Splinefunktionen II, J. Approximation Theory 10 (1974), 379-399.

6 Schaback, R., Interpolation mit nichtlinearen Klassen von Splinefunktionen, J. Approximation Theory 8 (1973), 173-188.

7 Schomberg, H., Tschebyscheff-Approximation durch rationale Splinefunktionen mit freien Knoten, Dissertation Münster 1973.

8 Schumaker, L.L., Uniform Approximation by Chebyshev Spline Functions II: Free Knots. SIAM J. Numer. Anal. 5 (1968), 647-656.

9 Schumaker, L.L., On the Smoothness of Best Spline Approximations, J. Approximation Theory 2 (1969), 410-418.

10 Werner, H., Tschebyscheff-Approximation mit einer Klasse rationaler Splinefunktionen, J. Approximation Theory 10 (1974), 74-92.

11 Werner, H., Tschebyscheff-Approximation nichtlinearer Splinefunktionen, in K. Böhmer-G.Meinardus-W. Schempp: Spline-Funktionen, BI-Verlag Mannheim-Wien-Zürich 1974, 303-313.

12 Werner, H., Approximation by Regular Splines with Free Knots, Austin, Symposium on Approximation Theory 1976.

APPLICATIONS OF PADE APPROXIMATION IN NUMERICAL ANALYSIS *

Luc Wuytack

Abstract

A survey is given of possible applications of Padé approximation in several fields of numerical analysis : solving nonlinear equations, acceleration of convergence, numerical solution of ordinary and partial differential equations, numerical quadrature. Special attention is given to the convergence behavior of the corresponding numerical techniques.

1. Definition and properties of Padé approximants

Let P_n be the set of polynomials with degree at most n for all $n \geq o$. The set of ordinary rational functions $r = \frac{p}{q}$ with $p \in P_m$, $q \in P_n$ and $\frac{p}{q}$ irreducible be denoted by $R_{m,n}$ for all $m,n \geq o$. Consider the following formal power series

$$f(x) = c_o + c_1 . x + c_2 . x^2 + c_3 . x^3 + \dots \qquad (1)$$

Definition of Padé approximant

The Padé approximant of order (m,n) for f is the unique element $r_{m,n} = \frac{p}{q}$ in $R_{m,n}$ such that

$$f(x).q(x) - p(x) = O(x^{m+n+1+k}) \qquad (2)$$

for some integer k, which is as high as possible.

Since $r_{m,n}$ is defined for every $m,n \geq o$ the following table can be constructed:

$$\begin{array}{ccccc} r_{o,o} & r_{o,1} & r_{o,2} & r_{o,3} & \dots \\ r_{1,o} & r_{1,1} & r_{1,2} & r_{1,3} & \dots \\ r_{2,o} & r_{2,1} & r_{2,2} & r_{2,3} & \dots \\ r_{3,o} & r_{3,1} & r_{3,2} & r_{3,3} & \dots \\ \vdots & \vdots & \vdots & \vdots & \end{array}$$

This table is called the Padé table for f. The elements of the first column of this table are the partial sums of (1). An element $r_{m,n}$ is called normal if it

* Work supported in part by the FKFO under grant number 2.0021.75

appears only once in the Padé table or equivalently $r_{m,n} = \frac{p}{q}$ is normal if p and q have degree exactly m and n respectively and k=o in (2).

Convergence of Padé approximants

The important topic of the convergence properties of certain sequences of Padé approximants has been treated in several books and papers, e.g. O.Perron (1929), H.S.Wall (1948), G.A.Baker (1965,1975). A survey of these properties has been given by C.K.Chui (1976) at the Austin Conference on Approximation Theory. From this theory it follows that the convergence domain of certain sequences of Padé approximants can be larger than this of the given power series (1). In a lot of cases this convergence can also be faster. Moreover Padé approximants can be used to approximate meromorphic functions succesfully. These are some of the most important features which make Padé approximants suitable for practical work.

Connection with Chebyshev approximation

Padé approximants can be considered as uniform limits of certain sequences of Chebyshev approximants. This connection was pointed out by J.L.Walsh in a number of papers. Recently C.K.Chui, O.Shisha and P.W.Smith (1974) proved the following result :
Let f be an element of C^{m+n+1} $[0,\delta]$ with $\delta > o$, and r_ε be the Chebyshev approximant from $R_{m,n}$ for f on $[0,\varepsilon]$ for every ε in $(0,\delta]$. Then the sequence $\{r_\varepsilon\}$ converges uniformly to $r_{m,n}$ on some closed interval $[0,\varepsilon_0]$ with $\varepsilon_0 \leq \delta$ as ε goes to zero. This property is called the locally best property of Padé approximants.

Algorithms for computing Padé approximants

Several algorithms exist for computing the elements in the Padé table. Some of them e.g. Baker's algorithm (1970) and Longman's algorithm (1971) make use of certain recurrence relations which do exist between neighboring elements in the Padé table. It is also possible to construct certain continued fractions whose convergents form certain sequences in the Padé table. Algorithms of this type are given by H.Rutishauser (1956), F.L.Bauer (1959), H.C.Thacher (1961), W.B.Gragg (1972), P.J.S.Watson (1973), G.Claessens (1975). To compute the value of a Padé approximant in a given point the ε-algorithm of P.Wynn (1956) or the η-algorithm of F.L.Bauer (1959) can be used. We refer to the paper of G.Claessens (1975) for more information concerning these algorithms.

Literature on Padé approximation

The topic of Padé approximation has been treated in a large number of books and papers. References to them will be found in the following

books : O.Perron (1929), H.S.Wall (1948), G.A.Baker (1975)

survey papers : J.Zinn-Justin (1971), J.L.Basdevant (1972), W.B.Gragg (1972), P.Wynn (1974), G.Claessens (1975), C.K.Chui (1976).

bibliography : C.Brezinski (1976).

proceedings of conferences and collection of papers : G.A.Baker and J.L.Gammel (1970), W.B.Jones and W.J.Thron (1974), P.R.Graves-Morris (1973), H. Cabannes (1975), D.Bessis, J.Gilewicz, P.Mery (1975).

The present paper is intented to be complementary to the survey papers mentioned above. We only will consider the applications of one-point Padé approximants in several fields of Numerical Analysis. A similar theory could also be given for the case of multipoint Padé approximation (where information of f at several points is used). We will mention applications of multi-point Padé approximation (also called rational Hermite interpolation) only briefly at the end of each section.

2. The use of Padé approximation in solving nonlinear equations

Consider a real-valued function f, defined in the interval $[a,b]$. The problem is to find an element x in $[a,b]$ such that $f(x)=0$. To solve this problem the following method can be used.

Let x_i be an approximation for x and r_i be the Padé approximant of order (m,n) for f at the point x_i. The element x_{i+1} is now defined as being a zero of r_i or, if $r_i=\frac{p_i}{q_i}$, $p_i(x_{i+1})=0$.

If the sequence $\{x_i\}$ is convergent to a simple zero of f and if r_i is normal for every i then the following relation can be proved (L.Tornheim, 1964) :

$$\lim_{i\to\infty}\frac{|x_{i+1}-x|}{|x_i-x|^{m+n+1}}=c<\infty \qquad (3)$$

This means that the order of convergence is at least m+n+1. Moreover it can be shown (L.Tornheim, 1964) that $\{x_0,x_1,x_2,\dots\}$ will convergence as soon as x_0 is chosen close enough to x. A similar convergence result is also given by G.Merz (1968) and J.S.Frame (1953) for modified versions of the above technique.

It is important to remark that the order of convergence is only dependent on the sum of m and n, which means that the order remains unchanged if elements on an ascending diagonal in the Padé table are used. The use of $n > o$ can be interesting since in this case the asymptotic error constant c can be much lower (G.Merz, 1968). Moreover it is important to note that the numerator p_i of r_i can be computed without having to compute q_i (e.g. using Baker's algorithm). This makes the case $m=1$, $n > o$ particularly interesting.

Remark that the case $m=1$, $n=o$ reduces to the classical Newton's method. It is also possible to compute all the zeros or poles of f simultaneously by using Rutishauser's qd-algorithm. See e.g. P.Henrici (1963), A.S. Householder (1970) and [13] for more details.

Applications of the use of multipoint Padé approximants to the above problem are given by P.Jarratt (1966,1970), J.F.Traub (1964), G.R. Garside, P.Jarratt and C.Mack (1968), D.K. Dunaway (1974), J.C.P. Bus and T.J.Dekker (1975).

3. *The acceleration of convergence using Padé approximation*

Let $\{a_i\}$ be a given sequence with limit A or $\lim_{i\to\infty} a_i = A$. The problem is to find A. Since the convergence of $\{a_i\}$ can be slow it is interesting in some cases to construct a sequence $\{b_i\}$ satisfying the following properties:

$$\lim_{i\to\infty} b_i = A \quad \text{and} \quad \lim_{i\to\infty} \frac{|b_i - A|}{|a_i - A|^\alpha} = o \quad \text{with } \alpha > 1.$$

In order to construct $\{b_i\}$ the following technique can be used. Consider

$$f(x) = a_o + \sum_{i=1}^{\infty} (a_i - a_{i-1}).x^i \tag{4}$$

Then $f_n(x) = a_o + \sum_{i=1}^{n} (a_i - a_{i-1})^*$ satisfies $f_n(1) = a_n$ for $n \geq o$. Let $r_{m,m}$ be the Padé approximant of order (m,m) for f. The sequence $\{b_i\}$ can now be defined as follows :

$$b_i = r_{i,i}(1) \qquad \text{for } i \geq o. \tag{5}$$

The elements b_i satisfying (5) can be computed easily by using the ε-algorithm of P.Wynn (1956). This algorithm works as follows :

Put $\varepsilon_{i,-1} = o$ and $\varepsilon_{i,o} = a_i$ for $i \geqslant o$

Compute $\varepsilon_{k,l} = \varepsilon_{k-1,l-2} + \frac{1}{\varepsilon_{k,l-1} - \varepsilon_{k-1,l-1}}$ for $k = 1,2,3,\dots$ and $l = 1,2,3,\dots,k$.

Then $b_i = r_{i,i}(1) = \varepsilon_{2i,2i}$ for $i \geqslant o$.

The convergence of $\{b_i\}$ depends highly on the properties of $\{a_i\}$. This relation has been investigated e.g. by P.Wynn (1966), C.Brezinski (1972), A.Genz (1973). The ε-algorithm can also be used if the elements a_i are vectors or matrices. See P.Wynn (1962), E.Gekeler (1972), C.Brezinski (1974,1975).
Padé approximation has also been succesfully applied for summing divergent series. See R.Wilson (1930), P.Wynn (1967). Applications of the use of multipoint Padé approximants for accelerating the convergence of a given sequence are given by R.Bulirsch and J.Stoer (1964), J.Oliver (1971), A.Genz (1973), L.Wuytack (1971).

4. *The numerical solution of ordinary differential equations using Padé approximation.*

Consider the problem of finding a solution for the following initial value problem $y' = f(x,y)$, $y(a) = y_o$ with x in $[a,b]$. Let $h = \frac{b-a}{k}$ for some integer k and $x_i = a+i.h$, for $i = o,1,\dots,k$. In order to find approximations y_i for $y(x_i)$ the following idea could be used : Let r_i^* be the Padé approximant of a certain order for $y(x)$ at the point x_i and take $y_{i+1} = r_i^*(x_{i+1})$, for $i=o,1,\dots,k-1$. A power series expansion for the solution $y(x)$ at x_i however is not known. But it is possible to consider the following power series in h :

$$y_i + h.f(x_i,y_i) + \frac{h^2}{2!}.f'(x_i,y_i) + \frac{h^3}{3!}.f''(x_i,y_i) + \dots \qquad (6)$$

Starting with y_o it is now possible to construct the sequence $\{y_1,y_2,\dots,y_k\}$ as follows : Let r_i be the Padé approximant of order (m,n) for (6), then define

$$y_{i+1} = r_i(x_{i+1}) \quad \text{for } i = o,1,\dots,k-1 \; .$$

It is not hard to see that this relation can also be written in the following form

$$y_{i+1} = y_i + h.g(x_i, y_i, h) \quad \text{for} \quad i = 0, 1, \ldots, k-1 . \tag{7}$$

The above technique can now be considered as a one-step method for solving the given initial value problem.

Applying a convergence result for one-step methods [56,p.116] we get : if $g(x,y,h)$ is continuous and satisfies a Lipschitz condition in y, then $\lim_{h \to 0} y_i = y(x_i)$. Moreover, using a similar argument as in [68, p.269], it is possible to prove that in the case of normal Padé approximants we have

$$y(x_i) - y_i = O(h^{m+n+1}) \tag{8}$$

In order to compute y_{i+1} as defined in (7) several techniques can be used. It is possible to construct explicit formulas for $g(x,y,h)$. See e.g. Z.Kopal (1959), J.D.Lambert and B.Shaw (1965). These formulas become fairly complicated for higher values of m and n. Another possibility is to use the ε-algorithm with $\varepsilon_{n,0}$ equal to the n-th partial sum of (6). See [49] and A.Wambecq [63].

Remark that the choice of n=o in the above technique corresponds to the Taylor series method in solving the given initial value problem.

Methods of the form (7) have the disadvantage that the derivatives of f must be known or computed. It is however possible to replace the derivations by linear combinations of values of f at different points, keeping a method with the same order of convergence. This replacement gives raise to nonlinear Runge-Kutta type methods. Some of the properties of these techniques are considered by A.Wambecq (1976). It is important to note that it is e.g. possible to derive nonlinear Runge-Kutta methods of order 5 using 5 evaluations of f, which is not possible in the linear case.

Formulas of the form (7) can also be used for solving systems of ordinary differential equations, see J.D.Lambert (1973) and [64].

Also multipoint Padé approximants can be used to solve initial value problems, giving raise to nonlinear multipoint methods. See e.g. J.D.Lambert and B.Shaw (1965, 1966), G.Opitz (1968), Y.L.Luke, W.Fair and J.Wimp (1975). Padé approximants for the exponential function play a very important rule to derive A-stable methods for solving initial value problems, see e.g. B.L.Ehle (1968, 1971), E.B.Saff and R.S.Varga (1975).

5. *Numerical quadrature using Padé approximation.*

The problem is to find the value of the definite integral $I = \int_a^b f(t).dt$. A first approach to this problem is as follows : approximate f by some rational function, e.g. a Padé approximant, and compute $\int_a^b r(t).dt$. The value of this last integral might not be easy to find and several difficulties can be encountered (see J.S.R. Chisholm, 1974). A second approach is based on the transformation of the given problem to the problem of finding the value y(b) where y(x) is the solution of the initial value problem $y'(x) = f(x)$, $y(a) = o$, with $x \in [a,b]$. Since $y(x) = \int_a^x f(t).dt$ it is clear that $I = y(b)$.

Let h, x_i and y_i be defined as in the preceding section, then the following technique can be used to find the value of y(b). Let r_i be the Padé approximant of order (m,n) for

$$y_i + h.f(x_i) + \frac{h^2}{2!}.f'(x_i) + \frac{h^3}{3!}.f''(x_i) + \dots$$

then y_{i+1} can be defined as follows

$$y_{i+1} = r_i(x_{i+1})$$

$$\text{or} \quad y_{i+1} = y_i + h.s(x_i,h) \qquad \text{for } i = o,1,\dots,k-1 \qquad (9)$$

for some function s(x,h). The formula (9) can be interpreted as a formula for approximate integration between x_i and x_{i+1} or

$$\int_{x_i}^{x_{i+1}} f(t).dt \simeq h.s(x_i,h) .$$

The convergence properties of the above technique follow immediately from the results in the previous section. We get : if s(x,h) is continuous then $\lim_{h \to o} y_k = I$ and $I - y_k = O(h^{m+n+1})$, in the case of normal Padé approximants. Again the ε-algorithm can be used to compute y_{i+1} in (9) or explicit formulas can be derived, e.g. in the case m=n=1 we get

$$s(x,h) = \frac{2.[f(x)]^2}{2.f(x) - h.f'(x)} .$$

Remark that the derivatives of f can be replaced by linear combinations of values of f at different points, keeping a method with the same order of convergence (see [68] for more details and some numerical examples).

Padé approximants can also be used for evaluating integrals having a singular integrand, see [69].

6. *The numerical solution of partial differential equations using Padé approximation.*

Consider the problem of finding a solution u(x,t) for the following boundary value problem :

$$\left.\begin{array}{l} \frac{\partial u}{\partial t}=\frac{\partial^2 u}{\partial x^2} \quad \text{for} \quad a<x<b \quad \text{and} \quad t>o \\ \\ u(x,o)=g(x)\ ;\ u(a,t)=\alpha\ ;\ u(b,t)=\beta \end{array}\right\} \qquad (10)$$

To find approximate values for $u(x_i,t_j)$ in certain points the following technique can be used (R.S. Varga, 1961). First spatial discretization of (10) is performed, with $h=\frac{b-a}{n+1}$ and $x_i=a+i.h$ for $i=o,1,\ldots,n+1$. Put $u_i(t)=u(x_i,t)$ for $t\geq o$ and $i=1,\ldots,n$, then (10) becomes :

$$\left\{\begin{array}{l} \frac{du_i(t)}{dt}=-A.u_i(t) \qquad \text{for} \quad t>o \quad \text{and } i=1,2,\ldots,n \\ \\ u_i(o)=g(x_i) \end{array}\right.$$

where A is a real symmetric positive-definite nxn matrix.
The exact solution of this initial value problem is given by

$$U(t)=e^{-t.A}.G\ , \qquad (11)$$

with $U(t)=[u_1(t),u_2(t),\ldots,u_n(t)]^T$ and $G=[g(x_1),g(x_2),\ldots,g(x_n)]^T$.
Let Δt be the stepsize in the t-direction then (11) can also be written as

$$U(t+\Delta t)=e^{-\Delta t.A}.U(t)\ . \qquad (12)$$

Let $r_{m,n}=\frac{p}{q}$ be the Padé approximant of order (m,n) for e^{-x} then (12) can be approximated by

$$U(t+\Delta t)=[q(\Delta t.A)]^{-1}.[p(\Delta t.A)].U(t) \qquad (13)$$

It can be proved that (13) is an unconditionally stable method if $n \geq m$. This property is based on the fact that $|r_{m,n}(x)| \leq 1$ for all $x \geq o$ if and only if $n \geq m$ (see R.S.Varga, 1961). The case $m = o$, $n = 1$ reduces (13) to a classical explicit method for solving (10). The case $m = 1$, $n = 1$ reduces to the Crank-Nicolson technique for solving (10).

The application of Padé approximation to solve more general parabolic partial differential equations can be found in R.S.Varga (1962).

7. *Other applications.*

In this section we only mention some other applications of the use of Padé approximation in numerical analysis :

- computation of Laplace Transform inversion, see I.M.Longman (1973)
- numerical differentiation, see [55] , [67] .
- solution of integral equations, see M.F.Barnsley and P.D.Robinson (1974), W.Fair (1974).
- analytic continuation, see [5] , J.L.Gammel (1974), J.Devooght (1976).

Of course Padé approximations can also be used to approximate a given function, see E.G.Kogbetliantz (1960), Y.L.Luke (1969), A.Edrei (1975).
Several other applications to related fields can be found in the literature given in section 1.

8. *Conclusion.*

It has been shown that Padé approximation can be applied to derive non-linear techniques in several fields of numerical analysis. These techniques have interesting convergence properties, similar to these for linear methods. Numerical examples show that nonlinear techniques can be more interesting than linear ones in the neighbourhood of singular points. Care must be taken in applying nonlinear techniques, due to the possibility of numerical instability during the computations.
Our experience shows that the use of Padé approximation is not better than the use of a linear technique in all situations. In those cases however where linear techniques give poor results or fail to converge it might be interesting to try a nonlinear technique.

REFERENCES

A. *Books, survey papers, bibliographies, proceedings of conferences on P.A.*

1 BAKER, G.A. Jr.: Essentials of Padé Approximants. Academic Press, London, 1975.

2 BAKER, G.A. Jr.: The theory and application of the Padé approximant method. Advances in Theoretical Physics 1 (1965), 1-58.

3 BAKER, G.A. Jr. and GAMMEL, J.L. (eds.) : The Padé approximant in theoretical Physics. Academic Press, London, 1970.

4 BAKER, G.A. Jr. and GRAVES-MORRIS, P. : Review on Padé approximation. In "Encyclopaedia of Applicable Mathematics", Addison Wesley, New York, announced to appear at the end of 1977.

5 BASDEVANT, J.L.: The Padé approximation and its physical applications. Fortschritte der Physik 20 (1972) 283-331.

6 BESSIS, D.; GILEWICZ, J.; MERY,P. (eds.) : Proceedings of the Workshop on Padé Approximants. Centre de Physique Théorique, CNRS Marseille, 1975.

7 BREZINSKI, C. : A bibliography on Padé approximation and some related matters. In [8] ,pp. 245-267.

8 CABANNES, H. (ed.): Padé approximants method and its applications to mechanics. Proceedings of the Euromech Colloquium, Toulon, 1975 Lecture Notes in Physics 47, Springer-Verlag, Berlin, 1976.

9 CHUI, C.K.: Recent results on Padé approximants and related problems. Proceedings of the Symposium on Approximation Theory, University of Texas at Austin, 1976. To appear.

10 CHUI, C.K.; SHISHA, O. and SMITH, P.W.: Padé approximants as limits of best rational approximants. Journal of Approximation Theory 12 (1974), 201-204.

11 CLAESSENS, G.: A new look at the Padé table and the different methods for computing its elements. Journal of Computational and Applied Mathematics 1 (1975), 141-152.

12 DONNELLY, J.D.P.: The Padé Table. In "Methods of Numerical Approximation" (Handscomb D.C. (ed.), Pergamon Press, Oxford, 1966), 125-130.

13 GRAGG, W.B.: The Padé table and its relation to certain algorithms of numerical analysis. SIAM Review 14 (1972), 1-62.

14 GRAVES-MORRIS, P.R., (ed.): Padé Approximants and Their Applications. Academic Press, London, 1973.

15 GRAVES-MORRIS, P.R. (ed.): Padé Approximants. The Institute of Physics, London, 1973.

16 JONES, W.B. and THRON, W.J. (eds.): Proceedings of the International Conference on Padé approximants, continued fractions and related topics. Rocky Mountain Journal of Mathematics 4 (1974), 135-397.

17 PADÉ, H.: Sur la représentation approcheé d'une fonction par des fractions rationnelles. Ann. Sci. Ecole Normale Supérieure 9 (1892), 1-93.

18 PERRON, O.: Die Lehre von den Kettenbrüchen, Band II. B.G.Teubner, Stuttgart, 1957.

19 WALL, H.S.: The analytic theory of continued fractions. D. van Nostrand, London, 1948.

20 WYNN, P.: Some recent developments in the theories of continued fractions and the Padé table. In [16], pp. 297-323.

21 ZINN-JUSTIN, J.: Strong interactions dynamics with Padé approximants. Physics Reports 1 (1971), 55-102.

B. *References on the use of P.A. in solving nonlinear equations*

22 BUS, J.C.P. and DEKKER, T.J.: Two efficient algorithms with guaranteed convergence for finding a zero of a function. ACM Transactions on Mathematical Software 1 (1975), 330-345.

23 DEJON, B.; HENRICI, P. (eds.): Constructive aspects of the fundamental theorem of Algebra. Wiley-Interscience, New York, 1969.

24 DUNAWAY, D.K.: Calculation of zeros of a real polynomial through factorization using Euclid's algorithm. SIAM J. Num. Anal. 11 (1974), 1087-1104.

25 FRAME, J.S.: The solution of equations by continued fractions. Amer. Math. Monthly 60 (1953), 293-305.

26 GARSIDE, G.R.; JARRATT, P. and MACK, C.: A new method for solving polynomial equations. The Computer Journal 11 (1968), 87-90.

27 HENRICI, P.: The quotient-difference algorithm. Nat. Bur. Stand.-Applied Mathematics Series 49 (1958), 23-46.

28 HOUSEHOLDER, A.S.: The numerical treatment of a single nonlinear equation. McGraw-Hill, New York, 1970.

29 JARRATT, P.: A rational iteration function for solving equations. The Computer Journal, (1966), 304-307.

30 JARRATT, P.: A review of methods for solving nonlinear algebraic equations in one variable. In [32], 1-26.

31 MERZ, G.: Padésche Näherungsbrüche und Iterationsverfahren höherer Ordnung. Computing 3 (1968), 165-183.

32 RABINOWITZ, P. (ed.): Numerical Methods for Nonlinear Algebraic Equations. Gordon and Breach, London, 1970.

33 RALSTON, A.: A first course in numerical analysis. McGraw-Hill, London, 1965.

34 TORNHEIM, L.: Convergence of multipoint iterative methods. Journal ACM 11 (1964), 210-220.

35 TRAUB, J.F.: Iterative methods for the solution of equations. Prentice-Hall, Englewood Cliffs, 1964.

C. *References on the use of P.A. in accelerating the convergence of sequences*

36 BREZINSKI, C.: Conditions d'application et de convergence de procédés d'extrapolation. Numerische Mathematik 20 (1972), 64-79.

37 BREZINSKI, C.: Some results in the theory of the vector ε-algorithm. Linear Algebra and Its Applications 8 (1974), 77-86.

38 BREZINSKI, C.: Numerical stability of a quadratic method for solving systems of non linear equations. Computing 14 (1975), 205-211.

39 BULIRSCH, R. und STOER, J.: Fehlerabschätzungen und Extrapolation mit rationalen Functionen bei Verfahren vom Richardson-Typus . Numerische Mathematik 6 (1964), 413-427.

40 GEKELER, E.: On the solution of systems of equations by the epsilon algorithm of Wynn. Mathematics of Computation 26 (1972), 427-436.

41 GENZ, A.: The ε-algorithm and some other applications of Padé approximants in numerical analysis. In [15], 112-125.

42 GENZ, A.: Applications of the ε-algorithm to quadrature problems. In [14], 105-116.

43 HOUSEHOLDER, A.S.: The Padé table, the Frobenius identities, and the qd-algorithm. Linear Algebra and its applications 4(1971), 161-174.

44 KAHANER, D.K.: Numerical quadrature by the ε-algorithm. Mathematics of Computation 26 (1972), 689-694.

45 OLIVER, J.: The efficiency of extrapolation methods for numerical integration. Numerische Mathematik 17 (1971), 17-32.

46 WILSON, R.: Divergent continued fractions and non-polar singularities. Proc. London Mathematical Society 30 (1930), 38-57.

47 WUYTACK, L.: A new technique for rational extrapolation to the limit. Numerische Mathematik 17 (1971), 215-221.

48 WYNN, P.: On a procrustean technique for the numerical transformation of slowly convergent sequences and series. Proceedings of the Cambridge Philosophical Society 52 (1956), 663-671.

49 WYNN, P.: The espilon algorithm and operational formulas of numerical analysis. Mathematics of computation 15 (1961), 151-158.

50 WYNN, P.: Transformations to accelerate the convergence of Fourier series. Blanch Anniversary Volume, Aerospace Research Laboratories, U.S. Air Force, 1967.

51 WYNN, P.: Acceleration techniques for iterated vector and matrix problems. Mathematics of Computation 16 (1962), 301-322.

52 WYNN, P.: On the convergence and stability of the epsilon algorithm. SIAM Journal on Numerical Analysis 3 (1966), 91-122.

D. *References on the use of P.A. in solving O.D.E. numerically*

53 EHLE, B.L.: High order A-stable methods for the numerical solution of systems of Differential Equations. BIT 8 (1968), 276-278.

54 EHLE, B.L.: A-stable methods and Padé approximations to the exponential. SIAM Journal on Mathematical Analysis 4 (1973), 671-680.

55 KOPAL, Z.: Operational methods in numerical analysis based on rational approximations. In "On Numerical Approximation" (R.E. Langer, ed., Univ. Wisconsin Press, Madison, 1959), 25-43.

56 LAMBERT, J.D.: Computational Methods in Ordinary Differential Equations. John Wiley, London, 1973.

57 LAMBERT, J.D.: Two unconventional classes of methods for stiff systems. In "Stiff Differential Equations" (R.A. Willoughby, ed.,1974), 171-186.

58 LAMBERT, J.D. and SHAW, B.: On the numerical solution of $y' = f(x,y)$ by a class of formulae based on rational approximation. Mathematics of Computation 19 (1965), 456-462.

59 LAMBERT, J.D. and SHAW, B.: A generalization of multistep methods for ordinary differential equations. Numerische Mathematik 8 (1966), 250-263.

60 LUKE, Y.L.; FAIR, W.: WIMP, J.: Predictor-corrector formulas based on rational interpolants. Int. J. Computers and Mathematics with Applic. 1 (1975), 3-12.

61 OPTIZ, G.: Einheitliche Herleitung einer umfassenden Klasse von Interpolationsformeln und anwendung auf die genäherte Integration von Gewöhnlichen Differentialgleichungen. In 'Numerische Mathematik, Differentialgleichungen, Approximationstheorie" (L.Collatz, G.Meinardus, H.Unger, eds., Birkhäuser Verlag, Basel, 1968), 105-115.

62 SAFF, E.B. and VARGA, R.S.: On the zeros and poles of Padé Approximants to e^x. Numerische Mathematik 25 (1975), 1-14.

63 WAMBECQ, A.: Nonlinear methods in solving ordinary differential equations. Journal of Computational and Applied Mathematics 2 (1976), 27-33.

64 WAMBECQ, A.: Rational Runge-Kutta methods for solving systems of ordinary differential equations. To appear.

E. *References on the use of P.A. in numerical quadrature*

65 CHISHOLM, J.S.R.,: Applications of Padé approximation to numerical integration. In [16], 159-167.

66 DAVIS, P.J.; RABINOWITZ, P.: Numerical integration. Blaisdell Publ., London, 1975.

67 WATSON, P.J.S.: Algorithms for differentiation and integration. In [14], 93-98.

68 WUYTACK, L.: Numerical integration by using nonlinear techniques. Journal of Computational and Applied Mathematics 1 (1975), 267-272.

69 WUYTACK, L.: Nonlinear quadrature rules in the presence of a singularity. In preparation.

F. *References on the use of P.A. in solving P.D.E. numerically*

70 VARGA, R.S.: On higher order stable implicit methods for solving parabolic partial differential equations. Journal Mathematical Physics 40 (1961), 220-231.

71 VARGA, R.S.: Matrix iterative analysis. Prentice-Hall, Englewood-Cliffs, 1962.

G. *References on the use of P.A. in various fields of numerical analysis*

72 BARNSLEY, M.F. and ROBINSON, P.D.: Padé-approximant bounds and approximate solution for Kirkwood-Riseman integral equations. Journal of the Institute of Mathematics and its Applications 14 (1974), 251-285.

73 DEVOOGHT, J.: Analytic continuation by reproducing kernel methods combined with Padé approximations. Journal of Computational and Applied Mathematics. To appear.

74 EDREI, A.: The Padé table of functions having a finite number of essential singularities. Pacific Journal of Mathematics, 56 (1975), 429-453.

75 FAIR, W.: Continued fraction solution to Fredholm integral equations. In [16], 357-360.

76 GAMMEL, J.L.: Continuation of functions beyond natural boundaries. In [16], 203-206.

77 KOGBETLIANTZ, E.G.: Generation of elementary functions. In "Mathematical Methods for Digital Computers" (A.Ralston, H.S. Wilf, eds., John Wiley, New York, 1960), 7-35.

78 LONGMAN, I.M.: Use of Padé table for approximate Laplace Transform inversion. In [14], 131-134.

79 LONGMAN, I.M.: Application of best rational function approximation for Laplace transform inversion. Journal of Computational and Applied Mathematics 1 (1975), 17-23.

80 LUKE, Y.L.: The special functions and their approximations. (Vols. 1 and 2). Academic Press, New York, 1969.

81 SHAMASH, Y.: Linear system reduction using Padé approximation to allow retention of dominant nodes. International Journal of Control 21 (1975), 257-272.

Luc Wuytack
Department of Mathematics
University of Antwerp
Universiteitsplein 1
B-2610 Wilrijk (Belgium)

Vol. 399: Functional Analysis and its Applications. Proceedings 1973. Edited by H. G. Garnir, K. R. Unni and J. H. Williamson. II, 584 pages. 1974.

Vol. 400: A Crash Course on Kleinian Groups. Proceedings 1974. Edited by L. Bers and I. Kra. VII, 130 pages. 1974.

Vol. 401: M. F. Atiyah, Elliptic Operators and Compact Groups. V, 93 pages. 1974.

Vol. 402: M. Waldschmidt, Nombres Transcendants. VIII, 277 pages. 1974.

Vol. 403: Combinatorial Mathematics. Proceedings 1972. Edited by D. A. Holton. VIII, 148 pages. 1974.

Vol. 404: Théorie du Potentiel et Analyse Harmonique. Edité par J. Faraut. V, 245 pages. 1974.

Vol. 405: K. J. Devlin and H. Johnsbråten, The Souslin Problem. VIII, 132 pages. 1974.

Vol. 406: Graphs and Combinatorics. Proceedings 1973. Edited by R. A. Bari and F. Harary. VIII, 355 pages. 1974.

Vol. 407: P. Berthelot, Cohomologie Cristalline des Schémas de Caracteristique p > o. II, 604 pages. 1974.

Vol. 408: J. Wermer, Potential Theory. VIII, 146 pages. 1974.

Vol. 409: Fonctions de Plusieurs Variables Complexes, Séminaire François Norguet 1970–1973. XIII, 612 pages. 1974.

Vol. 410: Séminaire Pierre Lelong (Analyse) Année 1972–1973. VI, 181 pages. 1974.

Vol. 411: Hypergraph Seminar. Ohio State University, 1972. Edited by C. Berge and D. Ray-Chaudhuri. IX, 287 pages. 1974.

Vol. 412: Classification of Algebraic Varieties and Compact Complex Manifolds. Proceedings 1974. Edited by H. Popp. V, 333 pages. 1974.

Vol. 413: M. Bruneau, Variation Totale d'une Fonction. XIV, 332 pages. 1974.

Vol. 414: T. Kambayashi, M. Miyanishi and M. Takeuchi, Unipotent Algebraic Groups. VI, 165 pages. 1974.

Vol. 415: Ordinary and Partial Differential Equations. Proceedings 1974. XVII, 447 pages. 1974.

Vol. 416: M. E. Taylor, Pseudo Differential Operators. IV, 155 pages. 1974.

Vol. 417: H. H. Keller, Differential Calculus in Locally Convex Spaces. XVI, 131 pages. 1974.

Vol. 418: Localization in Group Theory and Homotopy Theory and Related Topics. Battelle Seattle 1974 Seminar. Edited by P. J. Hilton. VI, 172 pages 1974.

Vol. 419: Topics in Analysis. Proceedings 1970. Edited by O. E. Lehto, I. S. Louhivaara, and R. H. Nevanlinna. XIII, 392 pages. 1974.

Vol. 420: Category Seminar. Proceedings 1972/73. Edited by G. M. Kelly. VI, 375 pages. 1974.

Vol. 421: V. Poénaru, Groupes Discrets. VI, 216 pages. 1974.

Vol. 422: J.-M. Lemaire, Algèbres Connexes et Homologie des Espaces de Lacets. XIV, 133 pages. 1974.

Vol. 423: S. S. Abhyankar and A. M. Sathaye, Geometric Theory of Algebraic Space Curves. XIV, 302 pages. 1974.

Vol. 424: L. Weiss and J. Wolfowitz, Maximum Probability Estimators and Related Topics. V, 106 pages. 1974.

Vol. 425: P. R. Chernoff and J. E. Marsden, Properties of Infinite Dimensional Hamiltonian Systems. IV, 160 pages. 1974.

Vol. 426: M. L. Silverstein, Symmetric Markov Processes. X, 287 pages. 1974.

Vol. 427: H. Omori, Infinite Dimensional Lie Transformation Groups. XII, 149 pages. 1974.

Vol. 428: Algebraic and Geometrical Methods in Topology, Proceedings 1973. Edited by L. F. McAuley. XI, 280 pages. 1974.

Vol. 429: L. Cohn, Analytic Theory of the Harish-Chandra C-Function. III, 154 pages. 1974.

Vol. 430: Constructive and Computational Methods for Differential and Integral Equations. Proceedings 1974. Edited by D. L. Colton and R. P. Gilbert. VII, 476 pages. 1974.

Vol. 431: Séminaire Bourbaki – vol. 1973/74. Exposés 436–452. IV, 347 pages. 1975.

Vol. 432: R. P. Pflug, Holomorphiegebiete, pseudokonvexe Gebiete und das Levi-Problem. VI, 210 Seiten. 1975.

Vol. 433: W. G. Faris, Self-Adjoint Operators. VII, 115 pages. 1975.

Vol. 434: P. Brenner, V. Thomée, and L. B. Wahlbin, Besov Spaces and Applications to Difference Methods for Initial Value Problems. II, 154 pages. 1975.

Vol. 435: C. F. Dunkl and D. E. Ramirez, Representations of Commutative Semitopological Semigroups. VI, 181 pages. 1975.

Vol. 436: L. Auslander and R. Tolimieri, Abelian Harmonic Analysis, Theta Functions and Function Algebras on a Nilmanifold. V, 99 pages. 1975.

Vol. 437: D. W. Masser, Elliptic Functions and Transcendence. XIV, 143 pages. 1975.

Vol. 438: Geometric Topology. Proceedings 1974. Edited by L. C. Glaser and T. B. Rushing. X, 459 pages. 1975.

Vol. 439: K. Ueno, Classification Theory of Algebraic Varieties and Compact Complex Spaces. XIX, 278 pages. 1975

Vol. 440: R. K. Getoor, Markov Processes: Ray Processes and Right Processes. V, 118 pages. 1975.

Vol. 441: N. Jacobson, PI-Algebras. An Introduction. V, 115 pages. 1975.

Vol. 442: C. H. Wilcox, Scattering Theory for the d'Alembert Equation in Exterior Domains. III, 184 pages. 1975.

Vol. 443: M. Lazard, Commutative Formal Groups. II, 236 pages. 1975.

Vol. 444: F. van Oystaeyen, Prime Spectra in Non-Commutative Algebra. V, 128 pages. 1975.

Vol. 445: Model Theory and Topoi. Edited by F. W. Lawvere, C. Maurer, and G. C. Wraith. III, 354 pages. 1975.

Vol. 446: Partial Differential Equations and Related Topics. Proceedings 1974. Edited by J. A. Goldstein. IV, 389 pages. 1975.

Vol. 447: S. Toledo, Tableau Systems for First Order Number Theory and Certain Higher Order Theories. III, 339 pages. 1975.

Vol. 448: Spectral Theory and Differential Equations. Proceedings 1974. Edited by W. N. Everitt. XII, 321 pages. 1975.

Vol. 449: Hyperfunctions and Theoretical Physics. Proceedings 1973. Edited by F. Pham. IV, 218 pages. 1975.

Vol. 450: Algebra and Logic. Proceedings 1974. Edited by J. N. Crossley. VIII, 307 pages. 1975.

Vol. 451: Probabilistic Methods in Differential Equations. Proceedings 1974. Edited by M. A. Pinsky. VII, 162 pages. 1975.

Vol. 452: Combinatorial Mathematics III. Proceedings 1974. Edited by Anne Penfold Street and W. D. Wallis. IX, 233 pages. 1975.

Vol. 453: Logic Colloquium. Symposium on Logic Held at Boston, 1972–73. Edited by R. Parikh. IV, 251 pages. 1975.

Vol. 454: J. Hirschfeld and W. H. Wheeler, Forcing, Arithmetic, Division Rings. VII, 266 pages. 1975.

Vol. 455: H. Kraft, Kommutative algebraische Gruppen und Ringe. III, 163 Seiten. 1975.

Vol. 456: R. M. Fossum, P. A. Griffith, and I. Reiten, Trivial Extensions of Abelian Categories. Homological Algebra of Trivial Extensions of Abelian Categories with Applications to Ring Theory. XI, 122 pages. 1975.

Vol. 457: Fractional Calculus and Its Applications. Proceedings 1974. Edited by B. Ross. VI, 381 pages. 1975

Vol. 458: P. Walters, Ergodic Theory – Introductory Lectures. VI, 198 pages. 1975

Vol. 459: Fourier Integral Operators and Partial Differential Equations. Proceedings 1974. Edited by J. Chazarain. VI, 372 pages. 1975

Vol. 460: O. Loos, Jordan Pairs. XVI, 218 pages. 1975

Vol. 461: Computational Mechanics. Proceedings 1974. Edited by J. T. Oden. VII, 328 pages. 1975.

Vol. 462: P. Gerardin, Construction de Séries Discrètes p-adiques. »Sur les séries discrètes non ramifiées des groupes réductifs déployés p-adiques«. III, 180 pages. 1975

Vol. 463: H.-H. Kuo, Gaussian Measures in Banach Spaces. VI, 224 pages. 1975.

Vol. 464: C. Rockland, Hypoellipticity and Eigenvalue Asymptotics. III, 171 pages. 1975.

Vol. 465: Séminaire de Probabilités IX. Proceedings 1973/74. Edité par P. A. Meyer. IV, 589 pages. 1975.

Vol. 466: Non-Commutative Harmonic Analysis. Proceedings 1974. Edited by J. Carmona, J. Dixmier and M. Vergne. VI, 231 pages. 1975.

Vol. 467: M. R. Essén, The Cos πλ Theorem. With a paper by Christer Borell. VII, 112 pages. 1975.

Vol. 468: Dynamical Systems – Warwick 1974. Proceedings 1973/74. Edited by A. Manning. X, 405 pages. 1975.

Vol. 469: E. Binz, Continuous Convergence on C(X). IX, 140 pages. 1975.

Vol. 470: R. Bowen, Equilibrium States and the Ergodic Theory of Anosov Diffeomorphisms. III, 108 pages. 1975.

Vol. 471: R. S. Hamilton, Harmonic Maps of Manifolds with Boundary. III, 168 pages. 1975.

Vol. 472: Probability-Winter School. Proceedings 1975. Edited by Z. Ciesielski, K. Urbanik, and W. A. Woyczyński. VI, 283 pages. 1975.

Vol. 473: D. Burghelea, R. Lashof, and M. Rothenberg, Groups of Automorphisms of Manifolds. (with an appendix by E. Pedersen) VII, 156 pages. 1975.

Vol. 474: Séminaire Pierre Lelong (Analyse) Année 1973/74. Edité par P. Lelong. VI, 182 pages. 1975.

Vol. 475: Répartition Modulo 1. Actes du Colloque de Marseille-Luminy, 4 au 7 Juin 1974. Edité par G. Rauzy. V, 258 pages. 1975. 1975.

Vol. 476: Modular Functions of One Variable IV. Proceedings 1972. Edited by B. J. Birch and W. Kuyk. V, 151 pages. 1975

Vol. 477: Optimization and Optimal Control. Proceedings 1974. Edited by R. Bulirsch, W. Oettli, and J. Stoer. VII, 294 pages. 1975.

Vol. 478: G. Schober, Univalent Functions – Selected Topics. V, 200 pages. 1975.

Vol. 479: S. D. Fisher and J. W. Jerome, Minimum Norm Extremals in Function Spaces. With Applications to Classical and Modern Analysis. VIII, 209 pages. 1975.

Vol. 480: X. M. Fernique, J. P. Conze et J. Gani, Ecole d'Eté de Probabilités de Saint-Flour IV-1974. Edité par P.-L. Hennequin. XI, 293 pages. 1975.

Vol. 481: M. de Guzmán, Differentiation of Integrals in R^n. XII, 226 pages. 1975.

Vol. 482: Fonctions de Plusieurs Variables Complexes II. Séminaire François Norguet 1974–1975. IX, 367 pages. 1975.

Vol. 483: R. D. M. Accola, Riemann Surfaces, Theta Functions, and Abelian Automorphisms Groups. III, 105 pages. 1975.

Vol. 484: Differential Topology and Geometry. Proceedings 1974. Edited by G. P. Joubert, R. P. Moussu, and R. H. Roussarie. IX, 287 pages. 1975.

Vol. 485: J. Diestel, Geometry of Banach Spaces – Selected Topics. XI, 282 pages. 1975.

Vol. 486: S. Stratila and D. Voiculescu, Representations of AF-Algebras and of the Group U (∞). IX, 169 pages. 1975.

Vol. 487: H. M. Reimann und T. Rychener, Funktionen beschränkter mittlerer Oszillation. VI, 141 Seiten. 1975.

Vol. 488: Representations of Algebras. Ottawa 1974. Proceedings 1974. Edited by V. Dlab and P. Gabriel. XII, 378 pages. 1975.

Vol. 489: J. Bair and R. Fourneau, Etude Géométrique des Espaces Vectoriels. Une Introduction. VII, 185 pages. 1975.

Vol. 490: The Geometry of Metric and Linear Spaces. Proceedings 1974. Edited by L. M. Kelly. X, 244 pages. 1975.

Vol. 491: K. A. Broughan, Invariants for Real-Generated Uniform Topological and Algebraic Categories. X, 197 pages. 1975.

Vol. 492: Infinitary Logic: In Memoriam Carol Karp. Edited by D. W. Kueker. VI, 206 pages. 1975.

Vol. 493: F. W. Kamber and P. Tondeur, Foliated Bundles and Characteristic Classes. XIII, 208 pages. 1975.

Vol. 494: A Cornea and G. Licea. Order and Potential Resolvent Families of Kernels. IV, 154 pages. 1975.

Vol. 495: A. Kerber, Representations of Permutation Groups II. V, 175 pages. 1975.

Vol. 496: L. H. Hodgkin and V. P. Snaith, Topics in K-Theory. Two Independent Contributions. III, 294 pages. 1975.

Vol. 497: Analyse Harmonique sur les Groupes de Lie. Proceedings 1973–75. Edité par P. Eymard et al. VI, 710 pages. 1975.

Vol. 498: Model Theory and Algebra. A Memorial Tribute to Abraham Robinson. Edited by D. H. Saracino and V. B. Weispfenning. X, 463 pages. 1975.

Vol. 499: Logic Conference, Kiel 1974. Proceedings. Edited by G. H. Müller, A. Oberschelp, and K. Potthoff. V, 651 pages 1975.

Vol. 500: Proof Theory Symposion, Kiel 1974. Proceedings. Edited by J. Diller and G. H. Müller. VIII, 383 pages. 1975.

Vol. 501: Spline Functions, Karlsruhe 1975. Proceedings. Edited by K. Böhmer, G. Meinardus, and W. Schempp. VI, 421 pages. 1976.

Vol. 502: Janos Galambos, Representations of Real Numbers by Infinite Series. VI, 146 pages. 1976.

Vol. 503: Applications of Methods of Functional Analysis to Problems in Mechanics. Proceedings 1975. Edited by P. Germain and B. Nayroles. XIX, 531 pages. 1976.

Vol. 504: S. Lang and H. F. Trotter, Frobenius Distributions in GL_2-Extensions. III, 274 pages. 1976.

Vol. 505: Advances in Complex Function Theory. Proceedings 1973/74. Edited by W. E. Kirwan and L. Zalcman. VIII, 203 pages. 1976.

Vol. 506: Numerical Analysis, Dundee 1975. Proceedings. Edited by G. A. Watson. X, 201 pages. 1976.

Vol. 507: M. C. Reed, Abstract Non-Linear Wave Equations. VI, 128 pages. 1976.

Vol. 508: E. Seneta, Regularly Varying Functions. V, 112 pages. 1976.

Vol. 509: D. E. Blair, Contact Manifolds in Riemannian Geometry. VI, 146 pages. 1976.

Vol. 510: V. Poenaru, Singularités C^∞ en Présence de Symétrie. V, 174 pages. 1976.

Vol. 511: Séminaire de Probabilités X. Proceedings 1974/75. Edité par P. A. Meyer. VI, 593 pages. 1976.

Vol. 512: Spaces of Analytic Functions, Kristiansand, Norway 1975. Proceedings. Edited by O. B. Bekken, B. K. Øksendal, and A. Stray. VIII, 204 pages. 1976.

Vol. 513: R. B. Warfield, Jr. Nilpotent Groups. VIII, 115 pages. 1976.

Vol. 514: Séminaire Bourbaki vol. 1974/75. Exposés 453 – 470. IV, 276 pages. 1976.

Vol. 515: Bäcklund Transformations. Nashville, Tennessee 1974. Proceedings. Edited by R. M. Miura. VIII, 295 pages. 1976.